THE SCIENTIFIC USE
OF FACTOR ANALYSIS
in Behavioral and Life Sciences

THE SCIENTIFIC USE OF FACTOR ANALYSIS
in Behavioral and Life Sciences

RAYMOND B. CATTELL
University of Hawaii

PLENUM PRESS • NEW YORK AND LONDON

Library of Congress Cataloging in Publication Data

Cattell, Raymond Bernard, 1905-
   The scientific use of factor analysis in behavioral and life sciences.

   Bibliography: p.
   1. Factor analysis. 2. Psychometrics. I. Title. [DNLM: 1. Factor analysis, Statistical. BF39 C368s]
BF39.C33   519.5'3   77-10695
ISBN 0-306-30939-4

First Printing — May 1978
Second Printing — June 1979

© 1978 Plenum Press, New York
A Division of Plenum Publishing Corporation
227 West 17th Street, New York, N.Y. 10011

Printed in the United States of America

To Karen
whose mathematical interests called me
to make concepts and experiences
in this domain explicit

Preface

It needs no great scientific insight to see that such multitudinously multivariate subjects as psychology, physiology, sociology, and history need multivariate methods. As this book may show, those methods—multivariate analysis of variance, regression analysis, typal and discriminant function analysis, multidimensional scaling, and factor analysis—belong to a single structural arch which bears up conceptual and causal understanding in all these subjects. But factor analysis is the keystone of that arch. Since factor analysis has itself developed fantastically in thirty years [my first small book (Cattell, 1952b) could almost cover the field in a few chapters!], this book confines itself to that subject, with only brief connecting asides on the related areas.

The purpose of a preface is to permit more personal comments and to explain why the design is what it is. Among the former the author often apologizes for writing in an already crowded library shelf, and, having confessed the crime, thanks his friends for their connivance. The area already enjoys a truly excellent array of books, from extremely good elementary introductions by Child, Henrysson, Lawlis, and Chatfield, through the workbook emphasis of Fruchter, and the intermediates of Guertin and Bailey and Comrey, to the comprehensive technical works of Anderson, Ahmavaara, Gorsuch, Harman, Horst, Lawley and Maxwell, Mulaik, Rao, Rummel, and Van de Geer, not to mention the undating books by Burt, Thomson, and Thurstone. Seldom has there been such a contrast between a feast of first-rate texts and the starvation level at which many departments keep their students in formal instruction. It seems by no means unusual for undergraduate majors to try to handle their psychology with no instruction in factor analysis but, for some reason, an excessive drilling in *analysis of variance* (ANOVA). Yet factor analysis is a gift of psychology to the life sciences, created by the needs of the subject, whereas ANOVA began with the needs of farmers.[1]

Factor analysis was fathered in psychology by Burt, Kelley, Spearman, Pearson, and Thurstone, and mothered in mathematics by Hotelling and the developments of the eigenvalue concepts going back to Jacobi. The offspring shows the more mathematical inheritance in Anderson, Ahmavaara, Harman, Horst, Lawley and Maxwell, Mulaik, and Rao, and the more psychological or social science inheritance in Burt, Gorsuch, Rummel, and Thurstone. This is not to say that the latter, in the least, lack any mathematical precision and, indeed, the excellent balance in Gorsuch and in Rummel came near to persuading me that a further book in the field is uncalled for.

However, nearly fifty years of discussing the difficulties both of students and of leading researchers in relation to statistics, and factor analysis in particular, have convinced me that a radical new approach is necessary. One must respect first the scientific interests of the beginner and only slowly and with patience reorient to mathematical concepts. The aphorism is as old as Herbart that good education moves from the known to the unknown and from the simple to the complex—and no psychodynamicist needs to be told that it must link to powerful interests the whole way. My title *The Scientific Use of Factor Analysis* means just this: that for the student the interest begins in his science as such; and for the researcher the demands of viable scientific models are given precedence over mathematical neatness, which can easily become pedantry. This radical change of emphasis from most existing texts connotes also an art of gradation and a policy of keeping the final center of gravity in the right place.

The gradation is shown in the fact that the levels of mathematical sophistication and precision demanded are quite different at the beginning and the end. I am assuming that in the one- or two-semester course (according to student ability) in which the book is completed the student will grow—in general attitude to mathematical models. One or two of my braver fellow authors have confessed that they are afraid to make simplified, two-thirds true or insufficiently qualified statements at the beginning and that the same high rigor of statement and formulation must apply from the first page. They are afraid of their peers, which means they are writing for their peers. The natural growth of an enquiring mind is quite different from that: it is willing to believe at first that water is H_2O, and

---

[1] My emphasis is only to redress a balance; for here as elsewhere I have always argued for a two-handed use of factor analysis and analysis of variance. I trace this to student years in which I shuttled across a little plot of grass between the laboratory where Spearman was developing factor analysis and the Galton Laboratory where Fisher was shaping with equal brilliance the analysis of variance.

talk of deuterium is at that stage distracting and offensive pedantry. If the reader is not too aware of these transitions I am happy to have had some success in the art that conceals art; but the procedure of "sketching in" and then returning to more precise details may be mistaken for repetitiousness (though good education requires planned repetition with growth). It has also required more cross reference than usual, back and forth, so that the reader explicitly recognizes amendments. I have, where possible, followed a further principle, one that, other things being equal, the historical order of development of a science is the best teaching order. I followed this order quite closely in my first book on factor analysis, but I have adhered to it only loosely in this book for reasons connected with some peculiarities in the later growth of the subject, which I have no space to explain here.

The book is divided into two parts and the teacher will have no difficulty in recognizing that the second part is significantly more difficult and involves a step from almost concrete exercises to abstract conceptions on the advancing fringe of mathematical psychology in factor analysis. Although the emphasis here is on the scientist's interests, so that mathematical derivations present in some other factor analytic texts are omitted, there has been no omission of the exact ultimate formulas by which a given mode of analysis or mathematical model is properly represented and conceived. If the reader is mathematically endowed, he will not lack the comprehensive statement he desires. If he is not so endowed he will get a verbal equivalent. Although reviewers of my 1952 book saw each, according to his specialty, various shortcomings, I was gratified to find that it was received as one of the clearly readable books they had encountered on factor analysis. I have attempted to live up to that here, though one result has been of some cost to the publisher, namely, that the book's higher ratio of explanations in reading to condensed formulas has made it consume more paper.

A further instance of adjusting to learning interests is the way in which I have brought in matrix algebra concepts in a smooth sequence, introducing them as they are needed, rather than starting with a solid block chapter devoted entirely to that topic. One consequence of my plan, which teacher and student must recognize, is that the chapters must necessarily be read in the order given. However, I have explained to the reader at one or two chapter beginnings that he may choose to skip all or parts of the particular chapter if he is reading for a first general perspective. Nevertheless, at no stage can he jump ahead very far—except in the last four chapters which might be picked up in a freer order.

The above must not be taken as any implicit criticism of the more purely mathematical presentations. Indeed, without the existence of these handsome and painstaking mathematical volumes (not to mention *Psycho-*

metrika, the *British Journal of Mathematical Psychology*, the *Journal of Mathematical Psychology*, and *Multivariate Behavioral Research*) what I have attempted here would be impossible. In part, I have written this book as a signpost, directing the reader to them at appropriate points. In particular, I have referred the reader to more detailed developments in Gorsuch and Rummel, the scientific emphases in which dovetail more readily with these chapters, but also to the mathematical–statistical overviews of Anderson, Harman, Horst, Lawley and Maxwell, Mulaik Van de Geer, and other well-known texts. On the other hand such short introductions as those of Child, Fruchter, Lawlis and Chatfield, and others mentioned above may function as introductions to the present work.

As to the final "center of gravity" at which the book aims to leave the student or researcher finally well balanced, it is best described as that of a scientist in his own field. I have watched over decades several recurrent well-meaning attempts to place the teaching of mathematical statistics where it seems logically to belong—in the mathematics department. With the exception of a few advanced graduate students who ended their careers as mathematicians rather than as psychologists, these transfers of courses—equally in psychology, economics, and physiology—from the substantive department to the department of mathematics, were a failure.[2] In the first place, psychology is a big enough subject in itself over

---

[2] Since some may wish more documentation of my meaning here I would say that historically when mathematicians joined the development of factor analysis they wanted to direct it according to the development of their own abstract systems rather than according to what I have frequently contrasted here as a set of more complex and flexible *scientific* models. The latter are equally precise, and largely mathematical (though they demand other properties too, in an analogical basis). However, these models respond to what the sensitively inquiring psychologist sees as "a best fit," not to the logically possible permutations and combinations of principles which the mathematician sees as the next logical development. For example, it is admittedly a simpler and cleaner job for the mathematician if he keeps unities instead of "guessed" communalities in the diagonal of the correlation matrix and if he keeps axes mutually orthogonal to rotation, whereas the scientific model has every indication of requiring something a bit more complex and even a bit messy. For some years, the mathematicians thus dragged the less perceptive psychologists from factors to the authority of components. They also called psychologists to the apparent mathematical purity of orthogonal factors and it has taken much argument, by the present writer and others, on behalf of psychological models to draw the pack of researchers from an unreasoning pursuit of "mathematical prestige" to what psychological research indicated as a more apt model.

On the one hand the pure mathematician's demands are a harsh intrusion, and on the other they sometimes invite a wild goose chase or irrelevant developments we do not need. For example, he may suggest that factor space be handled in Riemannian rather than Euclidean geometry. We should thank him for providing this possibility, but many psychologists seem to need to be reminded that we do not have to adopt it, just to be "up-to-date" mathematically, if it proves irrelevant. As a neces-

which to stretch a student's interests, without his having to master a new discipline from the ground up. Except for a few utterly rare instances, even with the greatest attention in teaching to the relevance of math–stat to substantive issues, the student's interest will move only slowly, over months and years, to a real understanding of the need for mathematical models in science, and a real enjoyment of their intrinsic beauty. If psychological substantive research can recruit a few hundred such men, it will be fortunate; but every psychologist, because he is in a field of multiply determined behaviors, should grasp the general principles of factor analysis and the behavior specification equation.

This book attempts a highly comprehensive survey in Part II of

---

sary illustration, though I trust not invidious, I would mention the developments around "alpha factor analysis," made by mathematicians among psychologists, and valid as a rejuggling of the possible ways of doing a factor analysis, but scarcely suggested by the needs of any psychological relations ever perceived. The juggler of models has a right to juggle, and perhaps our concern should be rather with the weakness of the average psychologist which leads him to jump on the latest bandwagon with no perception of the direction in which it is going. As one contemplates a certain *tour-de-force* offered by a certain kind of mathematical statistician to his brother psychologist one is reminded of the question of Macaulay regarding the powerful intellect of Dr. Johnson, "How it chanced that a man who reasonned on his premises so ably, should assume his premises so foolishly, is one of the great mysteries of human nature."

What we should like the gifted mathematical statistician to give us, in terms of helping us meet the most likely scientific model, can be illustrated by many urgent needs. After the initial breakthrough of MacDonald we need a thoroughly programmed nonlinear factor analysis, and we need a way of handling the *reticular* model instead of having to restrict to the *strata* model (p. 201, below). For twenty years we have called to mathematicians for a solution for oblique confactor rotation; and for more theoretically insightful significance tests for oblique rotated factor loadings as well as the difference between an obtained and a hypothetically stated factor structure (see, however, Jöreskog, p. 482). With all due gratitude for those things in which pure mathematical statisticians have turned aside to help us, the situation familiar to scientists remains: that generally we ask for a solution to one problem and get from the mathematician an answer to a different question, in which he has shifted the assumptions from those we must make. For example, we asked in the forties and fifties for a test of significance for the size of an oblique rotated factor but got an answer only for an unrotated factor, indeed, not even for a factor but for a principal component.

We have no right to demand of mathematicians that they be supermen, but better cooperation from both sides in defining goals would have been fortunate. Failing to get responses to their problems, psychologists have in fact resorted to solutions for their methodological problems which may be approximate and often look relatively clumsy to the mathematician. Sometimes (as in those publications by the present writer) they have formulated the real problem but have had to answer it by brute Monte Carlo methods, in order to get on with their work with some assurance of knowing roughly where they are statistically.

advanced mathematical statistical techniques, so that the psychologist knows what is going on. But it invites him to look over the fence rather than to get over it, knowing that psychology is a vast enough field to require him on this side of the fence. However, its contrast to some other leading texts is not just in this restraint. It lies rather in the far greater space given to scientific models, to the judicious choice of procedures in relation to experiment, to factor interpretation, to relations that broaden psychometric concepts, such as scaling, validity, and reliability, to higher strata models, and to the whole strategy and tactics of the use of factor analysis in research.

Finally, the book differs in venturing to convey, along with the explicit mathematical and statistical formulations, some values of a more intuitive kind arising from long and diverse experience in the field. In justification of this liberty it must be stated, at the possible risk of immodesty, that the Laboratory of Personality and Group Analysis at the University of Illinois is to have done factor analyses, in more diverse substantive areas (psychology, animal behavior, learning, socioeconomic trends, group dynamics, educational achievement, motivational states, and cultural history) and with more varied designs, than almost any laboratory in the world. Inevitably that brings to the recommendations and decisions certain perceptions and values not encompassed in the most thorough mathematical–statistical texts.

With this statement of the book's design and intentions the teacher and student can best judge for themselves its role in courses. The writer believes that it will function as an attractive avenue to the more statistically detailed treatments in the books indicated. Part I can stand on its own feet as a craftsman's introduction to the basic concepts and processes of factor analysis. Part II leads out on a new level to the more complex statistical treatments which can be pursued as far as desired in the given directions and with particular textbook aids.[3] Since the horizon of new

[3] If undergraduate senior courses are concerned I would advocate making a one-semester course from Part I (with associated illustrative examples and quizzes). If graduates, or if a class of highly selected undergraduates, then I believe it is practicable to cover the whole book in a semester. An alternative is to make Part I half of a first-semester graduate course in general statistics, in which ANOVA takes the other half, and to keep Part II for a second-semester course specifically in factor analysis. But there is also the lone reader to consider, in research or applied work, whose statistical and general background is more mature, and this important individual has often been in my mind, especially in Part II. Indeed, my dialogue after the first five chapters is often with the seasoned researcher and philosopher of science. In order not to talk down to him, I have attempted in the first five chapters (to some degree in the rest of Part I) a palimpsest, in which the writing on the surface will be easily followed by the student, but which hopefully will not offend the specialist inasmuch as he sees writing between the lines which has implications of a more subtle kind. Such a reader can be left to his own devices, but almost certainly in Part II most readers will need the guidance and exegesis of a good teacher.

devices is constantly growing today that pursuit is likely to end in the journals. The bibliography supplied here is a large one, produced with an eye to a stratified sample, and should start the researcher in whatever direction he wants to go. Finally, I would recommend to the teacher some attention to the standard notation in the Appendix, for unnecessary difficulties for the student arise from strangeness of languages.

Whatever differences may arise among multivariate experimentalists about the style of teaching, there are virtually none about the need for better education of psychology students in the whole area. There are good reasons why quantitative advances in psychology need these methods more than those in the physical sciences did. In the physical sciences, where intuition alone was soon able to settle on the basic concepts and dimensions such as space, time, temperature, and energy, which proved to be few, progress has repeatedly occurred through single acute perceptions of relations, as Lavoisier's perception of a change in weight on burning, or Boyle's observations on pressure and volume of a gas, or Becquerel's noticing of fogged photographic films near radium. Human behavior has been under observation so long that it is unlikely that any new observation at this "clinical level" will bring a radical discovery, and it is so complex that it is also unlikely that simple laboratory bivariate experiment will unearth the complex connections. Future discoveries in psychology are surely most likely to depend on methods capable of revealing relatively small effects, necessarily based on accurate psychometric measurement methods but *expressed in quite complex relationships*. For example, one might ask today whether the evidence shows that the *integrated* and *unintegrated* factor components in motivation strength interact with cognitive skill components according to an additive or a multiplicative model (Cattell & Child, 1975). The domain of finer measurement and analyses, to examine more complex models of multivariate determination, surely offers the virgin territory in which the greatest discoveries in psychology may be expected.

Although the psychologist has been the main client in my mind for this book—especially since I believe psychologists have not yet sufficiently recognized what the above professionally first-class textbooks mean to their science—I have also approached needs and illustrations from other life and social sciences. Biological scientists are beginning to utilize these more powerful methods, where previously they beat on closed doors with bivariate designs inapt to at least half of their problems. (This was brought home to me by many responses I received when I wrote two invitation articles on factor analysis in 1965 in *Biometrics*.) There are some area peculiarities but, as Rummel's book written mainly for sociologists and political scientists shows, the main multivariate problem and methods are very similar across the life and social sciences.

I have to thank the staff of the Institute for Personality and Ability

Testing and several former students there and elsewhere for putting increasing pressure on me to get this book out! I am greatly indebted for technical comment and shrewd constructive suggestion after reading the manuscript to Professor Ralph Hakstian and Professor Maurice Tatsuoka. Parts of the text also benefited from comments by Professor John Nesselroade, John Horn, James Laughlin, Samuel Krug, and George Woods. In more intangible ways I owe much to years of neighborliness and friendship with Ledyard Tucker, Jack Cohen, Herbert Eber, Jack Digman, Charles Wrigley, Carl Finkbeiner, Peter Bentler, Will Meredith, and others—particularly the members of the Society of Multivariate Experimental Psychology—too numerous to list here. Finally, I wish to thank Sanchia Foiles and Larry Sine for help with diagrams and tables; Professors Daniel Blaine and Walter Hudson for vigilance and professional insight in proofreading of formulas; and Harvey Graveline of Plenum Press for perspicacious editorial direction.

Contents

CHAPTER 9
Higher-Order Factors: Models and Formulas **192**

CHAPTER 10
The Identification and Interpretation of Factors **229**

PART II

PART I

CHAPTER 1

The Position of Factor Analysis in Psychological Research

SYNOPSIS OF SECTIONS

1.1. The ANOVA and CORAN Methods of Finding Significant Relations in Data
1.2. The Reason for the Salient Role of Statistics in the Biosocial Sciences
1.3. What Is an Experiment? Bivariate and Multivariate Designs
1.4. Dimensions of Experiment and Their Relation to Historic Areas of Research
1.5. The Relation of Experiment to the Inductive-Hypothetico-Deductive Method
1.6. The Relative Power and Economies and the Mutual Utilities of ANOVA and CORAN Methods
1.7. Summary

1.1. The ANOVA and CORAN Methods of Finding Significant Relations in Data

The student of psychology lives today in that exciting phase of the socio-biological sciences wherein genuine quantitative laws have at last begun to emerge. The prequantitative phase of shrewd literary observation and of penetrating clinical intuition, gave us some useful theories; but the age of quantitative experiment places psychology with the more mature sciences. It promises deeper understanding and the practical effectiveness of precise models and equations.

In that advance two main branches of statistical analysis, with their accompanying experimental designs, have been the major servants. They are the *analysis of variance*—ANOVA for short—and *correlational analysis*—CORAN for short. The former is taught in all elementary statistical courses, primarily as a method of examining the significance of the differences of mean score of control and experimental groups. The latter has diverse forms of expression and difficulty ranging from significance of ordinary "zero-order" correlations, to multiple and partial correlation, and so to factor analysis. The underlying kinship of these two approaches and the ways of functionally combining them in research are explained in many excellent statistical texts, and perhaps most fundamentally and briefly in the writings of Bentler (1976), Burt (1966), Cohen and Cohen, (1975), and Tatsuoka (1975). But in preliminary terms we can say that analysis of variance (ANOVA) tests or generates a theory by *observing if means of various experimental and control groups are significantly different*, while correlational methods ask *both about the significance of the relationship and its magnitude as such*. Factor analysis, which is the furthest logical development and reigning queen of the correlational methods, goes further and is capable of revealing patterns and structures responsible for the observed single relation connections.

In the simplest case in ANOVA we subject an experimental group to, say, a year's learning, comparing its level of performance relative to a control group not subjected to learning, asking if any significant "effect" exists. In correlational methods applied to the same problems we would take people with varying learning experience, from zero onward, and correlate their gain scores with the amount of their varied experiences of learning. In 50 people any correlation above 0.28 would be significant ($p < .05$), but it is of further interest to know whether it is 0.3 or, say, 0.7, as far as explanatory theories are concerned. Factor analysis, as a "second story with a skylight" on top of correlation, will tell us further whether the several measures used to assess the learning gains can, conceptually, be thrown together as a single kind of learning, or whether two or more distinct independent kinds of learning are going on, requiring separation of concepts.

1.2. The Reason for the Salient Role of Statistics in the Biosocial Sciences

Any student familiar with science as it is understood in the older, physical sciences is likely to be a bit puzzled on seeing the much larger role which statistical analysis, and such methods as factor analysis, seem to play in the newer, sociobiological sciences.

There are two main reasons for this. (1) The physical sciences can control and isolate phenomena manipulatively to a far greater extent than either ethics or circumstances will allow us to do with human beings. Thus, most of the extra, interfering influences which make the statistics of "error variance" necessary to us can often be ruled out in physics, and a single experiment on a single occasion, like Galileo's observation on the length of a pendulum, will suffice to settle matters. (2) The sheer number of possible variables relevant to theory is much smaller in the older sciences. Astronomers, physicists, and chemists had little difficulty in seeing that the important dimensions were distance, mass, temperature, time, etc., whereas psychologists and sociologists grope in a "big, buzzing, booming confusion" of endless possible variables to find the underlying dimensions of greatest theoretical potency. In personality study alone the dictionary offers some three or four thousand variables. And since there are about 50,000 registered psychologists in the USA alone the intuition, inspiration, or whim of one or another will sooner or later build theories of the signal significance of practically every one of them!

One important use of factor analysis is in finding the "significant dimensions" in such a jungle of variables. How far this is a mere reduction to mathematical predictors and how far an extracting of significant theoretical concepts we can debate later. A second major service of factor analysis is in exposing and counteracting that vice of the human mind which (as Francis Bacon pointed out in 1605) assumes that when there is one word there must be one thing corresponding to it in the real world. Psychologists have set out to investigate how "depression" changes under a drug or how "group morale" is affected by certain kinds of leadership. Factor analysis, in ways we shall see, tells us, however, that there are as many as seven distinct kinds of depression in the general area of depressive variables, and two independent dimensions of group morale. These behave quite differently in response to conditions of the patient, or of the small group, respectively, so premature manipulative experiment without first factor analytically determining what the real functional unities are, is wasted effort, trying to make laws about an accidental conglomerate of concepts, where the intelligible laws effectively apply to only the separate and quite distinct concepts.

1.3. What Is an Experiment? Bivariate and Multivariate Designs

When Wundt and his generation set out with admirable intentions to make psychology a peer for the older sciences, and gathered an impressive array of brass instruments around them, they naturally assumed that the existing methods of the physical sciences defined an experiment. That is to say, they did controlled, manipulative experiments in perception, learning, etc., which at most required only the development of ANOVA designs. This we may refer to as *bivariate design,* i.e., dealing with a *dependent* and an *independent* variable and all else ideally held constant. It is not possible in this compressed review to do justice also to history (see Cattell, 1966c), but we may note that the new methods more apt to the generality of real psychological problems came with Galton, Pearson, Spearman, and others interested in individual differences, correlational methods, and the recognition that what cannot be manipulatively controlled must be measured and allowed for. Thus the Wundtian and Galtonian traditions are two vital streams which converge in the broad river of method on which the traffic of modern psychology proceeds. But even today the proper notes and strategic uses of these methods seem not widely understood, and so it is the object of this book not only to explain factor analysis but to show where it fits into research activities. A man with a key to his house and another to his car sometimes fumbles with the wrong key. For all too long in the area of general process of perception and learning, for example, handled by Wundtian methods, there have been attacks on locked doors by traditional application of bivariate designs where novel multivariate designs could have opened them. The discovery of the number and nature of human drives as determiners of reinforcement is a case in point (Cattell & Child, 1975).

A two-handed use of ANOVA and CORAN (if we may so continue to abbreviate for the *correlation–analytic "factor" approach*), picking up each tool as needed, is the ideal in intelligent, flexible experimental progress. This means dropping the traditional narrow misnomer of "experimental psychology" in college course listings and embracing a broader concept and definition of experiment which may be given as follows.

An experiment is a recording of observations, quantitative or qualitative, made by defined and recorded operations and in defined conditions, followed by examination of the data by appropriate statistical and mathematical rules, for the existence of significant relations. It will be noted that the ambiguous term "controlled," as in, e.g., "under controlled conditions," is deliberately avoided here, as also is the question of whether we are using the relations to create or to test an hypothesis. Unfortunately, "controlled" sometimes means "manipulated," sometimes "held constant," and sometimes just "noted and recorded." Only the last is essential to an experiment, as indicated by "defined and recorded operations and . . . conditions" above.

The present definition of experiment is thus designed to include observation and measurement both of *a naturally occurring event* (like an eclipse, an hysterical paralysis, or the change in magnetic variation), or of *an event abstracted from its natural setting* and manipulatively repeated in the laboratory.

1.4. Dimensions of Experiment and Their Relation to Historic Areas of Research

Experiments in general have some six dimensions along which they can be varied. These are briefly listed in Table 1.1 leaving the student to

Table 1.1. The Logically Possible and Practically
Viable Dimensions of Experimental Design

Parameter title	Polar dichotomy
1. Number of variables	Multivariate to bivariate
2. Manipulation	Interfered with (manipulated) to freely occurring
3. Time relation	Sequentially dated to simultaneous observation
4. Situational control	Controlled (held constant) to uncontrolled
5. Representativeness of relatives[a] (choice of variables)	Abstractive to representative
6. Distribution of referees (population sampling)	Keyed to a biased sample (selected or "unrepresentative") to normal or "true"

[a]Here and in later use the word "relatives" is used for the set of variables
to be related and "referees" for the set of organisms or objects ("entries"
in a mean or correlation calculation) to which the various measures are
referred.

satisfy his curiosity more fully by reading about their implications in
the *Handbood of Multivariate Experimental Psychology* (1966, chap. 1
and 2). Therein he will find that six bipolar parameters yield $2^6 = 64$ types
of experimental design, so the difference on the parameter of *multivariate
versus bivariate* is not the only one which may separate various "factor
analytic" designs from various "classical experimental" designs. The
importance of Table 1.1, however, is that it reminds us that the use of
factor analysis in the past has built up an image of a "classical factor analytic" design which is as parochial and misleading as the "classical brass-
instrument" design. An enormous amount of multivariate, correlational,
factor analytic research has been done, in education for example, on individual differences rather than on processes, just as laboratory experiment
on learning and perceptual processes has confined itself largely to ANOVA.

Individual difference research concerned with the "character of the
organism faced with a standard stimulus" and process research, concerned
with "change in the response for the average person when the stimulus
changes" as in perception, learning, and developmental psychology, constitute the warp and weft of psychology. There has been an historical
habit, since Galton and Wundt, to apply CORAN, in the former and
ANOVA in the latter. This began as a convenience, in relation to rather
short-sighted objectives, but was unfortunately raised by textbook writers
to the dignity—and rigidity—of a tradition which for three generations has
blocked wider views. Let us hope the thought behind Table 1.1 will eman-
cipate us from these sterilities.

For example, it is an historical accident that factor analysis has commonly been used without the manipulation ("control of variables") customary in brass-instrument bivariate experiments. But as Cattell and Scheier (1961) pointed out, and as P-technique experiments for example, illustrate, there is no reason whatever why manipulation of an independent variable should not be included in multivariate designs. "Two versus many variables" and "occurring naturally *in-situ* versus manipulating" are quite distinct dimensions of experiment in Table 1.1. Parenthetically, one should be careful to distinguish between dependent–independent variables in terms of *experimental manipulation* and dependent–independent in *statisticomathematical analysis equations.* In the latter you are free to put whichever variable you like on the right of the equation as a "predictor." In manipulation, on the other hand, e.g., when you measure the amount of learning with different magnitudes of reward, you are committed to the latter as the manipulated independent variable. Actually, it is an especial virtue of factor analytic method (and CORAN methods generally) that it can proceed without the need for manipulation. In a classical experimental design on, say, the effect of a cold in the head on intelligence test performance you would hold all else constant and inoculate half the subjects with a cold in the head. In the nonmanipulative experiment you would not interfere with nature, but pick up cases with a cold in the head and equate for other differences by sophisticated statistics instead of "brute" manipulation.[1] Thus it is conceivable that the finding in Banting and Best's experiments on diabetes by manipulatively cutting out the pancreatic cells of dogs could have been reached by a correlational, factor analytic experiment using the naturally existing variations in pancreatic secretion as one variable and the various diabetic signs, e.g., sugar metabolism, as others in the same correlation matrix. Sometimes the nonmanipulative CORAN results are clearer, because the interference with the organism or group in manipulation is itself traumatic or productive of side effects.

Against the nonmanipulative experiment it is sometimes argued that it is incapable of isolating causal action—and causal explanation, admit-

---

[1] Minor claims are possible in both directions of advantage in respect to the manipulative vs nonmanipulative dimension of design. For the manipulative design it is argued that statistical equating of (or allowance for) influences not manipulated is never fully effective and precise. On the other hand, it is pointed out that manipulation often affects more than is intended and still leaves doubt as to the causal direction it is supposed to clear up. For example, in a study of the role of alcohol in arthritis of the arm, in which one group drinks and the other is forbidden to do so, perhaps the fact of being forbidden, or the elbow bending itself could be the cause of the observed difference. The nonmanipulative solution is to measure the amount of elbow bending, which need not correlate perfectly with liquor consumed, and partial it out, after including it as a measured variable.

tedly, is highly important in science. But, without entering metaphysical discussions on just what a "cause" really is, let it be said that this confuses the manipulative dimension in Table 1.1 with the time relation dimension. Causal connections can be established when it is shown that an invariable time sequence of A followed by B exists, and to show this it is not necessary to manipulate A: it suffices to know that B follows A, not A follows B. To recognize that a rain cloud is one necessary causal condition for a rainbow I do not have to create a rain cloud. It is true that *manipulation usually also means a time sequence, but not that establishing a time sequence requires manipulation.* Consequently, a factor analytic or correlational experiment in which the data gathering established the existence of a time sequence permits a factor analytic experiment to establish causal connections. A correlation of the amount of alcohol drunk on one day with severity of hangover on the next suffices to point to causality in one direction. However, here, as in manipulation, the possibility exists, pending further investigation, that some third, as yet unknown or unmeasured variable is causal to both.

1.5. The Relation of Experiment to the Inductive-Hypothetico-Deductive Method

So much for what an experiment is and what you can get out of it in the way of established relations. But what about the design of an experiment in relation to a general theory or a specific hypothesis? One can have a particular hypothesis on its own, or as an inference from a broader theory. When I directed a clinic in a large English manufacturing city I got the impression that the children of lower IQ came from larger families, and I set out to investigate this relation as a specific hypothesis. On the other hand, I could have investigated the hypothesis as part of a much broader theory, in writers like Gibbon, Spengler, McDougall, or Toynbee, that civilizations have cycles of rise and decline and that the breeding out of the inheritable component in higher intelligence is part thereof. In the latter there would be other subhypotheses, but the checking of each would hinge on investigating the statistical significance of some relationship.

This interplay of theoretical hypothesis and experiment is sometimes called the hypothetico-deductive method. But as the *Handbook of Multivariate Experimental Psychology* (1966) points out, it would be less misleading to call it the *Inductive-Hypothetico-Deductive (IHD) method.* For no one but a lunatic makes hypotheses out of thin air. There are always some previous inductive, even if unconscious, reasonings from

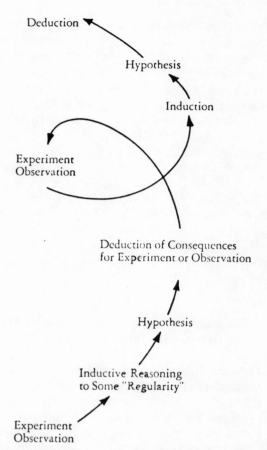

Fig. 1.1. The inductive-hypothetico-deductive spiral. (From R. B. Cattell, *Handbook of Multivariate Experimental Psychology*, Rand McNally, Chicago, 1966, p. 16.)

scattered observations, as in the instance above, which generate the hypothesis. Scientific work begins with data observation and returns to data observation, though the first encounter may be unsystematic. Consequently, the progress of research is really based on an IHD spiral as shown in Fig. 1.1. It is usual, in "playing up" the elegance of science, to put more emphasis on the deductive reasoning from a theoretical model to an expected result in a well-designed experiment. But in the actual history of science at least equal credit must be given, in terms of ultimate fruitfulness, to the initial inductive synthesis, whether from casual observation or more systematic data gathering, which gave birth to the theory. Darwin's ruminations during the voyage of the Beagle or Franklin's watching of thunderclouds and the rubbing of plates of resin, suffice to remind us of the importance[2] of the I in IHD.

1.6. The Relative Power and Economies and the Mutual Utilities of ANOVA and CORAN Methods

The empirically observable relation which is the father or child of a psychological theory can appear either from a bivariate experiment and ANOVA analysis of results, or a multivariate experiment and a CORAN factor analytic examination. Either design and method is both hypothesis creating and hypothesis testing. A convincing argument can be made, however, for greater economy and greater power, in both roles, for investigating relations by multivariate than bivariate designs (Cohen, 1968).

The question of power is best left until more has been explained about the method, but the question of economy can be quickly illustrated. Among, say, 10 different variables there are 45 possible paired relationships. Let us suppose it takes 10 hr to test and score 100 people on a pair of variables, i.e., 5 hr per variable. (Parenthetically, by rule of thumb, some ANOVA investigators would descend to 10 cases for one comparison, but this is irrelevant to the argument.) If each of 45 investigators interests himself in the relation presented by one of the above particular pairs, in bivariate experiment, then $45 \times 10 = 450$ man-hr will be expended (assuming investigators independent) in accumulating the data necessary to examining these relations. But if all are done in a multivariate experiment by one investigator only $10 \times 5 = 50$ man-hr suffice to yield all 45 relationships. If the 45 investigators got together, or if the study is pursued by one investigator as a multivariate design, then not only is there no duplication of data gathering but the relations are more accurately comparable, being all from the same sample, and the calculation of relations is quicker computer-wise through avoiding repeated entries. However, the far more basic argument for the multivariate design, as we shall see in the next chapter, is that it permits a more determinate answer about the *underlying independent factors* in the correlational relationships recorded. This more structured "basic-concept-generating" answer in turn rules out a lot of theoretical wild goose chases into which the examiner of relations one at a time could easily be led by the immense number of purely theoretical possibilities.

[2] It may, indeed, be argued that since the deduction has to be tested against actual data, the spiral finally ends again in induction and that IHD should be IHDI. There is no objection to this except that the final I is actually not inductive reasoning to a generalization, but only the checking of an existing generalization.

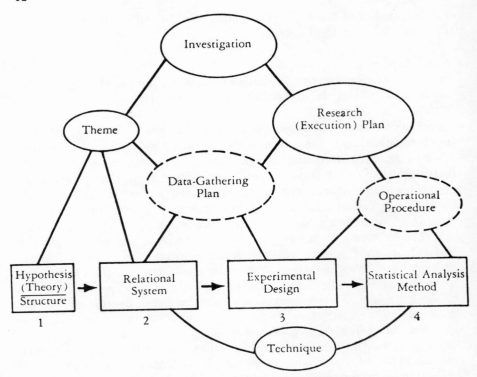

Fig. 1.2. Interdependence of four essential and other definable parts in a scientific investigation. Parts 1, 2, 3, and 4 are the essential parts of any investigation and it is for these that dimensions have been assigned in *The Handbook of Multivariate Experimental Psychology*. The remaining headings are organizational conveniences in distinguishing aspects of the total investigation. (From R. B. Cattell, *Handbook of Multivariate Experimental Psychology*, Rand McNally, Chicago, 1966, p. 52.)

What has been said above about the relation of experiment and theory to design in an investigation—as well as some further points expanded on in the *Handbook of Multivariate Experiment*—can perhaps be helpfully summarized and visualized in Fig. 1.2.

The reality one must never overlook in theorizing is that both the creation and the checking of a theory alike depend ultimately on a perceived *relation between variables*. The hypothesis in a theory describes or should describe exactly the expected relation; the research should check it. Thus, Darwin built the theory of evolution on relations perceived among species and traits, and Einstein's theory of relativity produced an inferred relation which was checked by turning the telescope on Mercury's transit. Figure 1.2 sets this relation checking in the broader context of design of a total investigation.

1.7. Summary

1. Correlational analysis, CORAN, of which factor analysis is the fullest development, and analysis of variance, ANOVA, are the two main instruments for examining the significance of relationships among psychological experimental variables.

2. The greater role of measurement and sampling error and the greater intrusion of variables irrelevant to the relation studied make statistics more important in life sciences than physical sciences.

3. An experiment may be defined as the collection of data under defined conditions, followed by an orderly and appropriate mathematico-statistical analysis to test the existence of systematic relationships.

4. The dimensions which cover all experiments, and by which they may be classified are six: *bivariate* to *multivariate*; *manipulated* (interfered with) to *freely occurring*; *sequential* and dated to *simultaneous* in observation; *abstractive* (specific hypotheses) to *representative* (general hypothesis) choice of variables (relatives); and *selective* to *representative* sampling of subjects (referees).

5. Research proceeds by the inductive-hypothetico-deductive method rather than the supposed hypothetico-deductive sequence, since all realistic hypotheses begin with inductive observation, even if the hypotheses are borrowed from other areas. Thus factor analytic research, for example, begins with inductively reached hypotheses that structure exists in the given area or that some particular structure exists, and proceeds in a spiral of emerging factors followed by testing of new variables inferred to belong to the hypothesized concept from the emergent pattern and so on.

6. Although both ANOVA and CORAN are indispensable common work-horses, there is a time and place for each to be strategically and tactically applied to the haulage. Strategically, many—indeed most—areas would be best first attacked by a concept-generating and hypothesis-sharpening factor analysis, followed by bivariate designs in which the factors become the variables. For example, an enormous number of bivariate researches, ambiguous and confusing in answer, have been performed on intelligence in the absence of a clear factorial understanding of the structure and measurement of intelligence among abilities. Apart from this sequence of "task forces" the multivariate approach is intrinsically far more economical in the number of relations tested for a given amount of data gathering.

7. The real relations and limits of multivariate and bivariate experimental designs seem not to have been generally envisaged, because it has not been realized that each can be combined with either alternative in the four remaining polar dimensions listed in (4) above. In particular, histori-

cal custom has assumed that only bivariate designs can be manipulative; that individual difference research can be multivariate but process research cannot; that causal inference requires manipulative design (whereas it needs only to be sequential) and that irrelevant variables are better handled by control to fixed values than by free variation accompanied by measurement. One aim of this book is to show the inaptness of these traditional restrictions, e.g., by demonstrating the indispensableness of factor analytic research in such process studies as perception and learning.

CHAPTER 2

Extracting Factors: The Algebraic Picture

SYNOPSIS OF SECTIONS

2.1. The Aims of Multivariate CORAN: Component Analysis, Cluster Analysis, and Factor Analysis
2.2. The Basic Factor Proposition: Correlation Size Related to Common Factor Size
2.3. The Centroid or Unweighted Summation Extraction of Factor Components
2.4. The Principal Components or Weighted Summation Extraction
2.5. Communality: Common (Broad) and Unique Variances
2.6. Checking Back from Factor Matrices to Correlation Matrices
2.7. The Specification and Estimation Equations Linking Factors and Variables
2.8. Summary

2.1. The Aims of Multivariate CORAN: Component Analysis, Cluster Analysis, and Factor Analysis

Typically, the factor analyst takes a careful choice of variables in the area of behavior in which he is out to test a theory or to explore structure, e.g., primary ability variables; marker variables for a supposed extravert temperament; measures which psychologists have thought to be indicators of strength of motivation; or representatives of an hypothesized learning gain pattern. As he looks at the square matrix of all correlations among, say, thirty such variables, he wants to know how many distinct independent influences—presumably decidedly fewer than thirty—can be considered responsible for the observed covariations in this domain of behavior.

15

In short the purpose of factor analysis is to find a new set of variables, fewer in number than the original variables, which express that which is common among the original variables. Whether these new variables are themselves to be allowed to be correlated or uncorrelated is a secondary decision. Also, whether these new variables are assumed to represent real scientific influences (determiners) or merely convenient mathematical abstractions for further calculations is left to secondary procedures.

Psychologists, however, have resorted to other principles besides factor analysis for finding sets of variables abstracted from the given variables, and we must take a side glance at these before proceeding with factor analysis itself. It will be only a side glance, because their natures will be more fully encountered in later developments. One is *principal components* analysis in which there is no reduction in the number of variables (*n* components from *n* variables), but in which the first few components account for much of the total variance of all the variables. It has properties useful to some mathematical procedures, but typically the components have no relation at all to defensible concepts of scientific entities. Moreover the components among *n* variables change as further variables are admitted to the correlation matrix, so they lack the constancy of a scientific concept.

The second alternative is *cluster analysis*, in which variables are grouped together according to the high correlations they have with one another. Such correlation clusters (in psychology the basis of *surface traits*, *surface states*, and some kinds of *types*) can be of any number from one and the same set of correlations, depending on where the experimenter arbitrarily decides to draw the line between a high and a low correlation. By contrast the number of factors can be fixed nonarbitrarily from the experimentally given values in the correlation matrix. Component analysis and cluster analysis will therefore be set aside from our main search for *determiners with nonarbitrary scientific meaning*, though we shall see more of their properties in other connections from time to time.

Even within factor analysis itself (sometimes called *common factor analysis*), however, the student should be prepared to encounter a diversity of subconcepts and of calculating methods. As regards concepts, some psychologists are content to think of a factor just as some "common property" among the set of variables, others, e.g., Burt, as a "principle of classification,"[1] others, e.g., Horst, as an economical predictor in place

---

[1] There are logical objections to using factor analysis itself as a system of classification, which become most evident in the description (p. 325) of Q technique. Briefly, in classifying persons, the classification into high and low in, say, intelligence is quite different from that on, say, anxiety. The process ends not in a classification but in *numerous* classifications each on one factor. The same is true of variables as of

of several variables, and others, as in the case of the present writer, as an underlying determiner, such as mass, oxygen pressure, temperature, anxiety, size of cortex, etc., of the covariance manifested in the set of variables concerned. This last aim and design in factor analysis is that of extracting the important concepts—the "significant variables"—which is the central purpose in the early—and often in the later—stages of all sciences.

The diversities in factor analysis are (a) of model and (b) of method of calculation, though minor features of the model itself are sometimes modified by the method of calculation. Cluster analysis, as we have seen, strictly belongs outside factor analysis. It often appeals to beginners because it is simple to pick out variables which correlate highly with one another and in this way keeps "close to the ground," but as will be seen (p. 44) it has a fatal Achilles' heel of arbitrariness in separating clusters, and we shall therefore set aside and study basically two models: *component analysis* and *factor analysis*. And since, as just indicated, component analysis is no good as a final scientific model it will be studied only as an adjunct useful in some stages of factor analysis.

Component analysis aims to finish with as many component "factors" as there are variables, each component presenting a particular pattern of correlation with all variables. That is to say each successive component takes a slice off the original variance of all variables, though the first components take much more than later ones, so that, in, say, a 20 variable problem the first three components might account for 70% of the variance–covariance of the variables. Thus an investigator contented with a quick and rough solution to his calculations might be prepared to substitute those 3 for the 20, though if he wanted to be absolutely accurate he would have to take out all 20 components. There is a sense in which factor analysis takes out fewer factors than variables—in contrast to component analysis—without losing the precision of its concepts, as we shall see.

Let it be said at the outset that the traditional designation "common factors" is a misnomer, for when the process which we describe later of reaching "simple structure" has been well carried out no factor is a common influence to all the variables. The correct expression is a *broad* factor, such as might cover, say, a third of the variables. The expression "broad factor" (essentially covering two variables or more) is constantly needed in factor analytic description to contrast the remaining kind of factor which is called a *specific* factor and is unique in its action to *one*

---

people. The student should also beware of supposing that seeking for correlation clusters of variables, in cluster analysis, has any relation to Q technique (transposed factor analysis). They are entirely different from one another and from the grouping of people in factor space described in finding *types* (p. 325).

variable. (Parenthetically, "broad" and "specific" are essential in psychology to distinguish from "common" and "unique" designating a trait factor common to all people as opposed to one unique to an individual personality.) In addition to the model of *components* and *broad* (*common*) *factors*, which account for *n* variables, respectively, by *n components* or *k broad and n specific factors*, there are some minor models of which alpha analysis is taken as a representative in (see Table 13.2). It is a form of broad (common) factor analysis which artificially does away with the specific factors and scarcely fits most expected conditions on real data, but deserves references in regard to possible models. We shall not meet it again until Chapter 13.

Although the beginning reader will not be able to get the detailed picture until certain mysteries are explained, it seems desirable to set out once for all in Table 13.2 (p. 403) what the varieties of abstractive analyses from concrete variables, by CORAN methods, can be today. At this point the reader need only note in the top rows of Table 13.2 the divisions having to deal with the primary distinction we have just made between component analysis and factor analysis. The columns are concerned with methods of calculation, such as centroid, principal axis, and maximum likelihood, which methods can become clear only as we proceed. In the next section we shall begin with the simplest calculation method: the *centroid method*. Except where experimenters happen to be restricted to desk or hand computers, it is the general practice nowadays, with large electronic computers available, to use the *principal axis method*. But as the history of a subject often reveals the best sequence for teaching it, so the *centroid* (or *unweighted*) method proves to be the best introduction to understanding the more complex *weighted, principal axis* method. As indicated, its relation to the principal axis method is that of a simple unweighted to a weighted composite and, as such, it gives the best insight into what happens in taking factors from a correlation matrix.

2.2. The Basic Factor Proposition: Correlation Size Related to Common Factor Size

In going through the centroid to the principal axis and so to the final factor model our aim is to show the reader, in general terms, how the factors which emerge are related to the observed correlations in the experimentally given correlation matrix. Actually, there are two ways in which this can be demonstrated (1) in arithmetic–algebraic reasoning and (2) in geometrical form. The psychology of individual differences tells us that some individuals will learn more easily through symbol usage and others through geometric visualization. (Even the leading exponents have their

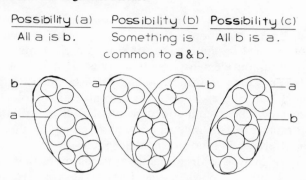

Possibility (a)   Possibility (b)   Possibility (c)

All a is b.   Something is   All b is a.
                   common to a & b.

Common elements of equal size shown by O's

Fig. 2.1. Three possible interpretations of a given correlation coefficient.

preferences. Thus, Thurstone visualized whereas Burt told the present writer this approach meant nothing to him and he felt it more natural to proceed algebraically.) Since there is no harm and much good in meeting the same ideas twice from different angles, we shall reinforce understanding with both approaches here, beginning with the algebraic.

Experimental research deals with correlations between variables, which it seeks ultimately to interpret in ideas about what is causing the correlations. Most students are familiar with the proposition that a significant correlation of a with b may mean one of three things: (1) that a causes some of b, (2) that b causes some of a, or (3) that a third factor causes some of both. This is illustrated in the "common elements" representation in Fig. 2.1.

What we mean by an element is a unit which can vary, say, from plus one to minus one, or zero to one, and thus contribute to the variance of the collection of units which constitute the variable. Thus in a collection of 10 pennies if we call the count of total number of heads the score, then each penny is an element which may be heads or tails—one or zero.[2] The variance of a variable, i.e., $\Sigma d^2 / N$, where the d's are deviations from the mean, is simply proportional to the number of (equal)

[2] In such a model the ten would be tossed separately, each coming down heads or tails, the heads being scored one and the tails zero. All ten coins would be tossed to get a person's score in variable a. Let us suppose that seven of these are supposed to be common elements with variable b. Then whatever score the tossing procedure gave to a person x on the first seven coins in a would be transferred to his b score, the remainder of that b score being made up by the toss on the remaining three coins. Only the last three coins would be free to give different scores in a and b. This is what we shall always mean by "common elements": they are values that will vary from person to person but stay the same for one person from one psychological variable to the other (or even to several others).

elements. It is one quarter, thereof, to be precise, if elements can each be zero or one. The common variance in two variables which correlate is simply proportional to the number of common elements. In an educational setting common elements might be illustrated by the correlation observed between scores in arithmetic and English, the assumption being that intelligence contributes to both. In laboratory psychology this common variance might be a common perceptual skill contributing to two learned performances. The independent part, equal to three elements, might be the difference in reinforcement schedules.

Now factor analysis is, in principle, nothing more than asking what the common elements are when one knows the correlation. We shall start discussing this for two variables and then extend it to several variables. The basic proposition, which Spearman (1904) stated almost a century ago (and also in Karl Pearson's earlier writings) is that when there are common elements g to two variables a and b, then

$$r_{ab} = r_{ag} \cdot r_{bg} \qquad (2.1)$$

That is to say, the correlation of a with b in Fig. 2.1 is the correlation of a with the factor common to a and b multiplied by the correlation of b with that same factor. Obviously, knowing as we do, from experiment, only r_{ab}, we cannot directly find r_{ag} and r_{bg} in this case.

The secret of how factor analysis can approach a solution from this basis is that instead of just two it takes a lot of variables into the calculation, using the correlations of a with b, with c, with d, and so on. The value we want to get r_{ag}, i.e., the amount of common elements in a, is

$$r_{ag} = \frac{r_{ab}}{r_{bg}} \qquad (2.2)$$

If we assume that the mean correlation of all the variables with the common element in all of them is

$$r_{mg} = \frac{r_{bg} + r_{cg} + r_{dg} + r_{eg} + r_{fg} + \cdots}{n}$$

and let the mean correlation of a with all the variables be written

$$r_{am} = \frac{r_{ab} + r_{ac} + r_{ad} + \cdots}{n}$$

then r_{ag} will be a function containing r_{am} divided by r_{mg}. We say "a function" because there are statistical niceties to observe and at this point we are concerned only to give a general logic. That general logic is that

$$r_{ag} = f \left(\frac{r_{am}}{r_{mg}} \right)$$

$$r_{bg} = f \left(\frac{r_{bm}}{r_{mg}} \right)$$

(2.3)

and so on for all variables. The function is to be discussed below, but for the moment let us take the common sense position that r_{mg}, the mean correlation of all variables with their common factor is a function of the mean of all their mutual correlations [of which there will be $n(n - 1)/2$] and call this r_{mm}. (Alternatively, since these correlations are likely to range from 0 to 1.0, one might very roughly take 0.5 as r_{mg}.) The value r_{mm} would be obtainable simply by summing the whole correlation matrix. With this assumption we can proceed to an actual calculation to estimate r_{ag}, namely,

$$r_{ag} = f \left(\frac{r_{am}}{r_{mm}} \right)$$

(2.4)

That is to say, the mean correlation of any variable a with all variables, divided by a suitable function of the mean intercorrelation of all variables among themselves, is an estimate of how much of the common elements (general factor) variable a possesses.

The centroid and the principal axis methods of factor extraction are based on this relationship derived from the full correlation matrix, though they differ in the algebra of weighting procedures. Also, they cannot simply take the straight means above, because it will be seen that we must add variances, and variances are related to the *squares* of the correlation coefficients.

This relation of correlations and variance contributions must be examined before proceeding, at the cost of a momentary divergence from the main treatment of the R matrix in factoring. It is a truism that the square of the correlation of any variable a with b tells the proportion of the variance (σ^2) of b predictable from a, or vice versa. Thus if we take Fig. 2.1,

$$r_{ab}^2 = \frac{\sigma_a^2}{\sigma_b^2} = \frac{7}{10} = 0.7$$

(2.5)

(Since common elements tell the fraction of the variance, so that r_{ab}^2, equals the *variance* ratio—predicted to total—not the *sigma* ratio.) The same is true if the actual relation is as shown in Fig. 2.1 (c), where all b is in a. To generalize to our factor situation with c common elements, as in Part (b) in Fig. 2.1, we would conclude that the correlation of the variable with the general factor is such that

$$r_{ag}^2 = \frac{\sigma_g^2}{\sigma_a^2} \tag{2.6}$$

But the extreme instances in Fig. 2.1 where all a is in b, or all b in a (which corresponds to a factor being all in a variable) are uncommon. If someone gives us a correlation of .6 of leg length with body length over 1,000 people we know where we are: leg length is 36% of the total body length variance. But we are not granted such prior knowledge of structure in psychological data and the safest assumption is model (b) in Fig. 2.1., where a and b have some common elements but each has something specific besides. In this case, for simplicity we have made (b) symmetrical— 7 in common and 3 elements specific to each. Incidentally, this is also the specific case of the reliability coefficient.

Note that whereas 7 elements in common in cases (a) and (c) yields a correlation of the root of .7, i.e., .84 [see Eq. (2.5)] it yields literally a correlation of .7 in case (b). This says

$$r_{ab} = \frac{\sigma_g^2}{\sigma_a^2} \tag{2.7}$$

but it does not deny the fundamental premise that the square of the correlation coefficient is the fraction of variance predicted. After all in (b) the 7 common elements, comprising σ_g^2, are not the whole variance of one variable any more. If a predicts σ_g^2/σ_a^2 of the variance of g and g predicts σ_g^2/σ_b^2 of the variance of b, then the squared correlation of a with b must be calculated as predicting $\sigma_g^2\sigma_g^2/(\sigma_a^2\sigma_b^2)^{1/2}$,

$$r_{ab}^2 = \frac{(\sigma_g^2)^2}{(\sigma_a^2\sigma_b^2)^{1/2}} \tag{2.8}$$

If a and b have symmetry, as in Fig. 2.1, b is then

$$r_{ab} = \frac{\sigma_g^2}{\sigma_a^2} \tag{2.9}$$

That is to say the common fraction is not predicted from the square of the correlation as in Eq. (2.6), but from the correlation itself, Eq. (2.9), which is why we stated in Eq. (2.7) that the square does not equal the common variance fraction of each. This, by the way, is the "catch" in making inferences from the reliability coefficient, as in Eq. (2.9). If as is commonly the case, however, the two variables have different size specifics, then Eq. (2.8) is the preferred formula. And if the elements in a and b in (b) above were 7, with 3 specific in a and 5 specific in b, the correlation would be

$$r_{ab} = \frac{7}{(10 \times 12)^{1/2}} = .64$$

It is because this division into underlying common and specific parts can never be made from a single correlation coefficient (except with inside information as in the legs example) and since one value is compatible with a host of different divisions of the elements, that we turn to another resource: factor analysis. (Numbers of elements fix the correlation; a single correlation does not fix the number of elements.) The basic principle of comparing the variable a's correlations with everything in the matrix, with each and every other variable's correlation with everything in the R_v variable intercorrelation matrix runs through all factor models and methods. The reader can probably see insightfully the essential correctness of this on common sense grounds, if he wants to get any variable's correlation with what is common to all. He can also see that the results will be more determinate (in converging on some stable value) the more variables we take into the R_v. What may bother him a little, if he is thinking around the problem, is that there may surely often be more than one factor in common. But let it be said that since common variances from different factors can be united in simple additive fashion this creates no difficulty in eventually finding the totality of common elements. As we shall see, methods are available for splitting the total communality into several independent common factor variances, if several exist.

Let us now introduce the more precise statements of the above general principle by accompanying the student through the actual calculating steps in the smallest convenient but representative numerical example. We shall start with the *centroid* ("center of gravity" or *unweighted summation*) method because, although as we have stated it is in these days of large computers far less used than the *weighted summation* (principal axis) method, it is an excellent learning introduction to the more sophisticated principal axis method, and there are in fact only trivial differences in the results of the two forms of calculation.

2.3. The Centroid or Unweighted Summation Extraction of Factor Components

The centroid method systematizes what we have already indicated above as the logical approach, namely, to sum the correlations of a given variable with all variables, down column a of the R_v matrix and divide that by the sum of all correlations of all variables with one another.

This should give us an estimate of the r_{ac} correlation (or *loading* as it is often called when a *factor* is one term in the correlation), i.e., the correlation of the variable with that portion of it which is common to all.

We shall work through a small example of eight variables. (We cannot cut down to, say, four variables because we want to illustrate the sequential extraction process through as many as three factors and four variables only would not permit this.) Table 2.1 represents the experimentally obtained correlation matrix R_v among 8 variables.[3] For ease of adding columns the triangular matrix is repeated in the top right half, and this will be done in all correlation matrices hereafter. The only oddity to be explained is the set of values down the diagonal. These values, not directly given by the experiment and called *communalities*, are defined as the correlation of each test with itself *due to the common elements part only*. Since we do not yet know what the common factors are we have to start with a good guess—or in polite terms an estimate. How to get an estimate for these values will be explained later; but let it be clear that it is not good, ideally, to put an r of 1.0 or a reliability coefficient in the diagonal since both of these are, with logical certainty, too high. The ninth row Σr at the bottom (Σ meaning "sum of") adds each column of r just as we advocated above.

The procedure is to total each column (including the communality) as shown at Σr in the 9th row (yielding for example 1.18 for V_1). The total value T for all correlations in the matrix is then found by adding across the Σr row. Since all loadings must never square to more than 1, these Σr totals are then divided by $\sqrt{T}$ (actually one commonly multiplies in the computer by $1/\sqrt{T}$). By doing this for each column we obtain as in the bottom (11th) row four figure answers which are reentered in the 10th row, rounded to two. These are according to our theory the estimated correlations of the eight variables with the common factor. These correlation values, .33, .39, .11, etc., will henceforth be called *factor loadings*.

Now if the correlation of V_1 with the common factor is .33 and that of $r_{2.c}$ is .39 then from formula (2.1) above the correlation of V_1 with V_2 (r_{12}) should be

$$r_{12} = .33 \times .39 = .13$$

But the actual experimental value in R_v is .20 (Table 2.1) so the first factor as thus estimated is not enough to account for all the correlation. It

[3] From this point on we shall begin using the word *matrix*, though the algebraic rules of multiplying matrices, etc., will be deferred a while. A matrix is simply a square or oblong "box" of cells in which numbers are placed according to some agreed rule. If a factor matrix has n rows and k columns it is called an $n \times k$ matrix. If we lay the same V (variable dimension) matrix on its side (and "over") instead of upright it becomes a $k \times n$ matrix and is called the *transpose* of the V matrix and written V^T.

Table 2.1. Correlation Matrix[a] of Eight Variables R_v

Test variables	V_1	V_2	V_3	V_4	V_5	V_6	V_7	V_8
V_1	(.30)	.20	.28	.36	-.09	.01	.06	.06
V_2	.20	(.30)	.00	.52	.36	-.04	.04	.04
V_3	.28	.00	(.30)	.04	-.36	.04	.04	.04
V_4	.36	.52	.04	(.90)	.66	-.02	.43	.39
V_5	-.09	.36	-.36	.66	(.90)	-.06	.27	.24
V_6	.01	-.04	.04	-.02	-.06	(.05)	.09	.08
V_7	.06	.04	.04	.43	.27	.09	(.80)	.72
V_8	.06	.04	.04	.39	.24	.08	.72	(.60)
Σr	1.18	1.42	.38	3.28	1.92	.15	2.45	2.17
T (sum of Σr)	12.95							
$\frac{1}{\sqrt{T}}(\Sigma r)$	.33	.39	.11	.91	.54	.04	.68	.60
	.3280[b]	.3948	.1056	.9118	.5338	.0417	.6811	.6033
$m = 1/\sqrt{T} = 0.278$								

[a] As soon as communalities are inserted in the diagonals it is called a *reduced* matrix, R, contrasting with the original *unreduced* matrix, R, which had ones.
[b] More accurate values, rounded in row above.

is at this point that we assume a second factor must be invoked to account for the .07 remaining. The subtraction of what can be accounted for by the first factor from what the actual correlations (R_v) are is done systematically for the whole matrix by constructing a first *factor product matrix*, as shown in Table 2.2, by making such a calculation as gave us .13 above for all cells, i.e., the loadings of all variables are multiplied together in pairs. The values are then subtracted cell by cell from the original R_v

Table 2.2. First Factor Product Matrix

Test variables	Loading on last factor	V_1	V_2	V_3	V_4	V_5	V_6	V_7	V_8
		.33	.39	.11	.91	.54	.04	.68	.60
V_1	.33	(.11)							
V_2	.39	.13	(.15)						
V_3	.11	.04	.04	(.01)			*a*		
V_4	.91	.30	.35	.10	(.83)				
V_5	.54	.17	.21	.06	.48	(.28)			
V_6	.04	.01	.02	.00	.04	.02	(.00)		
V_7	.68	.22	.27	.07	.62	.36	.03	(.46)	
V_8	.60	.20	.23	.07	.55	.32	.02	.41	(.36)

[a] These r's are the same as the r's in the lower left. As we do not need to add columns they are omitted.

matrix. To get the first factor product matrix we "bound" an 8×8 empty matrix with the first factor loadings as shown in Table 2.2. Then each cell is filled by the product of its row and column loadings. In this case the *communality* (see below) diagonal is also filled by products showing the communality due to the first factor (in parentheses).

Now we subtract these products from the original r systematically. This goes well till we get to the r of $V_1 \times V_7$ which equals $+.22$ and is greater than the original r of .06. What does this mean? It indicates that an r of $-.16$ remains to be accounted for by a subsequent factor or factors. In other words, the loadings of these two tests in the second factor, since their common possession of it caused them to be correlated negatively, are bound to be of opposite sign. Such negative residues will occur quite frequently. In fact, typically in about one half of the r's, as the *residual matrix* presented in Table 2.3 shows.

The fact that some variables can be positively and others negatively loaded in the same factor should occasion no conceptual difficulty. Among physical variables, for example, we might obtain a general factor of body weight and it is easy to see that body weight would influence some performances, e.g., wrestling, in a favorable way, but other variables, e.g., pole vaulting, negatively.

Can we now find the loadings in the second factor by repeating the column-adding procedure used for the first, aiming at obtaining the mean r of each test with all others? Any attempt to do so reveals the surprising fact that the columns now add to zero due to balancing of the positive and negative r. Indeed, when we consider later computational checking procedures (p. 34), we shall see that this addition precisely to zero is a proof of the correctness of the preceding step. In this example, due to rounding off loadings to two decimal places, the column totals do not

Table 2.3. First Residual Matrix[a]

Test variables	V_1	V_2	V_3	V_4	V_5	V_6	V_7	V_8
V_1	(.19)	.07	.24	.06	−.26	.00	−.16	−.14
V_2	.07	(.15)	−.04	.17	.15	−.06	−.23	−.19
V_3	.24	−.04	(.29)	−.06	−.42	.04	−.03	−.03
V_4	.06	.17	−.06	(.07)	.18	−.06	−.19	−.16
V_5	−.26	.15	−.42	.18	(.62)	−.08	−.09	−.08
V_6	.00	−.06	.04	−.06	−.08	(.05)	.06	.06
V_7	−.16	−.23	−.03	−.19	−.09	.06	(.34)	.31
V_8	−.14	−.19	−.03	−.16	−.08	.06	.31	(.24)
Totals	−.00	.02	−.01	.01	.02	.01	.01	.01

[a] $T = .07$.

sum exactly to zero, but even with this approximation, the mean of column totals is only +0.01.

In the centroid method of extraction, which we are here following, it is necessary now to make every column add up positively again by reflecting rightly chosen variables—in this case 1, 2, 3, and 4. We want them to add positively in order to take out as much variances as possible with each new factor, and the "rightly chosen" variables are those which best do this. This is quite legitimate: we are simply saying that as far as the residual matrix is concerned a variable that was, say, "sociability" now becomes "unsociability" and all the r's in its row and column change signs. The immediate loading will be for "unsociability" in the second factor loadings but to regain consistency with the first these four will have their loadings reversed. The second factor will thus have negative loading on sociability, i.e., it will operate to reduce a person's sociability.

Once the first residual matrix is thus "positivized" as a whole, as far as possible, the new (second) factor loadings are found by just the same adding, etc., as with the original R_v. Thereafter the process repeats itself: a new cycle begins and a new product matrix and a new residual matrix are obtained. In the present case after three cycles had yielded three factors the last residual appeared as shown in Table 2.4.

The residual correlations are now considered so trivial—many being zero to two decimal places—that it is not worthwhile trying to take out a fourth factor, and so the results are entered as a three-factor solution as shown in Table 2.5.

The reader may wish to see whether, as an exercise, he reaches these values beginning independently with the R_v matrix above. The intermediate details for this particular example are given in Cattell (1952, 1967), while excellent alternative illustrations are available for student exercise

Table 2.4. Residual Matrix After Taking Out the Third Factor[a]

Loadings on last factor	+.21	−.34	+.48	−.34	+.65	−.16	−.34	−.31
−.21	(.06)	.00	−.04	−.03	.00	.00	.00	.00
+.34	.00	(.08)	−.01	−.01	.00	−.01	−.03	−.05
−.48	−.04	−.01	(.07)	−.02	.02	−.01	−.01	−.01
+.34	−.03	−.01	−.02	(.08)	.01	.00	−.02	−.04
−.65	.00	.00	.02	.01	(−.02)	.00	−.02	−.01
+.16	.00	−.01	−.01	.00	.00	(.07)	−.02	−.02
+.34	.00	−.03	−.01	−.02	−.02	−.02	(.08)	−.01
+.31	.00	−.05	−.01	−.04	−.01	−.02	−.01	(.10)
Totals	−.01	−.03	−.01	−.03	−.02	.01	−.03	−.04

[a]Total before reflection = −0.16.

Table 2.5. Unrotated Factor Matrix, V_0

Test variables	Common (broad) factors				Unique factors							
	F_1	F_2	F_3	h^2	U_{s1}	U_{s2}	U_{s3}	U_{s4}	U_{s5}	U_{s6}	U_{s7}	U_{s8}
T_1	.33	−.48	.21	.38	.79							
T_2	.39	−.29	−.34	.35		.81						
T_3	.11	−.38	.48	.39			.78					
T_4	.91	−.20	−.34	.98				.13				
T_5	.53	.24	−.65	.76					.49			
T_6	.04	.07	.16	.03						.98		
T_7	.68	.47	.34	.80							.45	
T_8	.60	.44	.31	.65								.59

in Child (1970), Gorsuch (1974), Fruchter (1960), Lawlis and Chatfield (1974), and in the more advanced textbook by Harman (1967).

2.4. The Principal Components or Weighted Summation Extraction

The above simple summation method, as Burt (1940) calls it (commonly called the *centroid method*, meaning "center of gravity") will become clearer when we see the geometrical illustration below. "Unweighted summation" has the advantage over *weighted summation* (in which each column of correlations is weighted by the loading for that factor before proceeding) of simplicity, but it does not extract the factor variance as efficiently. This weighted summation is generally called the *principal axes* method, due to Karl Pearson and in other forms to some pure mathematicians such as Jacobi. Incidentally, one should beware of confusing *principal axes*, which is a computational method of *extraction*, due to Pearson, Jacobi, and others with *principal components* which as Table 13.2 shows, is a scientific model, due to Hotelling, Kelley, and others which puts unities in the diagonal instead of communalities and takes as many components as there are variables. The centroid and principal axes calculations yield essentially the same end results (after rotation), but the principal axes method pulls more variance out in the earlier (unrotated) factors. It does not alter the number of factors required, which is n, the number of variables, if we use the component model (putting ones instead of communalities as above), or k, if we agree to stop sooner because we have used communalities and reached a residual matrix which looks trivial. This somewhat greater efficiency along with other mathematical properties have caused principal axes to be preferred by mathematicians, while the avoidance of the "reflection" problem, which is tiresome in the centroid, adds to the attractiveness of principal axes to computer programmers. Incidentally, the same essential convergence has been shown by Digman and Woods (1976, in press), Tucker, Koopman, and Linn (1969), Harris and Harris (1973), and others for various special derivative methods (minres, alpha, etc., discussed here later).

In the principal axes method one takes the totals for the columns and multiplies each column (and corresponding row) by a fraction of this total, thus weighting each variable differently—giving "to him that hath." When this weighting and column totaling is repeated several times the result converges on a set of limiting values for the loadings. Residuals are then taken and the weighting process is repeated, taking out factor after factor, as in the centroid process. If the principal axes method is pursued to n factors making a solution on the principal components model we have a result corresponding to what mathematicians have pursued for other purposes and which is called *determining the latent roots* and *latent vectors* of a matrix. (These are also called *characteristic roots and vectors*.) Therein the vectors are columns corresponding to the factors, with their loadings on variables, and the roots contain the evidence of their sizes. It is desirable to point out this bridge to the world of mathematics while we are dealing with principal components, but it would be quite inappropriate to cross that bridge into complex formulas at the student's present stage of interest, though he would do well to note the names. There are two common algorithms (sequences of arithmetical steps arranged for the computer) in use, the Hotelling and the Jacobi method, but their differences are slight and irrelevant to the psychologist. An excellent review of principal axis procedures and of their expression in the principal components model can be found in Gorsuch (1974), Harman (1967), and Mulaik (1972). This principal axis procedure underlies, nowadays, as a beginning step, practically all minor and some major variations of factor analysis, which the student will meet under such names as alpha factoring, minres, canonical factoring, maximum likelihood, and so on.

The next problem concerns what to do in either the centroid or principal axes calculations, in the case that we aim, as in Table 1.1, at inserting communalities in order to use the factor rather than the components model, i.e., to stop at k factors (fewer than n variables). Then we have to decide when the residual matrix has really come to zero (short of n factors) and this will depend partly on what estimated communalities we put in. The important fact to keep in mind is that the communalities and the number of factors are mutually dependent values.

If the experimenter keeps ones in the diagonal he will end with n factors; if he makes communalities large he will end with more than k and if small with fewer than k (assuming we call the correct number k).

It will be seen later that nowadays one either decides on the number of factors by some ulterior means, e.g., what will be described as the scree test (p. 60), and then asks that the communalities be calculated exactly to fit that number, or one uses the maximum likelihood test. But in the traditional centroid method one goes at it the other way, i.e., one makes a best estimate of the communalities and then extracts factors till some statistical test says the last residual matrix is essentially full of zeros. The communality for a variable a_j, which is written h_j^2, is the fraction of its variance shared with the broad factors in all other variables. With a *single* common factor it is the square of the variable's correlation with the factor, as we would expect from the above discussion of variance. The reliability coefficient for the given variable has sometimes been used as an estimate of h^2, but it is always an overestimate because a variable agrees with itself through having other things in common besides the broad factors. The more common choice has been the highest correlation of that variable with any other variable. As Burt pointed out, if we arrange these highest loadings for each variable in its columns in declining order of the variables to n values, the higher values will underestimate and the lower will overestimate, so one can, with artistry born of experience, add about 10% to the higher r and take 10% off the lower ones to give the required communalities and get surprisingly close to the true values (see Cattell, 1966). Of course, 10% is an arbitrary but likely value.

With larger computer resources one finds the multiple correlation, R of all variables with the given variable, and squares this to get h^2. As Roff (1935) and Guttman (1940) have shown, this is an accurate determination of the *lower bound* for h^2. More will be heard about communalities below. We shall not go further here into determining when (with the h^2 put in by one of the above methods) the residual matrix has fallen to zero, since that approach is rarely used today. But let us assume these steps have been taken and that k factors are indicated, so that the experimenter possesses a matrix of n rows and k columns giving the correlation of every variable on the k factors. This is commonly called the unrotated factor matrix (Table 2.5) and though it is a complete statement of the factor variances, it is not the end of the story. For reasons given before it has to be rotated to become meaningful. That rotation process from V_0 the unrotated matrix to V_{rs} the rotated matrix we shall take up in the next chapter. Here, however, it is necessary to pause and examine more fully the nature of this unrotated matrix and its relation to the correlation matrix, which we shall do for unrotated factor matrices in general, without restriction to our initial approach by centroid extractions. For most psychologists nowadays will have given this R_v matrix to the computer, and, after specifying the number of factors, by peeping in at an intermediate step, will receive a final printed out V_0, the unrotated matrix.

2.5. Communality: Common (Broad) and Unique Variances

The above factor matrix is commonly called the *unrotated factor matrix* and is written "V_0" since rotation is still to come and will usually move by successive V_1, V_2, etc., approaches to a final rotated *resolution* V_{rs} (automatic programs can also handle some of this rotation). From the V_0 itself one calculates the communality finally reached, written as h^2 in the last of the common factor columns as shown in Table 2.5. This h^2 for each variable is the sum of the squared factor loadings for that variable, e.g., it sums the three values in each row, in Table 2.5, after squaring them. Now it has been pointed out above that when a correlates with b and we want to find how much of the variance ($\Sigma x^2/N$) of b is accounted for by a we have to square the correlation r_{ab}. Consequently, since the factors in a V_0 are *un*correlated among themselves, their variance contributions are simply additive: the communality h^2 is the sum of the squared correlations of the variable with the factors. It tells us how much of the variance of the given variable j is accounted for by the totality of broad, common factors, thus (when 1, 2 to k are the common factors):

$$h_j^2 = r_{j1}^2 + r_{j2}^2 + \cdots + r_{jk}^2 \qquad (2.10)$$

Contrasted with these broad factors running over two or more variables (as it happens in Table 2.5 over all but one, to any extent), are those that remain absolutely unique to one variable. Incidentally, as pointed out below in another connection, "broad" is really a better term than the traditional term "common" since most are *not* common to all variables in the matrix and experiment but only broad, in showing loading across two or more variables. Besides, the word "common" risks confusion with a factor pattern *common across people* as distinct from one *unique* to an individual, and it is only in this latter sense that we shall here consistently use "common."

Turning our attention now, by contrast, to the variance *unique* to a variable, which is usually written u^2, we see that

$$u^2 = 1 - h^2 \qquad (2.11)$$

Although this variance is mathematically unique to the variable, we should clarify the above-hinted truth to the psychologist, that psychologically this "unique variance" has less claim to correspond to any real entity or psychological concept than does the broad, common factor variance. One should not overlook that the loading of a so-called unique factor is nothing but a "leftover variance," calculated as what remains after the known broad factors have been taken out. Thus, for variable 1 (Table 2.5) the unique variance is $(1 - .38) = .62$, and the loading (correlation) of "factor U_1" on variable 1, is, naturally, the square root of that, i.e., .79, which is therefore entered into the unique factor part of the matrix—the diagonal part to the right of Table 2.5.

The verbal usage "unique factor" for this variance is misleading for two reasons: (1) In a scientific model we assume that it is actually a sum of some supposed real psychological specific factor, F_S, and error of measurement, e, thus:

$$u_j^2 = b_{sj}^2 + b_e^2 \tag{2.12}$$

Note that we are using F_{sj}, i.e., a factor specific to j, for the psychological factor and u_j for the statistical leftover uniqueness, which is F_{sj} plus error, b is a general term for a loading, and hence b_{sj} is the loading for the specific F_{sj}. (2) The specificity of u_j or F_{sj} is peculiar to the given factor matrix and experiment. If other variables are added to the matrix, which have some overlap with j part of u_j^2 will be transferred to the broad factor part of the matrix. Indeed the implication is that the so-called specific factor consists of some as yet undiscovered broad, common factors and perhaps of several specific factors thus:

$$F_{sj}^2 = F_c^2 + F_{sj1}^2 + \cdots + F_{sjx}^2 \tag{2.13}$$

where F_c is the new common factor. Thus it is, psychologically, less likely to mislead us if we speak of a unique variance rather than a unique factor. The latter is, in psychological discussion, better spoken of as "specific factor and error" variance, knowing that the specific is almost certain to be broken up further with more extensive research into new psychological entities. The broad factors, on the other hand, have no intrinsic reason to change their patterns in further research.

Incidentally, it has become customary to reproduce, in a published factor matrix, only the common part (since the unique part can always quickly be calculated from it). This is a good economy, provided one is not in danger of forgetting the existence of this unique variance and the specific factor and error which constitute it. One is forcibly reminded of the uniqueness, however, when one comes to estimate a person's factor scores (p. 238) or to make theories about what the specific factor means psychologically.

2.6. Checking Back from Factor Matrices to Correlation Matrices

Compared with most other calculations that a psychologist in quantitative experimental work has to make factor extraction is a long job. It begins with the scores of individuals on a series of tests and ends with the loadings of factors on those tests. This is a long trek—even though the modern computer may hide the distance—and when one considers the further steps yet to be taken (such as rotation) it is not surprising that, with many successive chances for error, different psychologists may reach rather divergent results. At least, they are sometimes astonished at one another's conclusions from the same data. More divergences occur at the later stages of analysis and can be left to later discussion, but let us point out here that there is at least a sure way of checking the path *up to the stage of the unrotated V_0 matrix.*

Let us suppose a_j and a_k are variables defining two adjacent rows on the V_0 factor matrix. To be concrete, let us take variables 1 and 2 from Table 2.5.

Then, since according to our basic proposition $r_{ab} = r_{ac} r_{bc}$, we can calculate the correlation r_{12} due to factor F_1 only as $r_{12(F_1)} = .33 \times .39 = .13$. This $r_{12(F_1)}$ is called an "inner product" of the two rows, when it is carried further to all factors. When we carry it further to F_2, the added correlation will be

$$r_{12(F_2)} = -.48 \times -.29 = .14$$

and from the F_3 loadings we similarly get

$$r_{12(F_3)} = .21 \times -.34 = -.07$$

When such sources of covariation are uncorrelated as they are in the unrotated matrix, they can be simply added, so that

$$r_{12} = r_{12(F_1)} + r_{12(F_2)} + r_{12(F_3)} = .20$$

If we turn to the original R_v (p. 25) we see that this figure agrees with the correlation of variable 1 with 2 in the original, experimentally given correlation matrix in Table 2.1. The reader may wish to check one or two other original r's in the R matrix by these inner products.

So far we have made no use of matrix algebra, but to anticipate the fuller instruction given therein in Chapter 5, we will point out that every correlation in R_v—computed by these *inner products* of two rows in the V_0—can be very quickly calculated in the computer. We instruct it to multiply V_0 by its transpose. As stated, its transpose V_0^T is simply V_0 turned over sideways as shown here in the first matrix equation we have introduced—Eq. (2.14). The regular rule in matrix multiplication is to

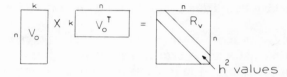

Fig. 2.2. Restoration of correlation matrix R_v from unrotated factor matrix V_0, by matrix multiplication. R_v will also have the communalities created, falling down the diagonal. It is therefore called a "reduced matrix" relative to the original R_v which had 1.0 entries down the diagonal.

multiply each and every row in the first matrix by each and every column in the second. Since we have arranged the matrix to come before its transpose, the computer following this rule will get the inner products of corresponding factors and add them, just as we did above. Doing this in all possible row and column pairs will fill all rows and columns of the R_v matrix, including putting communalities in the diagonal, as shown in general terms in Fig. 2.2 and Eq. (2.14):

$$V_0 \times V_0^T = R_v \qquad (2.14)$$

As regards labeling matrices, we shall throughout call the correlation matrix of the r's in the original matrix itself R_v and for the matrix of variable loadings on factors (which as individual values will be b's) we find V a convenient term. In any case it is common, as in Eq. (2.14), to write all matrices in capital letters (since they are vectors or collections of vectors) to contrast with lower case letters for *scalars*, which mathematically designates them as ordinary single numbers, having magnitude but not direction (as vectors have). This restoration of R_v is a useful check to see how close our communalities and numbers of factors have been estimated and how accurate all computation has been.

2.7. The Specification and Estimation Equations Linking Factors and Variables

The factor analysis initially gives us a breakdown of variables into factors and says nothing at that time about individual people. It is a general statement, as far as individuals are concerned, of *relations* existing in that population sample. However, it provides a basis for a next step in which the factor scores of individuals can be derived from their scores on variables or, conversely, the raw scores of any individual on his variables can be estimated from his scores (when known) on the factors. Let us consider the latter, under the title of the *specification* equation.

A correlation is a special form of a *regression coefficient* in which the same, standard-score metric is used in both variables. A regression equation, of one or more regression coefficients, is a means of estimating a person's score on b when one knows his score on a, and (in multiple regression) also on other variables. Clearly the factor matrix V_0 has regression coefficients which are the means of estimating a person's score on any variable in the matrix if we have obtained, say from some factor test batteries, his individual factor scores. Conversely, they are (but not as they stand) the means of getting his scores on factors if we have his scores on the variables. The details of the latter process are a little more complicated because the variables are mutually correlated whereas the unrotated factors are not. This makes the regression weights and the correlations two different things. The complication is slight but it is best to leave this transformation until later. In any case the former is of greater interest for psychological theory, for it shows how much a person's performance on some variable such as "rate of learning French" is a function of his score on, say, the general intelligence factor (F_1, below), a personality factor of his capacity to concentrate (F_2, below), his auditory memory (F_3, below), and so on. Let us imagine for the moment that variable T_7 in Table 2.5 (p. 28) is a test of achievement in school in French. Then we can take row 7 and write it as an equation:

$$v_7 = .68\,F_1 + .47\,F_2 + .34\,F_3 + .45\,U_7 \qquad (2.15)$$

This is called the *factor specification equation* for any given performance. For a particular individual we would take his standard score on F_1 and multiply it by .68, then on F_2 and multiply it by .47, and so on through however many common, broad factors are involved. The result will be only an estimate of his score on v_7 so long as we have no value for U_7. The needed scores for the broad factors F_1, F_2, and F_3 can be estimated in standard scores, for the given individual, from test variables T_1, T_2, T_3, T_4, T_5, T_6, and T_8, in a way soon to be described (p. 238). Alternatively, they can be directly measured (still as an estimate) from other tests existing as factor batteries, e.g., an intelligence test for F_1, Scale G on the 16 PF for F_2, and so on. Once the factor scores are known they can be used for estimation, by appropriate specification equations, not only of the set of variables first used to get the factor scores, but of quite a number of other variables as well for which loadings are known.

However, the reader will have noticed that though we can estimate scores for the common, broad factors from many variables at once, and so get a reasonably reliable value, there is nothing to estimate U_7 from except v_7 itself. This would be running in a circle and there is no point in doing it. If, miraculously, we did know the error and specific factor values in U_7 we would not be estimating v_7, with some degree of uncertainty, but precisely calculating it, by Eq. (2.15). Since we do not know this value it does not help us to know it should be weighted .45, so the fourth term on the right is dropped from the estimation of our predicted (estimated) score for v_7. This should remind us that estimation in both directions (factor to variables and variables to factors) is better as the h^2 values—communalities—get larger.

In this example, as stated, the unrotated factors are uncorrelated so their contributions can be simply added in the specification equation. However, in most rotated matrices the factors are correlated, so this makes special beta weights (regression weights, different, of course, from correlations) necessary for both directions of estimation. A point later made more systematically, is that the use of factor analysis in psychology is helped by keeping to a standard language of symbols, so we shall consistently use b for these factor *weightings* or *loadings* (this is convenient because they are "betas" both in statistics and psychology, meaning behavioral indices, i.e., statements of how far the given factor enters into a particular behavioral response or performance). What we shall, with equal consistency, call the *specification equation* (sometimes factor specification and also behavioral specification) is thus one special class of regression equation—one using factors instead of other predictor variables—and may be written in general symbols as

$$a_{ij} = b_{j1}F_{1i} + b_{j2}F_{2i} + \cdots + b_{jk}F_{ki} + (b_{js}F_{jsi} + b_{je}E_{ji}) \qquad (2.16)$$

Here a_{ij} is a measure of individual i's response or performance in act a_j. The b's are the weights, with subscripts showing they are peculiar both to j and to each factor, from 1 through k. (They are the numbers in the rows of Table 2.5.) It is assumed here that the factor scores, F, and the variable scores, a_j, are in standard score form. We settle here on the convention of writing the specific factor for a variable a_j as F_{js} and its b as b_{js}. The remaining part of the v_j variance is written as E_j indicating that error is specific to the measure of j. F_{js} and E_j are together in brackets to remind us that statistically they go together in U_j, the unique variance in a_j.

The reciprocal procedure we have spoken of—estimating factor scores—can be set out in the following equation, though the calculation of the w's must be left till later (p. 274):

$$\hat{F}_{xi} = w_{x1}a_{1i} + w_{x2}a_{2i} + \cdots + w_{xn}a_{ni} \qquad (2.17)$$

where $\hat{F}_{xi}$ is i's estimated score on factor x; the a are variables in standard scores and the w's are weights to be derived from the factor matrix. These also are beta weights, in general statistical terms, but it is helpful to distinguish these factor score estimation weights, by the symbol w, from the b in the behavior specification equation.

2.8. Summary

1. Multivariate correlational analysis methods (CORAN) have the main aims in most models of reducing the prediction from many variables to prediction from a few and, in all, the aim of reaching useful conceptual abstractions from the concrete variables used in experiment.

2. *Cluster analysis, component analysis,* and *common (broad) factor analysis* are the main generically different approaches (to which, with some hair splitting, alpha analysis may be admitted as a fourth genus). The first is rejected for further study here since, in the form of correlation clusters, its results are arbitrary and not useful for any superstructure of calculation. Component analysis and factor analysis differ in that the abstractions in the first are as numerous as the variables, whereas in the second the broad common factors (though not the specific factors) promise a useful scientific reduction of the number of determining concepts. The principal axis extraction method can be used, however, as the calculation avenue to both.

3. Within *factor analysis* (broad, "common" factor analysis) factors and variables can be given either uniformly unit variance, in what may be called standardized FA so far discussed, or, alternatively, differences of size in what is described later as real–base FA (Chapter 13).

4. One must clearly distinguish between *models*, as placed in their initial taxonomic relations in (2) and (3) above, and *methods* of analysis to reach results formulated in models, though differences in the latter will slightly modify the former. The main five reduction *methods* are the centroid, principal axis, weighted least squares, maximum likelihood, image components, and factoring of covariances methods, though various hybrids, e.g., rescaled image analysis, exist.

5. This chapter, introducing the student as by a first course, presents the centroid method of calculation in detail and the principal axis in outline, respectively, as the *unweighted* and *weighted summation* methods. These methods (as do all methods when aimed at the same model) give results so similar that for most purposes choice among them hinges on cost, convenience, and the user's predilections. There are slight advantages in the principal axis method, notably when used with the principal components goal, in that it has the mathematical properties associated with taking latent roots and vectors from a matrix, though maximum likelihood (yet to be described) ultimately supersedes it for statistical reasons. However, though the unweighted summation method is nowadays the standby only of those who have nothing but desk computers, it is followed in detail here as the best simple practical illustration of what factor analysis is doing.

6. The basic proposition in all, but most readily seen in the un-weighted summation (centroid), resides in the common elements' view of the correlation coefficient. The basic introductory formula is

$$r_{ab} = r_{ac}r_{bc}$$

where c is some common source of variance. It is shown that three different structures in the common elements—exhausting the possibilities in correlation—can yield the same correlation coefficient and that no solution of r_{ac} and r_{bc} is therefore possible from knowing a single correlation r_{ab}. Factor analysis makes headway, though at the cost of approximation, by taking many (n) variables correlated in all ways. The mean correlation of variable a with the ($n - 1$) others is a tolerable first estimate of the amount of c (common factor) in it. In the unweighted summation method the total of a's correlations with all variables is divided by the square root of the total of all variables' correlations with one another, to give the first factor loading, and so on for all variables.

7. After thus determining the correlation of each variable with the first extracted factor, a *product matrix* is formed showing the correlations among variables due to their loading on this first factor and this is subtracted from the original R matrix to see what correlation still remains to be accounted for. The *residual matrix* (after "reflections") is then treated like the original matrix. A series of cycles of product matrices and residual matrices is continued until a residual is reached which has essentially nothing left but zeros.

8. The correctness of the unrotated, $n \times k$ factor matrix thus achieved can be checked by back multiplying ($V_0 V_0^T$) to restore the correlation matrix R_v. It will be shown later that the estimate of the communality can be improved by inserting in R_v the h^2 obtained at the end of the first analysis and reiterating the analysis (p. 68). Since number of factors and size of communalities are mutually dependent, there are two possible directions of procedure. One can give careful attention to estimating communalities, by four methods mentioned here, and stop factoring when residuals are deemed at zero. Alternatively, one can use methods to decide the number of factors and let that fix the communalities. It can be indicated here that the latter is better recommended, later, except with use of the maximum likelihood method.

9. The correlations in each row of the V_0 (wherein factors are orthogonal) constitute the regression coefficients of the *specification equation* for estimating a person's score on the variable in that row from his factor scores. Reciprocally, the correlations of each column provide the basis for estimating the factor score from the variable scores, though they must first be changed to beta weights because the variables are correlated. This transformation from correlations to weights, as in calculating a multiple

correlation, must be made for both directions of estimation when, as in the advanced factor model, factors also go oblique (correlated). The beta weights in the *specification equation* are written "*b*" for *behavioral indices*, and those in the *factor estimation* equation "*w*," for weights of variables.

10. Besides the broad common factors, the factor model admits U, the unique variance for each variable, which breaks down into a specific factor F_s, and an error factor E. Thus, there are n plus k factors in all, and the F_s may actually split again. Because of this and because no variable normally correlates perfectly with a factor, both of the equations in (9) give estimates not exact answers. An exact calculation of a variable score by the specification equation would be possible only if the score for the specific factor were known, and this is normally unobtainable.

CHAPTER 3

Rotating Factors: The Geometric Picture

SYNOPSIS OF SECTIONS

3.1. The Correlation Coefficient Geometrically Represented

It has been pointed out in the last chapter that in factor analysis as in most scientific analysis systems, e.g., the mathematical structure of the atom, the same model can be presented either algebraically or geometrically. Having given in Chapter 2 the algebraic and arithmetic meaning of a factor common to several variables we now propose to reinforce it by a geometrical, graphic presentation and to use this as the clearest approach to the next step—that of rotation.

The student is used to depicting a correlation by two coordinates, one for each variable, and plotting the positions of people thereon by their scores. This "scatter diagram" then has two regression lines drawn on it, the mutual closeness of which gets greater as the correlation gets higher. The geometric representation now to be used is different (though related) and moves away from representing the two coordinates at right angles. Instead, the convention we are now to follow, geometrically represents the correlation as an angle between two lines (vectors) corresponding to the two variables. Further it does this among orthogonal coordinates

40

representing the dimensions of the factor space. To be precise r is made equal to the cosine of such an angle, θ, as shown in Fig. 3.1. Therein we are shown an r of +0.6 on the left, of about 0.8 next, of zero next (since a right angle has a cosine of zero), and one equal to a negative correlation (an angle greater than a right angle) on the extreme right. For a correlation of +1.0 the two lines O–A and O–B would lie tightly together, and for a full negative correlation (–1.0) these vectors would go in completely opposite directions from the origin, O.

The above is labeled "simplest" because when we come to factor space it is also necessary to let the length of the variable vector play a part. The correlation then becomes a function of the cosine and the standard deviation (length) of the variables, which, with unit length *factors*, becomes, for the variable, the root of the communality of each *variable h* in that space. Thus, precisely,

$$r_{ab} = h_a h_b \, \cos \theta_{ab} \qquad (3.1)$$

An example that will show in geometrical terms (as was shown in Chapter 2 in algebraic terms) how the correlations fix the number of factors is possible if we take the simplified case of no specific and error factors.

Let us suppose that in some experiment with three variables X, Y, and Z we obtained correlations between them of .56, .83, and .69 (and that we know there are no specific factors involved). Then let us cut out cardboard arcs with angles whose cosines corresponded to these r's as shown in Fig. 3.2. There are two possibilities when we fit these to a common origin: (1) that they can be fitted flat on the paper as shown in Fig. 3.2(a), or (2) that they "buckle" when we attempt to lay them flat because the angle XOZ is no longer just the sum of XOY and YOZ as it is in Fig. 3.2(a) (where $r = .0$ for XOZ would permit X, Y, and Z to lie in the flat) but is smaller.

In this case the only way to get them together is as a "horn" as shown in Fig. 3.2(b) which demands three dimensions. Now if test variables are represented as vectors and put together at the given cosine angles (from the R matrix given by experiment) they will tell us what factor space is necessary and the variables will appear as fixed points in that space. Thus if we had a dozen variables that fitted into three-factor space

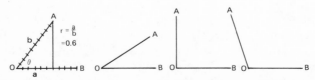

Fig. 3.1. Simplest geometrical representation of correlation coefficients.

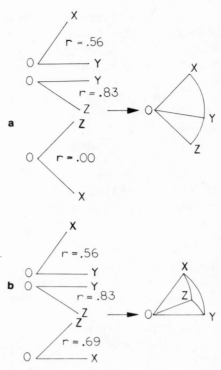

Fig. 3.2. Dimensionality of space as determined by correlations as cosines. Angles corresponding to three correlations among three personality variables. (a) Fitting angles permits a two-dimensional figure. (b) Fitting angles together creates a three-dimensional figure.

we could represent them by knitting needles stuck into a cork. However, to do that we would have to have an estimate of their communalities for as we have seen, the convention, to be consistent for calculation, would have to make each knitting needle equal in length to the h for the given variable.

3.2. Plotting Variables in Hyperspace from the V_0 Matrix

Since three dimensions are as much as the eye can handle and physical space allows, it is not practicable to extract the factors in any but two- and three-dimensional cases by cutting cardboard arcs and seeing how they fit together—even if we guessed the communalities correctly. Since it is thus not a working proposition to use this as a factor extraction procedure, we return in practice to the algebraic methods of Chapter 2. However, this geometrical approach is valuable as an illustration of what is happening, and, what is more, it is a real basis for the calculations in what we have called rotation procedures, once the V_0 has been reached by arithmetic-algebraic approaches.

The dimensionality of the hyperspace (space beyond three dimensions) has been fixed by the number of factors given us in the V_0, and we can now plot the positions of the test points (or vectors) in that space if we are content to take views seeing two dimensions at a time. In that case we take two factor columns from Table 2.5 at a time and plot the position of the eight variables (by their projections—loadings—on the factors) as shown in the three "shots" in Fig. 3.3. (The student would do well to check a few positions against Table 2.5.)

It should be noted that as we get into hyperspace and take just two-dimensional plots therefrom the length of the vector as drawn is no longer the total h but only the h value due to the two factors being plotted. Since we started in Figs. 3.1 and 3.2 with unit length (full standard deviation) test vectors, and since even when all common factors are covered the length of the test vector in hyperspace is still less than unity, the student may wonder where the rest of its variance has gone. The answer is that, as the factor matrix (p. 28) reminds us, there are also eight unique variances, which are all orthogonal (at right angles) to the three common broad factors and to one another. If we completed the pictures we should therefore have one special dimension into which each variable—and it alone—projects. There would thus be $8 \times 3 = 24$ two-dimensional plots for the 8 unique dimensions with the 3 broad ones, besides the possible paired comparisons among the 3 broad factors, namely, 3 such plots. However, only the latter are involved in rotation, so with k broad factors we can typically look at $k(k-1)/2$ two-dimensional plots.

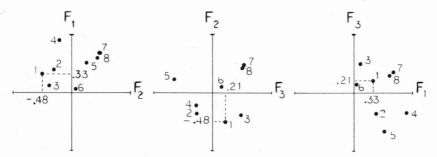

Fig. 3.3. The conception of variables as vectors in factor space. The factor loadings of test 1 are shown as projections to illustrate what the values in the specification equation on p. 28 mean in the spatial presentation.

3.3. The Relation of Cluster Analysis to Factor Analysis

Although our central concern is factor analysis, we shall pause a moment over these plots to explain more adequately the remarks made above about cluster analysis, since the plots present a good opportunity to do so clearly. For visual illustration it is best simply to take a two-factor example, as shown in Fig. 3.4. This is built up from a correlation matrix in which high correlations were found among variables 1, 2, 3, and 4 and again among 27, 28, 29, and 30, and so on, through clusters A, B, C, D, E, and F. The communalities happen to be high and the correlations such that all fall in the flat, i.e., in two dimensions requiring only two common factors, F_I and F_{II}, to account for most of the correlations. They are drawn oblique instead of exactly orthogonal to illustrate an important point to be met later.

A cluster is to be recognized as a "sheaf of arrows," i.e., a set of test vectors all highly correlating with each other, but little correlated with any others. There are no fewer than six such clusters—A, B, C, D, E, and F—visible in the plot, though there are only two factors, F_I and F_{II}. This fact that clusters and factors are neither identical in number nor in nature should never be overlooked, though there may often be a cluster close to a factor, as B is to F_I and C is to F_{II}. On the other hand it is possible for factors to exist where there are no clusters—as would be so if the 30 vari-

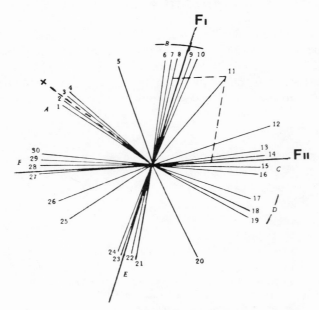

Fig. 3.4. The meaning of clusters in relation to factors.

ables in Fig. 3.4 were quite homogeneously distributed in space. Granted that a fairly high correlation is demanded for belonging to a cluster, as in this illustration, there will normally, however, be more clusters than factors. This point should be remembered when we later condemn certain computer programs for "chasing *clusters*" in attempting to place factors.

As mentioned, earlier, because of its apparent simplicity and closeness to the correlation matrix itself, the correlation cluster has had a perennial appeal to psychologists. That has happened especially in personality, where it is used to define some concept of temperament or some unitary pathological syndrome. Apparently, this happens mainly because a cluster is more concrete, immediately evident, and easier to understand than a factor. To find such syndromes all one needs is a correlation matrix among variables and an eye to pick out the subgroups of variables within which there are substantial mutual correlations.

The syndrome or correlation cluster does have some use in psychology, though as we shall see later it has ineradical weaknesses of definition. To avoid confusion the present writer introduced in the personality context the now fairly widely used distinction in personality psychometrics between a *surface trait* (correlation cluster) and a *source trait* (uniquely rotated factor). A surface trait is thus just an obvious "going together" of this and that, e.g., the educated vs. uneducated man cluster which rests on positive correlation of performances in English, arithmetic, etc., or the cluster of symptoms seen in the schizophrenic syndrome. But the source trait is an underlying contributor or determiner and in the correlation cluster just cited the positive correlation of English and arithmetic proves to be partly due to the fluid intelligence source trait, partly to a factor of years of education common to these and other school subjects and partly to a third broad factor—a source trait of superego strength which determines steady application in school (Cattell & Butcher, 1968). Similarly, the clinician recognizes that surface trait syndromes like schizophrenia, or obsession–compulsive neurosis are unitary only in appearances and that they must be understood as products of interaction of several distinct source traits and influences.

Actually, the correlation cluster has several disabling weaknesses as a psychometric concept and measurement tool, to which brief reference has already been made. Let us focus them a bit more sharply and then put cluster analysis aside and confine ourselves henceforth to factor analysis. [Except insofar as we return to discuss the danger of "cluster chasing" (p. 134) in factor *rotation*!] The first weakness of the cluster is that its boundaries—and therefore the central vector to the cluster (a "centroid") which is its essential basis of measurement—are arbitrary and hard to get agreement upon. For example, in Fig. 3.4 is 12 a part of cluster *C* along with 13, 14, 15, and 16? To draw a definite bound for a cluster one has to fix the lowest degree of correlation that will be accepted for admission of a variable. And this arbitrary decision results in one psychologist's fiat leading to different sets of clusters from another's. The centroid vector—X for cluster A in Fig. 3.4—will, of course, wobble from place to place with the changing boundaries of the cluster.

Secondly, with change of accepted "admission" correlation, two clusters in one research become one in another. Imagine that two new variables, 31 and 32, fell between variables 16 and 17 in Fig. 3.4. Then there would no longer be clusters C and D but only one sprawling cluster. Therein variable 12 would actually correlate rather low with variable 19. The problem of cluster search (for both variables and people) has been exhaustively examined by Cattell, Coulter, and Tsujioka (1966; chap. 9), with development of the necessary new concepts of *homostats* and *segregates*, and of the TAXONOME automatic program for objectively and efficiently finding them by computer in a correlation matrix. But in the last resort the problem remains inevitably that of deciding which to call separate clouds in a stormy sky. Thus even with the objective TAXONOME computer program the same result is reached by different investigators only if they plug into the computer program the same arbitrary values for lowest acceptable intercorrelation of variables and degree of overlap of homostats. These are but two of three or four weaknesses which should lead us to set aside, except for certain more loose descriptions of variables by "types," the appeal of the meretricious correlation cluster. Rather than deal with surface traits and states it is better—especially when an ambitious superstructure of further prediction and calculation is involved—to learn factor analysis.

3.4. The Effect of Factor Rotation on Factor Patterns

The geometrical view of factor axes and variables as vectors has a more constructive value, however, merely than that of persuading us to discard clusters. It introduces the important topic of factor rotation. Let us see what this means geometrically before we look at it psychologically. On looking at any plot we discover a curious fact, very disturbing at first to our concepts of factors as meaningful source traits. The psychological data—the correlations—fix the position of the *test vectors* themselves in a rigid configuration, but in a mathematical sense they do not fix the positions of the *axes*. On reflection it will be seen that we are free to spin the axes where we like. The projections (loadings) of variables on them will change as we spin, but all positions are equivalent in accounting for the correlations! That is to say by rotation we can reproduce the same experi-

Table 3.1. Different Rotations of the Same V_0,
Equivalent in Restoring R_v

	(a)				(b)		
Variable	F_1	F_2	h_2^2	Variable	F_1'	F_2'	h^2
1	.5	−.4	.41	1	.13	−.63	.41
2	.4	.2	.20	2	.44	−.10	.20
3	−.3	.7	.58	3	.24	.73	.59
4	−.4	−.4	.32	4	−.56	−.05	.32
5	.6	.6	.72	5	.85	.08	.73
6	−.5	−.3	.34	6	−.58	.10	.34

Note: The text states "the communalities remain unchanged." The slight changes
here are from rounding approximations.

mentally given correlations from quite different sets of factor abstractions.
Thus, let us for simplicity consider a factor matrix, as in Table 3.1(a) in
which just two broad factors (and six unique variances, not shown)
account for the correlations among six variables. Fig. 3.5(a) shows a plot
of these six variables as points in two-space. In Fig. 3.5(b) we have left the
vectors in the original, experimentally given correlational relation to one
another, i.e., the angles and their cosines (the latter representing the
correlations) remain unchanged. But now we have rotated the reference
vector or factor axes through 40°. What Thurstone aptly called "the *test
configuration*"—the angular relations among variables—remains the same.

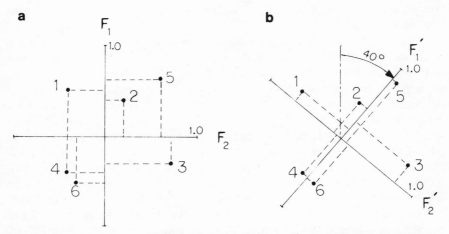

Fig. 3.5. Effect of rotating factor axes shown in a 40° clockwise shift, illustrating
invariant configuration with changing loadings.

However, the projections of the variables on these new axes, F_1' and F_2', are altered by the rotation, and if we measure them carefully we shall find the new projections in (b) come out as Table 3.1(b).

If one wishes to calculate the new projections, which we will call F_1' and F_2', from the prerotated projections F_1 and F_2, instead of measuring them on the graph, a well-known trigonometrical formula shows that

$$F_1' = F_1 \cos \theta + F_2 \sin \theta \qquad (3.2a)$$

$$F_2' = F_2 \cos \theta - F_1 \sin \theta \qquad (3.2b)$$

where θ is the angle of shift. If the reader cares to substitute .77 for cos 40°, and .64 for sin 40°, in this example of a 40° shift, he can check the values obtained from the graph. Note that the signs in Eqs. 3.2(a) and (b) are correct for a clockwise shift of F_1 and F_2, as in Fig. 3.5 (the trigonometric ratios always being positive). For a counterclockwise shift, the negative sign shifts from Eq. 3.2(b) to Eq. 3.2(a).

Incidentally, this orthogonal rotation calculation is presented in some texts in slightly different form. Since the sine of an angle is the cosine of its complement, we can write

$$F_1' = F_1 \cos \theta + F_2 \cos (90 - \theta) \qquad (3.2c)$$

$$F_2' = -F_1 \cos (90 - \theta) + F_2 \cos \theta \qquad (3.2d)$$

As will be evident, it is important to keep track of the sign of sine or cosine in the above equations in terms of the quadrant sign into which one is moving the reference vector. In later work, when we move into matrix multiplication, the four sine or cosine values in Eqs. (3.2a, b, c and d) will be set in a 2 X 2 matrix and called the transformation matrix, symbolized as lambda or L.

Although all the loadings alter, the new factor matrix in Table 3.1(b), is in most of its vital respects the equivalent of (a). For example, not only the configuration of variables but also their communalities remain unchanged. Not only do communalities remain unchanged (in orthogonal or oblique rotation) but the total variance of all orthogonal factors together remains unchanged (by columns). The variance contribution of any individual factor, however (squaring down the column), can change considerably. This can be seen algebraically by squaring the row of loadings, as usual, in Table 3.1(b) and it can be seen still more quickly geometrically by the fact that the test vectors retain their former lengths when axes are rotated about the origin. The graphically visible fact that the correlations (the cosines) among the variables remain unchanged can be checked algebraically by taking what are called the "inner products" of the factor loading of the two tests as we did earlier when restoring the R matrix. Thus, in Table 3.1(a) the correlation of the last with the second-to-last variable (5 with 6) is given by

$$r_{56} = (.6 \times -.5) + (.6 \times -.3) = -.48$$

In matrix Table 3.1(b), however, by our inner product rule, the correlation of variables 5 and 6 becomes

$$r_{56} = (.85 \times -.58) + (.08 \times .10) = -.49$$

i.e., except for rounding error it is the same. From pursuing this with other variables it will be seen that Table 3.1(a) and (b), the differently rotated matrices, always restore exactly the same correlation matrix and configuration. In short, the many possible rotations, giving different factor loading patterns, are equally consistent with the original correlation matrix R, and are therefore equivalent in an important sense to one another.

3.5. The Possibility of a Unique Rotation and the Need to Find It

Why, however, would one want to rotate? Since there are obviously an infinite number of positions at which the spinning axes can be stopped and all are equally good and equivalent mathematical "derivatives" from the original correlations (and capable of restoring those correlations), is not one as good as another? And have we not exploded at the outset the idea that factor analysis gives unique, scientifically meaningful factors? No, for it will be shown that one position is more meaningful than all others, and that this is not (save as an extreme coincidence) the position in which factors come "hot" out of the computer, i.e., the matrix position (a) in Table 3.1. That computer-delivered immediate position is certainly quite "accidental." Further, it can be shown that most others are equally indefensible as having any special meaning. Since the original accidental position, however, is "free" to be moved to a position of unique meaning, how do we find it?

In Figs. 3.3 and 3.5(a) the factors as they come in the unrotated matrix, V_0, from the computer are drawn as vertical and horizontal coordinates. It is correct to draw them orthogonal (at right angles) because, as the centroid extraction method reminds us, the successive factors are independent. But the verticalness is accidental and the true unique position of the ordinate that remains to be found could be anywhere. All that we know is constant and fixed is the restoration of the correlations from the projections of the variables on the various axes. Those correlations in R_v, including the h^2 values, will remain constant no matter how the projections change as we spin the axes. The next chapter, using both the geometrical view and its matrix algebra equivalent, will explain how we spin to the best position. For the technical issues in that process are so important as to deserve a chapter to themselves.

3.6. Summary

1. A correlation can be represented geometrically by vectors corresponding to the two variables and an angle of the right size between them. The length of each vector is represented by its standard deviation expressed graphically in units such that a factor is unit length. (This is the same as saying the length is h, when the communality is h^2). Then $r_{ab} = h_a h_b \cos \theta_{ab}$.

2. In the simplified case with no unique variances it can be seen geometrically, by cutting out angles for the given correlations as cosines, how the correlations determine what number of factor dimensions will be necessary to contain the test vectors.

3. In the ordinary case, with specifics, the vectors can be placed in space (if only two or three space, by knitting needles, of varying lengths, in a cork) by plotting their loadings given in the V_0 as projections on axes corresponding to the unrotated factors. The relations of the variable vectors to one another is called their *configuration*. When the configuration exceeds two space it can be graphically represented by a succession of two-dimensional graphical plots, of which "shots" there will be $k(k - 1)/2$ for k factors.

4. That the initial factor axis position coming out of the centroid or principal axis computer analysis is accidental and arbitrary can be realized from the fact that if we added a few extra variables, say x to the original n, the first centroid or axis factor (and therefore all subsequent ones) would have significantly different values for the same n variables. New company in the variables changes the center of gravity. (With rotation, incidentally, a position could be found to restore the n essentially to their original values: new company does not *finally* change the values, i.e., after proper rotation, except on any new factors it adds.)

5. Inspecting configurations brings home the arbitrariness of putting variables in distinct clusters, and the fact that clusters are generally more numerous than factors. This, and the poorness of the superstructure of calculations that can be built on clusters relative to factors, supports our preference for the factor over the correlation cluster model. Cluster analysis has often been—but should not be—confused with Q technique discussed later.

6. The factor axes to a configuration of variables can be rotated, the calculation giving new factor patterns for each position, but the configura-

tion (R_v) and communality h^2 remain constant and restorable from each different position. That h^2 remains the same can readily be seen in a two-space plot, as following from Pythagoras' theorem, since h is the hypotenuse of a triangle to which the loadings are the two sides.

7. Although all spins are in these respects mutually equivalent, and consistent with the original data, this multiplicity of possible *mathematical* solutions does not vitiate the conception of a factor as a real and uniquely definable determiner in scientific data. There is one unique position that satisfies, by further conditions, the requirements for a real determiner. The next task is to find principles for locating it.

CHAPTER 4

Fixing the Number of Factors: The Scientific Model

SYNOPSIS OF SECTIONS

4.1. A Return to the "Number of Factors" Issue

Before proceeding to the art of rotation, which gives the unique resolution ultimately needed as the solution to a factorial investigation, we need to return to base and tidy up some by-passed technical details of the factor extraction process itself. One may hope that the reader has, through the combined algebraic and geometrical approaches, got the conception of a factor as a common source and direction of variance contributing to the observed covariation of several psychological measures. He has, further, become acquainted with the computational process of extracting a series of successively smaller factor variances from the R matrix. The general outline has thus been given for the extraction process but there still remains thorny practical technique details about how we decide the number of factors, how we guess (or, to use a better term "estimate") the communalities, like those inserted in Table 2.1 and how we finally get to the most accurate form of the unrotated factor matrix. Also we need to note the slight distinctions between the factors from weighted and unweighted summation methods.

The answer to the number of factors might seem to be to go on extracting successive factors, as we did in the centroid example, until a residual is reached (p. 27) which is essentially emptied of all indications of covariance, i.e., in which the correlations have sunk virtually to zero. But they will never go exactly to zero, and experience shows that when more exact criteria are used there generally proves to be something left after the investigator declares he has wrung the R matrix dry. Whether for this or other reasons it is at any rate true, as one may notice in passing, that all but a few psychologists erred during the fifty years prior to 1970 in taking out fewer factors than we now know are required in the personality and ability data they examined and their theories have been accordingly misleading. Statistical methods have become available, notably those of Bartlett, Sokal, and ultimately Jöreskog for testing the significance of a last residual, but the end point is then still dependent on another influence, namely, on how well one has estimated the communalities originally placed in the diagonal of R. As Jöreskog points out statistically, and the present writer has noted in practice, by the time one gets to a residual that is by statistical test not different from zero the latest factors already contain an appreciable amount of error variance, and are not found to be accurately replicable.

Some psychologists have responded to this by urging that we retain the approach of successive extractions after inserting estimated communalities but sharpen the estimation procedures for getting the communalities. The only means to the latter goal that has a tight theoretical basis is the substantial undertaking of calculating the multiple correlation of each variable with every other, as proposed by Roff (1935) and developed by Guttman (1956). As will be argued later, though precise, this is theoretically wrong. For instead of giving us the most likely value for the communality this $h_j^2 = R_j^2$ value gives us the lower bound for that value.

Accordingly, the present writer (1958), Tucker and Lewis (1973), and others proposed some years ago the alternate approach of first deciding on the number of factors and then fitting such communalities to R as would make the residual go out exactly to zero when that number has been extracted. This will be described, precisely, below; but for the moment we shall pursue further the attempts to define an essentially zero residual. We shall do so not because it ends in what we would consider a really good approach, but because it throws some interesting light on the factoring process and on conceptions of significance of factors.

The question of whether a residual correlation matrix is significant can be considered alternately as that of whether the factor is significant that one would extract from that residual. Doubting the assumptions of the few theoretical attempts to develop formulas, Humphreys, Tucker, and Dachler (1970), Howard and Gordon (1963), Linn (1968), Horn (1965b), and others have resorted to concrete Monte Carlo methods, creating correlation matrices from random numbers (random normal deviates, to be exact, to simulate the usual normal distribution of scores) and extracting factors from them. If a correlation matrix is built up from a sample of, say, 100 "subjects" using variables which are random normal deviates, some correlations range into the .2, .3, and even .4 levels. The sigma of a zero r on 100 cases is .10, and factors with loadings ranging into .4 and .5 will consequently be obtained, as shown empirically by Humphreys, Linn, and others above. If values of unity are used in the diagonal (component factoring) there will, moreover, be as many factors as there are variables and the apparent final communalities of variables in these "random number" factors will be quite large. Humphreys makes an argument for stopping a factor analysis when one reaches down to factors no bigger than these factors given by random numbers. But there are restrictions, not to say fallacies, inherent in this argument. One gets, of course, quite large loadings when there are random correlations of 0.4 (since $r_{ab} = r_{ac}r_{bc}$ the loadings on two variables having such a correlation could each reach .63). What the straight random deviate factoring overlooks is that the random error variance, through error of measurement, is usually only a small fraction of the substantive variance in a normal experiment, and the former on its own does not replicate the real situation. Moreover, deviates from Z transformations are needed.

A rather common misconception at the outset about this whole business of deciding the number of factors present is that one is being "conservative" and demonstrating a superior scientific caution, if one takes out too few factors rather than too many. Thus we find, among well-known users of the method, Tupes and Cristal, and Norman stopping at six factors among personality ratings where tests applied by Barton, Dielman, Horn, Nesselroade, Schaie, Vaughan, and others show a dozen to twenty. (We find Eysenck even building a theory around the first three factors in data where most statistical tests show 12–20.) The attitude is a carryover from ANOVA, where one might conservatively decline to build a theory on a difference significant at the $p \leqslant .05$ level and, with good scientific sense, decline to assume a noteworthy relation until a $p \leqslant .01$ significance is shown. It is desirable to avoid a "type 1" error, and therefore $p \leqslant .01$ is more desirable, but science also loses when "type 2" errors are made instead from overcaution.

By contrast, the final conclusions of a factor analysis are *likely to be more distorted by underfactoring than overfactoring*.[1] The reason for this lies in the rotation process, and even though this latter has not yet been fully discussed the point can be made. Factors come out of the extraction process with variance sizes ("roots" or "eigenvalues," as later described here) in a steeply diminishing curve, as shown in Fig. 4.1(a). After rotation, however, it is commonly the case that the main substantive factors, when arranged in diminishing order, are more nearly alike in size, as in the first five factors in Fig. 4.1(b) contrasted with the steeper artificial "back staircase" effect in Fig. 4.1(a) which is an inevitable result of an efficient extraction process. The first seven rotated factors (assuming seven factors are taken finally in this example as the correct number), at the meaningful position, in Fig. 4.1(b) are quite *different entities* from the first seven extracted factors [Fig. 4.1(a)]. In the process of rotation the factors of the patterns of loadings on the variables are all reconstituted, becoming new mixtures (in various proportions) of the unrotated factors (see transformation matrix, p. 120). The new factor 1, for example, may have in it a third of the loading pattern of the original factor 1, a quarter of the unrotated factor 4, a quarter of the old factor 5, and so on. Consequently, if an investigator had decided in this case "conservatively" to cut off at three factors, his subsequent rotation would produce a very impoverished and misleading version indeed of the true first three factors. (This is the basic fallacy, for example, in Eysenck's three-factor account of human nature.) The new factor 1 would be a distorted version of the true one because vital parts of its variance lie beyond the point where such an investigator had "conservatively" cut off. For example, it would be missing the loadings needing to be contributed by unrotated factors 4 and 5 and it would be inflated by spuriously high relative contributions from the old factor 1. The distortion in the second rotated factor, when there are such omissions, is still greater, and so on increasingly down the line. That is why the agreement of rotations from an inadequate extraction with those from a correct extraction get poorer as one proceeds from the first factor. Note that the shadings in Fig. 4.1 are irrelevant to this discussion and belong to a later comment.

---

[1] Since factors decline in size, the "error" of adding an extra factor adds less error than the error due the lost variance from dropping the preceding factor. This is commonly admitted for orthogonal factors but a few have disputed it for ultimately oblique factors. In the present writer's opinion since all the needed variance is in the orthogonal factors, the question of rotation to obliquity is irrelevant to this issue of error.

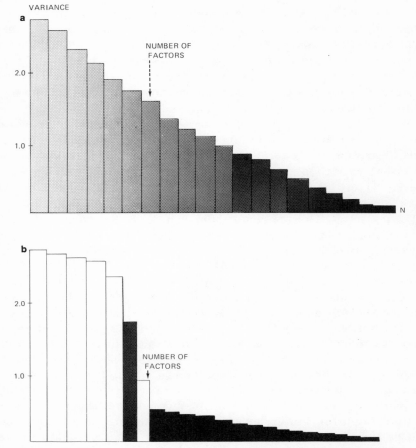

Fig. 4.1. Transformations of size and content between extracted unrotated and rotated meaningful factors with reference to cut-off point. (a) From extraction process—
before rotation. Shading indicates mixture of real and error variance in each factor.
(b) After rotation. Ideally hoped-for distribution (a few roughly equally large real
factors and comparatively small error factors) obtainable after exploratory stage, with
well chosen variables in a hypotheses-testing factor-analytic experiment.

4.2. The Alternatives of Statistical and Psychometric Bases of
Decision on Factor Number

Discussion on techniques for deciding factor number have done well
to contrast the *statistical* basis of approach, which takes account of
"blurring" of the population value by smallness of sample, and the *psychometric* approach which accepts the correlations at their face value

and asks what factor number gives the best fit. In approaching these alternatives, however, we run into a far deeper schism between a math–stat approach and a scientific model approach, which general issue has to be explained before the lesser issue can be intelligently handled.

The schism is most simply introduced by pointing out that according to the mathematician one cannot determine more than n factors or components from an R matrix of only n variables. The scientist, however, dealing with a total universe of possibilities, has to admit that (if influences are factors) the particular n variables he has chosen in one experiment may in fact have their relations determined by the impact of more than n influences; x more factors than n can of course be mathematically imagined, but with $(n + x)$ columns for a factor matrix of n variables, the resulting $(n + x) \times (n + x)$ factor correlation matrix acquires mathematically undesirable qualities. It becomes what is called a "singular" matrix, in which we have put more factors than there are dimensions, so that their vectors are forced into mutual correlation (some of them being superfluous), instead of being orthogonal.

The fact, however, that there could—according to the scientific model—be more than n factors (even if we cannot determine them properly from n variables) should be clearly realized and respected.

This can readily be seen if we take the 32 ball behaviors in the Cattell ball plasmode (p. 172), and, according to the evidence on factor number, take five factors—weight, elasticity, size of ball, length of cord, and distribution of weight toward the circumference. If we were to factor just a corner of the correlation matrix constituted by some four variables, e.g., height of bounce, time of pendulum swing, distance of rolling on a carpet, etc., these variables would still need to be accounted for, in scientific determination, by all five properties of the balls. Indeed, in the context of factoring all variables we can see from the V_0 that five factors enter these four variables. However, in factoring them alone we could only reliably separate and determine the nature of four factors at most.

As we move into discussion of matrices, let us note two new terms: that the *order* of a square matrix of correlations is the number of variables, and its *rank* is the number of factors that can be taken out of it. The logic of mathematics says that the rank of a matrix cannot exceed its order, and is, indeed, generally much less; and we have no quarrel with this. But if we are trying to infer the structure of the real world we have to keep in mind that our above four-variable factoring design does not prove that in fact there are only four factors at work. More will be said on this dilemma later; but for the moment we must note that we encounter an "artificial" limit to the number of factors as far as the scientific model is concerned due to the mathematical fact that k unknowns cannot be determined from less than k simultaneous equations.

4.3. Broader View of the Number of Factors Problem in the Light of the Scientific Model

If we go beyond mathematical considerations to consider the common sense probabilities of the real world, which our scientific model aims to fit, it seems likely that there are in most natural setting experiments at least as many factors as variables. However, we might also reasonably expect that most of them will be very small. For example, if we take 40 variables as used in researching on the Hakstian and Cattell (1976), *CAB* (*Comprehensive Ability Battery*) representing 20 primary ability factors, we may expect, in addition to the 20 ability factors, to see entering the actual test performance scores the effects of the 16 personality factors. These enter into any stressful performance in the examination room, and, in addition, we can expect a noise factor, affecting examinees nearer to a noisy street, and a bluntness of pencil factor slowing the performance slightly of those with blunter pencils, and perhaps a temperature factor because those at the back are rendered a little sleepier by the warmer air away from the windows, and a goodness of optical aids factor, since children with inadequate glasses may have slightly greater difficulty on certain tests. All these may extend the number of factors from 20 to, say, 40 (the number of variables) or more. But all except the 20 primary ability factors will be of an altogether lower order of size. After rotation, permitting the 20 real factors to emerge with distinctly larger variance as in Fig. 4.1(b), the remaining factors representing the above-mentioned small influences, will visibly be trivial. However, in the original extraction process the small species and large species will be intermingled in a smooth, relatively continuous descent [Fig. 4.1(a)].

These smaller "extraneous" species of factors (extraneous as far as the experimenter's planning and his choice of "important" variables are concerned) are, however, not the whole story of "factor debris." There is also the possibility of "error factors" and by this we mean *broad* factors such as Horn found, and not only specifics. Parenthetically, error in psychology is either *measurement* error or *sampling* error, and here, dealing with one sample as we are, our reference is only to the former. It is reasonable to expect that most measurements of variables will contain some error, small or large according to the accuracy of the instruments. Most models suppose that this error will be random and therefore essentially, i.e., *in the total population*, uncorrelated as between one variable and another. But as we have seen, in a sample as small as a hundred or so, such random normal deviates get correlated and we can extract factors from such correlations—as many as there are variables, though small in size, as seen in the studies of Humphreys, Horn, and Linn above.

As a matter of scientific faith, if we believe every event is determined, there are no such things as far as measures in nature are concerned as "random" numbers—despite the impressive tables of them provided in statistics books! But it is not worth going philosophical about whether there are correlations among errors simply because the sample is finite, or because the measurements of, say, both the GSR and reaction time were made for Smith on a humid day (slightly reducing their values) while for Roberts the same measures were all taken on a dry day, thus producing some correlations of experimental error. The correlations in a small sample set of random normal deviates, which should ideally be correlations of zero, may in fact be taken to simulate the effect of a large number of very small influences. Unless the instrumental errors are great, taking much reliability from the measures, these instrument factors we must remember will be of an altogether lower order of size than the substantive factors, e.g., personality and ability source traits. However, it will also sometimes happen that instrument errors are quite appreciable, so we cannot always assume a sharp break between large and small factors. The reader who wishes to delve further into the classes of instrument factors and their origins should study general *perturbation theory* (Cattell, 1961b; Cattell & Digman, 1964). But in connection with the number of factors problem, we need only note the existence, in general, of extra but usually quite small common factors due to measurement error. Of course, even when error is quite appreciable most of it will be quite *unique to the variable.* This has already been recognized in the specification equation [Eq. (2.4)] as E_j, and does not mess up the common factor structure. But in more sophisticated statements of the specification equation than we began with [see below (p. 67)] we must, consistent with what has just been concluded, also introduce terms for broad common factors of "correlated error." Sometimes only a single broad error factor is allowed for in factor extraction, on the assumption that all error is "swept up" into this one factor. Strictly, however, there may be several broad error factors and the "sweeping up" is therefore a convenient approximate procedure, with the limitations arising in all representations of the variance of several factors in one—chiefly that correlations cannot be exactly restored.

Now although measurement error and sampling error causes are distinct, they interact in certain effects. With smaller samples of subjects the error correlations can get larger and it is conceivable that one of these error factors will actually be larger than a smaller real, substantive factor, as shown in Fig. 4.1(b) for the sixth (shaded) rotated factor. In that case it will be likely to be included in the number of factors when we have cut off and thrown away what we designate as trivial factors, but this will be rare. However, it reminds us that error factors and trivial factors are not always identical. In Fig. 4.1 the darkness of shading is meant to indicate the amount of error variance in a factor. It will be noted that in the extraction series the error variance is not confined to the last factors, but spreads itself more evenly, and, in fact, in smaller degree contaminates even the first six extracted factors in Fig. 4.1(a). This is why it is fallacious to think that by underfactoring, by two or three factors, one is "cutting off the error." The error is everywhere. However, after rotation, as shown in 4.1(b) we may reasonably expect to have thrown most real variance into simple structure substantive factors and left most common error variance in the smaller (only roughly simple structure) factors, which consist, therefore, largely of pure error. (The largest of these, in a small sample experiment, however, may be larger than the smallest real, substantive factor, as shown in the sixth factor in Fig. 4.1.)

The only sure way to determine which of these rotated factors is error is to repeat analysis in a second factor experiment done with the same variables. Then the real substantive factors should show *matching* patterns, factor for factor, but the error factor patterns will be peculiar to each study. This is one reason why, if the indicator used to decide the number of factors leaves some doubt, it is better to take out one factor too many rather than one too few. For such extra variance is likely to be extruded in rotation from the simple structure system constituted by the true factors and thrown into the garbage of an error factor, though this may fail in an unskillful or automatic rotation. The other reason for risking a slight overfactoring is that the percent loss of true variance if one underfactors is, as mentioned above, greater than the false variance brought in if one overfactors, due to decline in size of successive factors [Fig. 4.1(a)] and rise in error ratio. It is true that taking out one more factor than is really present, in cases where there is really a definite number, i.e., where there is actually no "debris" of smaller factors, can lead in the rotation process to splitting a factor (see p. 168 below and Cattell, 1952, 1954, 1958, 1966a). But nearly always there is debris and in that case the rotation will correctly yield the final small substantive factor as such, even though it is too small to be of much importance, and will pile error into one error factor too few, without splitting a real substantive factor.

4.4. Practical Decision by the Scree Test and Maximum Likelihood

It might seem that this argument that there is virtually always an indefinite number of smaller and smaller factors, likely to extend the number of factors needing to be extracted up to at least the full number of variables present, should lead to a routine procedure of simply taking out as many factors as there are variables. (Even so one would still use communalities, not ones, as is done in component analysis.) Then one would trust to rotation to separate the wheat from the chaff. However, to dispense with applying some test for the number of factors would be to miss opportunities in that majority of experiments where there really are decidedly fewer factors than variables (frequently barely half as many). And even in the other cases the waste of time in rotating a lot of trivial factors argues against such a procedure, for the rotation labor goes up exponentially with the number of factors. [Our laboratory records show that the ROTOPLOT (see p. 143) convergence of a 10-factor problem to maximum simple structure may take 10 rotations and 10 days; but a 35-factor rotation can take four months.]

From the above account of the scientific model—of our best construction of what actually happens—what we briefly call *taking out the right number of factors* does not mean in most cases a number correct in some absolute sense, but in the sense of not missing any factor of more than trivial size (in the case of a psychometric approach) or one that is statistically significant (if we take a statistical approach, as in maximum likelihood, based on sample size). If, in the psychometric approach, the idea of these species—large substantive, trivial substantive, and error factors—seems a bit arbitrary, let the reader keep in mind that $p < .05$ and $p < .01$ are also arbitrary conventions. A conventionally agreed numerical size limit can, if needed, be just as readily assigned to the psychometric as to the statistical sense of trivial. Moreover, we can support this basis of separation in most cases, where samples are adequate, by the fact that the error factors are different in kind—a recognizably different species—from the substantive factors, by being unreplicable and showing a merely normal distribution of loadings instead of the more skewed distribution in substantive factor loadings.

One must recognize, nevertheless, as we have done above, that substantive factors may also be among the trivial ones, and in that case if we cut them off (preferably after rotation) we can only say they are psychologically too trivial to study at that stage of research. Just so such explorers as Captain Cook or Francis Drake did not stop in a first transit of the Pacific to study all the islands seen. How difficult this decision will be depends on the design of the experiment. If it is (a) exploratory and (b) representative in design, i.e., is compelled to work with a wide stratified sample of variables, then the transformation from large to small may be gradual, and in deciding the number of factors we may be compelled to recognize marginal uncertainties. But if it is (a) hypothesis testing and (b) aimed to polish findings in a field already explored, then a sharp decision on the number of factors is generally ensurable by attention to choice of variables and conditions. One chooses substantive variables such that they may be expected to give common factors with loadings altogether bigger than any due to error of measurement or to situation, instrument factors, of the kind discussed above. As will be seen when we come to study the design of factorial experiments, this condition is not unattainable even in exploratory researches, while in hypothesis-checking researches, "marker variables from past studies," i.e., two or three of quite high loading on each factor, can almost always be chosen to yield large substantive factor variances if those factors are there.

If a break exists, as will almost always be the case, between such larger substantive factors and the debris of error factors and factors largely outside the test variables, then the number of psychologically significant factors can be found, as shown in more detail later, by what is called the *scree test*. To use the scree test one has to begin by taking out successive components, i.e., one puts ones in the diagonal and runs to n components. One then plots their diminishing sizes (defined as the summed squares of their loadings, i.e., from the "latent roots"). Typically this plot line shows a distinct break between the "chute" of the larger factors and a much more gently sloping straight line running thereafter to the nth root (see Figs. 5.3–5.8). This latter runs at a constant angle, like the scree of rock debris at the foot of a mountain—hence the present name. Evidence will be given in Sections 5.2–5.4 that this break is the end of the real substantive factors.

The main alternative methods to the scree are (1) the Kaiser–Guttman rule of stopping when the last latent root falls below one, and (2) shifting the whole basis of factor extraction to the maximum likelihood method. The latter is expensive but good, though it needs to be watched, as explained when we study it later. The former—referred to as the K–G rule— has been extremely popular and is incorporated in countless programs because of its simplicity. Yet, we would suggest it is wrong, in principle, and erratic in practice as is shown below, and in various publications over the last decade. That is to say, where ulterior knowledge exists, to check the answers, the KG is in error about five times as frequently as the scree and the average magnitude of the miss is perhaps ten times greater.

The two main methods we shall develop—scree and maximum likelihood—are sufficiently different to offer just the balance the researcher needs, the *scree* being representative of the *psychometric* basis of evaluation and the *maximum likelihood* of the *statistical* evaluation, i.e., the former takes no account of sample size whereas the latter does. In most ranges of sample in which psychologists work—say, 200–400 cases— empirical evidence indicates that the two agree quite well; but the scree is, computerwise, quicker and cheaper, while the maximum likelihood is more complete in theoretical basis. To both of these methods we shall return in detail; they are introduced here to indicate the existence of practical solutions to deciding the number of factors.

4.5. The Contrasting Properties of the Principal Components and Factor Models

In this chapter, concerned essentially with rotation and in the more detailed developments of the next chapter, reference is made to the factor and component models requiring sophistication somewhat beyond that reached on these models in their first introduction in earlier chapters. Accordingly, in the present section we will look a little more closely at the principal components model and principal axis method of extraction and so bring out contrasts with the factor model. For although one may settle on the factor model, some use of principal components is needed from time to time, as in the above use of latent roots of components in order to apply the scree test.

It has been mentioned that in using the principal axis or weighted summation method of factor extraction one puts unities in the empty diagonal of the initial correlation matrix. Thus the correlation matrix one begins with will look like Table 4.1 instead of Table 2.1 which has entered guessed communalities filling in the diagonal. The sums of the columns will consequently be larger.

The important difference now is that instead of taking out fewer (k) factors than there are variables, which happens when communalities are used and one stops when the residual matrix is zero, the principal axis method, carried to the components model, takes out n factors (n being the number of variables). It will be found that with unities the residuals do not vanish till the nth factor is abstracted. Our introductory diagram (p. 403) setting out the varieties of (a) models and (b) methods for reaching underlying abstractions from covariance data was meant

Table 4.1. R Matrix Prepared for Beginning a Principal Components Analysis Correlation Matrix of Eight Variables[a]

Test variables	V_1	V_2	V_3	V_4	V_5	V_6	V_7	V_8
V_1	(1.00)	.20	.28	.36	-.09	.01	.06	.06
V_2	.20	(1.00)	.00	.52	.36	-.04	.04	.04
V_3	.28	.00	(1.00)	.04	-.36	.04	.04	.04
V_4	.36	.52	.04	(1.00)	.66	-.02	.43	.39
V_5	-.09	.36	-.36	.66	(1.00)	-.06	.27	.24
V_6	.01	-.04	.04	-.02	-.06	(1.00)	.09	.08
V_7	.06	.04	.04	.43	.27	.09	(1.00)	.72
V_8	.06	.04	.04	.39	.24	.08	.72	(1.00)

[a]This is a reproduction of the correlation in Table 2.1, above, as used for a centroid factor extraction, except that unities instead of estimated communalities are placed in the diagonals.

to help avoid the rather prevalent confusion of the principal components model and the principal axes method. Calling the latter *weighted summation*, as Burt does, might help but in any case we should note that one can use weighted summation with *communalities* entered and stop short of n factors (at k factors). But, alternatively, one can use full-length summation—which becomes component analysis—as a first step in factor analysis, cutting down later to k factors as shown below.

In accordance with the most common usage we shall speak of *component analysis* when using the principal axis method to produce as *many components as variables*, after entering ones in the R diagonal. The contrast of the two end results—from component analysis and factor analysis—is shown for our main example in Table 4.2.

In discussing Table 4.2, which has three common (broad) factors in one case and eight components in the other, let us nevertheless repeat that though ninety-nine times out of a hundred the factor analysis will have fewer factors than variables this is not the essential distinction. The essential distinction of the components and the factor model is that the components solutions *account for the variance of all variables in terms of common factors alone*, thus giving communalities (h^2) of unity, whereas factor analysis also *requires the aid of specific factors* to account for this total variance and thus has communalities of less than one. Clarity is gained in subsequent research and conceptualization if the student takes care to get at the outset this distinction between *component* analysis and *factor* analysis. To the psychologist the attraction of the factor model is the conceptual gain of dealing with fewer factors than there were variables— and factors to which, moreover, he can hope to give conceptual meaning. But the mathematician may remind him that in fact he finishes with more entities than in the components model. For in the example in Table 4.2 there are 8 components, but $3 + 8 = 11$ factors *if we include the specific factors*. Thus, in general, a factor analysis will have $n + k$ factors.

The psychologist does not worry much about these n specifics, being willing to set them aside as narrow sources of variance for subsequent investigation, while he concentrates on the broad common factors. But their existence does in fact point to some indeterminacies, since one cannot produce $(n + k)$ values from given n values in the exact way one can produce n components from n variables.

This shows itself, as we have clearly seen, in the practical reality that one cannot estimate a person's score on a variable from the broad factor specification equation exactly. A good (but imperfect) estimate can usually be made from scores on all the broad factors involved, but, since we are assumed not to know the score on the specific factor there is a gap. Conversely, there is also approximation in estimating a person's factor scores from his scores on the variables. In component analysis

Table 4.2. Comparison of Components Analysis and Factor Analysis from the Same Correlation Matrix

(a) Component analysis (unrotated general components)

Test	General components									% of total variance in successive components
	F_1	F_2	F_3	F_4	F_5	F_6	F_7	F_8	h^2	
1	.27	.56	.53	.01	−.56	−.04	.01	.11	1.00	32.2
2	.52	−.16	.60	.18	.34	−.44	−.02	.05	1.00	18.6
3	−.06	.80	.22	−.12	.47	.28	.02	.07	1.00	17.5
4	.89	.00	.28	.03	−.02	.22	−.01	−.29	1.00	11.9
5	.71	−.56	.05	.07	.01	.36	.05	.22	1.00	8.2
6	.02	.25	−.27	.93	.00	.04	.01	.00	1.00	6.1
7	.69	.26	−.53	−.14	.01	−.10	−.38	.05	1.00	3.5
8	.67	.27	−.53	−.15	.01	−.19	.36	.01	1.00	1.9

(b) Factor analysis by centroid method (unrotated)[a]

Test	Common (broad) factors				Uniqueness								% of total common factor variance in successive factors
	F_1	F_2	F_3	h^2	F_{s1}	F_{s2}	F_{s3}	F_{s4}	F_{s5}	F_{s6}	F_{s7}	F_{s8}	
1	.33	−.48	−.21	.38	.79								$F_1 = 50$
2	.29	−.29	.34	.35		.81							$F_2 = 22$
3	.11	−.38	−.48	.39			.78						$F_3 = 29$
4	.91	−.20	.34	.98				.13					
5	.53	.24	.65	.76					.49				
6	.04	.07	−.16	.03						.98			
7	.68	.47	−.34	.80							.45		
8	.60	.44	−.31˙	.65								.59	

(c) Factor analysis by principal axes method (unrotated)[a]

Test	Common (broad) factors				Uniqueness								% of total common factor variance in successive factors
	F_1	F_2	F_3	h^2	F_{s1}	F_{s2}	F_{s3}	F_{s4}	F_{s5}	F_{s6}	F_{s7}	F_{s8}	
1	.21	.11	.57	.38	.79								54.4
2	.41	−.29	.27	.33		.82							24.7
3	−.06	.34	.46	.33			.82						20.9
4	.92	−.19	.33	.99				.09					
5	.72	−.51	−.31	.87					.36				
6	.01	.14	−.01	.02						.99			
7	.65	.53	−.23	.76							.49		
8	.61	.51	−.22	.68								.57	

[a]This (c) is a principal axes extraction stopped after three factors. The centroid (b) and principal axes (c) look unlike, and show the unimportance of the initial extraction position; but they rotate into each other, apart from the slight differences shown in the communalities, h^2 values, and due to the different methods of setting them. The correlation basis for the above analyses is taken from Table 2.1. Note that in this computer extraction by centroid, independent of "hand" calculation in Table 2.5 above (page 28), the third factor is reversed in sign. This is equally compatible with the correlation matrix and reminds one that choice of factor direction is arbitrary.

the component scores will exactly restore the variable scores and vice versa, since there is no specific factor term in the specification equation.

Consequently, it is not surprising that for the mathematician, component analysis has greater appeal. There is no merely statistical estimation, and the logical cautiousness in dealing with n factors from n variables, instead of $(n + k)$ factors in factor analysis appeals to him as "cleaner" and more self-contained. For some years, and at least prior to Thurstone, the mathematician's taste tended to dictate the psychologist's practice. But most psychologists (and other social scientists and biologists) have long since come to recognize that what happens in most scientific fields of data correlation is better represented by the factor model than the components model. *Even in a large set of variables (say 100) we do not gather all the sources of influence that will actually account for the variance of every one of them.* The analysis cannot be treated, as the components model seeks to do, as a complete, self-explaining system. Each variable is likely to be affected by some influences not covered by its companion variables in the matrix. One might, for example, measure the growths and associated environmental features of thirty plants and include rainfall but not sunlight measures among these variables. However, what is missed out is principally the influences that make up the specific factors, for a specific factor is one that has found no companion variable to create with it a common factor and so is not included in the broad factor "explanation." Since all experiment is to some extent parochial in its horizons it is highly improbable that even three or four hundred variables will have encompassed all the factors in the universe— and yet all the factors in the universe can act on some small corner of it, as represented in the given matrix of variables.

Consequently, to accept the component model as a scientific model is to be deceived by a mathematical artifact—a kind of conjuring trick in which say, all of six variables (by the false addition of ones to the diagonal) are made to do duty for all the numerous extraneous sources of variance impinging on those six variables. The first disillusionment of those who claim that principal components can be real, constant, scientific influences arises when they take two overlapping sets of variables—$n + p$, and $n + g$— whereupon it will be found that the common factors across the n variables do not remain the same in the two contexts, and no matter how rotated they remain different. With factors, on the other hand, although the unrotated loading patterns may look different, the same factors will appear loading the n common variables after suitable rotation to unique simple structure. The reason for the difference in the first case is that what are really specifics have been forced into the common component space, and since the total specifics are different in the two cases—$(n + p)$ in one and $(n + q)$ in the other—the common components come out differently. Commons have been falsely made out of specifics.

On the other hand, though the factor model calculation puts aside what we really cannot interpret from our data into a "unique factor variance," this unique variance is sometimes taken more literally than we are entitled to take it. Actually, it is an enigmatic package consisting partly of as yet undiscovered common factors, and possibly of several psychological specifics rather than one, not to mention specific error factors. It is a statement of ignorance and a challenge to further research.

The final derived factor model to which we are brought as the best approach to the desirable scientific model will now be briefly formulated, as far as the decomposition of a variable is concerned, by the following more developed specification equation briefly formulated by

$$a_{ij} = b_{j1}F_{1i} + \cdots + b_{jk}F_{ki} + b_{je1}E_{1i} + \cdots + b_{jem}E_{mi} \qquad (4.1)$$
$$+ b_{js1}S_{j1i} + \cdots + b_{jsx}S_{jxi} + b_{je}E_{ji}$$

where the b's are loadings as before, with subscripts peculiar to the performance a_j and to the factor. F_{1i} through F_{ki} are the individual i's scores on the k broad substantive, e.g., psychological factor traits. E_{1i} to E_{mi} are his scores on m broad instrument (nonsubstantive) and error factors. By instrument factors, we mean, as described later, nonsubstantive factors artificially introduced by the properties of a measuring instrument. S_{j1} to S_{jx} are his scores on x specific factors; and E_{ji}, his score on an error absolutely specific to a_{ij}. In the ordinary use of factor analysis, the specific substantive and error factors cannot be separated and are lumped together in a single specific term, but indirect methods may later separate them. The actual factors, in a scientific sense, that are at work on a set of variables should not change when the variables have new company, though some new factors may be added. As Thurstone showed, in factor analysis, adding p to n variables and refactoring does not alter the patterns of the factors on the original n as stated above. The rate at which a brass bar expands with increase in the temperature factor should remain the same, regardless of whether I also include in the calculation data on the expansion of a copper bar in relation to the same temperature changes.

On the other hand, if one is not looking for entities which act as scientific concepts do, and wants only an immediate statistical gain within the private, isolated, self-contained world of a single experiment, then the seeming precision with which components, relative to factors, will calculate factor scores from test scores (and, in a circle, test scores from factor scores) may appeal to the pure statistician. But the scientist who wants factors corresponding to stable concepts, like weight, temperature, intelligence, or anxiety, that will reappear usefully in other experiments and contexts, will face the difficulties of the somewhat more complex model of factor analysis, and the problem of getting the truest communalities.

At this point the reader has hopefully gained perspective over the three main types of correlational analysis (CORAN) which aim to reduce concrete variables to abstracted underlying predictors, namely *factor analysis*, *component analysis*, and *cluster analysis*. The last we have dropped, and as a final goal rather than an immediate step we are about to drop also the component model. Component analysis, reached by the principal axes method, using unities instead of communalities, will yield exactly as many common "factors" as there are variables, namely, n; factor analysis is likely to yield decidedly fewer common factors k than variables, but will require n specifics; cluster analysis is likely to yield more clusters than there are factors and unlike components, they will be hard to specify precisely and quite variously intercorrelated. In cluster analysis we can estimate a score on any single variable from the score on the centroid vector (X for cluster A in Fig. 3.4) marking the central vector in each of the various clusters, or by a more powerful multiple correlation if we know scores on all the other variables in the cluster. The usefulness of this is not great and, in any case, we have had to reject the cluster search and cluster concept approach because of the insuperable difficulties of objectively identifying and determining the number and nature of clusters. Similarly, we now reject the component analysis approach as not having any necessary relation to the model of stable, identifiable, replicable influences or determiners across the natural world. However, we shall still have many dealings with both components and clusters in various transactions in factor analysis, though the structuring by factors is the primary aim of analysis and the basic reference system in what follows.

4.6. Summary

1. The number of factors and the size of communality estimates being mutually related it is concluded that except in the maximum likelihood method yet to be described the better course is to fix the number of factors and allow communalities to be fitted to that number. Broadly, the number of factors problem can be approached in a psychometric framework which calls for the number giving the best fit to the correlation matrix, according to its rank as a matrix, or on a statistical basis which considers the sizes of samples employed.

2. No matter whether psychometric (sample free) or statistical methods are proposed they must ultimately be compatible with the most probable scientific model. This needs to assume the possibility that there are more broad factors than can be mathematically determined (n). Although it is thus theoretically possible, scientifically, to have more

than n factors at work on n variables, we have to restrict ourselves in any given matrix of variables to $k < n$ broad factors and n specific factors.

3. Considering (2) and also the fact that in addition to real, substantive, and substantial factors there will be instrument factors and common error factors due to correlations of random error in samples short of the infinite population, one might suppose it is ideally desirable always to take out n factors—the greatest number possible—though with communalities short of unity. The last is required because these n broad factors would not, on the scientific model, be expected to account for all the variance of the particular set of variables in the experiment and n specific factors would also be necessary.

4. Discussion of this alternative forces us to anticipate the problems of rotation for simple structure. But in so far as that may be done we have to recognize that the greatest simple structure is achieved with n variables and n factors by having only one variable loaded on each factor, which is a degenerate, spurious solution. Only as a special tour de force can one sometimes achieve the rotational structure indicated for the scientific model if one ventures to begin by taking out n factors. (Parenthetically, the communalities can then only be fixed as at the lower bound, and roots often then go negative.) In short this assumption that there must be n factors, most of which will be extremely small, cannot in practice be handled by taking out n factors, but only by finding the k nontrivial factors, stopping at that point, and neglecting the trivial factors.

5. The determination of k is important, for over- and underfactoring distorts the patterns of the rotated factors finally considered. Underfactoring is worse for (a) due to the downward size trend in initially extracted factors the variance taken out is more in error in stopping short than running over, (b) earlier factor patterns in the rotated series are distorted by the lack of variance that should have been rotated into them, and (c) in underfactoring one does not chop off error—for that is almost evenly distributed among the unrotated factors. In going one beyond the true number of factors one can generally use it to rotate most common error into the last "garbage factor."

6. There is no exact mathematical solution to the rank of a correlation matrix (rank equals number of common factors) since the diagonal values are unknown. A solution called *minres* (p. 379) and good communality estimates will, however, come close to giving a nonarbitrary rank, while maximum likelihood and generalized least squares methods do better still. Approaches to fixing the number of factors (and therefore the communalities, which conform to the number) can be divided into those of (a) *psychometric* and (b) *statistical* natures. The former does not directly take the sample size and distribution properties into account,

though indirectly it pays some regard to sample in allowing for size of error factors.

The available examples of each on which, in the writer's experience, we can put most dependence are (1) the *scree test* (but not the Kaiser–Guttman rule of unit latent root) in the *psychometric* approach, and (2) the *likelihood ratio test* (an adjunct of the maximum likelihood extraction method) in the *statistical* approach.

7. Discussion of the number of factors problem leads to more discussion and definition of the different properties of the components and the factor models. The former extracts by putting unities for the diagonal values and gets n *common components* (there being n variables), whereas the latter puts in estimated communalities and extracts k common factors leaving n specific factors. Since the latter cannot be known exactly, the factor model gives only estimates of common factor scores, whereas the component model permits one to calculate the n common factor measures exactly as functions of n variable scores. As far as scientific constructs rather than mathematical abstractions are concerned, this "accuracy" is a complete illusion. For no ordinary finite collection of variables contains within themselves the information necessary fully to analyze and predict all the variance of each variable in terms of real influences. The components model thus has a "completeness" which is spurious.

8. In connection with this closer approach to the scientific model we should also note that the purely matrix-algebra-derived concept and term "a common factor" becomes incorrect. Virtually every so-called common factor—indeed, all of them when simple structure is reached—fall far short of being common to all variables in the given matrix. Such factors have many zero loadings and show significant or salient loadings only on a minority subset. They fall short still farther of loading all variables across a whole substantive domain. To avoid the distortion of thinking which the misnomer "common" encourages the present writer has increasingly recommended the term "broad" to contrast with "specific." This helps in keeping perspective in psychology as a whole.

Moreover, there is an additional objection to common, namely that we use *common and unique* to distinguish factor traits whose patterns are, respectively, *common to all people* and *unique to an individual*. Common and unique traits (operationally clearly defined separately by R technique and P technique below, p. 355) can *each* in fact be either broad or specific, as far as the span of variables is concerned. Thus broad and specific, common and unique make a fourfold taxonomic table for factors.

9. The number of concepts of structural unities one can extract from a correlational matrix depends on whether one uses a *component*, a *factor*, or a *cluster* model. From n variables the former yields n com-

mon components; factor analysis yields k broad factors and n specifics; and cluster analysis may yield $>n = n$ or $<n$, partly depending on whether one does or does not permit a variable to belong to more than one cluster. The next chapter, accepting as the most useful scientific model of *factors as determiners*, pursues the general introduction to factor analysis, above, into the actual operational procedures for fixing numbers of factors and sizes of communalities.

CHAPTER 5

Fixing the Number of Factors: The Most Practicable Psychometric Procedures

SYNOPSIS OF SECTIONS

5.1. Fixing the Number of Factors by Deciding the Communalities
5.2. Deciding Number of Factors by Latent Root Plots: The Scree and K-G Tests
5.3. The Empirical Support for the Scree Test from Plasmodes
5.4. The Empirical Support for the ScreeTest from Internal Consistencies
5.5. Ensuring Objectivity of Evaluation Procedure in the Scree Test
5.6. Findings in the Use of the K-G Test
5.7. Proceeding from Factor Number to Sizes of Communalities
5.8. Summary

5.1. Fixing the Number of Factors by Deciding the Communalities

The psychologist eager to get to the use of factor analysis may have fretted a little at the almost philosophical perspective of the discussion in the last chapter concerning what we mean by factors. But neglect of the firm basis necessary for the scientific model itself would have landed us here only in benighted squabbles among various trivial technical rules of thumb.

However, with this perspective we are ready to handle in complete technical detail in the present chapter the first major step in a practical factor analysis: settling the number of factors to be extracted. As stated broadly, two approaches have been recognized—the psychometric and the statistical. [Gorsuch (1974), who uses these apt terms, can well be read

72

for a fuller discussion of them.] Here we shall proceed with the psychometric approach, leaving to the more difficult Part II of this book discussion of the more subtle statistical approach and its distribution problems. Although there are some superiorities to the statistical approach, and it should always be brought in as an additional check when vital theoretical issues hang on the decision, yet we know from experience that most studies by most psychologists will be handled by a psychometric check only, as in the scree, the K–G (Kaiser–Guttman), the inference from the Guttman (1956) lower bound for communalities, or Sokal's test of distribution of residuals. These have been introduced in principle in the preceding chapter. We shall now run over them in more detail, concentrating finally on the scree test, on its justification as the main "work horse," and on details of its use.

Let us first reiterate the principle that *the number of factors one extracts and the size of the communalities one settles for are organically mutually dependent.* We have seen (Chapter 2) that if we "guess" very large communalities, then the number of factors needing to be extracted before the last residual sinks to zero is greater than if we started with small ones. Conversely—though this has not yet been discussed—if we start extraction (with some large enough communalities—or unities—in the diagonal so that their residuals will not easily go negative) and then decide by some good ulterior evidence to stop at a given number of factors (short of leaving a negative h residual), we can find communalities to put in that will agree exactly with the specified number of factors. That is to say, having got these appropriate communalities if we go back and insert them in the diagonal of this R matrix at the beginning of extraction, the residual will go right out—vanish precisely to zero—when we reach the number of factors we have decided to take out. Thus it is possible for each chosen value— number of factors or size of communalities—to fix the other, so the whole extraction problem can be settled by finding some dependable procedure that decides the number of factors, or one that fixes the communalities— or possibly some third basis which fixes both simultaneously.

Although we are not advocating as the best general practice that one first estimate communalities and stop on getting a zero or negative residual, yet there are times and situations where this is best and so we shall give this approach the necessary study. The two ways of estimating communalities that are most used may be called the *short approximate* and the *long precise* methods. The former is most handy in Sir Cyril Burt's adaptation of "the highest r in the column" method. The longer more exact method has been fully worked out on its statistical basis by Roff (1935) and Guttman (1956). The short method requires little more than a five-minute inspection of the correlation matrix. Burt arranges the variables in rank order according to the highest correlation each can claim with any other variable. This involves a glance along each row (or down each column), and a circling of the highest therein. It is then convenient to arrange the variables in order of size of their highest r along the base of a plot, as shown in Fig. 5.1(b)—the solid line being fitted as the rough best fit to what would be a slightly straggling plot of these highest r values. It is now required to estimate another line, as shown, in the form of the interrupted line which gives a raised value to the higher-than-average values and a reduced value for those below average. This seemingly Biblical principle of "to him that hath shall be given, etc." has a sound psychometric principle behind it, namely, that the variable with higher possession of the common factor variance will have a correlation with variables possessing less which does not do credit to its own communality. It is pulled down. The correlation of the higher with the lower ($r_{ab} = r_{ac}r_{bc}$) will be less than if the higher one were correlated with itself ($r_{ac}^2 = h^2$). In algebraic terms $r_{ab} = r_{ac}r_{bc}$ will be less than $h^2 = r_{ac}^2$ when r_{ac} is greater than r_{bc}. Conversely, a variable with low communality will get higher correlations than its communality.

The simpler practice, following Guttman, is to insert the highest r squared as communality. But although this relieves the psychologist of the arbitrary act in the "Burtian guess" method it is necessarily systematically poorer than a really good estimate modifying the highest r^2. The present writer has found that an approximate .10 increase and discount, as shown by the ends of interrupted line in Fig. 5.1 compared to the continuous line (actual best r value), yields good results. One inserts, therefore, each variable's h^2 the value above it on the interrupted line. As shown by consistency with other findings, e.g., the result of the scree or the longer Guttman method below, it is astonishing how close an experienced "Burtian guess" generally comes to the true values.

The more widely used Guttman method is to calculate the multiple correlation of each of the n variables with the other $(n - 1)$ variables which Guttman and Roff demonstrated to be the lower bound for the communality. If the reader wants to consider this in more detail he will find an extended discussion in Gorsuch (1974), who states that the method tends to give more factors than some other procedures in communality estimation. But despite its clear rationale, it is nevertheless open to the objection that, though precise, it gives a limiting value to h^2 rather than attempting the most likely value, as, for example, in Burt's method. The "lower bound" method actually has two alternatives within it. The first is as described above—$h^2 = R_{j \cdot 1, 2, \ldots, n}^2$—requiring the squaring of the multiple R of each with all other variables, and proceeding to factor until the residual goes to zero. The second directly infers the rank of the matrix (the number of factors). Within the second there is the conception of a weaker and a stronger lower bound to the number of factors and the com-

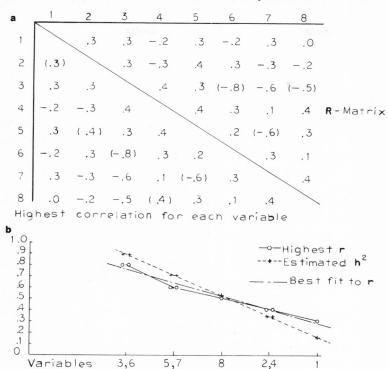

Fig. 5.1. Estimating communalities from the highest correlations in the column of each variable on the R matrix. (a) Correlation matrix inspected. The highest correlation for each variable is given in parentheses. (b) Plot of the highest r in columns modified to a new plot for a "Burtian guess."

munalities. The first is our familiar K–G test—put ones in the diagonal and stop when the roots fall below one, as discussed below. The stronger uses the above h^2 values and decides the rank of that matrix. Guttman's argument is that the latter must give a number of factors usually less and never more than the former.

If one is going to take the communality estimation approach to the number of factors his choice, broadly, between the short (Burtian) and long (Guttmanian) method must keep in mind that although the latter is accurate it gives a lower bound which, as pointed out above, is not the most likely value itself but is systematically (though only slightly) below it. Mathematicians tend to reject Burt's method, except as a quick approximation, because there is an element of doubt in it—but so there is in all scientific research. The choice will depend on how much computer time the investigator has (if any!), his skills, and the crucialness of the decision for theoretical ends.

5.2. Deciding Number of Factors by Latent Root Plots:
The Scree and K–G Tests

In any case, the present writer would suggest that the alternative approach—that of fixing first the number of factors rather than the communalities—has advantages in both theory and practice. In cases where theories hang crucially on the number of factors, it is a good thing to compare both approaches—through communality and through deciding the number of factors by the latent-root approaches now to be described. In latent-root approaches one puts unities in the R matrix, as in Table 4.1 and requests the computer to yield the latent-root sizes for all n of the successive principal components extracted. [At this stage in this book, where we still leave the detailed mathematics of weighted summation computation in the background, we need not define the term latent roots (or eigenvalues, in the original German derivation) other than as the sums of the squares of loadings for each factor in the unrotated matrix.] Essentially the roots thus correspond to the sizes of the successive components taken out by the weighted summation procedure, but a more technical statement is given on p. 412. As indicated, by the size of a component or root we thus mean a value reached by summing the squares of all of that factor's loadings on the n variables.

These latent roots in the typical computer output come with the V_0 matrix output, and are then plotted on a graph in the order they come in the V_0. They typically fall, as they come directly from the machine, in a curve as shown in Fig. 4(a) and in Figs. 5.2–5.8 below.

As we have seen above, Guttman pointed out that the lower bound to the rank (number of factors) in a matrix is given by extracting components (ones in the diagonal) until the size of the latent roots drops below 1.0. Kaiser pointed out additionally that another value—the alpha coefficient of homogeneity—ceases to be significant when this point is passed. The alpha coefficient may be briefly defined as the agreement of two estimates of factor scores from two random halves of the factor pattern weights, averaged over all possible random halves. This convergence of Guttman's and Kaiser's arguments on the latent-root of 1.0 as the point to stop factoring is theoretically impressive, but unfortunately we must argue from practice—and partly from theory—that the "root one" rule is not a reliable basis of decision. The evidence from practice will soon be given: that from theory turns on the fact that after rotation the size of the factors is quite different from that in the order of extraction. As Fig. 4.1 shows, it would be possible for the seventh factor to fall below 1.0 in the extraction but after rotation, which typically "evens up" variances, it would be well above that value. Of course, the decision to use it would depend on how far it goes above nonsignificance. Would one wish to use a factor pattern with only a 0.2 alpha generalizability? The direction in which one should lean is surely already given by the arguments above on the relative errors of over- and underfactoring, and the present writer would, in exploratory research, include such a borderline factor as surely as an early map maker would include an island, despite an uncertain outline, rather than tell the sailor that there is open ocean.

The main alternative psychometric approach is one that depends on the *slope* properties of the plot of latent-roots, rather than on absolute levels. The most explicitly worked out technique here is the *scree test*, which parallels some approaches made also by Tucker (see Thurstone, 1947) and by Horn (1965). Its criterion is not couched in terms of examining absolute size of the roots, as in the K–G method, but in terms of the angular change in the curve. In some hundreds of factor analyses it was the present writer's experience that the number of factors (judged by various independent evidences) coincided with the point on the roots plot where the downward descending curve relatively suddenly straightened into a much more gentle even slope. (It was given the name scree because it is like the straight scree of rock debris which a mountaineer often sees running out at a steady angle of debris stability from the foot of a mountain. For the theory is that it here also represents a debris of trivial factors.) The strong support for the scree test comes from extensive empirical evidence (Cattell & Vogelmann, 1976; Hakstian, Rogers & Cattell, 1978). However, a clear theoretical basis has also been given (Cattell, 1966b) in terms of many small factors appearing from real correlations appearing in the sample among specific and error factors that are normally independent and uncorrelated in an infinite population.

By this theory there should be either two or three successive screes, depending on the design of the experiment. With a choice of variables to mark big factors, the first scree would correspond to the substantive but trivial factors, the second to common factors produced by specific factors correlated in the sample (but not the population), and the third by similar correlations among what are in the population uncorrelated errors of measurement. Depending on the parameters, such as h^2, reliability, and size of sample, these would appear separately or fuse in a single slope. In the former case it would be the end of the uppermost scree that would mark the end of the significant substantive factors.

5.3. The Empirical Support for the Scree Test from Plasmodes

The empirical support for the scree comes from two approaches: (1) Plasmodes, i.e., cases where the number of factors is first known by a physical example (e.g., Thurstone's box problem; the ball problem of

Cattell & Dickman, 1962; Coan's, 1961 egg factoring, etc.), or by writing out a factor matrix with large and trivial factors and working backward from it to the correlation matrix. The R matrix is then in either case, given to someone to factor, to see how well the method advocated reaches the number of factors known to be put into it "backstage." (2) Comparing the results from applying the scree and other tests to real data experiments in comparisons where there is reason to believe the number of factors should be the same, e.g., two samples from the same population, or the same questionnaire items factored as *items* and factored as *parcels of items* (which would theoretically yield the same number) (Cattell, 1973).

A sufficient number of plasmodes has now accumulated to give a very wide testing of the scree criterion: eight in the Cattell and Jaspars monograph (1967), fifteen in the Cattell and Vogelman article (1976), and various single instances in Cattell and Dickman (1962), Cattell and Gorsuch (1963), Cattell and Sullivan (1962), Coan (1961), and Thurstone (1947). The most systematically planned comparison of the scree, the K–G, and other methods on a larger scale has recently been presented by Hakstian, Rogers, and Cattell (1978).

Illustrative instances of the outcome with the scree, when applied to plasmodes of known structure, are set out in Figs. 5.2, 5.3, 5.4, and 5.5. These are chosen to range from 2 to 11 as regards the number of factors, and from 11 to 40 in the number of variables.

Empirically and inductively, the best rule seems to be that the upper

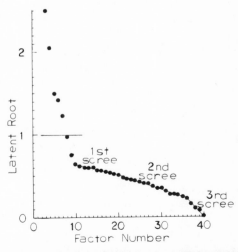

Fig. 5.2. Scree plot 1. Illustration by a plasmode, a case made up with 40 variables and 8 factors. (Note that the first root is not shown, being far above the graph.)

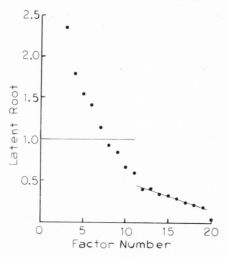

Fig. 5.3. Scree plot 2. Illustration by a plasmode, example made up to have 20
variables and 11 factors. (Note, first root not plotted on the graph.)

end of the scree marks the $(k + 1)$ factor when the true number of the
factors is k. It will be seen that in each of these varied plasmodes—the
first three (and also 5.6) made up on the mathematical model and the
last a concrete physical example, the "ball problem"—the scree test reveals
precisely the number of factors—8 out of 40 variables, 11 out of 20,
2 out of 12, and 5 out of 32—that are known to exist. As the reader who
studies the full roster of original instances in the above references will

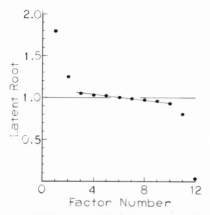

Fig. 5.4. Scree plot 3. Illustration by plasmode. This was given the structure of 2
factors in 12 variables.

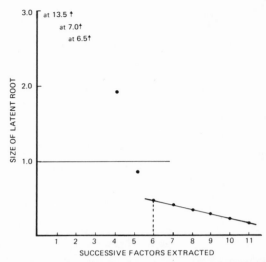

Fig. 5.5. Scree plot 4. Illustration by concrete plasmode of the ball problem. Last 21 variables not shown. From principles of physics, five factors should be involved.

find, this is a representative not an exceptional selection, though we shall recognize in a moment that the scree test is not infrequently "out" by one factor or so. One must also note that appreciable disagreements can exist among inexperienced users, in recognizing where the scree should be placed, such as do not arise among skilled users.

Experiments with plasmodes have also been done to show what error of measurement in the variables does to the scree. In Fig. 5.6(a) we have a scree for a plasmode of 30 variables and 10 factors in which the correlations and the scores of 50 individuals generated to fit the correlations should give exactly 10 factors. Into these scores random deviates have then been added to represent random error "blurring" the picture, and the factor extraction has been repeated from the new correlation matrix as shown in Fig. 5.6(b). The scree gives exactly the expected number, 10, in Fig. 5.6(a) originally, but after error is thrown in, it gives 12. The example in Fig. 5.3 above is also of interest in this respect, in that 10 ordinary simple structure factors were put in, followed by an error factor with a small and random loading pattern. Although this is not as thorough an exemplification of the error model as in Fig. 4.1, it agrees in showing that the scree will include one or two common error factors, over and above the main substantives, when the error is appreciable enough to generate large error factors. [In Fig. 5.6(b) because of the small sample (50), the common error was appreciable.]

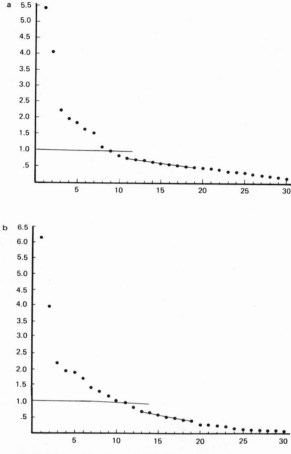

Fig. 5.6. Comparison of scree solution with and without appreciable error of measurement. Ten factors were actually inserted, as shown in the errorless example (a). Intrusion of considerable measurement error raises factors to 12 in (b), as expected on principles here discussed.

5.4. The Empirical Support for the Scree Test from Internal Consistencies

The alternative approach to testing the scree, by consistency among studies, is illustrated in Figs. 5.7 and 5.8. Figure 5.7 records an experiment in which there was comparison both over a change in method of factoring—items or parcels of those items—and also in sample size. The results are close in all four, but first let us look at the item–parcel comparison, shown in Fig. 5.7. The screes are close in conclusion—20 and 22

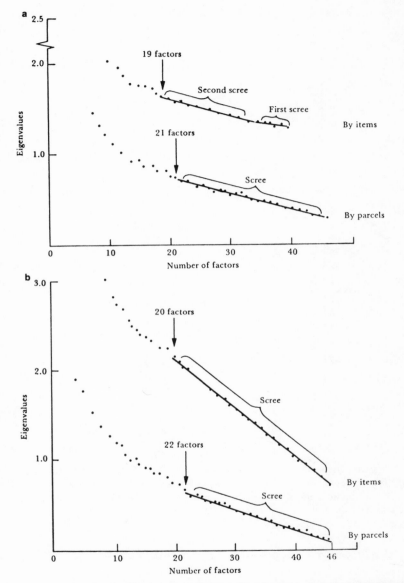

Fig. 5.7. Comparison of screes on same personality items (approximately 200) factored (a) as items and (b) as parcels on two different samples of people, showing convergence on 20–21 factors. (From R. B. Cattell, *Personality and Mood by Questionnaire*, Jossey-Bass, San Francisco, 1973, p. 298.)

factors—and this appearance of a slightly larger number—22—in the second has its reasons.[1]

Although consistency across researchers is not so unambiguous a test as checking against a known plasmode, for the general reason that even wrong methods can be consistent in their wrongness, without an ulterior source of knowledge, the scree test has certainly shown consistency across a diversity of comparisons. For example, it has steadily pointed to 8 or 9 factors in each of the twelve independent second-order personality factor studies aligned elsewhere (Cattell, 1973). At the level of first-order factors there have also been studies on the same 32 or 64 questionnaire scales (16 P.F., Cattell, Eber, and Tatsuoka, 1971) translated in several languages and applied upon population samples in the US, Britain, Japan, Australia, Germany, India, Italy, and Brazil, population samples, in which the scree has consistently indicated 18 to 22 factors (Cattell, 1973), whereas the K–G test has varied much more. A strong example with which to compare the decision in Fig. 5.7 is that shown in Fig. 5.8. Here the sample by Sells, Demaree, and Will (1970) was quite large—900—and the items, though equivalent in content were modified in detail (actually the original 200 items were increased to 300). The scree in Fig. 5.8 indicates 22 factors, essentially the same as Fig. 5.7. In this case it is interesting to note that, using a different "number of factors criterion," too questionable in nature for us to have discussed among the methods here, these investigators stopped at 11 factors.

5.5. Ensuring Objectivity of Evaluation Procedure in the Scree Test

Despite constantly accumulating evidence that the scree works very well when tested against concrete, definite criteria, as shown above, it has received criticism or two grounds: (1) that although experienced users generally concur quite well on the scree line, people untrained in its use, especially when faced with rather straggling plots from large numbers of items (as in Figs. 5.7 and 5.8) sometimes come to rather widely different

---

[1] See Cattell (1973) for discussion. The accumulation of variance on a certain factor when several items in that parcel share the factors can bring the parcel variable to a significant loading on a borderline factor when the separate items might not reach the same level. Thus borderline factors of small variance might become visible in the one case that do not come above the horizon in the other, despite the general proposition that factors in parcels and their items should essentially be the same. A difference in the opposite direction—fewer factors with parcels—is likely to occur at the limit where the number of parcels ($< 2k$) becomes insufficient to mark the k factors with at least two markers per factor.

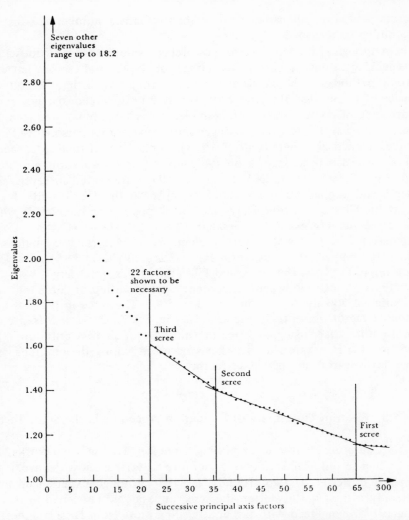

Fig. 5.8. Consistency of scree tested with same domain of items as in Fig. 5.7 (but not literally the same) on a very large sample of people and items. (From R. B. Cattell, *Personality and Mood by Questionnaire*, Jossey-Bass, San Francisco, 1973, p. 306.)

decisions; and (2) that this psychometric test lacks a theory as simple as the K–G criterion or the statistical tests. Whether or not the theory is adequate the reader must judge from the above brief account and discussion elsewhere (Cattell, 1956). As stated, the real support for the scree is empirical, and as surveys like the above will show empirical evidence is spread over more studies and kinds of situations than any offered for the alternatives among the brief psychometric methods. However, the question of how much training or experience is necessary to get reliable convergence of judgments from the scree as presented, deserves investigation and is getting it at the moment in the study by Hakstian, Rogers, and Cattell (1978). A tentative answer has already come, however, from the smaller study by Cattell and Vogelman (1976) where the average scatter (sigma) on 15 screes by 12 evaluators given only half an hour training with a set of instructions was 0.6 factors.

Steps will doubtless be taken to add techniques whereby all investigators will get the same result, from any scree test. In pursuing this objective, Dr. G. Woods has sought to get the computer to make a yes-or-no judgment, by calculating the successive tangent values as the line of points, for the scree is run from its early beginning (from the right), adding one point at a time. Wood's program calculates the sigma of the tangents about the mean slope at each addition until the added point gives a sudden increase in the sigma. The program works well and is in process of publication (Woods, 1978). While this aid is valuable, one must on principle point out in relation to such aids generally, e.g., in judging the plateau in hyperplane count (p. 143, below), that a process is not unscientific because it involves some human judgment. Especially where whole patterns are concerned, as in the scree, or in getting simple structure, or in examining a loading pattern to reach the best hypothesis for the next choice of significant variables, there is no substitute as yet for an intelligent human mind—though admittedly the operant word is "intelligent." In using the microscope or the telescope, training is expected, and trained bacteriologists, for example, can make decisions on inspecting a bacterial culture which the untrained cannot. It is true, on the other hand, that, like Schiaparelli seeing his canals on Mars, an investigator can fallaciously go too far beyond what others can see. To permit an objective judgment, therefore, scree data has been presented above from a range of variables, factors, simple structure and nonsimple structure, high and low communality, and orthogonal and oblique plasmode, and other examples. This has seemed worthwhile because it serves the purpose not only of permitting exhaustive examination of the validity of the scree, but also of giving the practitioner some feeling and experience in its use.

The key points in the ordinary visual use of the scree may be more explicitly and precisely stated as follows.

1. Three points in a row is an undesirably low limit for drawing a scree.

2. Some absolute limit to the sigma of points about an acceptable scree line, however straight it may be, must be kept in mind. The scree in

Fig. 5.3 and lower scree in Fig. 5.7 begin to approach such a limit and if we drew a line through the four points above the scree in 5.3 (imagining we have a scree there) it would not satisfy requirements.

3. A further condition is that the slope of a scree must not approach the vertical. Its angle will, of course, depend on the horizontal and vertical scales used. When each extra factor along the base is drawn to the same unit length as each movement of .10 in the size of the root (vertical scale), an angle above 40° from the horizontal should not be accepted. This 40° angle corresponds to a slope of tan = − 0.84, which becomes − .084 on our 10 to 1 scaling of coordinates, meaning that for each successive latent root on the horizontal scale we drop .084 in the size of the root on the vertical scale. Sometimes there are two or even three screes, but their angles should not change much and all should fall below the limit indicated. The first scree on the left is then the arbiter.

4. There is generally a relatively sharp, even if small, break in vertical level between the last point on the scree and the next point above.

The element of "art" lies in combining these conditions and judging by the sense of the majority of signs. For example, the two drawings in Fig. 5.6 do not meet condition (4), but no one would have any hesitiation in drawing the screes as there drawn.

5.6. Findings in the Use of the K–G Test

The second "quick" test of number of factors we have to examine, which until recently has been much better known than the scree and incorporated (rather uncritically because of its convenience) in many programs, is the Kaiser–Guttman limit, described above. This advocates that one drop all factors below a root size of 1.0. This is, of course, applied like the scree to an initial component analysis with unities on the diagonal. One can at once compare its verdicts with those of the scree by glancing at the horizontal line at 1.0 in all the above 9 varied examples.

On the positively known values in the four plasmodes (Figs. 5.2–5.5) the K–G gives 6 factors, erroneously, where the scree correctly gives 8; 7 where the latter correctly gives 11; 5 where the latter correctly gives 2; and 4 where it correctly gives 5. On the plasmodes, comparing without and with common error [Fig. 5.6 (a) and (b)], the scree gives 10 (known to be correct) and 12 (likely), respectively, and the K–G, 8 (erroneous) and 10. On the real data examples, where we are instead evaluating by internal consistency, by results on essentially the same data, the K–G method gives over a hundred factors with the 300-item variable study. The most disturbing inconsistency is where the same data is differently grouped for factoring (items and parcels) but on the same sample. Here the K–G gives 39 factors with items, and 11–12 for parcels, where the scree gives, respectively, 19–20 and 20–22 in the same two cases. This seems to suggest that Guttman's lower bound of 1.0 argument needs more examination, in regard to some wide framework of assumptions, than it has been given.

On the other hand, there are many known instances, in a fortunately chosen narrower range of conditions than we have comprehensively covered above, when the K–G and scree tests agree to within one or two factors (Cattell & Vogelmann, 1976). In terms of recognizing a range in which the K–G is relatively safe the present writer has in various places offered guidance to the K–G user that it is likely to overestimate the number of factors when the number of variables is large (Fig. 5.8) and underestimate when very few (Fig. 5.5). In the present data it happens that the underestimation persists in all plasmodes except that reproduced in Fig. 5.5, and fits the argument given above on p. 56 that, in the usual small factorings of many reported substantive researches, some factors excluded by the K–G from extraction and rotation would have positive alpha reliability if pursued through rotation. Indeed, in these cases where we are rejecting the K–G we know what factors are significant from the plasmode construction.

5.7. Proceeding from Factor Number to Sizes of Communalities

An "aside" which must in the name of realism be made to the student at this point concerns the discrepancy between what the psychologist needs and what the mathematical statistician has tended to give him especially in the field of computer programs. The situation has improved, but in the fifties and sixties the packaged programs, unawakened to the needs of the scientific model, gave him component analysis instead of factor analysis; a varimax or orthogonal rotation program instead of oblique simple structure programs like the Harris–Kaiser or MAXPLANE, and the K–G root-one criterion for deciding the number of factors instead of the scree. Program elements such as the K–G and VARIMAX are so brief, and so neat to the mathematician, that they have often been included tandem without any particular note on their presence and the numerous users of factor analysis who trust to the computer technician to do their factorings often did not know (or care about?) the basis of their conclusions concerning ability and personality structure, believing them "based on a factor approach." A crop of misleading results uncritically quoted in reviews has followed.

Even at this introductory stage, a warning should be issued that although this book suggests handing over as much as possible to computer programs, it does suppose that the researcher will only do so when he is clearly told what is in them, and that he will keep the design and strategy in his own hands to the extent that he will, in general, demand programs going beyond those adequate to the mathematical statistician alone. The psychologist has various extra conditions to meet in the scientific model, and he should understand them well enough to get the programs he requires.

As far as the decision on number of factors is concerned we have recognized three approaches: fixing the number and fitting communalities to it; estimating communalities and fitting the number of factors to it; and the maximum likelihood method which does both at once. The last has to be left for more advanced chapters. Even today, in terms of what is available and practicable to most computer users (and all nonusers) it is a "deluxe" method that few go to the expense of using. The second we have argued is inferior to the first, and consequently we have not gone in great detail into the question of deciding the significance of a residual matrix. Excellent discussions on this are found in Gorsuch (1974), Harman (1976), and Mulaik (1972), while the article by Sokal (1959) describes a brief and effective practical approach. (However, the essentials are also given here after we discuss significant tests in Chapter 14.)

As a matter of good tactics one is well advised to use mutually checking methods to an extent appropriate to the critical importance of the research, whether one is exploring new domains or in a well-staked-out area, and so on. As a minimum the present writer uses the scree and the K–G. For with the above allowances for number of variables in the latter it can give guidance as to the range in which the termination of the first scree can be expected to appear.

The procedure of fitting communalities to the number of factors decided upon is one of iteration. It is usual to start with unities in the diagonal of the R matrix. (Observation of computing time shows there is no advantage in beginning with a lower-bound estimate, and one may be troubled with negative roots when proceeding to the designated number of factors.) Either centroid or principal axes extraction can be followed, though virtually all programs offered nowadays advantageously follow the latter.

The communality immediately reached at the given factor number is then fed back into the diagonals of the R matrix, replacing the old values, and the factoring is repeated. Usually, with a matrix of 40–100 variables, four to eight repetitions of such cycles suffice for the communalities to converge to a point where no further change occurs at the second decimal place. Most programs today have an adjustable "default" parameter. Thus if one fixed this at, say, .002, the program will stop in default if any change greater than this fails to occur in any variable at the last rotation.

One limitation to such convergences is stated in a classical theorem of Albert (1944) that a unique set of converged values will be reached only when the number of factors is no more than half the number of variables ($n/2$). Wrigley and others have checked this with a range of empirical instances and found it substantially correct. Although a delayed convergence can almost always be reached when k actually exceeds $n/2$ the values reached are found to depend somewhat on the particular initial guessed communalities put in as starting point. In this case, therefore, it is best to begin with the best possible estimates by Burtian or Guttman approaches, hoping this will keep the converged values in the right neighborhood. Fortunately, the majority of studies that proceed to fix the communalities to the number of factors chosen *do* have more than twice as many variables as factors, for in a good design, this would be a limit one would not, for other reasons, want to go beyond.

A variant of communality estimation one sometimes meets is to start with one in the diagonal and stop at some decided number of factors, retaining (without iteration) the communalities reached at that point. These will, on an average, definitely be overestimates and the later factors, in particular, will have roots that are too large. It is not to be recommended, for it is in a sense a hybrid of a factor model and a "truncated components" model, lacking the clear assumptions and consequences of either.

5.8. Summary

1. It is possible to decide the number of factors by inserting the best possible estimates of the communalities in the R matrix and extracting factors until the last R residual is deemed negligible, or until a negative latent root appears.

2. Alternatively, one can decide the number of factors on an independent basis, and complete the factor analysis by discovering the communalities which exactly fit that number, given the particular experimental R matrix. Communalities and factor number are thus mutually determining. The third possibility, that of converging at the same time on both, known as the maximum likelihood method, is complex. It can be deferred to a later chapter because the methods in the present chapter are likely, in all but "expensive" work, to be those more used.

3. Estimation of communalities, for the procedure of determining factor number from communalities, is recommended, if one wants (a) a short approximate method by the "Burtian estimate" or if a mathematically exact method by (b) Guttman's stronger lower bound. Though exact in calculation, the latter is a *lower bound*, while Burt's is a guess at the

most likely real value. There are other, less recommended methods, notably Guttman's $h^2 = r^2$ (r being the highest r in the column) and his weaker lower bound which, however, first requires a decision on the number of factors. All these are psychometric rather than statistical methods; i.e., they do not take size of sample into account.

4. With communalities fixed the decision as to when enough factors have been taken out is sometimes made by observing when the first negative root appears in extraction. More sensitive tests of completeness of extraction, hinging on when a residual is negligible, have been given by Bartlett, Burt, Sokal, and others on a theory, and by Humphreys, Horn, Linn, and others on a Monte Carlo basis. What has been called the minres method (Harman, 1976) is also available as a special case of the minimum residual—hence, minres—approach. But since the more appropriate residual testing methods are statistical rather than psychometric, their discussion is deferred to Chapter 14. In any case, this approach from communalities is not recommended, compared to the direct method [(5) below] for determining them from factor number, though using the former may be inevitable in some situations.

5. Determining the number of factors can itself be approached in a psychometric basis, directly by the (a) scree or K–G methods, or (b) on a statistical basis as in the maximum likelihood method (or indirectly by the above tests for significance of a residual matrix). This chapter considers the psychometric approach which in the form here recommended, examines the plots of sizes of successive components, by methods such as the scree (which has alternatives in the Tucker and Horn methods) and the Kaiser–Guttman methods.

6. The scree has had the most extensive theoretical analysis and practical testing among these methods depending on properties of the curve in the plot of descending root sizes. It concludes that extraction should stop where the steeply descending curve turns into a gently sloping straight line—the scree.

Its recommendation rests on extensive empirical checking on examples (plasmodes) with known numbers of factors. But it has a theoretical basis: that the straight line represents a debris of small factors of different origin from the main factors, namely, from (a) correlations in the sample among specifics, (b) error broad factors from measurement errors that are uncorrelated in the infinite population, and (c) from trivial but numerous "instrument or remote influence" factors. Additional to the above empirical checks by plasmodes its validity rests on the finding that it gives consistent answers in real data when items and parcels from the same domain are factored, and in other situations where internal consistency can be examined.

7. A criticism of the scree is not that it is incorrect in principle, but that different observers, unless skilled and experienced in the method, may not agree. Rules to give greater objectivity to decisions have been put forward and a computer program has been invented by Woods which makes the decision automatic.

8. The Kaiser–Guttman test of stopping at a root of 1.0 has been shown in extensive empirical trials on plasmodes to underestimate when there are few variables and more seriously to overestimate with many. In so far as it is based on the alpha coefficient of homogeneity a theoretical criticism also exists. The scree test, by a ratio of 3 hits to 1, is more dependable than the K–G test. Consequently, automatic use of the K–G in package programs, except as a "second opinion" with allowances for the above tendencies, is risky.

9. At present it is best to bring evidence from two or three sources to bear on the decision on number of factors. As there is occasionally ambiguity as to which is the last scree, the K–G may supplement by indicating the general range in which to look for the scree's end.

In connection with "completeness of extraction" one must digress to point out that beginners sometimes read on computer outputs figures labeled "percentage of variance extracted" and take them as meaning how much they have taken out of what should ideally be extracted. Actually, these figures commonly give the fraction of the total *component* variance, taken out with each component, or of the decided total *common factor* variance, not of some ideal amount that should have been taken out with perfect knowledge of the number of factors. Thus, the value has little more relevance than being told the distance to New York when one is traveling from Chicago to Boston. In brief, a factor analysis is not more successful if it maximizes this percentage of total component variance. It might help avoid this and related confusions if the symbol $\underline{V}_0$ were used for the component and V_0 for the factor matrix.

10. With the number of factors decided the resultant variable communalities which exactly conform, i.e., which make the last residual exactly zero, are found by an iteration of factor analyses from the original R matrix. This is a successive approximation, substituting the newly reached communalities in the diagonal at each new cycle. Convergence to stability at a unique solution is possible when $k < n/2$, and such convergence, to two or three decimal places is somewhat quicker with certain initial choices of communalities. If $k > n/2$ convergence usually still occurs but at values depending on the starting "guessed" values.

CHAPTER 6

The Theory of Unique Rotational Resolution by Confactor, Procrustes, and Simple Structure Principles

SYNOPSIS OF SECTIONS

6.1. The Properties and Limitations of the Unrotated Dimension Matrices, $\underline{V}_0$ and V_0

Having determined the number of factors, and holding in his hands an unrotated factor matrix V_0 with consistent communalities, the investigator stands at the most perilous phase of a factor analysis. However, excellent the work up to this point, the way in which the next step is carried out—that of finding the uniquely meaningful rotation—will decide whether a conclusion is drawn that is enlightening, or somewhat misleading, or positively absurd.

The unrotated matrix is, in essence, of order $n \times k$, i.e., of broad factors only, since the n specifics are not to be rotated. That matrix we shall symbolize as V_0 (or $V_0 + U$ with an added matrix for the specific factor loadings.) Henceforth, we shall code all *variable-factor relation* matrices as capital V, short for VD or *variable-dimension* matrices . The general term *dimensions* is needed here because not all dimensions in transformed matrices are *factors*. The columns may, for example, be *reference vectors* (p. 108), to be explained shortly.

Seen in a wider context, the V_0 matrix is thus strictly not $n \times k$, but $n \times (k + n)$ since to be complete according to the model, it must contain the n specifics in the above U matrix. The statement that the $n \times (k + n)$ $\underline{V}_0$ matrix is cut to $n \times k$ because the unique ($n \times n$ diagonal matrix) cannot be rotated should perhaps be stated as that they geometrically could be, but the loadings would not change in rotation. For the specific for each variable j is $(1 - h_j^2)$ and h_j^2 does not change since all rotations are *in a space orthogonal to that of the n specifics.*

Now, Chapter 3 has explained that an infinite number of factor matrices, corresponding to possible spins of the geometric factor axes, can be found that are equivalent to the extraction-given broad and specific factor matrix. All are equivalent in (1) restoring the original correlation matrix, and in (2) maintaining the communalities—the h^2—unchanged. Since, as we have shown, the addition or subtraction of variables from an experiment affects the loadings—in the original extraction—even on the variables that remain constantly in the set, the initial V_0 position can be regarded as more or less accidental and in no way of "sacred" meaning. This necessity of proceeding to rotation adds to the gulf between the meanings of component analysis and factor analysis, since from a component analysis concerned with purely mathematical properties, there is something special (though not psychologically) about the initial components extraction position and nothing particularly to be gained by rotation. One such special mathematical property is that the loadings in the columns of a components extraction, or of a principal axes extraction preparatory to a factor rotation, are uncorrelated, which ceases to be true once it is rotated obliquely or orthogonally. This property is not an ideal to be pursued in factor analysis. We expect the patterns of some factors to have some chance or systematic resemblance to others for psychological reasons. For example, the loading patterns of fluid and crystallized intelligence are appreciably similar and therefore positively correlated. However, other properties of the unrotated position, such as the loadings of a given variable on the first component are local in the sense of being specific to the particular set of variables used. Indeed, in this connection it may be helpful to show geometrically in Fig. 6.1 what has been said above about overlapping sets, namely that the direction and composition

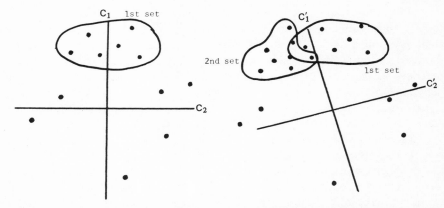

Fig. 6.1. Effect of choice of variables on factor positions. (a) First component going through the center of the first set of prominent variables. (b) Change in first component when another set of prominent variables is added to first in a new experiment.

of the first centroid (or first principal component) extracted are a "center of gravity" of the variables used and are therefore at the mercy of the experimenter's particular choice of a set of variables to experiment upon. On the set as shown in Fig. 6.1(a), the axis comes out through the middle of the most heavily represented behavioral area, shown in the first set. In Fig. 6.1(b), through a substantial addition of new variables to the experiment, the center of gravity of the test vectors, which the axis follows, is swung over to the left. This changes the loadings on the first set. Thereupon the second component C_2, to keep orthogonal, has to follow to C_2', and, all subsequent component patterns are changed too. The unrotated matrix, centroid or component, therefore looks quite different for the same variables in different company.

On the other hand, a correct rotation of two such different initial unrotated matrices can bring the shared set of variables to the same identical end pattern regardless of company, except in extreme cases. Incidentally, it was this situation which puzzled early factorists, when debating Spearman's theory of the uniqueness of the general intelligence factor. They argued that the nature of the general ability, "g" which he distilled from cognitive performances must depend on what particular handfuls of cognitive variables went into the flask, and, as often happens in scientific debates, there were subtle senses in which both contenders were right.[1]

---

[1] Although we must eschew unnecessary trips into history, this has some interest for basic concepts. Spearman's configuration could not, by his method of test choice, yield a picture like Fig. 6.1 for he introduced the special condition that one must throw out tests which upset the smooth "hierarchy" in the correlation matrix and which therefore failed to meet his famous "tetrad difference criterion." This business

6.2. The Rationale for Rotation by Hypothesis-Testing and Hypothesis-Creating Principles

The fact that the initial V_0 extraction position, i.e., the particular dimension patterning given, is local to the particular set of variables and partly accidental to the method of factor extraction, at least justifies our rotating to some new but equivalent position if we so wish to find constancy. But unless rotation is to be to a no more meaningful position than that of a weather vane in a thunderstorm, we must find a basic guiding principle. Here the scientific model becomes our guide, for according to our postulates a real influence in nature will show up as a factor and therefore one of the possible factor positions can be that of an important, unique, real influence. This means that one particular set of factors from among the infinite possible sets from the "spins" of rotation should be discoverable and will match these real influences. All other positions, including our extraction position, are merely mathematical transformations of the pattern originally cause by these influences. The problem is, therefore, to recognize some peculiar properties of the unique "factor as an influence" rotation that will enable us to distinguish it from all other positions.

Let us note in passing that the assertion that an influence will appear as a factor is not merely an *a priori* postulate. Experiments have been done in which known causal determiners are at work on known variables and where the ensuing factor analysis yields patterns corresponding precisely to the effects of the determiners known to be at work. Such experiments are the Cattell and Dickman (1962) ball problem; the Coan egg study (1959); the Cattell and Sullivan factoring of cups of coffee (1962); Thurstone's box problem (1947); and Barton's factoring of chemical compounds. These empirical justifications show, however, that the expression influence must be broadly interpreted as sometimes an ordinary cause (with a temporal sequence of cause and effect) and sometimes as a determiner without temporal sequence. For example, in the 32 ball behaviors measured in the Cattell–Dickman ball problem, rotation to simple structure brought out such influences as weight, size, and elasticity, as in factors determining covariance of the behavior of the balls. Conceivably some philosopher might balk at calling weight or elasticity a "cause" of the

---

of hierarchies and tetrad differences will not be understood except by those who have studied abilities, but it is enough to recognize that it did ensure that he had only one uniquely positioned common factor. Nevertheless, in principle, a second and a third, etc., broad factors could be imagined as later additions. Once a three-dimensional system arose there could be no guarantee that the axes for the second and third factors, though kept orthogonal to his "g," would be in a unique position. The reader wishing to follow up this issue should see the distinction between *univocal* and *uni-tractic* test batteries in Cattell (1973, p. 381).

ball's behavior, or still more, at the calling the length, breadth, and height factors in Thurstone's (1947) factoring of boxes "causes" of their volume. As pointed out in Chapter 1, temporal sequence is involved in cause, and a cause must be recognizably present before its consequence.

A whole series of possible philosophical concepts in the general area of "determiners" is covered by the entities which emerge in uniquely rotated factors. They may vary from length of boxes as a determiner of volume, through time and temperature as determining the products of a chemical action, to such causes as degree of childhood inhibition determining super ego level. Granted that only one direction of rotational resolution will lead us to land on such real influences, by what principles do we find this direction?

There are broadly two general uses of factor analysis in science, as described earlier—*hypothesis testing* and *hypothesis creating*—and these necessarily give us straight away two different approaches to rotation. In hypothesis testing we set up a *statement of hypothesis*, which, to be adequately concrete and foolproof, must be given in the form of loading values in a rotated factor matrix as they would be expected from the hypothesis. With this "target" or "test" matrix we can see if the given V_0 can be rotated to fit it. In hypothesis creating, on the other hand, we seek a rotation possessing certain *general properties* (stated below in simple structure and confactor principles) which should hold only when the special conditions appear in which the factors are true influences.

Let us begin here with hypothesis testing approaches. The *hypothesis matrix* described above must state both the number (k) and the pattern of loadings (V_{fp}) which the hypothesis requires. Using what we shall describe in more detail later as the PROCRUSTES program (p. 97) or a maximum likelihood approach the investigator rotates the V_0 as close as it will go to this *target matrix*. Such a position is uniquely the best possible, but whether it fits the hypothesized pattern well or poorly is a separate question and requires tests of goodness of fit.

For completeness we give here in Eq. (6.1) the formula for the PROCRUSTES solution, though, since matrix calculations have not yet been discussed the reader may not yet fully comprehend it. In any case, the whole use of PROCRUSTES hypothesis testing procedures will receive special discussion later (p. 116) and so will not be pursued further here. However, its essence is here briefly indicated by Eqs. (6.1) and (6.2). If L is a rotational transformation matrix to give a best possible fit to a special hypothesis matrix V_h, then we ask that

$$V_0 L \text{ approximates } V_h \qquad (6.1)$$

i.e., that it reaches V_h to an acceptable degree of approximation. Whence we solve for the unknown L by

$$L = (V_0^T V_0)^{-1} V_0^T V_h \qquad (6.2)$$

This formulation by Hurley and Cattell (1962) is also the basis of Ahmavaara's (1954) transformation analysis, approaching from other assumptions. The formula itself was used in another connection very early by Mosier (1939b) though this was unknown to Hurley and Cattell at the time and presumably also to Ahmavaara. There have since been valuable extensions by Schönemann (1966b). It has remained the main avenue to hypothesis testing by factor analysis until recently, when Jöreskog's factoring of covariances by "confirmatory" likelihood principles has provided a further tool (see Gorsuch, 1974; Jöreskog, 1969).

Although there is greater flexibility in Jöreskog's proofing or confirmatory maximum likelihood, such as using a target with only some of the hypothetical values fixed and others free to adjust (see below), it also has more complexities than many experimenters will be prepared to face, and in the latter situation the PROCRUSTES has appeal and is likely to be in frequent use. It has certain virtues in that, for example, as Tucker and Korth have shown (1975) the rotation position reached with the real data yields the maximum congruence coefficients possible for that data with the target factor matrix written out to define the hypothesis [V_h in Eq. (6.2) above]. Since the congruence coefficient is the most used, and one of the best indices for deciding whether a match of two factors from different researches is significant, this is a useful property. However, as will be discussed in some later connections, PROCRUSTES is able to achieve good factor pattern matches at the cost of reaching rather extreme correlations among the factors, and the testing of a hypothesis should therefore require that the hypothetically correct correlations among the factors should also be set out and compared. In any case it is a good procedure to finish off a PROCRUSTES rotation by a (visual) ROTOPLOT adjustment of angles among factors which can commonly be achieved with very little loss of simple structure hyperplane count. It should be noted, in passing, that in a much investigated field, where we know just where the zero loadings for the simple structure normally lie, the PROCRUSTES program is often a quicker and cleaner move to simple structure than a general simple-structure-seeking analytical program.

Hypothesis-testing use of factor analysis is on the increase, as the more precise theories which multivariate research has helped to develop have begun to crystallize in various fields. But during the past 50 years factor analysis has been used far more widely as an instrument of discovery in new fields, and a generator of hypotheses. In this role it is more powerful than ANOVA, as we have seen in Chapter 1, and thus may long keep hypothesis testing as its most valuable use. For example, it is factor analysis that has generated the clearest and most testable hypotheses about the structure of abilities, about the dimensions of culture patterns, about the nature of the human primary drives, and the patterns of culturally acquired motivation structures. It is now penetrating to new concepts in learning theory, behavior genetics, group dynamics, language structure, psycho-physiological state analyses, etc. (The reader may follow these in the chapters of Bereiter, Bock, Coan, Cohen, Horn, Hake, Nesselroade, Mefferd, Osgood, Pawlik, Royce, Sells, Thomson, Tucker, and others in the *Handbook of Multivariate Experimental Psychology.*)

Although CORAN and ANOVA methods need to go hand in hand, as discussed in Chapter 1, there are phases of research in which science is advanced better by a relatively long exploration by a "task force" of experienced factor analysts. In a chaos of speculation, such an approach is able relatively reliably to determine significant variables and to generate the most likely hypotheses, upon which ANOVA methods can then be brought to bear for testing "effects," especially where the latter are likely to be slight, since ANOVA with large samples can be very sensitive. A reference here is mainly to tactically coordinating CORAN and ANOVA in research; but in recent years some illuminating writing has been done coordinating them in their statistical meaning and equivalence, which can be seen notably in Cohen and Cohen (1975) and in Tatsuoka (1971b). Meanwhile, since the exploratory and hypothesis-producing role of factor analysis has long been and will long continue to be its more unique contribution to scientific research it is on this that we shall now first concentrate.

To use this hypothesis-creating virtue fully we must now ask "What general principle can be found in rotation that will locate the operative determiners—the unitary influences—regardless of their particular natures?" Two general principles which should apply to all unitary influences, whether they be abilities, genetic components in temperament, drives, cultural forces, emotional response patterns, or causes fixing the behavior of physical objects have been suggested for hypothesis creation: (1) simple structure, and (2) confactor rotation.

Since it is appropriate here to discuss the second relatively untried method only briefly, whereas the first needs all the detailed instructions for the care of a long-tried, standard work horse, we shall discuss and dismiss this still only partly perfected confactor method first, and then go to everyday concrete procedures with simple structure.

6.3. Hypothesis-Creating Rotation: By the Confactor Principle

Let us imagine three influences, such as intelligence (as IQ), age (in years), and physical strength (as pull on a dynamometer), operating on a collection of variables such as performance in arithmetic and English, the distance one can throw a baseball, number of miles one has traveled in one's life, as instanced in the hypothetical data loadings on Table 6.1. The design calls for the six variables to be measured on two groups of people, deliberately chosen so that they are likely to have quite different variances on the same three factors. In fact, if the reader compares the corresponding factors in (a) and (b) in Table 6.1 he will see that I has $\frac{3}{2}$ times more contribution in (b), II has twice as much, and III has $\frac{5}{6}$ th as much. The columns of loadings in the two studies are in fact parallel and proportional; The reason for the smaller loadings in the college men on intelligence will be that they are more selected for intelligence, i.e., less scattered. (Note we do not say they are higher in mean, but only cut more "to a type" than the businessmen.) Thus differences in intelligence among the businessmen will play a larger part in determining their differences in arithmetic and English performance than they will among the college men. On the other hand, let us suppose the businessmen have not cultivated their athletic abilities as much as the students. Consequently, this factor has become less important, with a reduced variance, shown by a drop of loadings from .6 to .5. (As a psychological reality note that intelligence is also given a small loading on throwing, since there may be a "knack" as well as physical strength.) Since the businessmen have an age range twice that of the college men (10 years versus 5 years), it is natural that the loadings of age in determining hearing acuity (which is negatively related because it drops over these years) will be greater and the number of miles traveled should be larger. Consequently, the loadings of (b) on II are in fact doubled.

If it is in the nature of things, then, that where there are real unitary influences in the world, *they will act with greater or lesser variance in different populations, and demonstrate parallel, proportionally increased or decreased loading patterns, as in Table 6.1.* The question will immedi-

Table 6.1. Confactor Rotation: Parallel Proportional Loading Profiles of Two Groups

Tests	(a) College men, 18–22 years			(b) Businessmen, 22-32 years		
	I[a]	II[b]	III[c]	I[a]	II[b]	III[c]
Arithmetic	.4	0	0	.6	0	0
English	.4	.1	0	.6	.2	0
Miles traveled	.1	.2	0	.15	.4	0
Hearing acuity	0	-.2	0	0	-.4	0
Throwing baseball	.1	0	.6	.15	0	.5
Long jump	0	0	.6	0	0	.5

[a] I = intelligence.
[b] II = age.
[c] III = physical strength.

ately arise if we want to use this property: "If a position can be found in some rotation of each of two studies from their V_0's such that the proportionality for each factor exists over all variables, is it unique, or are there several such parallel positions?" This vital initial question the present writer was fortunately able to answer, in proposing the method, by an initial proof that such a position is unique and not to be repeated (1944d). It is as unique as the position of the two barrels of a combination lock which permit the lock to open.

It is therefore well worth setting out to discover this position. But how do we do so? If this position had to be found by trial and error, rotating all possible positions of each unrotated matrix until they coincided it would take a lifetime. Fortunately, an analytical solution was reached (Cattell & Cattell, 1955), ten years after the statement of the confactor principle itself, which permits an immediate analytical solution by matrix multiplication on a computer. Although matrix algebra methods have only been explained to readers so far at an elementary level, we must proceed to slighty more sophisticated use at this point.

If we indicate the two experiments by subscripts A and B the statement of the confactor principle is that

$$V_A = V_B D \tag{6.3}$$

where V_A is some special rotation of the unrotated matrix V_{0A}, and V_B comes similarly from V_{0B}, such that V_A and V_B stand to each other in a relation like that in Table 6.1, i.e., D stretches or contracts the factor column values. Then we need to find the transformation matrices—L's— that will rotate the V_0's to these unique positions, thus

$$V_A = V_{0A} L_A \tag{6.4a}$$

$$V_B = V_{0B} L_B \tag{6.4b}$$

D is what is called a diagonal matrix, consisting of k values expressing the proportionality of variance of each factor in A to the corresponding factor in B, at the one and only position where such simple proportionality holds. L is the usual matrix for rotating an unrotated matrix to any given position, the discovery of the values for which we shall explore in a moment. After algebraic transpositions it can be shown that the required solution is first to discover a matrix K such that

$$K = (V_{0B}^T V_{0B})^{-1} V_{0B}^T V_{0A} \tag{6.5}$$

Obviously K is derivable as is the L in Eq. (6.2) from the data we *have*, namely, V_{0A} and V_{0B} which are found from a straightforward unrotated factoring of the experimentally given correlation matrices R_A and R_B. The desired L_A and L_B matrices for shifting V_{0A} and V_{0B} to the unique proportionality position are then derived from K by getting the principal components ("latent roots and vectors") of a matrix KK'. The D values are the square roots of the k latent root values of the KK' components and the L_B matrix (and by reciprocal procedures the L_A) are obtained as the component entries themselves.

This is a very condensed statement, oriented to what one does rather than the detailed steps of the argument for doing it, which only a minority will wish to study at this time. The fuller explanation is given later here and in Cattell and Cattell (1955). The mathematically inclined should also compare the above with important propositions by Meredith (1964 a, b), Gibson (1962, 1963), and Evans (1971) which take other approaches to the problem of factor transformations among studies, though not oriented to the requirements of the scientific model here posited. The confactor method as described above has two limitations.

1. Because we have not yet had opportunity to describe factoring of covariances (p. 407), as distinct from factoring correlations, we have treated the relation as if it held in ordinary factor loadings. The relations we have discussed could hold exactly only for loadings based on factoring covariance rather than correlation matrices. With the latter the relation could only be approximate, for reasons given later.

2. The solution given above is restricted to an *orthogonal* rotation, and our following of the scientific model should eventually convince us that this is an undesirable, because unreal, restriction. The present writer has tried several approaches over the last decade to achieve a satisfactory oblique solution, and through work jointly with Brennan, Meredith, Schönemann, and Hakstian a workable approximate procedure is now in sight (Cattell and Brennan, 1977).

Neither of these restrictions, however, rationally justifies the comparatively small use given the confactor method, since with certain problems conforming to its assumptions it would be apt. The lack of experience with it is probably due more to the complexity of its matrix calculations— though trivial in time demand on a computer—and especially to the planning required in getting two experiments completed with the same variables on two appropriately different populations. To test its practicability where some conditions are not met, plasmodes have been set up with moderate departures from obliquity and it has been shown that the orthogonal approximation has come out acceptably close to the true solution. As to the need for covariance factoring, the reader can see even if he is not familiar with covariance factoring why ordinary factoring runs into difficulties. For example, if two factors each loading a variable .7 in experiment A were to increase their variances, as real influences, by, say 50% each in experiment B, the correlations would have to go above unity and the communality (on the two factors above) to an impossible value of $(.7 \times 1.5)^2 + (.7 \times 1.5)^2 = 2.21$. Ordinary factor analysis supposes unit variance factors and variables, and, when one factor increases in loading, others, even though they do not actually change in the direct magnitude of their effect, have to reduce their loadings, as individual children have to take less space when more get into the bed. Covariance factoring (p. 407) avoids this, and it is in covariance equivalents of the given matrices in Eqs. (6.3) through (6.5) that confactor calculations must be made.

6.4. Hypothesis-Creating Rotation: By the Simple Structure Principle

The next principlelike confactor judges the structure on its merits in regard to the general properties expected in any determining real influence and not in terms of some specific theory. It is called simple structure, and will initially be described as originally proposed by Thurstone. Thurstone prescribed finding a rotation which maximized the zero loadings separately considered by columns and by rows in the factor matrix. This would mean that no factor loaded more than a minority of all variables and no variable was accounted for by more than a few out of all the factors. This is a clear operational statement of what one should hope to reach by rotation, but the present writer would substitute a single goal, not two, stating at the same time an explicit principle which requires that goal. The principle, discussed below, is that no factor should affect (significantly load) more than a minority of a wide array of variables, because this would be the property we would expect of a true influence. This would agree directly with Thurstone's first principle, but would be indifferent about the second. The latter tends to be required only because one cannot maximize zeros in the columns of a given matrix without also maximizing them when summed across the rows! However, though by our principle not even a single factor should fail to have sufficient zeros in its column, this does not connote that there should not be some rows with no zero. Such a difference of emphasis may seem trivial but becomes important when fixing an objective criterion for simple structure or designing automatic programs to find it.

To return to the rationale of getting a simple structure by obtaining many zeros in columns we should note that it rests simply and solely on the common sense expectation that any real influence in nature *will tend to affect only a limited fraction of any widely chosen set of variables.* The student will think of many instances in the physical world of such influences—light, gravity, temperature, pressure, humidity, oxygen level, altitude, etc. Factors or components as they come out of the extraction process generally load all variables. The later factors will have small loadings, it is true, but they will be scattered over most variables. Consequently, if we can find a rotation position from this V_0 which shows each factor loading a few variables decidedly, and leaving the rest alone, it is likely that we have landed on influences behaving in the above way. Of course, at the outset of experiment, we must be cautious, because it is logically possible that there are other causes of simple structure than independent influences. All influences may appear as simple structure factors but not all simple structure factors may represent determiners. The empirical success of simple structure rotation in actually picking out such influences, e.g., producing such recognizable factors as weight, volume, and elasticity in the ball experiment (Cattell & Dickman, 1962), however, and the failure of critics to suggest any other systematic source, point to the correctness of the main assumption. One can reason to it also from what is known about most influences. For example, in Table 6.1 it is reasonable to expect that intelligence will appreciably affect vocabulary and the mathematics of arithmetic, but matters like the distance one has daily traveled to work, or the goodness of one's hearing, or how far one can jump, are in a different domain. It seems unlikely that intelligence will have anything to do with them. At most it may load them almost insignificantly in that intelligence may help in the skill of throwing a ball or (as indicated in our suggestion of a loading of .10) students may have traveled, in going to college, a little more than others of their age. Similarly, there is no reason why the physical strength factor should affect one's performance at arithmetic or English, or the distance one has traveled in the world. The resolution position illustrated in Table 6.1 is in fact consistent both with simple structure and with the confactor resolution conditions, and this agreement would also ordinarily occur.

Let us next clearly recognize what simple structure means in geometrical graphical terms. In two-factor plots it means, as shown in Fig. 6.2, that the points at the end of vectors representing variables will tend not to be

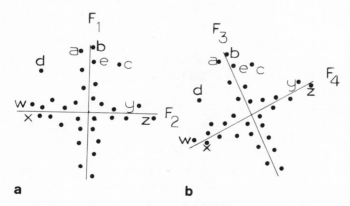

Fig. 6.2. Simple structure: (a) orthogonal and (b) oblique.

distributed evenly over the space but will accumulate in denser zones as shown. For if, in Fig. 6.2(a) we dropped perpendiculars from the points on the axis of factor 1 to obtain the loadings of all variables as projections on that factor, a minority of variables, such as a, b, c, d, and e, would have substantial loadings, but w, x, y, z, and the numerous other vector ends in that kind of position would have virtually zero loadings in the rotated factor matrix. Parenthetically, such uninfluenced variables would not be expected to be exactly zero, because of errors of measurement. Just as there is a set of points loading zero in factor 1, so there is another set to give us the position of factor 2, and so on for all factors.

There is no reason why these two lines of dense points should come out as orthogonal as the lines of the graph paper, as in Fig. 6.2(a). And in fact we often find that when plotted in the original orthogonal factor or component space (i.e., from the V_0 matrix) the points group themselves as shown in Fig. 6.2(b). Thus Fig. 6.1(a) is just a special orthogonal case among oblique factors: the rare case when they are at right angles. Since we know that the cosine of the angle is the correlation (at unit length) this obliqueness means that the factors in general are correlated—positively in the case of Fig. 6.2(b).

The whole question of correlated factors will be discussed in Chapter 8, but in passing we may note that it makes sense for factors to be correlated rather than represented artificially in rigid orthogonality, because influences in the real world *do* get correlated. For example, if I took twenty measures of plants, terrain, etc., which depend on the temperature and the rainfall at 100 US weather stations I should almost certainly find the rainfall and temperature factors negatively correlated. Similarly, in the ball program (p. 95), though size and weight of balls affect many behaviors differently, and hence are different factors, these two factors are substantially correlated (Fig. 7.4). The verdict that oblique factors versus orthogonal simple structure are normal, however, must be given by nature and empirical studies and in fact they give us, regularly, obliqueness like Fig. 6.2(b) not Fig. 6.2(a).

If plots like Fig. 6.2 were plotted from a three-factor instead of a two-factor matrix of V_0 projection (loading) values, there would be three plots—1–2, 1–3, and 2–3—instead of one. And what looks like a line of points would be actually a disk of points seen on edge, as if a thin ellipse were drawn around points in Fig. 6.2(a). In four and more factor-dimensional plots the plane or disk exceeds our world of three dimensions and becomes a *hyperplane*, i.e., a plane in more than three dimensions, which we cannot visualize. In general terms, therefore, we speak of simple structure as something which produces dense collections of points lying in hyperplanes. The extent to which this structure exists can be assessed by counting the number of zeros in a factor column (actually within say ±.05 to allow for error). This is called the *hyperplane count* for that factor.

Some users of factor analysis have had difficulty in accepting the reality of such hyperplanes, but Cattell and Gorsuch (1963) experimented by comparing a natural data matrix (the 16×14 ball problem, p. 95) with two artificial random matrices matched for the same factor variances, and the same communalities for each variable but with random values within that framework. (These comparable conditions are necessary because the small correlations from random normal deviates *do* yield factors mimicking simple structure, at least to the extent that they yield a normal distribution of loadings. It is necessary therefore to start with random loadings not random scores.) To quote "The results show that *one* position in the *natural* data is significantly superior in hyperplane count to any random [spin] position [in *that* data] and to [*any*] random [spin] positions in the [random] *unstructured* data." (Bracket terms added for clarity.) In short, the existence of distinct nebulae of variables in the correlational configurations from natural data is no more an accident than the existence of nebulae or the Milky Way in the sky. Laws of nature brought such structures about in both cases. Whenever we measure variables subject to real, shared influences—and what natural variables are not?—the configurations of simple structure will be found among them.

Granted that simple structure is inherent in real world data, the problem remains as to how we are best to find that structure. In a two-factor case it requires only that we plot the points from V_0 and move the axes to the denser positions, as in the results in Figs. 6.2(a) and 6.2(b). And with three factors the present writer has had colleagues who built a candelabra with corks and knitting needles and clambered over the furniture and on step ladders to find the positions from which the planes of the nebulae could be seen on edge in three dimensions! But in the general case, with four or more factors, such acrobatics is unnecessary, for there are several technical methods available for passing from the given V_0 by rotation to a V_{rs} (reference vector structure) at a maximum simple structure position. These techniques are important enough to be postponed for intensive discussion in the next chapter.

6.5. Illustration of Simple Structure Properties in Real Data

Before studying the aids to simple structure rotation, however, the reader may advantageously gather some experience of what varieties of simple structure await these methods, as can be shown by a brief but representative sampling of varied actual case plots. Figures 6.3 through

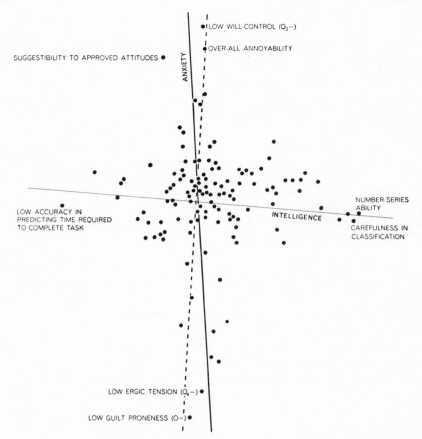

Fig. 6.3. The intelligence factor plotted against the anxiety factor in objective tests. This is a plot from an experiment with 500 airmen measured on some eighty behavioral variables. Two well-known factors, out of several located, have been plotted against each other. The anxiety factor is marked partly by such objective tests as annoyability and suggestibility and partly by questionnaire (16 P.F.) responses. The hyperplanes to the two factors are unmistakable and the intelligence source trait is fixed along the thick (near horizontal) axis. It loads most highly the test of series and of classification and negatively, low accuracy in estimating time. The identity of the unlabeled points can be found in the research indicated (Cattell, 1955) since the main purpose of this figure is to give a concrete view of hyperplanes in intelligence definition. (From "The nature and measurement of anxiety," R. B. Cattell. Copyright © 1963 by Scientific American, Inc. All rights reserved.)

6.9 represent most varieties of (a) measuring device media—questionnaires, objective tests, ratings, laboratory measures—over (b) most varieties of domains—ability tests, personality measures, motivation–interest, and national culture dimensions—over (c) several samples from different ages

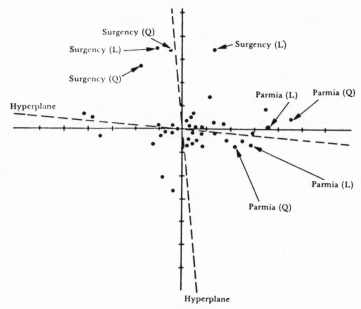

Fig. 6.4. Personality structure in joint rating and questionnaire media: Agreement illustrated by surgency and parmia source traits. Agreements of Q data with L data factors. From a factor analysis of combined Q data and L data on 181 boys and girls from 7 to 8 years of age (Coan & Cattell, 1958). The study was undertaken to answer this question: Can one personality dimension be found simultaneously loading, in a combined simple-structure plot, the variables chosen to mark what are hypothesized to be factors of the same meaning in the separate media simple-structure resolutions? These and at least two other factors confirmed the reality of such common structures (48 variables, 24 factors, 181 subjects). The obliquity of the factors is small ($r = 0.05$; angle = 87.5), but the H hyperplane is not unambiguous and could yield more positive correlation of F and H (parmia). (From R. B. Cattell, *Personality and Mood by Questionnaire*, Jossey-Bass, San Francisco, 1973, p. 265.)

and cultures and (d) data from two different experimental techniques— R technique as so far discussed and P technique to be discussed.

Selected so broadly for representativeness it is not to be expected that they will, every one, present an ideal simple structure, while their styles of presentation will alter somewhat with the years in which they were published; but it will be informative briefly to discuss each of the seven stratified sample instances set out.

In Fig. 6.3 the sample both of people and variables is large and the simple structure is relatively clear. Obliqueness is obviously mandatory, and the result is a slight negative correlation of intelligence and anxiety such as commonly exists in the general population. It is noteworthy that questionnaire and objective test markers for anxiety come together.

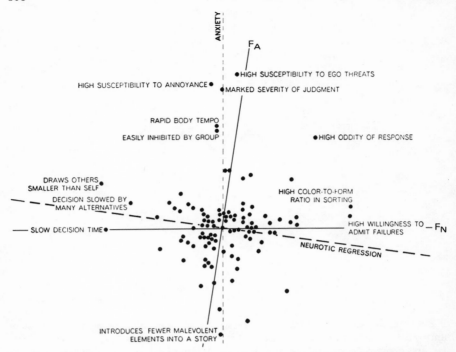

Fig. 6.5. Personality factors in objective, miniature situation tests: anxiety and regression. Hyperplanes are dashed lines. Factors are actually reference vectors. (From "The nature and measurement of anxiety," R. B. Cattell. Copyright © 1963 by Scientific American, Inc. All rights reserved.)

The Fig. 6.4 study was undertaken to check the agreement of questionnaire and observer rating personality factors. The interrupted lines are considered the best fit to the points, the continuous lines being reference vectors normal to hyperplanes. (The precise distinction of factors and reference vectors is postponed to the next chapter.) The hyperplane for parmia is appreciably wider than the ±.10 we like to set as a limit, but its position is clear. Obviously, questionnaires and ratings are here defining the same personality factors.

Figure 6.5 is the first illustration of one personality factor against another when both are measured by *objective* (laboratory) devices. Again hyperplanes are shown by the *interrupted* lines with the reference vectors as *continuous* lines dropped vertically (normal to) the hyperplanes. Indeed, this is the definition of the *reference vector*—that it is the perpendicular at the origin to the hyperplane at its best discovered position. For the present the reference vector may be thought of as the first representative of the factor itself—the axis on which we measure the loadings. But later

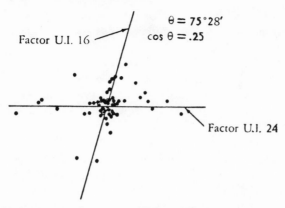

Fig. 6.6. Objective personality test in another culture: ego strength (U.I. 16) and anxiety (U.I. 24) factors. The factor traits here are competitiveness or assertiveness (U.I. 16) and anxiety (U.I. 24). The data is from objective tests (HSOA battery, given to 13-year-olds) translated into Japanese and administered to high school students in Japan. (From R. B. Cattell, *Handbook of Multivariate Experimental Psychology*, Rand McNally, Chicago, 1966, p. 185.)

a slight difference will be defined between reference vector and factor, yielding slightly different values in the V_{rs} (reference vector structure) and the V_{fp} (factor pattern) matrices. Our plots at present are from the V_{rs}. It will be noted in Fig. 6.5, that most variables, as usual, cluster around the origin, being markers loaded distinctively only on other factors and loading neither of these two we happen to have drawn. For the rest of the

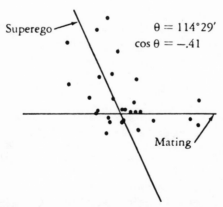

Fig. 6.7. Objective motivation measures: superego and mating erg dynamic structure factors. Data from American adults on the Motivation Analysis Test (IPAT, 1965). (From R. B. Cattell, *Handbook of Multivariate Experimental Psychology*, Rand McNally, Chicago, 1966, p. 185.) Note an appreciable negative correlation of sex and super ego.

points a hyperplane width of $\frac{1}{4}$ in. on either side of the central zero hyperplane, presenting "best fit" boundary lines would rope in most of the loaded variables.

The study shown in Fig. 6.6 was undertaken to give a check, as it does, that the same objective behavioral variables load the same personality factors (assertiveness and anxiety) across cultures—Japanese and American. (The comparable American data is elsewhere: Cattell, Schmidt, & Pawlik, 1973.) The position of the U.I. 24 (anxiety) hyperplane (vertical) would benefit from more points, but it is the nature of anxiety to have some role in many behaviors, and therefore to have a sparse hyperplane.

In Fig. 6.7 factor analysis was brought to bear on a hitherto untouched domain—motivation measures. The hyperplanes in this early study are not as neat and narrow as in, say, the much-investigated field of abilities, though with future improvements in the reliability of measures they may become so. Meanwhile one nevertheless has no difficulty in

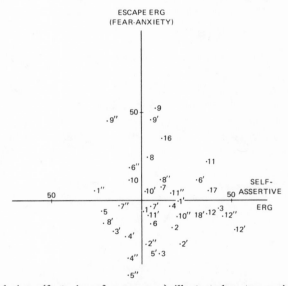

Fig. 6.8. P technique (factoring of one person); illustrated on two ergic structures: the escape erg and the self-assertive erg. Several numbers are in duplicate because two studies are set out together to show extent of agreement. Index to prominent attitude vector points as follows.

11. I want never to do anything that would damage my sense of self-respect.
 9. I want America to get more protection against the terror of the atom bomb.
12. I want to increase my salary.
 4. I want to make love to a woman I find beautiful.
16. I want to avoid ever being a patient in a mental hospital.
17. I want to become proficient and outdo my colleagues in my chosen profession.
18. I want to be smartly dressed with a personal appearance that commands admiration.

placing the best fitting hyperplanes. The result is a negative correlation of strength of sex erg and strength of superego structure, as has since been found in other research.

The novelty in Fig. 6.8 is a shift from ordinary factor analysis (later to be defined as R technique) to single person factoring (P technique). The points are numbered and the attitude variables involved as markers are defined below. However, to avoid prejudiced rotation the variables are never named to the experimenter in the actual rotation process but left as purely anonymous points. This is often called "blind rotation," which is a curious name for a purely visual process! "Abstract" or "unidentified" rotation might be better. The hyperplanes are quite as clear here as in ability or general personality factors and the degree of obliquity happens to be a negligible departure in this case from exact orthogonality.

Both small groups (as used in group dynamics) and nations have had their performances factored to give the dimensions of their *syntalities* (Fig. 6.9). Uniquely rotatable dimensions are found, as shown in the typical simple structure resolution here. This is a study of 82 variables on 120 countries. Here the hyperplanes are shown by continuous lines. This plot is not final, but is taken at the 12th overall shift, where the nature of the dimensions is already becoming clear, as indicated, namely, factor 4 as national wealth and factor 8 as educational level of the population.

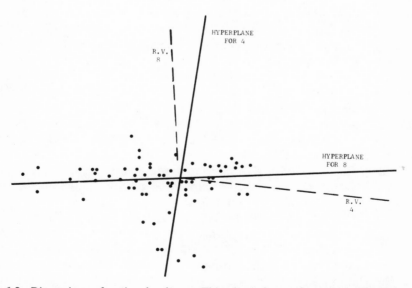

Fig. 6.9. Dimensions of national cultures. This plot is from a factoring of 82 variables on 110 countries. The hyperplane for factor 4 is not well defined. It is adequate for essential rotation but not for determining the precise correlation of the two factors. (Cattell & Wolliver, 1977.)

6.6. Some Problems of Research Design and Interpretation in Simple Structures

A really effective use of the simple structure approach cannot be expected without attention to the design of the experiment in which it is to be used. *Confactor* rotation can, in principle, be used with any choice of variables, though it requires a careful choice of subjects or situations, but *simple structure* requires a foresighted choice of the sample of variables. It is obviously absurd to expect to get any determinate rotation of a certain factor X if we have so chosen the variables in the study that probably all of them will have some significant loading on X. For X will then have no "hyperplane stuff" to rotate by. If in researching on, say, the nature of the inductive reasoning factor, as known hitherto by its highest loaded performances, we took 10 tests expected to load it and only two—say, a spelling and a vocabulary test—expected to lie in the hyperplane, the definition of the hyperplane by only two tests would be poor, and consequently the basis for orienting the reference vector for the induction factor would be vague. The definition of the parmia factor in Fig. 6.4 above comes perhaps dangerously near this with only five or six effective hyperplane variables. It is important to remember that the position of the reference vector is *not defined by the cluster of variables which load it highly, but by the hyperplane of variables that are quite unaffected by that factor.* (What we defined earlier as cluster may, as a composite surface trait, lie well away from the pure factor position.) Rotating to put reference vectors through clusters, incidentally, is a grievous misunderstanding of rotation, for the reference vector is quite different from its own or any hyperplane. Even variables far out on the hyperplane will have zero loadings on the factor (or reference vector) one is seeking to place. The distinction that should remove all risk of confusion in this area is the technical one that the hyperplane is the k-1 dimensional space which does not include the factor (and therefore the reference vector) dimension.

Since many published researches undoubtedly show the disabling weakness of planning without regard for determining hyperplanes it is well to remind the reader of two remedies for such a defect:

1. Put in hyperplane stuff. That is, when investigating domain "A" deliberately introduce several "irrelevant" variables you have reason to believe have a "not-A" nature. Hopefully this can be done with some previous knowledge of other common, broad factors that are "not-A" so that pairs of markers can be found for them that will yield factors standing out clearly and largely orthogonal to the factor or factors in the domain you are interested in. Incidentally, this fits the more general

philosophy of research that you don't know the nature of A unless you also know the nature of not-A.

2. Need for a special design of this kind occurs in what for lack of a better term we may call "fan-structure," as required in the intensive investigation aimed at some particular factor or source trait. The investigator is interested only in, say, a thorough investigation of the nature of the super ego source trait, or of even two such dynamic or ability factors and their exact correlational relationships, say of fluid and crystallized intelligence, g_c and g_f. He will want to use probably two or three dozen variables carefully chosen to explore the nature of behavioral expression of the one (or two) factor(s), i.e., a sufficient set of markers having high loadings on them. However, all this care will go to waste unless a precise orientation can be given to the main factor vector itself by an ulterior hyperplane. (As we have seen it is no use taking the center of the *cluster of chosen marker variables* if he wants to define that vector at all precisely.) In such a design he needs to give brief representation—say two really good markers—to each of as many factors as possible at the same stratum level as the factor intensively studied. (Stratum level is defined in Chapter 9. It can be taken for the moment as "equally important.") These factors will "fan out" (hence the above expression) into a hyperplane (in k-1 space) around the vector for the central factor studied and enable the loadings of its many expressions to be determined more precisely through a really reliable rotation. Countless factorial researches concentrating on some special concept have failed to contribute really dependable knowledge, through ignoring the need for "fan" design in choice of variables.

3. Aim to work with *as many factors at the same time* as you conveniently can—up to a dozen or so at any rate. In addition to the gain just mentioned through their mutually providing hyperplanes, as required for good rotation, there is a second gain through avoiding confusion of factors. It is statistically possible for two factors, even when orthogonal, to load the same set of variables in just the same pattern, up to loadings of 0.70 ($0.70^2 + 0.70^2 < 1.0$). Consequently, there would be times when one would not know, in a research with just those variables only, which of the two factors one has got hold of. They can be distinguished only with the use of some extra variables on which, in fact, they sharply differ. Thus, a study with more variables and more factors is always more clear and indubitable in findings than one which confines itself only to a narrow area. Furthermore, it is always a good plan to provide marker variables for an old "established" factor in a given area when one is searching to discover new factors. It was by such planning that Pieszko (1967), for example, was able to show that the clinician's concept of a "sensitivity" trait which burgeoned in the late sixties was actually nothing other than the

standard general anxiety factor (U.I. 24). Similarly, a wild goose chase after some new armchair invention, such as an alleged unitary source trait of "impulsivity," is likely to be avoided if markers for carefully established factors in the previously broadly mapped personality sphere are included, as instanced in Wardell and Royce's (1975) discussion of impulsivity.

In discovery (hypothesis-producing) factor analysis the idea of representative design, as stressed by Brunswik (1956) and by the present writer in the *personality sphere* concept (1946a, b, 1957) becomes important. It leads, in practice, to steps not unlike the above, but its principle is different. Though this chapter on rotation is not the place for a general discussion on research design (see Chapter 12) the rotation requirements can be seen to support other arguments for taking as many and varied variables as one conveniently can—"convenience" usually being settled by the total hours subjects can be persuaded to sit for a variety of tests or the availability, in social psychology, of sufficient sources of recorded data. In any case, in hypothesis-testing factor analysis as much attention to choice of presumed hyperplane and marker variables is required as in hypothesis checking, though with a slight difference of emphasis and aim.

The question sometimes arises in this connection whether our statement that there is a structure in natural data (but not in artificial data, Cattell & Gorsuch, 1963) needs to be supplemented by an admission that there may sometimes be more than one structure, i.e., alternative possible simple structures. When the variables in a study are poorly sampled, or unavoidably few, two equally tempting positions for simple structure undoubtedly sometimes appear. A "classical" case is among the eight main second-order personality factors in questionnaire data, where Q_{IV}, the independence factor, sometimes locks over into a characteristic alternative to its normal position, presenting ambiguity notably in the plots against exvia, Q_I, and cortertia, Q_{III}. In this case the problem was eventually solved by advancing to a representation of nearly 30 primaries, beyond the typical studies characteristically done with 16, whereupon, with the benefits of a larger sample, the original choice in the most careful studies was confirmed.

But there is a wider and scientific sense in which alternative structures may be said to exist, and which cannot be understood until Chapter 9 on second-order factors has been grasped. Briefly, however, the problem can be instanced by noting that in practice it is uncommon, but not unknown, for second-order factors to appear as alternatives among first-order factors. The second-order factors, however, do exist as real determiners, and are thus legitimate simple structure alternatives not constituting any denial that simple structure leads to real entities.

There are stages in science where it needs to tolerate and entertain, as almost equally promising, two, or even more, apparent alternative explanations, while further investigation proceeds. Sometimes, as in the chemical issue of whether atomic weights should or should not be exact multiples of that of hydrogen, discovery from new investigation decides. In others, as in the alternatives of a point quantum or a wave (De Broglie) we may have to live with dual thinking. If competing positions for simple structure genuinely appear, then it is important to report and study them, and the above instance of first- and second-order hyperplanes in the same item data appears to be such. But the vast majority of uncertainties are not of this nature and are due either to lack of thoroughness in simple structure search procedures, poor techniques, or inadequate design in regard to choice of variables.

In the latter one must include some inadequacies that are unavoidable, as in the above instance of trying to tie down second-order hyperplanes, if it should occur in a situation in which one has available, say, only eight primaries to go as points into the drawing of the plots. It is not easy, in terms of sampling of variables, to be sure that the best fitting line is correctly placed when only, say, four points out of eight can represent the "disk nebula on edge" that we need to locate, for four points is scarcely a swarm.

The inherent unavoidable limitations of simple structure are evident partly in its relative failure, in those cases where only few variables can be involved in a study, and partly in total failure arising from the real possibility that early in the game our choice of variables may have accidentally included factors which actually extend over all variables used, and therefore have no hyperplane. However, since all science, as Tennyson said, consists in "creeping on from point to point," these odd cases would in time be understood from other contexts of evidence. Meanwhile, where variables are few, the simple structure method needs to be supplemented by the confactor method, even though the latter may still be limited to orthogonality.

6.7. Summary

1. The principal components matrix $\underline{V}_0$, or the factor matrix V_0, as they appear at the positions when newly extracted from the R matrix, have few properties (except the noncorrelation of the columns) that make them, as they stand, of particular interest to the psychologist (as distinct from the mathematician). The loadings are "accidental" in the sense of being peculiar to a local choice of variables. However, any such matrix can be transformed by rotation into an infinite number of equivalent matrices, equivalent in being able to retrieve the original R matrix, in having constancy of communality, etc., and one of these is uniquely meaningful. The

problem is to find the rotation producing this unique matrix in which the factors correspond to determiners in scientific data, with properties of general interest.

2. The scientific model in multivariate experiments leads us to expect that real influences (causes—past or present—or determiners) will create unitary factors corresponding to them. But in the extraction process these particular factors have become transformed into some mathematically equivalent set of factors. The problem just cited can therefore be stated as one of finding the way back from the "accidental" V_0 as obtained by the extraction process from the R matrix, to a unique resolution obtainable by rotation, and this requires clear guiding principles.

3. There are broadly two principles of procedure: (a) to seek the unique position just described by applying criteria characteristic of determiners generally, or (b) to set up a "target" matrix expressing some hypothesis in the experimenter's mind, to see if a possible rotation position exists which will fit that target. These may be called "hypothesis-creating" and "hypothesis-checking" forms of rotation, and they should properly be accompanied by differences in the design of the factor experiments from the beginning.

4. The hypothesis-checking rotation can be carried out by a computer program called the PROCRUSTES transformation, of which there are orthogonal (Schönemann, 1966a, b) and generalized oblique (Hurley & Cattell, 1962; Browne, 1967; Gower, 1975) forms or by a form of maximum likelihood factoring. Freedom to go oblique is, however, scientifically essential, and the particular correlations arising among the factors in going oblique offer an additional basis for testing the soundness of match, as well as the foundation for reaching higher-order factors. More detailed technicalities, e.g., indices for assessing the goodness of such matches, are left until later.

5. For hypothesis-creating resolutions we need to meet conditions that apply to determiners in any kind of data. Two principles are available: (a) confactor rotation and (b) simple structure rotation. The former has the most theoretically positive features, but the procedures for confactor rotation ("parallel proportional profiles") are completely worked out only for the orthogonal case, and are as yet "experimental" for the oblique case. This design requires two experiments with the same variables, but on samples significantly differing in the variances within each pair of corresponding factors.

6. The form of hypothesis-producing rotational resolution on which psychologists have almost entirely depended in practice is that of simple structure. Simple structure, like confactor rotation, involves a set of required general relations (i.e., criteria of fit not tied to any specific hypothesis) derived from the scientific model of factors as influences.

The criteria involved in confactor and in simple structure are independent as operational statements, but since they derive from different paths of inference from the same concept of properties of determiners they should (a) be compatible and (b) in fact lead to the same resolutions.

7. The extent and nature of the approach in real data to the ideal simple structure are illustrated by experimental data plots from a wide range of data in personality, motivation, and social group syntality studies, with both R and P technique methods. Discussion of the techniques for reaching simple structure is deferred to the next chapter.

8. The use of simple structure rotation to complete a factor analysis requires prior attention to the design of the experiment, especially in the choice of variables. The main desirables are: (a) variables sufficient in number and kind to provide "hyperplane stuff" and (b) a sufficiency of factor dimensions to orient each factor with respect to several other dimensions, preferably some "established" ones. In the intensive study of one factor it is important to carry sufficient markers for other factors (even if only briefly and roughly measured) to "fan out" in a hyperplane sufficing for more exact orientation of the factor studied.

9. The possibility, and possible meaning of alternative hyperplanes being found in simple structure rotation is discussed. Mostly, supposed alternative rotations—and therefore explanations—arise from using too few variables; poor design; or running into what are later defined as "geometer's hyperplanes" (p. 169). However, the possibility of two equally good positions, with associated distinct theoretical explanations, only to be resolved by later advances and ulterior evidence, cannot be excluded.

10. Although we have barely touched on sample problems, in general, to this point, the question may be raised by the student whether simple structure on a sample is, for moderate samples, an acceptable approximation to that in the population. The issue has some complex parameters, e.g., the reliability of the measures *per se*. But otherwise the simple structure in a moderate sample should be as good as that in a large one, though the sampling of the factors may give different variances and correlations among the factors relative to one another and the population. The sampling of variables produces greater problems than the sampling of people, when seeking a correct simple structure.

CHAPTER 7

The Techniques of Simple Structure Rotation

SYNOPSIS OF SECTIONS

7.1. Transforming the Unrotated to a Rotated Matrix
7.2. The Definition of a Matrix and Matrix Multiplication
7.3. The Nature of the Transformation Matrix in Hyperspace
7.4. The Rationale for Oblique, Correlated Factors, and the Computation of Their Correlations
7.5. Rotation by Visual Inspection of Single Plane Plots
7.6. Analytical and Topological Automatic Rotation Programs
7.7. Comparative Studies of the Strengths and Weaknesses of Various Automatic Programs
7.8. The Tactics of Reaching Maximum Simple Structure
7.9. ROTOPLOT: The Calculations in Successive Shifts
7.10. Summary

7.1. Transforming the Unrotated to a Rotated Matrix

The preceding chapter has stated the general aims and styles of rotation. In this chapter we shall concentrate entirely on the practical techniques of obtaining a resolution on the simple structure principle, because, until confactor rotation is perfected, it is likely to remain the sole dependable principle in hypothesis-producing factor analysis. In any case, where the obtaining of suitable mutually related groups for confactor is not possible, and wherever for any reason one is confined to a single experiment, simple structure remains the one dependable everyday method of reaching the hidden unique resolution.

At this point, in making each of the steps toward simple structure, we encounter the need (a) to decide on what angular shift to make from

118

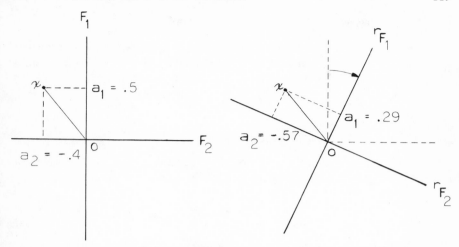

Fig. 7.1. Measuring new projections after orthogonal rotation.

the initial "accidental" axis positions in the single plane plots to a new "simpler" position, and (b) to work out what change that angular shift produces in the loadings on the factor or factors involved in the shift. This could be done geometrically by measuring the projection of the points on the new axes on the graph itself (see changes for x in Fig. 7.1); but it is altogether more convenient and accurate to do it by calculation. What in fact one wishes to do is to transform the $n \times k$ V_0 matrix to an $n \times k$ V_1 matrix, where V_1 is the first exploratory step in the search for the most perfect simple structure matrix V_s. For, except by complex methods explained later, one cannot go directly to maximum simple structure, but approaches it through several steps moving to increasingly dense hyperplane nebulae. One should keep in mind the basic principles set out in Fig. 3.3, that the *configuration of variables*, i.e., the angular correlation relations among them, remains unchanged and rigid under rotation. Only the axes change position, and therefore the projections of the variables upon them. Or, since all is relative, one can imagine the axes as standing still and the configuration being spun like an armature—but in any and all directions—within this framework of coordinates.

The simple basic and easily proven trigonometrical proposition was stated in Eqs. (3.2a) and (3.2b) that if one wants to get the new projections of a point such as x in Fig. 7.1, respectively, on F_1 and F_2 when the F_1 axis is shifted clockwise (toward F_2) through θ degrees, then

$$\underline{a}_{1 \cdot r} = a_1 \cos \theta + a^2 \sin \theta \tag{7.1a}$$

$$\underline{a}_{2 \cdot r} = -a_1 \sin \theta + a_2 \cos \theta \tag{7.1b}$$

where $\underline{a}_{1\cdot r}$ and $\underline{a}_{2\cdot r}$ are the rotated (hence r) projections of variable a on the new axes 1 and 2 and the a_1 and a_2 are the old projections of a on axes 1 and 2. Note that if we want to write this in cosines only, cosines conveniently having the meaning of correlations, we can easily (since $\sin \theta = \cos 90 - \theta$) write the multipliers of a_1 and a_2 in Eqs. (7.1a) and (7.1b) as

$$\cos \theta \qquad \cos (90 - \theta)$$

$$- \cos (90 - \theta) \quad \cos \theta$$

If the axis for F_1 had been shifted away from that for F_2, i.e., in a counterclockwise rotation, the signs of the cosine values would be

$$\cos \theta \qquad - \cos (90 - \theta)$$

$$\cos (90 - \theta) \quad \cos \theta$$

Without getting entangled in further discussion of the sign questions, let it be said that it so works out that when the student looks up cosines in a table or enters correlations he will actually give the numbers signs as immediately above for a rotation away from the other reference vector and as in the block above for a shift toward it.

We shall now show how the four numbers are entered into a matrix form, to permit the computer to do the calculations of 7-1(a) and (b) for us. The four values above are put into a 2×2 matrix, as shown in the middle of Table 7.1. If we had three factors to shift on, it would be a 3×3 matrix, giving the direction cosines of each of the new factors, in a column, to the three unrotated factors, according to the row x column multiplication rule of matrix multiplication stated earlier (and below), thus carrying out the calculation in 7.1(a) and (b) to get the new projections.

Let us suppose we have measured on the graph the angle, say 25° clockwise, through which it appears the axes should be rotated to improve simple structure. That is to say to move to where more points lie in the

Table 7.1. Illustration of Matrix Calculation of the New Factor Matrix in an Orthogonal Rotation

Variable	F_1	F_2		$\underline{F}_{1r}$	$\underline{F}_{2r}$		$\underline{F}_1$	$\underline{F}_2$
$(x =)$ 1	.5	−.4	$\times$ F_1	.91	−.42	=	.29	−.57
2	.4	.2	F_2	.42	.91		.45	.01
3	−.3	.7					.02	.76
4	−.4	−.4					−.53	−.20
5	.6	.6					.80	.29
6	−.5	−.3					−.58	−.06
	V_0			L			V_r	

hyperplane. We find from tables that cos 25° equals .91 and sin 25° equals .42. Now we need to carry out the calculations for the new projections not only for the variable a, but of course for all the points in Fig. 3.5 and Table 3.1. It will be realized from Eq. (7.1a) that to get the new F_1 loadings now called $\underline{F}_1$ loadings of all points we multiply down the *whole matrix*, with the same two numbers above, namely, cos θ and cos $(90 - \theta)$— which equal .91 and .42—keeping each to the factor column concerned. Note that a column, such as $\underline{F}_{1r}$ is giving us the direction cosines of the new, rotated factor $\underline{F}_{1r}$ to the framework presented by the original unrotated factors, $\underline{F}_1$ and F_2, respectively. To get the computer to do this calculation for us we have to arrange the unrotated values in one matrix and the two multipliers in another, as shown in Table 7.1, by the $\underline{F}_1$ column in matrix 1.

It will be seen that by Eq. (7.1a) the new loading of a_1 on factor 1, which can be written $a_{1r} = (.5) \times (.91) + (-.4) \times (.42) = .287$ (rounded = .29) and so .29 is entered as the new projection for a_{1r} under $\underline{F}_{1r}$ in the V_r matrix (r here standing for rotated). The computer will carry this combined multiplication and addition all down the rows of the unrotated matrix, producing a whole new column for $\underline{F}_1$. Then it will do the same for $\underline{F}_2$, using the multipliers in the second column of the transformation matrix L to give the numbers in the $\underline{F}_2$ column. (Parenthetically, it is a special case for two factors and an orthogonal case, as here, that the numbers in the second column in L happen to be those in the first column reversed.) Thus, in all cases except some illustration like that given initially in Fig. 7.1 we do not depend on geometrical procedures and ruler to measure the new projections but arrange, as here, for a matrix multiplication and insert the new points on the new graph from the new projections thus calculated.

7.2. The Definition of a Matrix and Matrix Multiplication

Our policy, stated earlier, has been to introduce matrix conceptions gradually, as they are needed, but at this point some more explicit formulation of what has been learnt is desirable. Some regular knowledge of matrix procedures will help the student not only in communicating with the computer technician, but also in neatly conceptualizing the various psychological models in terms of which much multivariate experimental psychology proceeds.

The present rotational procedures are a natural point at which to introduce some formalizing steps regarding matrices, though other steps in matrix algebra will be made at suitable junctures later. If the student wishes to devour the subject in one meal, there are some very good treatments oriented to psychologists, notably Chapter 3 in Gorsuch (1974), Chapter 4 in Rummel (1970), Chapter 2 in Tatsuoka (1971b), parts of Green and Carroll (1976), or for an extensive treatment, the books by Horst (1963) and Hohn (1958). However, the gradual development throughout these chapters will hopefully suffice for the reader following the main factor analytic arguments here *per se*. But in the end the student may wish to push on to the more thorough treatments indicated. For wherever multivariate experimental methods are used—and we have seen in Chapter 1 that they are vital to a real breakthrough in conceptions in many areas of psychology—it is likely that the analysis will use matrix algebra methods implemented by the computer. Indeed, it is a fortunate accident of history—like that of the invention of printing coinciding with new translations of the Bible—that the electronic computer appeared in the mid-20th century, as the social sciences recognized the power and indispensability of multivariate methods.

Now we have just said that the shifting by rotation of n points on k axes from a V_0 to a V_1 is most easily carried out by a *matrix multiplication*, which is a device to do systematically over many variables and factors what we have calculated for one variable and two factors in Eq. (7.1) above. But what exactly is a matrix and by what rules does one matrix multiply another? By definition, a matrix is nothing but a box—square or rectangular—of numbers (called "elements") arranged according to rules to ensure that some desired model of computation be carried out. (Besides square and rectangular we talk of triangular and diagonal matrices, the latter being square matrices in which nonzero numbers exist only below or above the diagonal or in the diagonal, respectively.) Now in multiplication of matrices of any kind, as in Table 7.1, the rule is that a row of the first matrix is to be multiplied by a column of the second, as shown by the arrows in Table 7.1. This $R \times C$ rule can be remembered by thinking of Robinson Crusoe. The numbers are so entered that each element in the column of the past-multiplying matrix multiplies the corresponding rank order element in the row and the products are added by the computer and entered in the new matrix at the point where the given row and given column intersect. For, of course, the row by column multiplication is carried out for all available rows and columns and puts a number in every cell of the column matrix.

Among the terms the student should handle at this point are *scalar*, *vector* (row vector and column vector), *conforming*, *order*, *transpose*, and *inverse*. What we have been accustomed to call "a single number" such as forms an element in a matrix, must now, for differentiation, be called a *scalar* quantity. It has size but not direction. By contrast, a row or column of a matrix, considered as a unit (an $n \times 1$ or $1 \times n$ matrix on its own) is called a *vector*, because it has both size (length) and direction. This is obvious if one looks at Ox in Fig. 7.1, for the two number elements in its row (Table 7.1) give its projections on axes, which numbers define both its length [as $(a_1^2 + a_2^2)^{1/2}$] and its direction. A vector is thus nothing but a series of numbers, but by geometric implication they are "arrows" having projections on axes in hyperspace. Because any element a belongs both to a row and a column vector, it is written a_{ij} where i is the row and j the column in which it stands, i.e., it stands at the intersection of a row and a column vector.

To perform certain operations with two matrices they must conform one to the other. That is, the second matrix must have as many rows as the first has columns. In factoring, the first unrotated matrix is described as $(n \times k)$, i.e., it has n rows (for variables) and k columns (for factors), illustrated by a 6 × 2 in Table 7.1. The $n \times k$ must, therefore, be multiplied by a "k by something" matrix, and in the case of transformation to k new factors it is $k \times k$, transformation matrix, which is 2 × 2 in Table 7.1. Thus $(n \times k) \times (k \times k)$ will produce $(n \times k)$.

If matrices conform and there are a series of multiplications, the first and last values give the order of the new matrix. Thus

$$(n \times k) \times (k \times p) \times (p \times y) = (n \times y)$$

The *order* of any matrix is defined by stating the numbers of rows and columns—$n \times y$ in this case.

By putting quite large matrices into a computer, with the program to fit the rules, we can commonly get in seconds what would take hours or days of computing on a desk calculator. Matrices can be subjected to the processes of addition, subtraction, multiplication, division, and the calculation of a reciprocal, i.e., X^{-1}, and other powers of a matrix. In fact the inverse of a matrix X is $1/X$, so that $X \cdot X^{-1} = 1$. (The 1, however, is a diagonal matrix of 1s.) Although the same processes can be carried out as with ordinary scalar numbers, the rules turn out to be different. An important difference is that what is called the *commutative law* does not hold, i.e., in matrices $A \times B$ is not the same as $B \times A$, so we must be careful to multiply in the order needed in the given case. And in division of A by B we commonly multiply A by the inverse ("reciprocal") of B, thus: AB^{-1}.

In symbolism, except where special confusions might arise, we denote matrices by capitals and scalars by lower case. (A matrix can be multiplied by a scalar, representable as bA, which means that every number in A is multiplied by b.) Thus the rotation calculation above represented in rectangular boxes can be written (with the conformity shown by the orders below each matrix) in capital letters as follows:

$$V_0 L = V_1 \tag{7.2}$$

$$(n \times k)\,(k \times k) = (n \times k)$$

where L is the transformation matrix of cosines recording rotational shifts.

Typically, a matrix is drawn, if it is not square, in the "upright" position. Thus a score matrix called S, setting out the scores of N people on n variables (N being $> n$) is written ($N \times n$), as we commonly see it in, say, a school class list setting side by side the scores in each of several areas of work. Similarly a factor loading matrix is written in the upright position ($n \times k$) since there are more variables n than factors k. If we wish to indicate the other "lying down" position we call it a *transpose* and write it A^T. Thus S^T has school subjects as rows and persons as columns, and V_0^T is lying on its back and is ($k \times n$) in order. Actually when, in transposing, a row is made to become a column the matrix both lies down and flips over. A few other rules of matrix algebra will be mentioned as we meet them, and the student who wishes to read systematically will find a good elementary treatment in Hammer (1971), an exhaustive one in Horst (1963), and a factor analytically convenient one in Gorsuch (1974, Chap. 3), and many others of which Tatsuoka and Horst have been mentioned above.

7.3. The Nature of the Transformation Matrix in Hyperspace

In the example of two-dimensional rotation (p. 120) we have taken the value of k to be 2 and of n, 6. But if, as usual, the number of factors required throws us into hyperspace, k in most experiments will lie between 4 and, say, 40. The L, *transformation matrix*, for, say 6 factors would then be 6×6, as shown in Table 7.2(a). The values in the cells of a transformation matrix represent, as we have seen above, the *cosines* of the angles (the correlations) of the new factor axis positions to the old, unrotated factors. Thus column 1 in the central matrix in Table 7.2(a) represents the new factor 1 and the L matrix states that it is to be placed at an angle of cos .5 to the old 1, an angle of cosine $-.3$ to the old 2, and so on. Thus each column in L defines the correlations (cosines) of the new factor with all the old ones, with a positive or negative sign according to direction of shift.

It will be noticed therefore that in the L matrix, as it usually stands, each column applies to a *rotated* factor and each row to an old *unrotated* factor. The values in a column as stated give the new factor's relations, in terms of cosines, to each of the old unrotated factors, and conversely for the rows. (Note that the descriptive "bordering numbers" around a matrix identify the factors. They are set out for clear explanation here in L and V_{rv} but would not go into the computer matrices.) A glance at F_{11} in L (the cell at the top left corner), which is 0.5 in this example, will tell us at once, if the value should be large, that the rotation has not moved much from the initial principle axes position. (A small angle has a large cosine.)

Table 7.2. Matrix Calculations in Deriving a Rotated Matrix and the Correlations among Rotated Reference Vectors

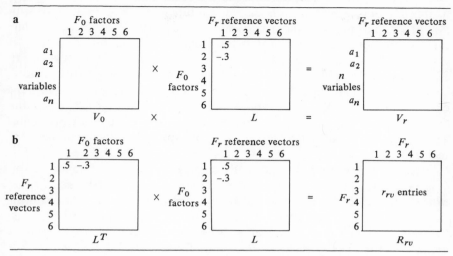

Incidentally, in good rotations the cosine at the end can be quite small, showing radical change from the V_0, but where little trouble has been taken with rotation a large value here (and down the diagonal) is one symptom that the rotation position has moved little—and probably not enough—from the original.

In Table 7.2 is shown another calculation that is regularly made with the L matrix. The main calculation is to get the new rotated loadings, as shown by $V_0 L_1 = V_1$ in Eq. (7.2) above. But if the simple structure shifts have made the axes go oblique then we want to know the angles (cosines, correlations) among them. The calculation to do this—multiplying transpose L^T by L is shown by a partial numerical illustration in Table 7.2. In matrix terms it is

$$L^T L = R_{rv} \tag{7.3}$$

where R_{rv} is a matrix of the correlations—r_f's—among the reference vectors.

Up till now we have mutually substituted, with some indifference, the terms *factor* and *reference vectors*; but the time has come for a first degree of definition of this rather subtle difference. The reference vector has been sufficiently operationally defined as the perpendicular to the discovered hyperplane. So long as all shifts are kept orthogonal the reference vector and the factor are the same. But if we go oblique they part company. Perhaps the simplest description is to consider the corner of a room to be the origin for a three dimensional plot. In the ordinary orthogonal room the perpendicular reference vector at the corner coincides with the intersection of the two walls. But in the attic room of a house with a mansard roof the two walls slope inward and their line of intersection is no longer the line of the perpendicular to the corner. The factor corresponds to this line of intersection: it is the intersection of the hyperplanes, of factors other than the one under consideration. The full consequence of this for the difference of the factor pattern matrix and the reference vector matrix of "loadings," with which we have dealt in the plots to this point, can be properly discussed only in the next chapter. But we should note here that obliquity requires the calculation of the angles among the reference vectors, as in R_{rv}, for proper plots, like those in Figs 6.3, 6.4, etc., above. Further, from these correlations among the reference vectors in R_{rv} Table 7.2, we shall later derive the corresponding angles among the factors.

Rotational changes are normally carried out in the reference vector system, because the calculations are simpler, and only at the end of the process does one change results to be in terms of the factor. The loadings on factors and reference vector are in any case simply proportional (see Chapter 8) by columns. However, it is because rotation operates with references vectors that we have R_{rv} instead of R_f in Table 7.2 and Eq. (7.3).

7.4. The Rationale for Oblique, Correlated Factors, and the Computation of Their Correlations

Various methods, graphical and algebraic, of getting the best final L matrix will now be described, but before doing so some other points need to be made. First, it should be noted that the squares of the cosines down a column in L must sum to 1.0 in k space just as they do by the familiar trigonometrical expression $[\cos^2 \theta + \cos^2 (90 - \theta) = 1]$ in two space. Thus any intermediate numbers we get from any process for the new L must be normalized before use. That is to say the sum of the squares of the values down the column must add to 1.0. Any unit length vector purporting to stand in a space of orthogonal unit length coordinates must (by Pythagoras' theorem or the trigonometrical statement above) have projections (which are cosine values) that square and add to 1.0. This has to be true for all the columns of the L matrix. If values cease to have this property, as they do when we invert the L matrix, or make changes on it by a shift matrix, then one must normalize the column again. This is done by adding the squares of whatever existing figures have been reached and taking the square root, thus $P_3 = (a_{13}^2 + a_{23}^2 + \cdots + a_{k3}^2)^{1/2}$ for column three. If the existing values in column three are then divided by P_3 it will be found that the new values have the normal property. The student should note a new term here for describing a matrix, namely a *column-normalized* or *row-normalized* matrix and an *orthonormal* matrix. An orthonormal L matrix is one in which not only the columns are normalized but in which the calculation upon them, as in Table 7.3 above, yields an *orthogonal* R_{rf} matrix, i.e., one with ones down the diagonal and zeros everywhere else. The initial L matrix in an oblique rotation is column normal but not orthonormal, i.e., there will be nonzero values all over the R_{rv} matrix. But after L is operated on in various ways it is for most purposes necessary to renormalize columns before proceeding, yielding 1.0s down the diagonal.

Let us scrutinize a little more closely what the calculation $L^T L$ does. It is in fact getting the "inner products" of all pairs of columns in the L matrix, and thus yielding the angle (as cosine or correlation) of any two new factors to one another. In orthogonal rotation these r's will all be zero, but the correlation of a column in L with itself will of course be unity (because normalized), and these unities will fall down the diagonal in what is called an *identity* matrix. Indeed if the small 2×2 example in Table 7.1 is worked out it will give a matrix for orthogonal factors:

$$R_{rv} = \begin{array}{c@{\quad}cc} & \underline{F_1} & \underline{F_2} \\ \underline{F_1} & 1 & 0 \\ \underline{F_2} & 0 & 1 \end{array}$$

(Actually, the calculation will yield 1.0045 for the 1.0 values here.) Thus we can generalize that so long as we rotate orthogonally, i.e., moving axis 2 as much and in the same direction as we move axis 1, the new angles between the factors will remain as in the original orthogonal unrotated, and the correlation matrix among the new factors will be a square matrix of ones down the diagonal with all zeros in the off diagonals. And as indicated above, regardless of whether the R_{rv} outcome is to be orthogonal or oblique the matrix calculation for what we have here just described in ordinary terms requires, by the Robinson Crusoe rule, that we must place L on its side and "flipped" (its transpose) and get it multiplied by L. That is, we tell the computer $L^T L = R_{rv}$. Now as mentioned and illustrated above, the relation of one factor to another at the meaningful resolution, i.e., by simple structure, is rarely orthogonal. Consequently, the typical calculation does not produce an R_{rv} which is an identity matrix, i.e., has ones in the diagonal and zeros everywhere else. Instead it gives ones in the diagonal and typically a majority, say, 80% to 90%, within a +0.3 to -0.3 range of correlations. Thus, moderate correlations among the reference vectors and the subsequent factors are the rule. These correlations of factors are of importance because they tell us about the psychological relations of source traits and they also provide the foundation for locating second-order factors.

The calculation complications with oblique factors have induced some noted psychometricians, e.g., Burt, Edwards, Guilford, and Horst (and Eysenck initially) to use only orthogonal factors. Also a host of less eminent users of factor analysis have been persuaded by the cheapness and convenience of the VARIMAX rotation program (discussed later), which is rigidly orthogonal in its product, to turn out orthogonal solutions, though in half of these cases it is evidently done in ignorance of the issue rather than by deliberate intent. If there is a real need, in the scientific model, for oblique factors, however, the scientist cannot afford to be deterred by complexity. The arguments put forward by the present writer 35 years ago and since (Cattell, 1945, 1946a,b, 1947b, 1952) remain as potent now as then, and, possibly have helped to convince an increasing proportion of researchers. The reasons begin with the fact that we should not expect influences in a common universe to remain mutually uninfluenced and uncorrelated. To this we can add an unquestionable statistical argument, namely, that if factors were by some rule uncorrelated in the total population *they would nevertheless be correlated (oblique) in the sample*; just as any correlation that is zero in the population has a nonzero value in any sample. And whenever we impose orthogonal axes on factors that are oblique we distort the meaning of the factors, so that the sample would yield mixtures of the true factors. A clear instance of this confounding of the real factors by an orthogonal solution is given in the VARIMAX solution, p. 137 below. One wonders whether the theoretical clinging to orthogonal factors by some system builders, e.g., Guilford's system of ability and personality structure, and by the users of Kaiser's VARIMAX, is not some philosophical confusion over the meaning of Plato's ultimate archetypical "ideas." Our concepts should admittedly be clear, distinctive, and independent in our minds. But one does not need to confound logical, ideational independence with statistical independence in the domain of empirical observation. Energy or power and mass, for example, may be quite independent concepts (at least until Einstein), yet in any example, e.g., a galaxy of stars or one hundred automobiles, they will generally be correlated. Energy radiation being related to star size and power to car size since demand calls for a heavier car to be designed with more power. In all such cases the pure concepts of energy and mass are reached only by accepting the obliquity of their real patterns of expression. The sad fact for research is that constraint to an artificial orthogonality destroys both the correctness of the pattern discovered and its constancy from one research to another.

7.5. Rotation by Visual Inspection of Single Plane Plots

In the search for simple structure from the extracted V_0 position, Thurstone, Guilford, the present writer and colleagues, and others used to proceed in the 1930s and early 1940s by trial and error visual plot shifts which, after perhaps 10–20 overall shifts, would attain to a unique, unimprovable simple structure. Various mechanical aids were developed, but the process was inherently a slow one. Through the work of a handful ᵒᶠ resourceful factorists—Browne (1967), Cattell and Muerle (1960), Carr (1957), Eber (1966), Gorsuch (1968), Hakstian (1971), Hakstian and Abᵢ (1974), Harris and Kaiser (1964), Kaiser (1958), Saunders (1961, 1962 Schönemann (1966a,b), Hendrickson and White (1964), Wrigley and Neᵢ haus (1955), and some others, a series of automatic computer program for rotation has now become available that will carry psychologists mosᵢ (but not all) the way to simple structure resolution. It is important to stress most, rather than all, for in the exuberance of a prolific invention of these factory aids the hard truth has been overlooked that nature cannot be perfectly reproduced by mechanical aids, that the most experienced human judgment is still needed, and that programs are no wiser than those who make them. We still stand, like the quick-view tourist, to face the fact that in the ascent of a mountain possessing some roads, though a car will take one much of the way, the last crags have usually to be ascended on foot, the equivalent of which here is visual plotting. This last step is necessary both for a scanning around to see that the automatic program has not done something completely foolish (as a computer program sometimes can) and also to effect the most refined possible improvements. The assertion here is one unpopular to automatic program users that gains in hyperplane count, in regard to an emerging hyperplane of a breadth prescribed by the sample, can be made by an experienced rotator and that they often noticeably—sometimes crucially—alter the psychological interpretation offered by the loadings at the end of the automatic program.

The true analogy to maximizing the hyperplane count (the percentage of the variables in the whole study matrix loading less than an agreed, conventional value, say that of lying within ±.10) is finding the top of a mountain in the dark. For it is *"de rigueur"* that rotation be what is called "blind" (though visual!) in the sense that the theorist shall not know, e.g., by any labeling of the variables on the graph, which positions of which variables are likely best to suit his own theory! The only criterion of importance, therefore is *maximizing this percent in the total hyperplane.* And when we think we have climbed to a maximum plateau in the plot of this percentage (see hyperplane history, p. 143 below), the only check that we have approached perfection is to move in some yet new direction and see what happens. Like a man believing in the dark that he has reached the top of the mountain the only proof is to walk in several directions and check that all go downhill.

Three common myths about visual rotation persist in depriving researchers entering factor analysis of the best results from a proper combination of automatic and visual plotting methods.

1. That visual plotting is still terribly time consuming. This is no

longer true. The ROTOPLOT program does all the work of plotting the points and the plots permit the investigator quickly to call shift angles to an assistant who enters them in the computer. Thus ROTOPLOT admittedly requires more time than pushing a button for a self-performing program yet it is 100 times quicker than the original visual methods.

2. There is a repeated myth that visual rotation is somehow cheating, allowing the investigator to move axes to fit his own theory whereas automatic methods are "objective." Even Gorsuch speaks of visual shifts as being decided by the visually "most compelling" structure as if compelling depended on something less objective than counting. First, let us reject "cheating" by reminding the researchers again that true visual rotation must be blind. And let us insist on its objectivity by pointing out that the criterion of what is correct is precisely the same as that in the better automatic programs: a demonstrated maximum hyperplane count, i.e., the final count of percentage of variables in the hyperplane. Critics here have succumbed to the slip of believing that "objective" is synonymous with "untouched by hand." But, one must first point out that a computer can be no more correct in its choice of a criterion of gain than the human who has written its program. And, secondly, one has to recognize that the real criterion of an objective resolution is whether it meets statistical significance in hyperplane count, regardless of whether the series of steps to gaining the maximum are done by desk calculator, visual oscilloscope, paper plot, or electronic computer.

3. Finally, there is the misplaced faith that automatic programs, being couched in terms of mathematical perfection are bound to give the correct result. This depends on the definition of perfection that the program is told to set up. If the real need is to maximize the count of variables falling within a ±.05 band, then all presently known automatic programs on the analytical model (see below) are incapable of doing this. The MAXPLANE automatic program, built on a topological model is intrinsically capable of working to a desired hyperplane width (though it may sometimes fail) and so also is the ROTOPLOT just described. Since both are more demanding of computer time than most analytic programs a practical procedure we would advocate is to get a first position by a cheap analytic program and then use these to polish (beating the first hyperplane count). Just as a Rolls-Royce or other top car manufacturer brings in hand craftsmanship for the last refinements of fit, so here ROTOPLOT can advantageously follow and polish a simple automatic program.

7.6. Analytical and Topological Automatic Rotation Programs

Before describing final processes and judgments on the combination of visual ROTOPLOT and automatic programs, let us briefly but systemati-

cally describe the two varieties of the latter just mentioned and their principles of action. What are best called *automatic, electronic computer
programs* are of two kinds:

1. Analytic programs in which some *continuous function* derived
from the projections of every variable point in the given study enters into
the final choice of position.

2. Topological programs, so called because their aim is to bring as
many points as possible into some definitely bounded subspace, the hyperplane, relative to other defined topological subspaces.

The discussion of these programs could easily lead us into extreme
complexities, so we shall give essentials in as simple language as possible.
The square of factor loadings on all factors must remain constant with
orthogonal rotation at a value

$$J = \sum^{j=n} h_j^2 = \sum^{i=k} \sum^{j=n} a_{ij}^2$$

where a is the orthogonal loading for n variables and k factors, but the
fourth power (or some higher power) can vary with the rotation. And if,
by calculus, we use a device to maximize the expression for the fourth
powers what we are doing is to find the position which increases the variance of loadings, i.e., which places factors so that some loadings are quite
high and others quite small instead of remaining approximately normally
distributed about zero as they normally are in the V_0. For the average
factor a high fourth power relative to the squared value does just that.

Figure 7.2 is an attempt to show in simple visual terms what (b) an
analytical program of this "powered" nature (as virtually all are) and (c) a

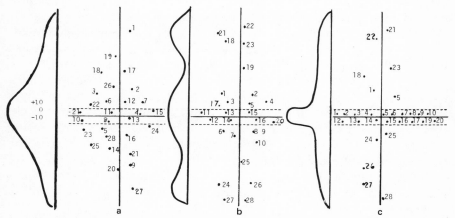

Fig. 7.2. Comparison of loading distributions of unrotated and rotated factor matrices
by analytical and topological programs. (a) Unrotated; (b) rotated by analytic principle rotation; and (c) rotated by topological principle.

topological program does to the distribution of loadings as they originally appear in the unrotated factor at (a). No attempt here is made to show what simultaneously happens on the second factor. And the number of points is kept small, though the ideal target of the distribution is shown as a smooth-drawn curve to the left of each. The effect of an *analytic* program is to shift the random or normal Gaussian distribution in (a) to a form like (b). The effect of the *topological* program is to maximize the number *specifically in the hyperplane range only*, as shown in (c) (the hyperplane boundaries being assumed set by the dotted lines).

Variables like 1, 2, 3, . . . , 10, low but not in the hyperplane in (b), should not really count as a gain in simple structure, because they are not really zero loadings (a zero loading width stretches at most to the boundaries shown). Yet by the analytical type of program such reductions are counted as gain. That is to say, they throw in some weight, even if not as much as do variables 11–20, in deciding that the rotation shall be settled in that particular position. In other words, the analytic program does not distinguish in the "vote" on the final rotational position between variables that are really in the hyperplane and a frequently more numerous group the members of which are merely "rather low" in loading. At the same time, in doing this it throws up as many variables as possible into high loadings, like 22 and 28. This is impressive if one likes at once to see high loadings on a factor, before much research has been done to find such variables; but it is quite likely that their values on a rotation to the true hyperplane are somewhat lower, as shown by the change in 18 and 23 from (b) to (c). That is to say, the analytical solution prefers positions where some variables are very high (their fourth powers being very influential), and this aim is strictly an extra goal independent of that of finding the hyperplane. Like any worshiping of two gods, it is apt to cause trouble. In effect, it commits analytical programs to some degree to *chasing clusters* with reference vector which, as we shall argue below, may often appreciably sidetrack the rotation from finding the maximum number in the hyperplane as such, which is the only defensible criterion.

By contrast, the solution at (c) recognizes two distinct classes of variables by a topological criterion—those in the hyperplane, as defined by a hyperplane ±.10 wide (or whatever statistics indicates) and those outside it. In seeking to maximize the number in the hyperplane space it is not biased at all by what happens to those outside since the magnitude of their loadings do not enter into the calculation. Consequently, (c) is the desired position and it does not matter whether a maximum in the hyperplane space is found by an automatic program—MAXPLANE—or by visual trial and error—ROTOPLOT.

Neither these exposures of fallacious assumptions in analytical programs, nor mention of this slight tendency to the vice of cluster chasing, should, however, divert one from the final practical compromise of using suitable analytical programs followed by MAXPLANE or ROTOPLOT polishing to develop and correct. But it *is* necessary to warn against the presently widespread fashion of putting blind faith in analytical programs alone, and the belief that they ensure flawless perfection. It is important, moreover, to understand the principles at issue, so that choices can be made among alternative program procedures in various rotational situations.

To these questions we shall return as soon as we have clarified a bit more the consequences for evaluation of oblique simple structure rotation of that difference between the reference vector and the factor, initially brought in the limelight a few pages back. All we need recognize at this point is that the shift from the V_0 matrix to simple structure on the references vectors V_{rs} (*rs* for *reference* vector *structure*) has to be translated at the end to a corresponding loadings on a second, factor matrix, called the factor pattern matrix V_{fp}. The transformation matrix L found to be right for the reference vectors can be regarded as changed to one for the factors by

$$L_f = L_{rv} D^{-1} \tag{7.4}$$

where D is a matrix of k diagonal values, which can be regarded as tensors ("stretcher" or "contractors") of the columns in the reference vector matrix. Thus substituting Eq. (7.4),

$$V_{fp} = V_0 L_f = V_0 L_{rv} D^{-1} \tag{7.5}$$

and

$$V_{fp} = V_{rs} D^{-1} \tag{7.6}$$

One value in the D^{-1} matrix (the source of which is described later) enlarges in the same ratio all "loadings" in one reference vector column, to values perhaps about 20%–30% larger in the V_{fp} matrix. The principal consideration this brings from a rotational point of view is that it enlarges the width of the hyperplanes too; so that if we are set to ±.10 in the V_{rs} we may be working to ±.12 in one factor and ±.14 in another since the d's in the D matrix differ in size. Nevertheless for the time being this must be tolerated, because (a) one can raise doubts later whether the hyperplane width should be the same width on different factors, and (b) the reference vector system is easier to work in. Decidedly more calculation is needed to make plots and shifts in the V_{fp} system. However, some researchers blessed with generous computer facilities are nowadays giving themselves the luxury of working on rotations directly in the factor system. With this note on a more complex development, we return to evaluating different rotation programs in the V_{rs} system, the evaluators applying, however, to both.

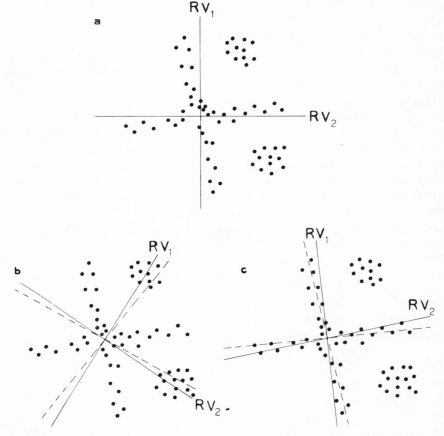

Fig. 7.3. Contrast of rotating hyperplanes to topologically true simple structure positions versus rotating reference vectors to go through clusters (as in some analytical programs). (a) Unrotated plot. (b) Rotated to analytical program criterion. (c) Rotated to topological program criterion. Reference vectors are shown as continuous lines, hyperplanes as dashed lines. Note that there are actually fewer points in clusters than in hyperplanes.

Let us now imagine an oblique case as in Fig. 7.3 which, as is not unusual contains two clusters besides the two (slightly oblique) hyperplanes, as they would appear as points are plotted in (a) from the given unrotated projections. The tendency of an analytical program to chase clusters, i.e., to give credit to distinctly high loadings may cause it to take the position shown at (b) putting reference vectors through clusters. In this essential false position, the hyperplane counts are poorer than they need be, since there are fewer points in clusters than in hyperplanes (when the ±.10 cut is made through them). It will be seen that the reference vectors are pulled to the centroids of clusters and correlate somewhat negatively. The topological (MAXPLANE) or ROTOPLOT search, on the other hand, would finish at the position in (c). Here the reference vectors are at right angles to the true hyperplanes and are in substantially different positions from (b), being now, moreover, positively correlated. And the clusters—judged straight measures of factors in (b)—are now seen to be factorially composite surface traits. For a comparison of MAXPLANE and some alternative rotations see Cattell (1966a), Hakstian and Abell (1974), Dielman, Cattell, and Wagner (1972), Burdsal and Bolton (1977), and Taylor and Coyne (1973).

It is objected, especially by those who have not absorbed the difference between a correlation cluster surface trait and a factor source trait, that the alternative position (c) may doubtless result in a good hyperplane, but lacks the presence of sufficiently high loaded variables [relative to (b)] to interpret or measure such a factor. This is true; but the important purpose is to locate the number and general nature of true influences, not necessarily to fully interpret and measure them at the first exploration. And it is no use deceiving ourselves at that stage of research that we have now got good practical measures of it if we do not. Variables higher in loading on a factor once well located as to a hyperplane can always be introduced later. When Leverrier inferred the existence and position of the planet Neptune he did not need to undertake to describe its atmosphere or temperature as part of the immediate job of demonstrating its position.

The objection—apart from missing true hyperplanes—to rotation methods putting axes through clusters is, among other things, the inconstancy of cluster outlines and positions already mentioned here in Chapter 1. Although the cluster is in k space it requires only the easy manufacture of a lot of similar items to create an artificial, unnatural cluster. (For example, a set of questionnaire items like "Do you like the look of President X?" "Do you like the way President X talks?" "Do you like the way President X walks?" can be depended upon to make a cluster of item variables as extensive as one wishes—and quite artificial.)

The rejoinder that if subjective, arbitrary, or stray influences can "artificially" make a cluster they can also make a hyperplane is quite wrong. A hyperplane is a very peculiar structure marked by a collection of variables that in general have good loadings on ($k - 1$) factors but none on the kth factor itself, whose hyperplane we are considering. It would tax the constructive ingenuity of the best personality psychologists living to make up 50 items alike meeting this pattern of requirement for, say, a 20-factor matrix. In short, to quote a familiar verse, clusters "are made by fools like me but only God can make" a hyperplane. And it is because it is such an inherent expression of nature that we pursue it in seeking natural structure.

In view of the results of Hakstian (1970, 1971), Hakstian and Abell (1974), Dielman, Cattell, and Wagner (1972), and others discussed below

showing in general quite good agreement in practice of different automatic programs it may seem that we have made too much of the above
defects of certain methods. It has been objected, for example, that deflec-
tion by clusters is much greater through the diagonal or the multiple
group methods of factor extraction than what is likely to happen from
any rotation process. But note we have not advocated these questionable
extraction methods here. It may be objected, again, that the number of
variables in the hyperplane region is so much greater usually than those in
the cluster that they can suppress cluster-chasing distortion. All this is
true, but the emphasis is on "generally" in such talk of evidence of adequacy of agreement of program results. There are special or slightly unusual cases, still important, which in the writer's experience have led to
distorted conclusions through the above defects in analytical programs.

7.7. Comparative Studies of the Strengths and Weaknesses of Various Automatic Programs

The distinction within *automatic* rotation programs between (a) the
analytic and (b) the *topological* principles of program design has been
brought out as sharply as possible in the above illustration, though in
practice the differences are on an average small. Nevertheless, it is worthwhile noticing the relative time, costs, reliability, effectiveness, and
specific vices and virtues of different programs. One should also remember that in the last resort the criterion of simple structure is topological,
i.e., the count is made strictly in a narrow, fixed-width hyperplane band,
which analytical programs are not fitted to recognize. However, good
standards can be reached by going most of the way with an analytical
program and then polishing with ROTOPLOT or MAXPLANE.

At the present moment there is a host of analytical automatic
programs VARIMAX, OBLIMAX, a series of OBLIMINS, QUARTIMIN, BIQUAR-
TIMIN, BINORMAMIN, PROMAX, EQUAMAX, HARRIS-KAISER, VARISIM, etc.
(see Cattell and Khanna, 1977). But there is as yet only one topological
automatic program, namely, MAXPLANE (Cattell & Muerle, 1960; Eber,
1966). The shortage of the latter type is unfortunate, for ingenuity could
surely produce another form sufficiently independent of MAXPLANE (perhaps avoiding its tendency to collapse factors) to give a rotational check
in the topological domain on analytic programs. The use of an orthogonal
program, like QUARTIMAX (Wrigley & Neuhaus, 1955; Wrigley, Saunders &
Newhaus, 1958), VARIMAX (Kaiser, 1958), or VARISIM (Schönemann,
1966a), except for quite special purposes, or as a preliminary to an
oblique, is rarely appropriate or needed. Moreover, even if hyperplanes
happen to be orthogonal, an oblique program will stop appropriately at

the orthogonal special position. On the other hand, if they are oblique, and if an orthogonal program is used, the orthogonal program will have a "neurosis," oscillating between the incompatible satisfaction of one hyperplane and another actually oblique to it, commonly ending in the meaningless compromise shown in Fig. 7.4.

For these reasons the indiscriminate and unreasoning installment of the (admittedly quick and cheap) VARIMAX program during the last decade, as the only program in a great number of push-button factor analysis programs used by the unwary, has had results as disastrous to advance in substantive concepts as the automatic inclusion of the quick

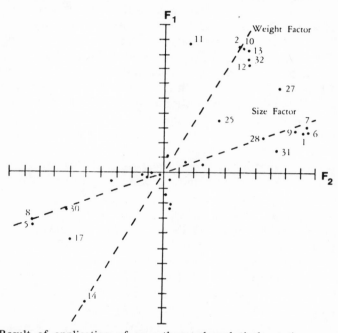

Fig. 7.4. Result of application of an orthogonal analytical rotation program to an intrinsically oblique simple structure. Identification of variables by number: (1) diameter, (2) weight, (5) number of rotations rolling, (6) distance of ball from eye required to cover circle, (7) diameter of shadow case, (8) rotations to one rotation of tire, (9) number of squares covered by ball, (10) distance of weight from ball on liner, (11) collision displacement, (12) paddle wheel rotations, (13) springboard depression, (14) angular momentum, (17) distance sent by croquet blow up an inclined plane, (25) inclined plane carpet friction, (27) weight of splashed out of water (28) inclined plane impulse distance, (30) winding necessary to use up string, (31) size of eclipse shadow, (32) impact displacement by pendulum. Note that by this VARIMAX solution F_1 and F_2 are mutually contaminated measures of the true oblique size and weight factors. (From R. B. Cattell, *Handbook of Multivariate Experimental Psychology*, Rand McNally, Chicago, 1966, p. 187. Concrete plasmode: Cattell–Dickman ball problem, 1962.)

"latent root of one" decision for number of factors as the sole factor number check in many programs.

A less definite weakness of VARIMAX and a number of other analytic programs is some degree of inability to get away completely from the first component—which is in some cases essentially a large cluster. This tendency of the first big factor not to split up as it should is quoted by several users, but does not always happen. It shows itself in the L matrix in the first (and to a lesser degree, the second) rotated factors having large correlations in the column opposite component 1, and results in an unduly large variance in the early rotated factors relative to later ones. The possibilities of some hitch on the way to final simple structure exists in analytic programs because "local maxima" exist which may be mistaken in some programs for the final maximum on the curve. One way tried in our own laboratory to get away completely from the sometimes steeply declining principal axis series to an evened distribution is to spin the principal axes by a special program which mixes first with last and so on, or by the useful Landahl transformation (Landahl, 1938). Another program with a similar aim is EQUAMAX (Saunders, 1962), the inventor of OBLIMAX (1961), but these, though they have done good work, are, as Gorsuch says (1974, p. 195), "seldom available or used."

Those who employed, during the fifties and sixties, the exuberant productions of analytical programs—QUARTIMAX, VARIMAX, OBLIMAX, the OBLIMIN series, BINORMAMIN, EQUAMAX, TRANSVARIMAX, PROMAX, VARISIM, the H–K ORTHOMAX approach, QUARTIMIN, BIQUARTIMIN, etc.—and the topological MAXPLANE program, learned certain advantages, dangers, and vices in each. The structure of these programs is sufficiently set out by Gorsuch (1974) and Rummel (1970), also in evaluative surveys by Cattell and De Khanna (1976, Chap. 9), Dielman et al. (1972), Hakstian (1971, 1976), Hakstian and Abell (1974), Katz and Rohlfe (1974), Guilford and Hoepfner (1969), and others. Although, as indicated above, they have their various economies, vices, and special suitabilities, it is possible to indicate some order of overall goodness, or at least to make some recommendations for use.

Dielman et al. (1972) compared five—MAXPLANE, OBLIMAX, PROMAX, Harris–Kaiser, and VARIMAX—across four types of data—questionnaire, objective test, a physical problem, and a mathematical plasmode. The results were compared for (a) goodness of simple structure (hyperplane count) and (b) factor invariance (congruence coefficients) across the different solutions. Unfortunately, the second was computed as a "democratic" agreement with all studies instead of with the external criterion of the true values available in the plasmode and thus provided no real test of the hypotheses that (1) oblique methods and (2) topological methods would provide greater invariance. The result on the given basis was that no significant difference existed among methods. But in simple structure, though results varied across areas, the best general results were reached with ROTOPLOT-finished MAXPLANE, PROMAX, and the Harris–Kaiser program, in general, in that order. With two different data problems, which in the present writer's opinion were more artificial and less representative, Hakstian (1971) found the H–K best, and then PROMAX, OBLIMAX, and MAXPLANE. Experimenting with classes of H–K algorithms, Hakstian and Abell (1978) found differences according to relative numbers of salients and nonsalients (1974). Hakstian and Abell (1974) compared the H–K, BIQUARTIMIN and direct OBLIMIN, finding BIQUARTIMIN the least effective; and concluded, however "no *single* . . . procedure can be expected to lead to optimal solutions for *all kinds of data*" (p. 444). Burdsal and Bolton (1977), on questionnaire data, found "the relative efficiency in terms of ±.10 hyperplane count" to be "VARIMAX, 71.7%; QUARTIMAX, 72.0%; OBLIMIN, 72.9%; QUARTIMIN, 14.3%; COVARIMIN, 28.1%; BIQUARTIMIN, 43.6% and ROTOPLOT, 80.8%."

The actual research examples used in the comparisons of methods of rotation, as described in the last paragraph, typically covered the middle range of number of factors—from 4 to 14. It has become evident more recently in Burdsal and Bolton (1977), and in as yet unpublished reports by Cattell, and Kameoka (in press) that analytical programs become more uncertain in action as the number of factors increases. Kameoka was unable to obtain agreement on a 31-factor problem among two or three well-known simple structure programs. She rotated by PROCRUSTES (see p. 97) to the position which had held in the given domain across earlier studies and, after applying a few ROTOPLOT shifts, reached the very high and significant simple structure hyperplane count of 89%, not reached by the automatic programs. The reasons for this unreliability of analytic programs with more than, say, about 24 factors are not yet clear. It may have to do with the increase in number of spurious "geometer's hyperplanes" (p. 169) or with the small ratio of high to low loadings when factors are numerous. But meanwhile the practical researcher should be warned about it. Since in all but absolutely new fields the positions of some factors will be known, the problem can often be solved by rotating these by PROCRUSTES, and then applying, say, MAXPLANE to the partly rotated matrix (the remainder being unrotated), finally tidying by ROTO-PLOT.

So many are now working on these analytic (but few on topological) programs that whatever is written in detail is likely to change in a couple of years. Some risk of not spreading the first factors adequately has been mentioned in PROMAX and its class of relatives; we may add that OBLIMAX has a tendency to make factors more correlated, and OBLIMIN (and its relatives) less correlated than the eye later sees as necessary in the ROTO-

PLOT graphs. Other risks such as the danger of responding to highly loaded clusters almost as much as to hyperplanes and of mistaking rather broad bands for the true hyperplanes manifesting a ±.05 band have been described here as defects of the analytical approach. In regard to such comparisons, one statistician complains of OBLIMAX that investigators find the factor correlations larger than is profitable; that MAXPLANE devours a rather substantial amount of computer time, and so on. Some of these defects are being remedied.[1] By taking higher powers of the loadings than formerly, Hakstian has got analytical programs to aim at narrower hyperplanes. Browne (1974) has brought gradient methods to bear on analytic rotation, and Katz and Rohlfe (1974) have improved on MAXPLANE, as shown by several examples.

Further discussion of rotation principles and calculations, and a setting out of flow charts and subroutines for the MAXPLANE program will be found in Cattell and De Khanna (1977).

7.8. The Tactics of Reaching Maximum Simple Structure

In this proliferation of programs some indication of choice for everyday is needed, and the present writer would vote, if the choice is to be among analyticals, for H–K and PROMAX (respectively, for very exact

---

[1] An instance of proposed improvement in the H–K is the proposal for "polishing" (Kaiser & Madow, 1976). This proposed improvement in the program stops at a Harris–Kaiser analytical solution, looks at the y variables in the, say, ±.10 hyperplane, and then takes subsets of them that look most promising, or, more objectively, all possible subsets of $(y - 1)$ or $(y - 2)$ and so on. The same was independently proposed by Digman in his "Mai Tai" program. Thurstone provided calculations (1937) to get a reference vector that will put a certain subset of variables as closely as possible into a zero loading hyperplane. By the Kaiser and Madow use of this method one goes through a trial and error process with various subsets to see which will "lie down" in a hyperplane, by testing the significance of simple structure achieved with each. As described in the discussion of significance of simple structure in Chapter 8, the method used here is the usual F test for comparing two variances. Shifting from the unit length factors in a sphere to unequal length eigenvectors in an ellipsoid, which give the same variable variance in all directions, Kaiser compares the F ratio for the reference vector (or the unstandardized primary) in the positions shifted to in thus letting various subsets constitute the hyperplane. A possible criticism is that (a) this may capitalize on chance in any single sample and (b) that it avoids the issue of having to set a likely width to the hyperplane. Indeed (see Chapter 9) it gives a changing width to the hyperplane from factor to factor. If the attempt is made to shift the *whole* set of variables loading, say, ±.15 into an exact hyperplane then we know it is almost certainly doing a misleading job, for, by the nature of things, several will not really belong. If it tries numerous subsets then it is essentially a wasteful procedure, which one would not dream of trying as a main method, but in special cases it may be worth trying on a set already picked out on other grounds as likely for a particular factor.

work, using Hakstian's improvements, and for a quick inexpensive program). However, he would vote for MAXPLANE, with recent improvements (Katz and Rohlfe, and the Brennan program) as a topological check or final tidier. The central fact is that the programs just mentioned are about equally good, but that they have their peculiar unsuitabilities, not easily known in advance. Consequently, the important thing today is good tactics more than hairsplitting among programs, and central in good tactics is to use a check always by two good programs, preferably one analytic (H–K, PROMAX, etc.) and one topological (MAXPLANE, ROTOPLOT), with a ROTOPLOT finish.

In our own laboratory for the last five years the general rule has been to apply (1) an H–K or PROMAX and (2) a MAXPLANE geared at first to a relatively quick solution, and then to follow either or both by a ROTOPLOT continued long enough to demonstrate a climb to an indubitable plateau. As to the H–K and PROMAX programs per se the reader is referred for a brilliant mathematicostatistical analysis to Hakstian (1970, 1971) and for programs and practical use to Cattell and De Khanna (1976) and Gorsuch (1974). As to MAXPLANE, and the above reference to "gearing" it to particular purposes the reader is referred below. Here we shall assume the programs applied and an L (transformation) matrix reached shifting the V_0 to an approximate, reasonably good simple structure V_{rs}. We propose now to deal with the tactics, in the next step, of applying ROTOPLOT.

First, however, a word is needed to those who continue to assume the standard push-button analytical output must be the best possible, because of its mathematical precision, and, therefore, needs no polishing. Unless some breakthrough in analytical design is imminent the fact remains, as we seemingly must repeat, that the analytic principle has at least two faults—some slight pulling over by clusters, and an insensitiveness to what the real boundary of the hyperplane should be. MAXPLANE, into which the designated hyperplane widths can be "plugged in," theoretically is the complete solution, such that no ancillary program need be used, but besides the slight disadvantage of its cost it has a weakness of occasionally missing one or two real hyperplanes completely if it is set to too narrow a band at the outset. If it is more safely set to a broader approximate hyperplane (say ±.12 with an N of 300) as proposed here, then it also needs ROTOPLOT finish. In any case there is everything to be said for the researcher taking a look after either the analytic or the topological programs have done what hopefully is 95% of the job—and reached a good mutual agreement—to see in the ROTOPLOT plots just what has happened.

To the practiced investigator the ROTOPLOT graphic output is always illuminating, if only about the general "texture" of the result, and a good researcher would no more want to stop with the automatic output of the preceding programs than a doctor would want to diagnose by phone without seeing the patient. Two kinds of improvement are likely from successive ROTOPLOTS: (a) an examination and removal (by an improved simple structure) of rather excessive or bizarre angles among the reference vectors and (b) an improvement of the count in a suitably designated narrow hyperplane band, which we have called "polishing." To those who are incredulous that man can "improve on mathematics" in the ROTOPLOT polishing the answer has already been given above and it may be added here that the improvement is small—say from 74% to 78%—compared to the improvement—say from 34% to 78%—if one starts from V_0 or a very rough program instead of from an H–K or MAXPLANE as above. Nevertheless, it sometimes is crucial to decision between two theories.

Typically, in the ROTOPLOT, the investigator looks at the graphs and draws hyperplane lines to give a best "least squares" fit, visually, to whatever "one edge nebula" of points he believes he can see. He will normally see such improvements only on about half the graphs, and when he shifts both reference vectors he will avoid shifts that would leave them at mutual angles of 50° or less. The shifts are written on a shift matrix as described in the next section and the new V_{rs} matrix, as the result of the multiple overall shifts comes back each time from the computer. The investigator counts the new hyperplane total each time and plots his progress on a graph which may be called a history of hyperplane plot. This procedure is illustrated by two actual cases, with prolonged rotation over 23 to 27 overall rotations in Fig. 7.5.

Since an immediate *a priori* decision of how wide the hyperplane should be (a function, as discussed elsewhere, of the smallness of N and of the reliability coefficients of the variables) one needs to see in what width the count in fact shows a tendency to go on improving. Commonly, we have taken counts simultaneously in the two bands ±.10 and ±.05. [Toward the end one may like to take ±.05 and the penumbra ±(.05–.10) as a separate count, since the latter may actually decrease slightly in good rotation.] In any case the percentages of variable loadings within the chosen bands are counted and plotted for successive overall rotations. The percentage is for the whole RV (reference *vector*) matrix, the assumption being that the best position for each factor is not entirely to be evaluated on its own but by the best count for all together. This percentage typically climbs, with various setbacks as one happens to make poorer gambles, shown by zig–zag progress, to a plateau, as in Fig. 7.5. As explained in our analogy of finding the top of the mountain in the dark, it is desirable to make three or four further trips when one believes one is already at the top, to confirm that fact, and this is shown by a (somewhat uneven) plateau being reached, as in Fig. 7.5, at the end of the history of hyperplane.

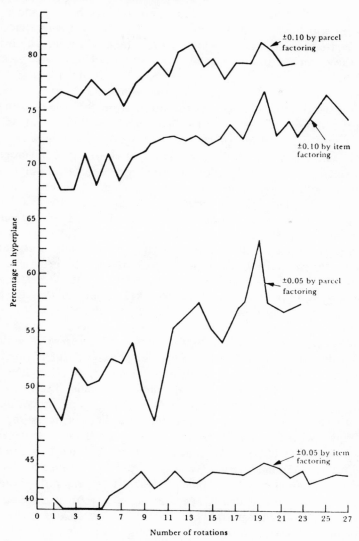

Fig. 7.5. "History of hyperplane" plots, climbing to plateaus, illustrated on real data. (From R. B. Cattell, *Personality and Mood by Questionnaire*, Jossey-Bass, San Francisco, 1973, p. 290.)

In the case of proceeding by MAXPLANE, either as tidying or from the beginning, one must note that this program has two or three kinds of parameters that can be plugged into it, determining its behavior according to the nature of the experiment and data. The mechanism of MAXPLANE is that on each factor it makes a succession of small angular sweeps through space counting the discovered number in the hyperplane at each. It settles on whichever of these, say 5°, shift positions yields the highest and records that position as the beginning for the next round which begins after all the other factors have been singly shifted. The second parameter that can be "set," over and above the first (modification of the original 5° shifts) is the width of the hyperplane in which the count is made. The third is a weighting for distance of the point from the origin, which will encourage pulling in a distal point rather than a proximal one. A fourth possibility—present in some of the class of MAXPLANE programs only—is to weight, within say a ±.10 band those captures within ±.05 more than those between ±.05 and ±.10. For example, a more central point, within ±.05, might count as two compared to an outer one.   In the use of MAXPLANE by a skilled operator—and, by the way, the above comparisons of outcome with the H–K are not made with MAXPLANE in the hands of an expert but an average beginner—these parameters can be adjusted to the nature and stage of the matrices. They are roughly automatically adjusted to the stage of the rotation. For example, the program at first prescribes wide sweeps—up to 45° either way by 5° steps—and bases the count on a wide hyperplane—±.15 or even ±.25 (since at first the hyperplane will make its appearance as a blurred nebula). But as the simple structure position is approached its visible features become more crystallized, and MAXPLANE does as the human searcher by ROTOPLOT would do. It narrows the width of hyperplane in which it expects the points to be— say down to ±.12 or ±.10—and it makes smaller, finer angular shifts. These features make MAXPLANE more demanding of general scientific sense and skill, in adjustment to the experiment and goals concerned, and also somewhat more costly in terms of machine time. The former is important. For example, with fewer factors the convergence from wide to narrow sweeps can come faster, and with larger samples the final target in hyperplane width can be made narrower. The "stops" to be pulled also include, as stated, giving greater or lesser relative weight to points farther out from the origin, as one does with the eye. Indeed, ROTOPLOT is a simulator of an experienced human rotator. Like ROTOPLOT, MAXPLANE proceeds till three or four successive shifts bring no further improvement in the total percentage count (the "plateau") usually at the second decimal place. Such possibilities of adjusting to sample size, reliabilities of measures, e.g., simply do not exist in analytical programs. The lack of freedom explicitly to define the hyperplane width to be maximized is, in principle and ultimately in practice, a defect of analytical programs relative to MAXPLANE and ROTOPLOT. All programs are likely to improve a little in the next few years, but it remains to be seen whether the improved analytical programs will finish better than MAXPLANE with an experienced user able to "play" the instrument. In any case it is hard to sympathize with the practice, when serious issues are involved, of stopping with an analytic push-button solution without asking to see a ROTOPLOT output.

Some closer discussion is called for at this point on just how wide the hyperplane sought should be—since ±.10 has so far been followed only as "conventional." It is true, as shown in Fig. 7.5, that the shifts which increase the ±.10 count are up to a point doing the same at the ±.05 width. But if the standard error of a zero loading were, say .02 in a given large sample, then 98% of the loadings that are truly zero would lie between ±.05. In that case with factors whose nonzero loadings range from, say, 0.01 to 0.40 a maximizing of the ±.10 band would be free to finish at a somewhat different position from that given by a demand for maximizing ±.05. Indeed, some instances have been tried by the present writer with large samples where a series of rotations can be found improving the ±.05 count by about 50% while leaving the ±.10 scarcely increased, and in the experiments of Finkbeiner (1973) immediately below it has been shown that—regardless of what theoretical statistical disagreements may exist about the width of hyperplane—it is possible to maximize the count only within a certain width in a given sample, and that width he found to be less than has conventionally been used in most factoring.

In regard to the general tactics espoused here for a really thorough rotation, namely, (a) beginning with an automatic program (analytical, H–K or PROMAX, or the topological MAXPLANE program initially set for a relatively rough ±.12 band) and then (b) viewing and polishing it in ROTOPLOT, we have sometimes found it time-saving yet effective to curtail the length of the human polishing by ROTOPLOT by use of with what has been called "hobbled MAXPLANE." The appropriate time for hobbled MAXPLANE is when the automatic output viewed in ROTOPLOT is seen as essentially sound in general structure and in general correlations among factors, and only in need of tidying, i.e., reduction of hyperplane widths to a desired narrowness appropriate to the sample, etc. One then reenters (assuming it used in the initial automatic phase) with MAXPLANE with parameters altered so that is is "hobbled" to quite small sweep steps (1° or 2° at a time) and weighted to accept only points close to the estimated hyperplane widths. Such a tidying may cut out the last 5–10 ROTOPLOT shifts.

As pointed out, the main advantage, in principle, of the topological MAXPLANE program is that it can be set to maximize in a hyperplane of prescribed width. The question of what this prescribed width should be we have postponed in a fundamental sense to Chapter 15 on distributions and significance problems generally, but a contingent answer can be given by referral to the work on standard errors of factor loadings, particularly the series of contributions by Jennrich (1973, 1974), Jennrich and Thayer (1974) dealing with both the ordinary oblique factor situation and the standard errors of maximum likelihood factor loadings. Obviously, because of the diagonal matrix transformation, if standard errors were the same for the variable correlations over all reference vectors they would not be so for all factor loadings, and vice versa. But the refinement of using a different hyperplane width for each factor, which would have to be recalculated at each shift, is not for busy people, but only for deluxe programs and absolutely crucial issues.

The recent as yet unpublished empirical studies of Finkbeiner and Cattell show that by laboriously rotating for a long time to be sure that all are in the hyperplane that could possibly belong the boundaries of this best—presumably true—hyperplane can be made to "stand out," by a high histogram distribution band from the general distribution of real loadings. In the examples so far examined it would seem that in typical studies (say $N = 300$, $n = 80$, $k = 20$) the conventional and arbitrary $\pm.10$ bound is too wide. Both this empirical approach and the theoretical calculation suggest a standard error of a loading at about 0.02–0.03 and then 19/20ths of the variables should therefore lie within $\pm.06$ rather than $\pm.10$. To keep tabs on the general nature of the ROTOPLOT advance we have continued with the practice of recording both $\pm.10$ and $\pm.05$ counts in the "history of the hyperplane" but the final effort has focused, in studies with a decently large experimental basis, at maximizing in the $\pm.05$ or $\pm.06$ band. This tends to agree with Harris's formula results (which, however, differ from factor to factor). It is also to be noted that at the time of Gorsuch's book (1974), when no single widely accepted expression existed, Gorsuch decided on indirect grounds that a value in the 0.5–0.9 range (say 0.7) is generally indicated for the hyperplane boundary.

7.9. ROTOPLOT: The Calculations in Successive Shifts

The above discussion has necessarily had to range widely, and meanwhile the student has been left without the necessary detailed treatment of the ROTOPLOT procedure itself. Parenthetically, he has also had to take unavoidably on faith conclusions on issues the ultimate basis of which can only be tackled in Part II of this book. However, it is better to be aware of all that bears on the problem, even if discussion in depth is not immediately possible.

The geometry, and algebraic calculations, of the first shift from the V_0 position have been explained, and the latter have been set out above both in ordinary and matrix form (see p. 120 for review). It is our purpose now to proceed from that point, explaining the mechanics of the series of shifts on the ROTOPLOT that converge on the best possible simple structure. In ROTOPLOT the plotting of the graphs and the calculation of the new matrix resulting from the visual decision shifts of the investigator are made by the computer. Incidentally, the convenience of this program is such that with important projects the present writer has several times begun with it from the V_0 rather than an automatic program jump much of the way—with the Harris–Kaiser, PROMAX, course MAXPLANE, etc., as above.

In describing ROTOPLOT it is necessary mainly to describe how the *shift matrix S* is set up by the researcher from his decisions on looking at the plots, and to explain how this shift matrix operates on the transformation matrix L with which the reader is already familiar. ROTOPLOT has to start with the V_0 and *some L* matrix for its fodder. If one is kicking off from the V_0 without any lift by an automatic program then the L will be an "identity matrix," i.e., a $k \times k$ matrix with ones down the diagonal and zeros elsewhere. One looks at the identity L matrix plots, makes shifts recorded on the S matrix, then lets this operate on L (as shown below) to produce a new L called L_1, and applies the latter again to the original V_0. But more frequently one will be starting from an automatically reached jumping off point. He will then already have an L matrix given by the program used, and this, with an identity shift matrix (since a shift matrix of some kind must go into ROTOPLOT; see below), will plot the picture at that point. From the shifts then made (which we need not describe again since, as usual they seek the hyperplane nebulae) a new S matrix will be constructed as shown below in Table 7.3. From that point on the same

Table 7.3. Shift Matrix: S

	1^a	2	3	4	5	6	7	8	...	19	...	24
1^b	1.00											
2		1.00								+.04		
3			1.00									
4				1.00								
5					1.00			-.16				
6						1.00						
7							1.00					
8					+.11			1.00				
.												
.												
.												
19		+.14								1.00		
.												
.												
.												
24												1.00

[a] Reference vectors shifted.
[b] Reference vectors with respect to which shifts are made.

cycle is repeated. There will be (a) a graphic output, (b) a series of pencil lines, drawn by the person rotating, on the graphs where the thicker hyperplanes can be seen, (c) the writing of the tangents of these lines in the right places in the shift (S) matrix, (d) the multiplication of the preceding L (transformation) matrix by the new S, and (e) the application of the new normalized L to the V_0 to get the next V_{rs} as graphic output. To repeat, in computer steps, as summarized in Fig. 7.7 below, after each set of plots a new shift matrix, S, will be punched in, which will operate on the preceding L matrix (transformation) to produce a new one as shown in Eq. 7.9 below. The new L matrix, however, will always *go back to operate on the same, original V_0* as shown in the fourth step of Fig. 7.7 below. Incidentally, in the plotting (a) if two or more variables should fall on one point this is indicated by the best ROTOPLOT programs by printing numbers instead of points (mostly 1's; sometimes 2's, 3's, 4's, etc.), and (b) orthogonal axes are used for drawing convenience despite the reference vectors actually being correlated. The distortion from it is slight and noncumulative.

This last deserves comment because it bothers some users—both theoretically and practically—unless explained. The problem is simply that if reference vectors were drawn oblique and points plotted accordingly, the oblique plots would have points that would often far overshoot the bounds of the paper. And since errors are not cumulative (a new L always goes back to the old V_0) this approximation creates no calculation problem. It also creates no practical visual problem with an experienced rotator, who makes allowance according to the angle he finds recorded at the top of the plot. Figure 7.6 will illustrate the procedure in finding the angle to choose for insertion in the shift (S) matrix.

In Fig. 7.6(a) the hyperplane of 5 has been shifted counterclockwise to better fit (by least square estimate made visually) the distribution of points in the upper left but particularly the lower right. This is a positive shift of the reference vector of 5 toward 8 and is entered in the shift matrix S (Table 7.3), in the 5th column and the 8th row as a tangent of $+.11$. The second shift in Fig. 7.6(a) is of the hyperplane of 8, also counterclockwise. But this takes the reference vector for 8 away from the 5 reference vector. Hence the *negative* sign is placed before the tangent $-.16$ in the column 8 (RV shifted) in the row 5 (Table 7.3) (5 being the RV "against" or "on" which it is shifted). The *size* of the tangent is read off on the edge of the graph (not shown in Fig. 7.6, but shown as a side scale on graph paper). In Fig. 7.6(a) the result is that the two RVs which began as essentially orthogonal remain essentially orthogonal. In Fig. 7.6(b) they begin negatively correlated ($-.231$ shown at top) and will be brought to greater orthogonality since both shifts are plus, namely, $+.14$ of 2 on 19 and $+.04$ of 19 on 2 (see Table 7.3).

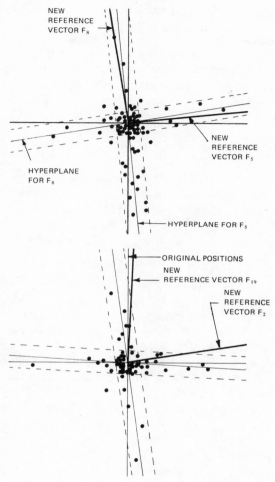

Fig. 7.6. Two plots from ROTOPLOT showing shifts made toward simple structure. Dashed lines are hyperplane boundaries around thin central lines for center of hyperplane. Thick continuous lines are new reference vectors resulting from the hyperplane shifts. Initial angle of F_2 and F_{19} has cosine of $-.231$, not shown since drawing is always initially orthogonal.

After similar treatment of all $24 \times 23 \div 2 = 276$ two-dimensional plots, the shift matrix in Table 7.3 would have perhaps 50% of its cells filled with shift values. The unities down the diagonal show that apart from some such given shifts the RV stays where it was. The rationale of the use of tangents in the shift matrix can be seen most easily by referring to a single shift of RV_x on RV_y where (see columns of S in Table 7.3) the existing $\sin \theta$ and $\cos \theta$ in L get multiplied by $(1 + \tan x)$ thus:

$$\sin \theta \times 1 + \cos \theta \tan x = \cos \theta' \tag{7.7}$$

This produces the new $\cos \theta_2$ for the new angle of RV_x to RV_y, whence it is evident that the tangent is the required entry in a shift matrix. This value, however, remains to be rendered part of a normalized column, and more correctly (but perhaps initially obscurely!), Eq. (7.7) should be written as

$$\sin \theta \times \frac{1}{(1 + \tan^2 x)^{1/2}} + \cos \theta \frac{\tan x}{(1 + \tan^2 x)^{1/2}} = \cos \theta' \tag{7.8}$$

In general, however, a whole column of tangents would be squared and added in the expression in the denomination. Alternatively, the normalizing can be done when the new L matrix is reached.

In matrix terms, extending the above calculation to second and later columns, the calculation in Eqs. (7.8) and (7.9) becomes

$$L_1 \times S_1 = \underline{L}_2 \tag{7.9}$$

followed by a normalizing of the columns of the new transformation matrix to L_2. In Eq. (7.9) L_1 is the "given" transformation from V_0 to $V_{rs.1}$ from which the plots were made and L_2 is the transformation to $V_{rs.2}$ (granted the step of $\underline{L}_2$'s normalization by columns to L_2 as just described). This is to keep the axes unit length, and the reader is reminded that in normalizing we square the values in the column, sum the results, take the square root of the sum and divide each number in that column by it. (The computer does this converting $\underline{L}_2$ to the required L_2.) As Fig. 7.7 reminds us we then go back to the original V_0 and postmultiply by the new L to get the new V_{rs} (actually $V_{rs.2}$). This is then plotted, new shift decisions made, a new shift matrix S_2 constructed, and so on around the cycle. From a tolerably good (a) automatic or (b) PROCRUSTES initial approach to a good position it may take 4 to 24 overall ROTOPLOT rotation cycles to reach satisfactory simple structure, the number increasing with the number of factors involved. The hyperplane percentages, counted and recorded by ROTOPLOT at each cycle, will be entered in the history of hyperplane plot to guide the rotator on his pilgrimage and tell him when he is essentially at his goal.

Experience in cooperative laboratory work, ranging from that with graduate research assistants to that with eminent colleagues enterprisingly exploring factor analytic usages for the first time, endorses the judgment that the degree of detailed instruction given in the last dozen pages is necessary and desired. Furthermore, there is urgent need in this area of factor analytic technique for apprentice-like training situations, in which skills not easily described in books can, like surgical skills, be learned. Even if the researcher is henceforth going to depend largely on a computer

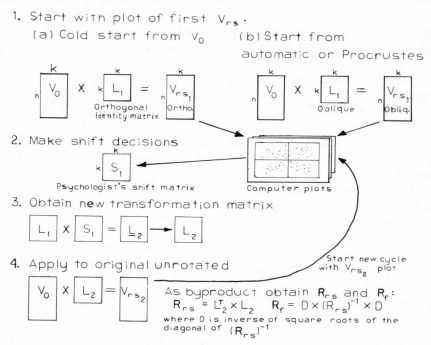

1. Start with plot of first V_{rs}.
   (a) Cold start from V_0 (b) Start from automatic or Procrustes

2. Make shift decisions

3. Obtain new transformation matrix

4. Apply to original unrotated

As byproduct obtain R_{rs} and R_f:
$$R_{rs} = L_2^T \times L_2 \qquad R_f = D \times (R_{rs})^{-1} \times D$$
where D is inverse of square roots of the diagonal of $(R_{rs})^{-1}$

Fig. 7.7. Calculation steps in visual, graphic rotation, as in ROTOPLOT program, etc.

package, with only occasional ROTOPLOT, first-hand treatment of rotations, what is learned through ROTOPLOT will give him altogether greater insight into the factor resolution problem by any instrumentality and its impact, for example, for higher-order factoring. The rationale for "dirtying one's hands" with graphical experience is analogous to the rationale of sending future steamboat captains to a brief apprenticeship in sail, which makes better seamen through a more sensitive contact with the elements!

7.10. Summary

1. This whole chapter focuses on various methods of rotating for simple structure, since research still depends almost solely on this principle—rather than the confactor alternative—in the hypothesis-generating use of factor analytic experiment.

2. Whenever rotation of one factor in a framework presented by itself and others is carried out, the projections of the variables on the new factor can be calculated by a simple trigonometrical formula from projection on the old factor and the angle of shift from it, of the reference vector toward or away from other reference vectors.

3. When carried out simultaneously for many factors each taking a new factor position with respect to an old factor position, i.e., in a resolution in hyperspace, the transformation from V_0 to V_1 (and so to V_{rs}, the final simple structure) is most economically and systematically performed by matrix multiplication. Accordingly, the elementary rules for these matrix multiplications are introduced here.

4. Unquestionably, if factors are to meet the conditions for determiners in a scientific model they must be allowed to go oblique when so indicated by simple structure. This brings some complications in calculation and requires that we distinguish between the *reference vector* (the perpendicular to the hyperplane) and the *factor*. Rotation is initially done in the reference vector system and is so explained here, though relatively complex computer programs can now be built to refer at every step to the corresponding factor matrix.

5. If factors (RVs of course) in the original orthogonal V_0 are rotated independently, each to its own maximum simple structure, the factors will almost invariably become oblique. The calculation of the obliquities (correlations of reference vectors) is made utilizing the transformation matrix L, describing the movement each factor has made from its original V_0 position. The correlation of reference vectors (R_{rv}) is then

$$R_{rv} = L_r^T L_r$$

and the correlation of factors can be simply derived from this as

$$R_f = D(L_r^T L_r)^{-1} D$$

where D consists of tensors derived from the diagonal of $(L^T L)^{-1}$. The d entries are the inverses of the square roots of the corresponding diagonal entries in $(L^T L)^{-1}$.

6. The calculations for successive trial-and-error visual shifts toward maximum simple structure have been incorporated in a program called ROTOPLOT (or ROTOGRAM) which greatly reduces the labor of the older visual shift procedure. It multiplies each successive transformation matrix by a shift matrix S, recording the tangents of the shifts the experimenter has made. Thus

$$\underline{L}_{(t+1)} = L_t S_{(t+1)}$$

where (á) $(t + 1)$ is the new and (t) the old L, and (b) the columns of $\underline{L}_{(t+1)}$ are renormalized in the final $L_{(t+1)}$ which is the transformation actually applied to V_0. The new transformation matrix then goes back to the original unrotated factor matrix thus:

$$V_{rs(t+1)} = V_0 L_{(t+1)}$$

so that any errors are not cumulative but corrected at each rotation.

7. Although reference vectors become oblique after the first shift, they are drawn orthogonal in the single plane plots on which the experimenter works, because it is impracticable to do otherwise and the approximation builds up no systematic error. Normally such rotations from plots are done "blindly" in the sense that points are not identified for test meaning, thus leaving the researcher's results free of conscious or unconscious bias toward supporting a particular theory.

8. The skill and labor needed for "hand rotation" when properly done being very great, a dozen or more automatic programs for computer resolution have been invented in the last two decades. They are of two kinds: (1) analytical programs which maximize or minimize some function of the loadings of *all* variables. Typically they result in maximizing the scatter of loadings, forcing some toward prominent salients and others toward zero. But no fixed width can be assigned to the near-zero band, which is a defect. (2) Topological programs which maximize the percentage of variables lying in a hyperspace slice, the width of which can be set to be appropriate to the sample and the reliability of the data. That width is best set to include 95% ($p < .05$) of the variables whose true loading is zero. Great progress has been made with analytical programs, but in principle the topological approach is sounder, though at present represented only in MAXPLANE, and the improvements by Eber (1966) and the "function plane" of Katz and Rohlfe (1974).

9. A comparative study is briefly given of the virtues and vices of various automatic programs. For example, some systematically overestimate and others underestimate the true angles at which factors should stand for maximum simple structure; PROMAX and other, and some VARIMAX-derived programs, as well as OBLIMAX, can hitch at local maxima, and the former may fail to spread the large first component sufficiently. The Harris–Kaiser approach, using the radical trick of rotating in the eigenvalue system rather than in the final V_0 factors, makes a substantial advance theoretically, among analytical programs, but cannot be altogether free of the moderate tendency in all of them to "chase clusters." MAXPLANE avoids this but has its own vice of occasionally erratically collapsing factors or leaving high angles and, of course, shares with the majority of programs the slight weakness of obtaining simple structure in the reference system and on the same band width for all factors. In the reasonably extensive comparisons of programs to this point, putting H–K, PROMAX, and MAXPLANE high, but in varying orders, it must be pointed out that the comparisons have not been made with a sophisticated use of MAXPLANE, employing the various parameter adjustments which it makes possible for a skilled user.

10. The above differences should not be overstressed, even though theoretically interesting, because extensive trials by Dielman, Cattell,

Hakstian, and others show that, in general (with the exception of programs like VARIMAX and VARISIM locked to orthogonality), the various programs recognizably hit the same factors in perhaps 95 out of 100 cases. Nevertheless, "recognizably" does not mean "exactly," and there is considerable evidence that a skilled use of ROTOPLOT can improve 9 times out of 10 on any automatic program output (except MAXPLANE) as to hyperplane count. Accordingly, it is recommended to go as far as possible by an automatic program, H–K and MAXPLANE preferred, and then to "polish" by ROTOPLOT in all important work.[1]

11. ROTOPLOT is recommended partly because the investigator should look at the general texture and angular appearances in the automatic output, as "quality control" and also because the test of objectivity of a simple structure resolution is not whether it is done by a machine (with a human program however!) or by human visual rotation, but whether a "history of hyperplane" plot shows that a hyperplane count plateau, unimprovable by any further trial and error, has been reached. Even then, the maximum reached must be tested for statistical significance by tests to be described.

12. Except for the recent contributions of Jennrich and others, there has been little basis or experience regarding means of fixing the proper hyperplane width in different conditions, i.e., the band within which maximization of count should be sought. But empirical tests on maximized hyperplanes show that with typical N's (about 300), n's (about 40), k's (about 15–20), and reliabilities (about 0.7–0.9)—the principal determiners of hyperplane width—the break between the hyperplane distribution and the usual distribution of true loadings lies at about $\pm.05$–$\pm.07$ (corresponding to a standard error of about .025 rather than at the arbitrary conventional $\pm.10$). Although ROTOPLOT can only improve on analytical rotations slightly and rarely to a substantial degree as regards the $\pm.10$ band, a skilled rotator with ROTOPLOT can often improve on the count in the $\pm.06$ or $\pm.05$ width, leading to changes in factor loadings that are only moderate, but psychologically important.

13. The method of calculation in successive shifts in the ROTOPLOT program (or in general visualization) is set out above in detail. The calculations made in automatic programs are more complex and are deferred for later study. Illustrations are given of ROTOPLOT use. The possibilities of transforming at each shift from an RV to a factor output can be readily set up by any programmer, but it is doubtful whether this is worthwhile in most studies, and the final transformation to factors is enough.

---

[1] A MAXPLANE program is available from the Institute for Personality and Ability Testing, Champaign, Illinois.

CHAPTER 8

More Refined Issues in Rotation and the Use of Oblique Factors

8.1. More Refined Tactics from Experience Necessary in SS (Simple Structure)[1] Resolution

The last chapter has "blocked in" the main concepts in simple structure rotation: the calculation of transformation and shift matrices; the reasons in the model for freedom for oblique factors and reference vectors; and the use of automatic and visual Rotoplot rotation programs. But it has by no means exhausted the conceptual intricacies, or completed the neces-

---

[1] From this chapter on we shall generally abbreviate *simple* *structure* as SS.

sary formulas, or reached out into the tactical skills needed to discover simple structure with greatest precision.

The two direct causes of failure and confusion in factor analytic research are wide misses of the number of factors to extract, and the attainment of only a very crude SS, and, of these, the second may well be responsible for the greater damage. The fashion in the last decade of inserting the V_0 and pressing the button of some blitz program, built as an arbitrary and unalterably linked "assembly line" of procedures has devastated many researches. The procedures as to number of factors and phases of rotation need rather to be judged separately, made open to permit appropriately adjusted parameters, e.g., in rotation to adjust to the "width" most suitable for hyperplanes in the given study, and, above all conceptually well understood by the computer user in relation to his research purposes. Accordingly the aim of this chapter is to carry the procedures and tactical skills in the pursuit of SS discussed briefly in the last chapter to higher levels of understanding. To complete this phase of study we shall then turn to a more complete mathematical–statistical formulation of the end product of this unique, oblique resolution of a factor analysis.

In this phase of a factor analytic experiment, as in certain methods of research in domains of the physical and biological sciences, it remains true that some skills and qualities of judgment are needed which are most effectively learned in an apprenticeship relation. The present chapter going beyond the bare statements of mathematical models and statistical formulas, commonly found in mathematical texts divorced from the actual activities of research, will deliberately infuse the presentation with some experiential, if not even artistic, judgments.

8.2. Three Tactical Principles in Controlled Rotation

Let us begin by setting out three principles in rotation, developed partly from logical analysis, but resting their cogency also on successful experience.

1. *Preservation in major shifts of approximate orthogonality.* The larger angular shifts that need to be made are often made early in the search for simple structure. In ROTOPLOT it is desirable to see if a shift to some extent coordinated with that made on the first reference sector can then be made on the other axis, to keep most axes marching together in approximate orthogonality—not unlike (a) in Fig. 7.7. Extreme angles bring confusion of factors. Indeed, two highly correlating factors may even climb upon one hyperplane, and an unjustified collapse of the total space losing one dimension can then occur.

2. *Changing policy in regard to outlying points with maturing of rotation*. Initially one is likely to rotate to bring the more outlying points into the hyperplane because they can be brought in with less angular movement and little loss of points already near the hyperplane but close to the origin. However, as the true hyperplane begins finally to crystallize as a relatively narrow (say ± .05) band, it may not "point up" exactly in the direction of such outlying variables and at that stage one should if necessary freely relinquish these outliers giving more authority to the more "compelling" line of the majority. One must remember that there is no logical reason why (in exploratory studies, at least, where old markers are not carried) a variable of high h^2 (communality) should be factor pure. Even a variable of $h^2 = 1$ need not be a pure measure of any factor. An assumption that a high h^2 variable must be factor pure (not will probably be) is one of the vices of analytical programs, especially as they proceed to rotate the RVs toward clusters rather than away from hyperplanes. Incidentally, this may be one of the reasons for the delay among psychologists in the ability measurement field in recognizing that the g factor is actually a mixture of fluid and crystallized general intelligence factors (Horn, 1965c, 1966).

3. *Reducing each separate apparent required shift magnitude when several shifts are to be combined*. Some investigators, especially some years ago, advocated shifting a given factor only on one other factor in any round of overall rotation. The alternative, which an experienced and more confident rotator is likely to use, is to shift on several other factors, not rejecting any really promising shift. It is certain that reaching simple structure, as a maximum hyperplane count across the whole V_{rs} matrix, is substantially quicker if at each output of graphs, i.e., each overall rotation, improvements of hyperplane for any one factor are simultaneously made on several factors. However, two risks are then faced which must be watched. First, when larger shifts (30°–40°) are being made, as typically at early but sometimes also at late stages in the process, it is possible for two shifts on the RV hyperplane of factor X actually to be aimed, unknowingly at two different hyperplanes, one being that which will ultimately settle as the hyperplane of another factor Y. The hoped-for clearer hyperplane for X at the next graph may then be found to be neither the one nor the other but be "blown to pieces" in a formless cloud of points, and totally lost. Many rashly large rotations thrown together may also cause RVs to play musical chairs and change their places and identities.

Parenthetically, this experience of, say factor 7 changing its character on rotation so that it becomes like the original 11, and say, 11 then becoming something else, should remind one that there is also no *lineal descent* of meaningful identity of say, number 7 in the final rotation from the original unrotated number 7. As the column for 7 in L shows, 7 is a linear combination of all unrotated factors, and there is no reason why the unrotated 7 should be the largest component therein. If it is, and the same parallelism holds for other factors (putting large values down the diagonal of L), the most likely explanation is either that the method of factor extraction, e.g., the multiple group (mentioned but not discussed in Chapter 3), has from the beginning oriented itself to clusters, and therefore fairly often to factors, or that the rotation has been so perfunctory and inadequate as not to have emancipated itself from the original position.

Returning to the rotational implications of combined shifts, the practical solution in this matter of possible confusion through two or more large shifts is (a) to keep large shifts single (alone in the column of the S matrix) in the first stages of rotation. Secondly, (b) one must recognize a combinatory principle of reducing the component shifts. For even when different shifts are happily moving toward the same hyperplane their effects will tend to add up to an overshift, so that each singly should be carried to a lesser magnitude than seems indicated by the particular two-dimensional graph being viewed. Thus, in Fig. 7.6a factor 8 might be taken away from 5 a little less than the 0.16 indicated. Some missing of the target in joint shifts is also due to overlooking (in the oblique reference vectors drawn orthogonally in the graph) the fact—shown at the top corner—that they are correlated, so that a positive shift of X on A and on B should be made less on each if A and B happen to be positively correlated. But even if they were completely orthogonal there is some probability that a point p, brought from, say, a loading of +.20 to one of .00 (hyperplane) on RV for X on one graph is also being brought from +20 to +.10 on another graph of the RV for X. These two attempts to get into the X hyperplane might thus end by giving it a substantial negative projection on X instead of leaving it lying on its hyperplane. In blind rotation we do not know which point is which, and even if they are identified by number the calculation of effect point by point would be very time consuming (MAXPLANE does this, but by computer program). Of course, it is possible for one shift to make a point go more negative than the hyperplane goal requires and another to leave it more positive and this is the usual expectation from probability, and the reason for aiming at a least squares fit to the swarm. Provided we have reason to believe that combined shifts are moving toward the same hyperplane the best procedure would be to let the shift line thus be a least-squares fit to the perceived set of points in each of the graphs being combined. However, a more complex geometrical argument than can here be given does show that in combining many shifts it is in general desirable to reduce each below the best estimate of what the tangent should be if one were shifting on that graph alone. A rough rule is so to reduce all initially indicated shifts that the sum of their squares does not exceed 1.0, i.e., "to normalize" them if they add up to more than 1.0; though a total of 2.0 can generally be tolerated.

8.3. Discussion of Matrix Inverses and Transposes in the Calculation of Correlations among Reference Vectors and Factors

Once the SS resolution is complete as such, there are several useful pieces of information contained in the matrices calculated in describing the final position. One wants to know, first, the correlations among the factors (hitherto—except at the conclusion of Fig. 7.7—we have calculated only those among the reference vectors) and then the *loadings* of the *factors* on the variables (the procedure so far deals with *correlations* of the variables with the *reference vectors*) and finally, other things, such as the significance of the simple structure reached. In the latter part of this chapter a thorough discussion will be given to this somewhat mysterious difference of reference vector and factor, but while our main concern remains with rotation we shall simply explain the actual calculation to be made in going to factors from the intermediate "scaffolding" of the reference vectors and justify the basis of the calculation later.

From L_s the last of the series of transformation matrices leading from V_0 to the simple structure reference vector matrix $V_{rs.s}$ it has been explained (p. 151) that we get the correlations among the reference vectors by a simple matrix multiplication. Since L_s gives the angle of each rotated factor to the unrotated it may be intuitively evident that if we multiply the column relating factor X to the original orthogonal framework by that relating Y to the same reference axes, we shall get the cosine (angle) relating X to Y. To multiply each column in any transformation matrix L by every other column, it will be seen that according to our Robinson Crusoe rule, we must put L on its side and postmultiply it by itself. A small three-factor example which the reader can check easily is given in Table 8.1, while a six-factor example was given in Table 7.2.

The transpose of any matrix M is written M' (or, in recent works, with less risk of confusion as M^T) and this newer convention we shall follow here. Each row in the original M becomes a column in M^T and the order left to right becomes top to bottom. Whenever we want to get products among columns within a matrix we write M^TM, and if among rows of a matrix, then MM^T. The R_{rv} matrix, as shown in Equation (8.1), has unities down the diagonal (correlation of vectors with themselves):

$$L_r^T \times L_r = R_{rv} \tag{8.1}$$

because, as explained earlier (Chapter 7), the columns of L are always normalized (made to have the squares of elements add to unity). The latter means, as can be seen from Pythagoras' theorem, that the new rotated RVs are unit length in the space defined by the original unit length V_0 factors. The off-diagonals in the R_{rv}, which are symmetrical about the diagonal, give the correlations (cosines) of the reference vectors with one another.

Now it is shown below (Fig. 8.1) that in a two-space plot the angle between two reference vectors, 1 and 2, is the supplement (180° minus the angle) of the angle between the corresponding factors, 1 and 2. Perhaps from this two-space case the reader can intuit the essential conclusion that the matrix of correlations among factors (in the general case that of hyperspace) must come from the inverse of the matrix among reference vectors, R_{rv}.

This brings us to a new concept in matrix algebra, only briefly introduced before—the inverse of a matrix M, written as M^{-1}. (Parenthetically, take heed that beginners sometimes confuse the inverse, M^{-1}, with the transpose, M^T.) The inverse of a scalar quantity n (any ordinary number) is $1/n$ which can be written n^{-1}. It then follows that $n \times n^{-1} = 1$. The same multiplication defines the inverse of a matrix M, namely that $MM^{-1} = I$. Here I is called an *identity matrix* and has ones down the diagonal and zeros everywhere else.

The calculation of a matrix, M^{-1}, such that when it multiplies M it will give I is not a simple matter. Even, say, a 10×10 sized M inverse would take hours on a desk calculator. And even on a computer the inversion of a 100×100 matrix can take up to an hour (depending on the model) and be expensive, so inversions are sometimes avoided when an alternative algebraic formulation can get to the same end. (Baggaley & Cattell, 1956, present such a means, involving inverting an R_f instead of the larger R_v.) The inversion R_{rv}^{-1} does not immediately give the correlation

Table 8.1. **Multiplying an L Matrix Transpose by L to Get Reference Vector Correlations**

.69	−.61	.40		.69	.60	.83		1.0	.09	.37
.60	.74	.31	×	−.61	.74	−.03	=	.09	1.0	.31
.83	−.03	−.55		.40	.31	−.55		.37	.3	1.0

L^T	×	L	=	R_{rv}
$(k \times k)$		$(k \times k)$		$(k \times k)$
3×3		3×3		3×3

Table 8.2. Matrix Calculation of R_f from R_{rv}

.93				1.16	.03	-.44		.93				1.0	.03	-.36
	.95		X	.03	1.11	-.36	X		.95		=	.03	1.0	-.30
		.89		-.44	-.36	1.27				.89		-.36	-.30	1.0

D	X	R_{rv}^{-1}	X	D	=	R_f
$(k \times k)$		$(k \times k)$		$(k \times k)$		$(k \times k)$
3×3		3×3		3×3		3×3

among the factors R_f because the inversion has upset the normalization of the columns. This is shown by the diagonal on R_{rv}^{-1} no longer containing unities. However, it is easy to find the multiplier of the columns and the rows that will bring the R_f matrix back to normalized columns. If $\underline{d}_x$ is the diagonal value shown in R_{rv}^{-1} for a given factor x, then if we multiply x's row and column by $1/(\underline{d})^{1/2}$ (which we will rename d), the diagonal value will be restored to 1, i.e., the factors will become unit length.

To carry this out in matrix terms we set up a *diagonal matrix, D,* with these d values down the diagonal. A diagonal matrix (not to be confused with a triangular matrix, filled only above or below the diagonal) has values down the diagonal and zeros everywhere else. If the reader will follow the usual multiplication rule he will see that the above required column and row multiplication can be achieved by double use of the diagonal matrix as shown in Table 8.2:

$$D \times R_{rv}^{-1} \times D = R_f \tag{8.2}$$

Thus, if we imagine a case with .64 appearing at the junction of row 3 and column 3 in R_{rv}^{-1} (instead of 1 as in R_{rv}), then its root is .8 and the reciprocal is 1.25. When by the first diagonal matrix .64 is converted to .80, the second (column) multiplication by 1.25 will bring the final value to 1.00. In passing the reader should note that though one would be inclined to write $DR_{rv}^{-1}D^T$, the transpose in the second is unnecessary because the *transpose of a diagonal matrix is the same in its action as the original.*

8.4. Reaching Correct Primary Factor Correlations in SS Rotation

The printout of the above R_{rv} and R_f values which typically comes with the ROTOPLOT output is of use in each graphic representation in the rotation series itself and is not usually left until the final simple structure position. As pointed out above, since the graphic plots though representing oblique reference vectors are kept orthogonal on the graph paper, one must have guiding information about the real angle and, as mentioned in Fig. 7.6, the graph will show at the top right both the correlation (cosine) between the RVs and that between the factors. These help guide the direction of movement if we want to keep to the most orthogonal positions compatible with SS. Watching the r appearing in the R_{rv} and/or the R_f is also important in avoiding collapse of two factors into one, through their axes approaching so close that they begin to move toward using the same hyperplane. It has already been pointed out that the best estimate of these angles is needed for second-order factoring which is increasingly and properly considered an essential part of comprehensive factor research.

Correlations among factors can, when all data are available, also be obtained in another way, by correlating people's actual factor scores. But for most purposes and basic theory the initial R_f from rotation is preferable provided the direction cosines (which are the correlations) are obtained with care. The point needing to be clearly made here is that if the investigator knows his results are likely to be taken as a basis for higher-order work, he should work to carry the rotations to the utmost possible SS precision. For instance he may persevere, for the sake of accurate factor correlations, through 6 to a dozen more overall rotations than are usually felt adequate merely to define the essential loading patterns of the primaries as such. It is in this more reliable determination of angles, based partly on maximizing within the ±.05 level (assuming a good N) rather than within ±.10, or whatever vague range an analytical program is maximizing, that MAXPLANE and ROTOPLOT show a superiority to analytical function programs. This criterion of the precision of angles has not, incidentally, been taken into account in the first empirical studies, listed in Chapter 7, on comparison of soundness of rotational programs (p. 148). The difference in factor loadings between maximizing the hyperplane count within ±.10 and within ±.05 is often appreciable when the angles among factors for the two solutions are compared, and, as the studies by Finkbeiner and Cattell suggest, most good factor studies ($N = 200$ or, say, 300; reliabilities = .8) if they maximized within ±.06 would get more dependable values for the angles.

Upon inversion of an R_{rv}, even when it has shown quite moderate intercorrelation of *reference vectors* (say, between +.3 and −.3), it is not unusual to find *correlations of factors* in the resulting R_f which are surprisingly large, starting sometimes at values like 0.8 and 0.9. Parenthetically, such high factor intercorrelations in the matrices sometimes appear in using the PROCRUSTES rotation to an hypothesis target, and these high values then often represent the contortions of the PROCRUSTES program in desperately trying to reach some intrinsically unreachable match to an investigator's hypothesis. Any claim to have obtained a good fit to hypotheses by PROCRUSTES therefore needs to be examined in the light of what the R_f reveals, though an exact evaluation of acceptability in twilight zones is a difficult issue postponed to Chapter 12 discussion of confirmatory (hypothesis-testing) factor analysis.

Before proceeding with what to do with initially reached grotesque correlations of factors on inverting R_{rv}, let us ask what correlations may be reasonably expected. There may be true correlations among substantive factors, both at the primary and the second-order level, that reach quite high values. For example, in the High School range, and in good cultural neighborhoods, fluid and crystallized general abilities, g_f and g_c, will typically correlate .6, and in the personality domain an experimentally well-based result with a large sample of women showed the superego (G) and self-sentiment strength (Q_3) factors reaching a simple structure correlation of 0.77. The expected correlations depend on the domain and the accumulation of experience in domains. Nesselroade, Curran, Cattell, and other researchers in the field of psychological states have repeatedly found decidedly larger correlations among primaries in state than in trait factors, due to time-coordinated ambient situation changes imposed on the former. For example, anxiety and depression characteristically correlate in dR technique about 0.7, because the dynamics of situations often cause them to occur together. Correlations will also tend to be larger with a "finer grain" in the domain of variables and therefore tend to occur more at lower than higher orders [see comparison of L–Q and T data factors (Cattell, 1973)].

The troubles we speak of in inversion of the R_{rv} matrix, however, are correlations that can be shown to be *erroneously* high, through rotational technical failures. The first of these is poor SS search. The second is the effect of inversion on resulting poor SS positions. Regarding the latter experience will show at once that these high correlations become more frequent with larger matrices. In fact, as one moves to 25–35 factors the amount of work involved in eliminating them is considerable. Presented with r's in the R_f standing obstinately in the .80s and .90s most researchers naturally attack the problem by rotating the originally no more than moderate r's in the R_{rv} matrix, down to really small values. Indeed, the writer has seen investigators despair, after 20 overall rotations of a 30-factor matrix, of ever reducing obliquities enough, and brutally putting the problem into an orthogonal simple structure program! Of course, this throws away the baby with the bath water: the orthogonal R_{rv} "inverts" to an orthogonal R_f (indeed, since they are the same in the orthogonal case one does not even have to invert it). But thereby all the information about the psychologically interesting correlations of factors, and all the higher-order factors, are lost.

The way to a true solution, out of the trap of a spuriously high R_f, is neither by orthogonalizing nor by resort to some other automatic program (for it is usually the automatic rotation program that has, in its imperfection, yielded an R_{rv} that inverts to a bizarre R_f). One must first recognize the underlying cause of the disease, which is that with more dimensions there are more chances of one vector lying nearly in the plane (hyperplane) of two (or more) others. (See the next section, on singular matrices.) One must have confidence—though often sorely tried by the slow progress of successive rotations—that when a number of factors test has indicated k factors, there exists a real simple structure of hyperplanes in which one reference vector is never a mere composite of others. If one finds that simple structure, the problem will vanish. However, it helps to recognize the fact that when the correlations among factors are moderate, and of the same size in a larger as in a smaller matrix, the correlations of the reference vector will be smaller in the larger than one is accustomed to in the smaller. Starting then at the best automatic (but not orthogonal) program output, one must be prepared to undertake patiently a long and skillful ROTOPLOT series of shifts. Therein, one aims generally to reduce R_{rv} correlations, but always in pursuit of the true immanent simple structure.

One should therefore reduce values only legitimately by choosing, when two apparently equally good alternatives toward simple structure present themselves, that which has the smaller cosine. Our experience has shown repeatedly that not mere reduction of all R_{rv} cosines but *unbiased pursuit of the best simple structure as such* will most quickly get large unwieldy matrices out of their seemingly intractably high R_f correlations. This is because the true position of the natural influences are consistent with k space, while most other positions produce correlations in the maxtrix inconsistent with k space, k being the rank of the second-order matrix. Absurdly high correlations in R_f are indeed usually a sign that R_{rv} is ceasing to be a Gramian or positive definite matrix, or is going singular. (These terms are explained immediately in the section now beginning.) Let it be repeated, however, that one should persist in the procedure of reducing most R_{rv} correlations in a large matrix, because one knows that truly (to give the same r as occur in a smaller R_f matrix) they should be smaller than in a small R_{rv}. But the real goal, with priority, is the obtaining of the best possible simple structure, and faithful pursuit of this in itself actually guarantees getting the investigator out of his "bind."

8.5. Singular and Non-Gramian Matrices and the Collapse of Factors

In discussing the number of factors problem (Chapter 5) we omitted one possible contributing criterion—that from rotation—because it involved matrix conceptions not easily clarified at that point. The psychology student even at this point cannot be expected to know in depth the mathematical meaning of a Gramian matrix, of the rank of a matrix, or singularity of a matrix, and so on. But since he will meet these terms he should know at least their implications for his matrix computations.

An "ill conditioned" R_{rv} matrix such as we have been talking about may, when inverted, as in Eq. 8.2, give absurdly high R_f values—say above .9 everywhere. But this is only an intermediate stage to a worse condition when the computer declares the R_{rv} matrix to be uninvertible! Such a matrix is said to be *singular*, and it implies in general that its rank is really less than its order. The rank, it will be remembered, is the number of factors required to explain the number of variables—the order of the matrix.

It should be remembered that the inversion of the R_{rv} is that of a correlation matrix with unities down the diagonal. If we worked, instead of with such an *unreduced* matrix, with a matrix of communalities down the diagonal, which is called a *reduced* matrix, it would not be so surprising to find it uninvertible because the communalities are usually deliberately chosen to bring the matrix to a smaller rank k instead of a rank n. However, even with communalities less than unity it is possible, as we saw in the number of factors problem, to end with n factors, i.e., the matrix will have n nonzero roots. However, the difference is that such a matrix with communalities is not what is called *positive definite* or *Gramian* because some of the roots will be negative and some positive, presenting us with values uninterpretable in experimental concepts. Technically, the term Gramian defines a matrix with either positive definite or semidefinite roots, i.e., some positive, some zero, non-Gramian connoting the existence of negative roots. These properties can be illustrated in the very small scale instance in Table 8.3.

In case (a) in Table 8.3 we have the correlation matrix with unities as set for the typical principal components solution. This matrix is Gramian, i.e., it has "positive definite" roots, and it is nonsingular, i.e., it will invert, as shown. In (b) with reduced ("communality estimate") values in the diagonal, it is still of rank 2, but the second root is negative and it is described as *indefinite* and non-Gramian (+ and − roots) but in this case it is still possible to find an inverse. However, when the rank is less than the order, as here, the inverse is not unique.

It is possible for a matrix to be uninvertible (by ordinary means, so as to yield a unique inverse) for other reasons than that its rank is less than its order. A matrix derived from an $L^T L$ in which the usual independences were lacking would be such. However, in the practical problems that arise in rotation, as above, a matrix is usually described as having gone singular and become uninvertible because its rank has become less than its order. And as it approaches singularity we run into the large r's in the R_f matrix encountered above.

Table 8.3. Positive Definite and Indefinite Correlation, Matrices and the Problem of Singularity

Unreduced R matrix

		1	2
$R =$	1	1.0	.5
	2	.5	1.0

Matrix is nonsingular
and positive definite

Roots 1.5 and .5

It inverts[a] to

		1	2
$R^{-1} =$	1	1.$\dot{3}$	$-.\dot{6}$
	2	$-.\dot{6}$	1.$\dot{3}$

Same R reduced

		1	2
$R =$	1	.25	.5
	2	.5	.25

Matrix is nonsingular
but not positive definite.
It is indefinite, non-
Gramian

Roots .75 and $-$.25

It can be inverted[a] to

		1	2
$R^{-1} =$	1	$-4/3$	$8/3$
	2	$8/3$	$-4/3$

[a] Inverses have been left in fractions for derivation to be seen.

The cause of these large r's and ultimate singularity is commonly due to two columns in the L_f matrix which generates the primaries (and corresponding columns in L_{rv}) becoming too much alike. One can often see this in looking at either L matrix and, of course, one can see mathematically that this similarity of values would produce a large correlation between the two factors concerned. Actually we are usually working on rotation in the reference vector system with the L_{rv} which produces R_{rv} ($=L_{rv}^T, L_{rv}$); but, as stated, one's concern is for the direction cosines of the factors as such, though these we do not usually calculate. In either case we have recognized that high r between two factors arises from the two columns being nearly alike. But in more general terms the condition of singularity is what the mathematician calls *linear dependence* developing among two or more columns of L. That is to say, some linear combination of the values in, say, columns 2, 3, and 5 of the L matrix comes close to reproducing the values in column 14. The R_f matrix (with unities in its diagonal) is then no longer of rank k, but becomes $(k - 1)$ or even $(k - 2)$ and as such will not invert by the ordinary program (or, indeed, by any program) to give a unique inverse. If one is a visualizer, the situation can be seen in three dimensions as one of having moved a certain reference vector right into the plane made by two other reference vectors. The problem rarely arises in small matrices, say up to 12 factors, where rotation can be poor without the experimenter being pulled up sharply by evidence of this roughess, in the form of a singular matrix. But it happens fairly often when many factors—say on a 35 × 35 matrix—are being obliquely rotated, and the demand for skilled and insightful rotation then becomes paramount. It can perhaps be perceived intuitively that where there are many factors, i.e., when k is large for the R_{rv} matrix, it becomes increasingly easy to slip into a position where the given factor lies in the subspace of certain other factors and is reproducible from them.

This experience of inverting to high R_f values, and finding, indeed, that some poor further shift may bring complete singularity, raises the question of whether one should sometimes let this "collapse of a factor" happen, and accept the fact that one must reduce the unrotated matrix to $(k - 1)$ factors, where one thought one had k. This would be an improvement if perhaps one made a mistake by originally taking out one factor too many. As a practical advice, Thurstone for one, advocated the policy we have here advocated: that if one is to make an error it is better to take out one too many factors than lose real information by taking one too few. The extra factor will "rotate out" as described in a moment. Before it does so it may cause rotational trouble as above. If there is increasing evidence that there is a factor too many it can be squeezed out in rotation in one of the two ways described below. But to finish the job one should then go back and delete the last column of the V_0 (and adjust the communalities) to $k - 1$ factors, and polish the rotation afresh from that new V_0.

A superfluous factor is likely to be "squeezed out" when demanded by good rotation in two ways. First, as just indicated, it may increasingly yield up its variance to other factors (often by being rotated negatively on large factors) until it has scarcely any loading left outside the hyperplane. To do this it has to move until it is almost orthogonal to all unrotated factors and has its largest value in the L matrix on the last and smallest of the unrotated matrix. It cannot be actually rotated to fall outside the unrotated factor space but it can shift to taking a very small dimension of it. And when the analyst finally decides it is superfluous, the last unrotated factor, or whichever really small V_0 factor is most involved with this smallest rotated factor, can be struck off as a column of the V_0. Some small readjustment is then needed by reestimating communalities and doing a tidying rotation on the renormalized L columns (all now lacking a last number as the bottom row is left out).

The second way in which a superfluous factor gets dropped is by its angle to another factor getting smaller and smaller (the r_{rv} cosine becoming larger and larger) until one finds one really has only a single factor. Like a game of "musical chairs" in which there is one fewer chair than there are players, there is actually one fewer hyperplane than there are reference vectors, due to the original factor extraction error. The collapse means that two reference vectors are being forced to sit on one hyperplane. The implication, of course, is that in real data there are only as many hyperplanes as there are factors, and this implication will be focused more sharply in the next session in using it as a cue regarding the number of factors. One must note, in passing, that this kind of sidelight—and indeed any real knowledge of what is happening to the simple structure—would be denied to anyone who restricted himself to a packaged automatic program and who failed to get the "feel" directly given by ROTOPLOT outputs. This is perhaps why Kaiser, in contrast to Thurstone and the present writer, concluded that overfactoring is invariably disastrous (1963, p. 165). There is an aspect of truth in this overstatement, for, as the present writer has pointed out (see next section), with *substantial* overextraction "factor fission" occurs, but just one or two factors extra do not seem immediately to produce this fission, since in real data there are always common error factors to occupy the extra dimensions.

As far as rotational resolution is concerned the point to note is that a factor may essentially disappear by one of the above two paths. The rotator should be warned, in passing, that with slightly careless rotation this disappearance can occur even when there is not a superfluous factor. (If that happens one must reinstall a kth reference vector by a new column in the L matrix roughly orthogonal to the others.) On the other hand if the kth factor is superfluous in the sense here of never finding anything like a hyperplane to "sit on," then one must as indicated above, strike out a factor (typically the last and smallest) from the unrotated matrix, and return with a shortened L matrix to a little more rotating on the $(k - 1)$ column V_0 matrix to sharpen the hyperplanes again.

When an investigator keeps tabs on individual rotated factors, which tend to retain their essential identity over the last few rotoplots as the hyperplane count settles to its plateau, it will often happen that though the last rotation, which we will call the fth, has the best total count, an individual factor x will have its best count on the $(f - 2)$th and y on the $(f - 5)$th rotation. There is then no objection to taking these better columns for x and y from the L matrix from these earlier rotations—and indeed to lifting any column from any L matrix that is the best in SS for a given factor—and putting them together in a final composite L matrix. This was done, for example, in the recent study on personality structure in objective, behavioral test measures (Cattell, Schuerger, Klein, & Finkbeiner, 1976), where the separate, independent ROTOPLOT pursuits of total structure, by the three investigators concerned with rotation covering some 20 overall shifts by each, were sometimes, *for a particular factor*, at a slightly better hyperplane count in the Klein, or the Finkbeiner, or the Cattell reference vector matrix. The combination of separately picked-out best L column across the three, for each factor, assembled in a single L, obviously yielded the best total solution by any criterion we applied later. Naturally, one must quickly check the R_f from the new composite L, since hitherto unseen abnormally large correlations could appear then (or the R_{rv} could even be uninvertible). But if the L columns chosen really are the true best SS positions such abnormal cosines are most unlikely. Only if the fth and $(f - x)$th rotations are so remote that the given factor loses its identity between, so that one has been pursuing convergence on what are actually different hyperplanes, could this occur.

8.6. Geometers' Hyperplanes and the Use of Rotation as a Check on Factor Number

The theoretical implication of this and some other findings touched on here is that the hyperplanes lie in the space of the test configuration in as real and substantial a form as cleavage planes in some crystal masses, waiting to be found. The theory also supposes that except where some alternates exist they are exactly equal in number to the dimensions of the factor space. The assumption that there are no alternative positions and that the number of hyperplanes is precisely the same as the dimensionality has been doubted and contested by some very able factorists, who claim to have found alternate sets of k hyperplanes lying in the same space. The present writer's response to this claim is (a) that the alternative hyperplanes are generally so inferior to the first choice as not to be considered genuine alternatives, and (b) that these nebulae are not factor hyperplanes at all but chance alignments of an unusual number of points in what has been called a *geometer's hyperplane*. The latter is defined below as a special kind of artifact, and (c) some real, organic alternative hyperplanes may exist because of higher-order factors operating on the same variables, or other real structures in cases where what we shall call the *reticular model* (p. 201) to produce a second species of hyperplane.

Since the third possibility—second orders—will be duly met later, we will deal here only with geometer's hyperplanes (Cattell, 1966, p. 209). In any set of n points scattered randomly in k space one can always find $(k - 1)$ of them lying in a hyperplane. This can be most quickly understood by reflecting that in a number of points in 3 space one can always find sets of two points in a plane with the origin. Indeed, in k space where k is of the size typical of most researches one can find a substantial number, the number being the number of possible combinations of $(k - 1)$ things out of a total of n (n being, as usual, the number of variables) namely $n!/(n - k + 1)! (k - 1)!$. Since these hyperplanes merely follow from the rules of geometry we have called them *geometers' hyperplanes* to distinguish them from the organic hyperplanes defined in the scientific model and produced in connection with the action of real influences, as already plentifully empirically demonstrated. A calculation for a typical multivariate experiment, say with 100 variables and 41 factors will show that there are an enormous number of such 40 item geometers' hyperplanes lying around in the factor space (in fact, $1.374623415 \times 10^{28}$ of them!). Even if we imposed the restriction of taking only those roughly orthogonal to one another and therefore, like true hyperplanes, not sharing more than 80%, say, of their variables, we would still have a large number.

However, no expert seeker for simple structure would be ready to accept a ±.10 hyperplane with only 40% of his variables in it, such as would exist in the above typical geometers' hyperplane. (Although the student who looks carefully at published studies of say personality structure will find that substantial theories have sometimes been based on alleged simple structures reaching little or not at all above 40%!) If, as with most ranges of variables and factors in good designs, something from 60% to 80% is required at ±.10 for $p < .01$ significance, no 40% geometers' hyperplane is going to deceive us. But as one gets to the number of factors where $k > n/2$—which can easily happen in exploration, though in well-informed planning one likes to avoid Albert's difficulty (p. 573), i.e., that arising at $k > n/2$—the danger of confusing geometers' and organic hyperplanes becomes very real. When variables are unavoidably few, as in higher-order studies factoring primary factors, a 50% hyperplane count ($n/2 = k - 1$) sometimes definitely occurs for a true organic hyperplane. Moreover, one must remember that some geometers' hyperplanes are likely to actually include more than ($k - 1$) points, for ($k - 1$) is only the count for the points exactly on the hyperplane, and if we permit ±.10 width we shall catch some more. Just as in three space there will be some two-point lines that for all practical purposes pick up a third and fourth point (assuming a cloud of say 40 points) so there will commonly be—assuming a ±.05 hyperplane width—a distribution running to a number of ($k + 1$), ($k + 2$), and more geometers' hyperplanes. It is not surprising that with values of k and n in certain not unusual ranges, and with hyperplanes allowed to go flabby to a width of, say, ±.15, careful investigators may claim that the space contains hyperplanes for several alternative simple structures!

To argue, as the present writer does above, that on theoretical grounds there will essentially only be k real hyperplanes in k space (except for our later concessions to possible higher order and reticular structures), is not necessary immediately to provide a remedy to the problem of escaping false hyperplanes generated as above, any more than the proper diagnosis of any disease brings an immediate remedy. However, good design is a prophylactic. On grounds of communality estimation we have already seen the desirability of n being more than $2k$. On present grounds of rotation we would raise this to $3k$. The geometers' hyperplanes would then contain around 34% of the variables and be safely out of the range of confusion with the 50% plus required in a significant true hyperplane.

Let us now turn to the possibility briefly introduced above of using the rotation process to check on the number of factors decided by the earlier processes. Such a practical possibility appears as a lemma on the writer's theorem above that there are as many hyperplanes as factors. It has been mooted on the grounds by experienced factorists such as Thurstone (1947) and Gorsuch (1974) and has twice been given a demonstration at a rough level (Cattell, 1960, 1966a; Veldman, 1974).

It has been pointed out above that an extra factor or two beyond the indicated number by the scree or maximum likelihood will commonly fail to get a hyperplane of its own with any worthwhile variance on it. But it can also happen that the extra factor dimension will split the hyperplane of a legitimate factor (unless there are large error factors) creating two spurious factors, as described below, in referring to factor "fission." Table 8.4 shows (from Cattell, 1966a) the effect on a well-known physical plasmode (the Cattell–Dickman ball problem; Cattell and Dickman, 1962) as discussed below.

Further examination of this example shows also (not on data presented here), as one would expect, that the hyperplanes are thicker when one rotates an underextracted V_0. This follows because the hyperplane in a k space has to be projected on $(k-1)$ or $(k-2)$, etc., space. One can visualize this by imagining a disk in three space being projected into two space, whereupon, in anything but an end on position, it will appear as a fairly broad ellipse.

The second rotational effect of a wrong number occurs in underfactoring. With gross underestimates the distortion is devastating, for the massive variance in the early components has nowhere to go. However, one gets impressive high loadings, it is true, as when Eysenck or Peterson takes out three personality factors where there are probably 15 to 20! The spuriousness of the high loadings has been well brought out by Digman (1973). With overfactoring, on the other hand, though one extra factor may easily rotate out, as above, yet with two or more a degenerative process typically sets in exhibiting what has been called *factor fission*. In Table 8.4 the true number is 5 factors, and one can see the beginning of fission (notably of factors 4 and 5) even with one extra (6 factors). Fission means that a factor with a, b, c, and d as salients and e, f, g, h, i, and j in its hyperplane is likely to split into one factor with a and b as markers, and c, d, e, f, g, h, i, and j in the hyperplane and another with c and d as markers and a, b, e, f, g, h, i, and j in the hyperplane. As this false excess of factors is carried toward n, more degeneration occurs, and finally we have n factors

Table 8.4. The Degenerative Process of Factor Fission as Revealed by the Rotational Test for Number of Factors

V_{fp}	4 factors: insufficient number for the substantive (physical) factors				5 factors: number according to scree test					6 factors: first extraction of definitely excessive number of factors					
	1	2	3	4	1	2	3	4	5	1	2	3	4	5	6
1				.65				.59	-.11		.60		.36	-.11	
2		.56				.62				.43					
3	.81	-.23		.10	.85				.14						
4			.99				.98					.96			
5				-.50				-.38	.47					.48	
6		-.12		.67				.60	-.11				.37		
7				.64				.59					.37		
8				-.50	.10	.11		-.39	.10					.49	
9		-.12		.62				.63	.11				.48		
10		.56				.61					.59				
11		.75		-.30		.63		-.35			.62		-.30		
12		.50				.41			-.19		.39			.17	
13		.52				.54					.51				
14		-.56				-.61			-.11		-.59				
15	.71	-.15			.77				.22	.43					
16	.81	-.23		.11	.84	-.10		.13	.10	.42					
17	.59	-.16		-.34	-.56	-.20		-.31		.26	-.13		-.29		
18	.86				.86										.37

	21	15	22	14	20	20	23	18	15	27	22	23	22	26	28
	1	2	3	4	1	2	3	4	5	1	2	3	4	5	6
19	.85	.16			.83	.11		-.17							.38
20			.98	-.14			.97					.96			
21			.97	-.11			.96					.94			
22			.98	-.13			.97					.96			
23			-1.00				-.98					-.97			
24			-.94				-.95					-.93			
25	.54	.25			.48				.11		.15		-.31		
26	.80				.80				-.37				.42		
27		.19		.12		.22		.33	-.12	.31	.22		.12		
28	.52			.39	.54			.31					.27		
29	.78		.11	.36	.75										
30			.58	-.39			.54	-.25	-.15			.54		-.36	
31		-.13	-.42	.54			-.43	.47	.35			-.42	.31		.28
32	.14	.47	.32		.16	.47	.29				.44	.28			.44

Number in hyperplane

R_f	21	15	22	14
	1	2	3	4
1	1.00			
2	.31	1.00		
3	-.06	-.03	1.00	
4	-.37	-.81	.03	1.00

	20	20	23	18	15
	1	2	3	4	5
1	1.00				
2	.34	1.00			
3	-.10	-.17	1.00		
4	-.29	-.56	.11	1.00	
5	.22	.60	-.13	.03	1.00

	27	22	23	22	26	28
	1	2	3	4	5	6
1	1.00					
2	.44	1.00				
3	.17	-.03	1.00			
4	-.40	-.41	-.08	1.00		
5	-.24	.26	-.14	.65	1.00	
6	-.88	-.27	-.23	.40	.40	1.00

each with only one variable out of the hyperplane. At this *reductio ad absurdam* the factors from the R_{rv} will correlate like the variables in the original correlation matrix, i.e., $R_f = R_v$. From this it is clear that we would expect the goodness of hyperplane—as the percent in a band of say ±.05—to continue to improve moderately till the true factor number and then to climb quite rapidly. Thus in Table 8.4 the mean for the successive factors as the extracted number goes from 4 to 5 to 6 is 18, 19.2, and then 24.7 after which it climbs inordinately.

In seeking to help to decide number of factors from SS rotation properties the present writer proposed to combine the above two signs with two others: (3) the magnitude of the loadings on the factor relative to the communalities, which should drop rapidly on factors as they begin to split their markers, and (4) the angles among the factors, which would increase past the true point because variables are commonly correlated more than factors, and because, at first fission, there is a real reason (see Cattell, 1966a) for the two fission factors, because they share in part the same hyperplane, to be much correlated. This increase can be seen comparing the mean r for the 5 and the 6 factor R_f in Table 8.4 (from the R_f's at bottom).

It was the writer's suggestion that all four of these signs be combined though it is not possible to suggest any definite weighting of indices. However, Veldman (1974) has recently nearly shown empirically that the third criterion alone will give dependable evidence. He takes for each factor the expression b^2/h^2 for each variable on a given factor, and then takes the Σ^2 of this value over the column. This is of a similar value as indicated above and is typically maximized in analytical programs in their "spreading" of the loadings between the extremes of hyperplane lows and marker highs. Veldman found that this value summed over the k factors in the study typically advanced monotonically to its highest value at the true number of factors and then fell, at least for a while. The present writer would hazard that this index would be better still if he had used an oblique rather than the VARIMAX program in getting the SS positions. The value thus behaves as one would expect the hyperplane count to behave if divided by the statistically expected random count for the given numbers of factors (see significance of hyperplane, Section 8.7 below).

One should thus recognize that there exists, beyond such reasonably effective tests of factor number as the scree, and the negative root with Guttman–Roff communalities, as well as such theoretically sound tests as the maximum likelihood method, a fourth, independent and quite valuable test in an index derived from these four rotation properties. When vital psychological issues are at stake, such as whether the number of personality factors in the broad questionnaire domain of normal behavior is nearer 3–4 or, as the present writer believes, 23–28 or so, there is today no need for doubt to persist, if these resources are properly used. A helpful further discussion of hyperplane properties and definition is given in Gorsuch (1974) and Van de Geer (1971). Let us note also that the advent of "proofing (confirmatory) maximum likelihood" methods for testing hypotheses (Section 13.6) has resulted in a new approach to evaluating simple structure and factor number. It consists of stating in a target matrix, as in PROCRUSTES, whose loadings should and should not be zero, and testing the goodness of fit of factors obtainable from the given R matrix (Jöreskog, 1966). Some limitations on the proofing method are discussed later (p. 400).

8.7. The Evaluation of a Simple Structure

The initial test of having attained whatever simple structure exists lies in the history of hyperplane plots (Fig. 7.5) and evidence of gradual but real approach to an unimprovable plateau. This is not given by an automatic program with confidence (or even by an iterative topological program) but only by ROTOPLOT. Indeed, reasons have been given for finally subjecting any automatic output to a ROTOPLOT examination. We have also pointed out that since analytic programs, unlike the topological MAXPLANE, do not define specifically the hyperplane width in which they are attempting to maximize, the application of ROTOPLOT, though it may not improve on the automatic program's broad ±.10 or ±.15 count, is normally likely to improve the ±.05 count. Although the change may be slight it can recognizably alter the factor angles, lead to loading changes altering the interpretation, and bring definite changes in the ensuing second-order factor structure.

However, the final judgment needed on a rotation is not only an anwer to "Is this the demonstrated *maximum* hyperplane count obtainable?" but also to "Is this a statistically *significant* simple structure?" For many years only one solution has been offered for this determination—that of Bargmann (1954)—and its tolerable adequacy perhaps accounts for lack of further proposals. In principle it is simple, though in calculation, complex. It counts the frequency of variable points in a "slab" in hyperspace constituted by the hyperplane represented by the ±.10 (or other) cut, in the given example. It then compares this with the count expected if all points (variables) were randomly distributed throughout the sphere, with the given communalities, and assigns a P value to the given frequency. This expected number depends on the number of points, i.e., n, and the number of factors k, and for that reason, as the set out of the Bargmann test tables in the Appendix indicates, a separate graph is needed for each combination of parameters.

The Bargmann test in practice has turned out to be a severe critic, and it is no exaggeration to say that about half of the investigators publishing factor studies would have their conclusions wiped out thereby— which is perhaps why a test of the significance of the obtained simple structure is so rarely published. Nevertheless, by any standards of true scientific procedure an SS significance test is an essential part of a factor analytic experiment, and is one of the six conditions which Vaughan (1973) and the present writer (1973) have advocated for evaluating the conclusions of a factor analysis. In an attempt to improve on the Bargmann test or supplement it with independent evidence the present writer has experimented with (a) substituting random normal deviate loadings rather than random numbers over the range in Bargmann's test itself, (b) using Harris' value for the standard error of a loading (this, however, founders, as a practicable measure, on the hyperplane then being different for every variable and factor combination), requiring excessive calculation, and (c) taking the V_0 of many actual experiments and spinning them randomly to obtain, by Monte Carlo methods, the chance distributions. The last is quite the firmest if not the most elegant basis the psychologist can use today, and tables for this are offered by Vaughan (in press). Of course, these tables (Cattell–Vaughan) need to be expanded to other ranges as soon as psychologists appear with resources to do so. A test on essentially the same principle has very recently been proposed by Kaiser and Madow (1976). Instead of determining the distribution by random spins, their approach depends on the fact that if a factor extraction is arrested at the *eigenvector* stage and these values are not multiplied by the diagonal matrix of roots to give the usual V_0 (see Chapter 13), then any vector in any rotational position in that space will have projections of variables upon it that will stay constant as to variance (not constant as single projections, of course). Consequently, one does not have to create tables of distributions by rotating to all kinds of different positions. To test whether some given position has a significant hyperplane one takes either a cut at the width of the statistically indicated standard error of zero, or one takes a subset of variables decided on earlier (in the case of Kaiser's approach) and compares their variance with the general variance of variables in that study, which, we have seen, stays fixed for all positions in the eigenvector space.

The hyperplane seeking and the hyperplane testing virtues and drawbacks of the Kaiser program should be evaluated without prejudice of one to the other. The former has been examined in the previous chapter, expressing some misgivings and the latter can be examined here. The advantage of the Kaiser–Madow relative to the Cattell–Vaughan method is that one avoids either (a) the approximation of comparing a given study with an average result from several studies of roughly the same h^2, $n \times k$, or (b) having alternatively to apply the random spin program to determine tables for one's own special study. The advantage of the Cattell–Vaughan relative to the Kaiser–Madow method is its empiricism. One may also ask whether the latter may not introduce such increasing exaggeration on dimensions after the first (in the comparison of factor and eigenvector) that in some directions the natural hyperplane, of say ±.08, would have to be represented by a value many times larger—how many times one does not know—to allow for the stretching in the method. The method compatible with our basic theoretical model would regard the standard error of a zero loading as the same in all factors, and this would suggest the Cattell–Vaughan method to be less distorted than the Kaiser–Madow approach, in principle too. As suggested in other procedures, if the study is one of great theoretical importance we would advocate applying both tests. One would give greater weight to the K-M where its principles are most applicable to a given research, and weight to the C-V where freedom from distortion of hyperplane widths is essential.

However, if the investigator is prepared to go to the trouble of applying a random rotation program on his own V_0 matrix instead of taking the general C-V tables from "average, typical V_0," values he would get the most reliable result of all. No factor study of any importance should be left without a test of significance of simple structure. (Use the Kameoka–Sine tables, Appendix 6.) One should note, however, the special sense in which an SS significance test is a statistical inference. In the Bargmann test it is statistical only in regard to a sampling over the test domain, i.e., it is what we called a psychometric rather than statistical test, if we restrict the latter term to subject sampling matters. If the Cattell and Vaughan Monte Carlo approach were carried further, systematically over samples of variable sizes, and samples of people or occasions (and also over reliabilities), we should have distributions permitting a statistical inference with due reference. A problem for critical surveyors of research conclusions is that many results are reported in the literature where the best possible simple structure is believed to be reached, but some or even all factors, when tested, fall far short of SS significance. What should be concluded in such cases? In three out of four such instances neither demonstrated maximal simple structure nor significant simple structure has been reached. (Commonly, the study has stopped at use of a poor analytical program.) These studies cannot be given any weight in psychological conclusions. In others a demonstrated maximum, unimprovable SS even if not significant, should be accepted contingently on the assumption that the choice of variables has perhaps not permitted well-developed hyperplanes in *some* factors.

8.8. Reference Vectors and Factors, Loadings and Correlations Distinguished

Granted the checked attainment of a maximized and reasonably statistically significant simple structure (which will normally be found to be a position with primary factors to some degree correlated), the investigator is on the threshold of interpretation of results. This he does either by developing hypotheses to fit the loading patterns found or by checking the fit to a preconceived hypothesis stated as a pattern. But what exactly are the "patterns" about which we are talking here?

In an orthogonal system they would be either correlations between factors and variables, or the weights of variables for the best estimation of factors. With correlated factors, however, he has another choice besides the weights for estimating factors: namely, that between *correlations* of the factors with the variables and the *loadings* of the factors on the variables—which, as we shall see, are usually distinctly but not vastly numerically different. The loadings are values derived from the correlations of factors with the variables *when the correlations among the factors themselves are allowed for*. They are just the same as beta weights in multiple correlation—the weights if we consider the factors as predictors and the variable as the predicted criterion. They are therefore the weights used in the basic *specification equation* (p. 67) for estimating scores on a variable from a person's scores on the factors.

From the rotational process itself, however, the concluding matrices set out are neither factor correlations nor factor loadings on variables. They are simply *correlations with the reference vectors*. It is true that all three of these matrices come out at *very roughly* the same values, so that the salient, prominent variables by which one would recognize or interpret a given factor would differ little from taking the highest on any one of the different matrices. However, accurate work requires the necessary conversions, and, for reasons now to be given, the most meaningful expression of the final outcome is the *factor pattern* matrix. To reach this we must begin by illuminating the difference of reference vector and factor touched on operationally above (p. 133).

As mathematical statisticians we have to define the *factor vector* (the factor itself) as something distinct from the *reference vector*. As stated in introducing this difference above, the factor vector is, in geometrical terms, the line of intersection of all hyperplanes other than that of the factor itself and it will be roughly orthogonal to this factor and reference vector hyperplane. The corresponding reference vector, as we have seen, is the exact normal (perpendicular) to that factor's hyperplane. Now algebraically, in further calculations, it will be found that it is the *factor* rather than the *reference vector* that corresponds to what we called a "determiner" in the scientific model.[2] That is to say, the variance on vari-

---

[2] "Primary factor" has often been used to contrast the factor with the reference vector. This common but undesirable use has been avoided here, because when we come to higher-order factors (Chapter 9) the term "primary" is properly used and needed in distinguishing first-, second-, and third-order factors, respectively, as primaries,

ables needs to be finally described as the sum total of contributions from unit variance factors. Since we want to keep this and other transactions in terms of the model we have to keep to the factor rather than reference vector "system." Indeed, as mentioned before, reference vectors are best regarded as a temporary piece of scaffolding around the developing building, needed for convenience particularly during the handling of the rotation job.

The essential difference of factors and reference vectors can best be seen diagrammatically, although this causes a little distortion, perhaps, because graphic representation means taking the extreme case of just two dimensions. The hyperplane of a reference vector appears in two dimensions not as a plane but a line, at right angles to the RV, though one can imagine a third factor sticking out of the paper. For in three dimensions there would be a plane there and the intersection of the planes for RV_1 and RV_2 in Fig. 8.1 with that paper plane would be the hyperplane lines we see. Then, in this case, the thick line which is the hyperplane of reference vector 1 in Fig. 8.1, is actually factor 2 and conversely the hyperplane (a line vector) to reference vector 2 is factor 1. It will then be seen from simple geometry that the angle between the primary factors is the supplement $(180 - \theta_{12})$ of the angle θ_{12} between the reference vectors. (This two-dimensional illustration we invoked before in indicating why the correlation matrix between factors is derived from the inverse of the matrix of correlations among reference vectors—for the inverse is the k-dimensional generalization of the geometric supplement relation here.) However, Fig. 8.1 is also introduced to show the relations of *correlations with the reference vectors* to *loadings on the factors*. Correlations, spatially, are as we have now long seen, cosines of angles between unit length vectors. (If two vectors are not unit length, but have standard deviations—here taking the form of communalities—of only h_1 and h_2, then $r_{12} = h_1 h_2$ cos θ_{12}.) The perpendiculars from variables a_j and a_k in Fig. 8.1 on RV_1 thus give projections which are in fact (when multiplied by their communality roots) *their correlations with the reference vector 1*. These are the values we are accustomed to in the V_{rv} matrix directly resulting from our oblique rotations. Their projections on the *factors*, on the other hand, give the *loadings on the factors* (not the correlations with the factors). Let us note the mathematical rule, however, that when one works with oblique coordinates the lines of projection *do not fall vertically on the factors*. Instead, the magnitude of the projection is defined, in such oblique graphical work, as the cutoff on the factor axis, when the lines are

secondaries, and tertiaries. In contrasting RV and factor, "primary" is quite superfluous, "factor" being enough. Indeed, among secondaries we also have to speak clearly of reference vectors (RVs) and factors, and "primary" would then introduce confusion.

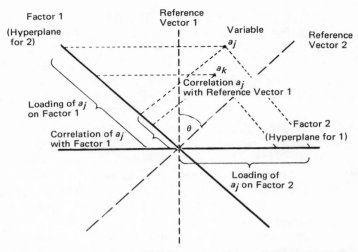

Fig. 8.1. The relation of factor and reference vector correlations and loadings. It will be seen that the loadings of a series of variables, a_j, a_k, etc., on the factors shown here by cutoffs of intercepted lines on Factor 1 are proportional to their correlation with the corresponding RVs shown (proportionally) by the right-angle projections on Reference Vector 1. (From R. B. Cattell, *Handbook of Multivariate Experimental Psychology*, Rand McNally, Chicago, 1966, p. 181.)

dropped *parallel to the other factor* (inspect Fig. 8.1 carefully). It will be seen from the upper part of Fig. 8.1 that the *loadings on the factors will therefore always be simply proportional to the correlations with the reference vector*, for any given pair (factor and RV). This will be true for all variables on that pair, but the fixed ratio of correlations to loadings, for all variables, will be different on each other paired factor and RV.

In the orthogonal case, as has been mentioned earlier and as will be evident from collapsing RV and F together in Fig. 8.1, there is no difference of factor and reference vector, and no difference, therefore, of correlation and loading. But in the oblique case, it becomes very important to grasp the distinction and also to be able to calculate the relation which then arises between a loading and a correlation. In matrix terms it means nothing more than that we have an $n \times k$ *factor pattern matrix* (the matrix containing *loadings*) and an $n \times k$ *reference vector structure matrix* (containing *correlations*) in which a certain ratio—a constant multiplier—transforms all values down the first column of one into those in the first column of the other. (A possible mnemonic to remember that structure rather than pattern means correlations is that there are two *r*s in structure and two *r*s in correlation.) A second multiplier (called *d* because it falls in a diagonal matrix) does the same to bring simple mutual proportionality to all values in the second pair of columns across the two matrices, and so on.

Further illustrations as we proceed will make this a familiar and simple fact of oblique rotations. However, in the now passing generation the concept has seemingly brought confusion and alarm to some psychologists. In spite of the sound reasons for expecting factors to be oblique, some users of factor analysis have retreated to the seeming security of orthogonal factors familiar in the nursery stage of factor analysis. Since most psychology students become familiar with partial and multiple correlations before they meet factor analysis, the best introduction here is perhaps to point out, as done above, that loadings are nothing but beta weights in a multiple regression equation. They tell how much each predictor variable (in this case, each factor) should contribute to the score estimate for the predicted variable (in this case, any variable in the R matrix) for maximum prediction. When predictors are uncorrelated, we know that the contribution of each is given directly, simply by squaring its correlation with the predicted variable. This is true of orthogonal factors, where beta weight and correlation, i.e., loading and correlation, are the same. But oblique factors are mutually correlated, and what we call their loadings are what they contribute when the correlated effects of other factors are excluded. [The final contribution to the variance of each variable (not to the score itself) is, however, as in multiple correlations, the product of the loading and the correlation of the given factor with the variable. As will be seen later, both the pattern and the structure matrices are therefore involved in that calculation.]

We do not intend to go into all aspects of oblique calculations until the reader has digested the above. More precise analysis and formulation is deferred to page 184. However, from what has just been said, it will be evident that getting the loading (factor pattern) matrix is formally identical to calculating the weights for a multiple regression. For it is an attempt at the *best estimate* of a person's variable score from his known scores on the oblique factors. The student familiar with graphical representation of the partialling out of one variable from another will recognize that that is what we are doing in Fig. 8.1 in making the projection of a_j on Factor 1 run parallel to Factor 2.

As at similar stages in other concepts we shall now move from handling the problem in terms of a couple of factors, and on familiar elementary statistics course regression computations, to the matrix formulation appropriate to actual factor analytic practice with many variables and many factors. The above has explained the terms *factor pattern matrix* and *factor structure matrix*, as introduced and stabilized, incidentally, to the benefit of clear discussion, by Holzinger and Harman (1941) and

Thurstone (1947). Now corresponding terms can be used for the two forms of the reference vector system and in both, *pattern* will mean *loadings*, and *structure* will mean *correlations*. Since there are two forms of relation and two kinds of vector, there are thus actually *four* possible matrices, as shown in Fig. 8.2.

The reference vector *pattern* is given only for mathematical completeness since it has no role of any importance in a scientific model, but occurs only as a transient expression in calculations and indeed is rarely used. The reference vector *structure* is also no part of the final model, but we handle it so frequently in the rotating processes that its domestic familiarity gives it a certain concrete reality. Illustrations have already been given (p. 151) of how at the end of the rotation process (a) the reference vector structure V_{rs} is converted into the final factor pattern V_{fp}, and (b) the correlations among the reference vectors R_{rv} is converted into correlations among the factors R_f. For both of these calculations we require a certain diagonal matrix D, which is $k \times k$. It will be remembered that (p. 161) we obtain R_f from R_{rv} (rs among the reference vectors with which we have been working) by

$$R_f = DR_{rv}^{-1}D$$

This D is defined and obtained as the diagonal matrix formed by the reciprocals of the square roots of the diagonal values in R_{rv}^{-1}. Its purpose is to restore the diagonal values in R_f to the unities existing in R_{rv} before it was inverted: for correlations of unit length factors with themselves must be unity. Closer scrutiny of these relations will now show that this D is

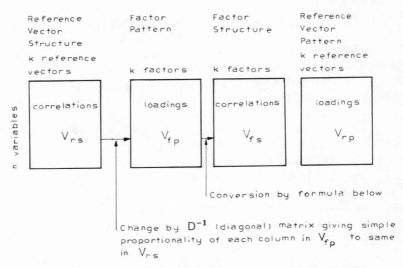

Fig. 8.2. The four interrelated reference vector and factor matrices.

the same (but inverted) as is applied to the transformation of columns of V_{rs} to get V_{fp}. The algebraic demonstration of this identity of D values will be obvious from equations below (such as that $R_v = V_{rs}R_{rv}^{-1}V_{rs}^T = V_{fp}R_fV_{fp}^T$) and can be looked up in Harman (1976). The application of D to getting the V_{fp} from the given V_{rs}, however, is as the inverse D^{-1} so that the columns of V_{rs} are "stretched" giving proportional but usually larger values in the V_{fp},

$$V_{fp} = V_{rs}D^{-1} \tag{8.3}$$

The derivation of the factor structure matrix V_{fs} (correlations with the factors) from V_{fp} is next obtained by the calculation

$$V_{fs} = V_{fp}R_f \tag{8.4}$$

or from the reference vector matrix directly

$$V_{fs} = V_{rs}D^{-1}R_f \tag{8.5}$$

The results deriving from (8.3) require, granting that starting point, only the inversion of the small $(k \times k) R_{rv}$ matrix (not the much larger R_v, as in some later calculations) and can quickly be performed for the psychologist on any computer.

8.9. Which VD Matrix Should Direct the Simple Structure Search?

Since in the writer's laboratory experience there has been a constant demand for working out transitions among the four main VD matrices, as well as those for the V_{fe} (factor score estimation matrix; see Chapter 11) which is also often in demand, it is evident that psychologists should be very clear about the different natures and functions of these variable-dimension matrices. The V_{re} matrix (reference vector "score" estimation) is practically never used but is included in Table 8.5 for symmetry and completeness. The purpose of the Table 8.5 is to permit calculation of each V from a variety of possible starting points and resources.

Four matrices relating variables to dimensions have thus been defined above. As a class—along with some other additions to the category— they may briefly be called VD (*variable-dimension relation*) matrices. Diverse uses exist for these VD matrices, e.g., the V_{fp} is the basis for interpretation, and for estimating scores on variables; the V_{fs} (*factor structure*) is the basis for getting to factor scores via the V_{fe} matrix and so on. These uses will be pursued in due course, but in this chapter on rotation we must keep to the question of which of them best directs pursuit of simple structure per se.

The crucial question is "should we look for simple structure (maxi-

Table 8.5. Summary of Mutual Derivations Among Important Factor Matrices[a]

Matrices derived	Matrices as Sources of Derivation					
	R_v	V_0	V_{rs}	V_{fs}	V_{fp}	V_{fe}
R_v		$V_0 V_0^T$	$V_{rs} R_r^{-1} V_{rs}^T$	$V_{fs} R_f^{-1} V_{fs}^T$	$V_{fp} R_f V_{fp}^T$	$V_{fs} V_{fe}^{-1}$ (gives $\underline{R}_v$)
V_0	$R_v V_{rs}^{-1} L_r$		$V_{rs} L_r^{-1}$	$V_{fs} R_f^{-1} DL_r^{-1}$	$V_{fp} DL_r^{-1}$	$R_v V_{fe} R_f^{-1} L_r^{-1}$
V_{rs}	$R_v (V_{fp}^T)^{-1} R_f^{-1} D$	$V_0 L_r$		$V_{fs} R_f^{-1} D$	$V_{fp} D$	$\underline{R}_v V_{fe} R_f^{-1} D$
V_{fs}	$R_v (V_{fp}^T)^{-1}$	$V_0 L_r D^{-1} R_f$	$V_{rs} D^{-1} R_f$		$V_{fp} R_f$	$\underline{R}_v V_{fe}$
V_{fp}	$R_v (V_{fs}^{-1})^T$	$V_0 L_r D^{-1}$	$V_{rs} D^{-1}$	$V_{fs} R_f^{-1}$		$\underline{R}_v V_{fe} R_f^{-1}$
V_{fe}	$\underline{R}_v^{-1} V_{fs}$	$\underline{R}_v^{-1} V_0 (L_r^T)^{-1} D$	$\underline{R}_v^{-1} V_{rs} D^{-1} Rf$	$V_{fs} R_f^{-1}$	$R_v^{-1} V_{fp} R_f$	

[a] The notation here is that adopted throughout the book, namely: R_v = reduced, $\underline{R}_v$ = unreduced correlation matrix of variables; V_0 = unrotated factor matrix; L_r is the transformation from V_0 to the reference vectors; V_{rs} is the reference vector structure; V_{fs} is the factor structure; V_{fp} is the factor pattern; and V_{fe} is the factor estimation matrix.

The D matrix is defined by $DR_r^{-1} D = R_f$, where R_r is the reference vector correlations, and R_f the factor correlations, where the converse derivation is $R_r = DR_f^{-1} D$. A transformation from the V_0 to the V_{fp} parallel to that from V_0 to V can be written such that $V_{fp} = V_0 L_f$. The correlation among factors in L_f's then $R_f = (L_f^T L_f)^{-1}$, and the relation of the factor transformation to the reference vector transformation is that $L_f = L_r D^{-1}$. Note that as stated in the text, some books used D^{-1} for what we have here called D (because in fact it is first calculated as a D, from $R_f = DR_r^{-1} D$).

In the above columns there would, of course, be other equivalent formulas for deriving the values on the left from those above. For example, $V_{rs} = V_0 L_r$ could, from the last paragraph, be $V_{rs} = V_0 L_f D$. So the stipulated source is usually combined with whatever usable matrices are commonly most conveniently at hand.

It will be noted that some formulas above call for the inversion of matrices that are not square. This can only be done by using the generalized inverse. For example, in row one, $V_{fe}^{-1} = (V_{fe}^T V_{fe})^{-1} V_{fe}^T$, the expression in parentheses being a square matrix. A lack of uniqueness in the inverse sometimes causes difficulties, but here we are concerned to show the essential relationships of major matrices, rather than calculation methods.

mum count of zeros in columns) in the V_{rs} (on which we initially rotate) the V_{fp} or the V_{fs}?" The first two are not alternatives: the zeros in the V_{rs} and the V_{fp} will fall in the same places, because one is the same as the other multplied only by the D values. (There arises only the minor issue between them of the differing widths of the standard error of those "zero" hyperplanes.) The real issue is between the V_{fp} (or V_{rs}) and the V_{fs}—between zero loadings and zero correlations as the mark of simple structure. Since the loadings represent the influence of a factor determiner purely on its own action on the variables (other factors being partialled out of the V_{fs} correlations) and our model supposes the factor to be a determiner, then the zeros which mark the hyperplane, indicating a complete lack of relevance of influence, should logically be made to appear in the loadings, i.e., in the V_{fp} and V_{rs} matrices. For the loadings indicate what a factor does *in and of itself* when other factors are partialled out; whereas the observed correlations of factor and variable are partly due to other factors correlating with the given factor and "supporting" it. [For instance, if intelligence correlates .6 with school achievement this may be partly due to the fact that superego strength, which also helps school achievement (Cattell & Butcher, 1968), correlates positively (though quite moderately) with intelligence. The real effect of the latter, i.e., its loading, or beta weight might therefore turn out only to be .4, and its contribution to variance in achievement (.6 X .4 = .24).]

If simple structure should be in the V_{fp} then it resides also in the entirely parallel V_{rs}, and we have been right in the standard practice of seeking it in rotational operations in the V_{rs}. The only problem which now arises regarding the general practice of seeking simple structure in the V_{rs} is that due to the D multiplication ($V_{fp} = V_{rs}D^{-1}$ or $V_{rs} = V_{fp}D$) the width of the hyperplane will not be the same in both. If the stretching value in the D^{-1} matrix is for factor X equal to 1.5 then in using a ±.10 width for the hyperplane in the V_{rs} we have actually been lenient—or careless—in factor terms by accepting everything with ±.15 on that factor in the V_{fp}. Thus if, as seems at present reasonable as a rough rule (though Harris's formula gives a different standard deviation of a zero loading for each factor), that a uniform ±.10 should be adopted for all factors then it might seem we should do rotations in the V_{fp} rather than the V_{rs}. For equal hyperplanes in the former would not be equal in the latter, as we now treat them. The issue needs more discussion than it can be given here, for some tentative formulas for the standard error of a loading introduce the correlations among factors which would complicate the calculation further. Furthermore, if one is courting perfection regarding hyperplane widths it would seem that one should go to real base factor analysis (p. 406) which involves factoring covariances. Meanwhile, for the present, the arguments for keeping in the V_{rs} plots are practical rather than theoretical but cumulative, namely, that R_{rv} (the correlations among RVs) would have to be inverted and carried into the factor pattern calculation at every plot; that oblique plotting of factor projections is difficult on most machines and that our visual habits seem better to fit dealing with reference vectors drawn perpendicular to hyperplanes.

The end of the rotation process, however, is the factor pattern V_{fp} matrix and this we shall use, to compare, and interpret in discussing the final inferences from a given factor analysis. Furthermore, the ultimate test of significance of simple structure discussed above should finally be carried out in this V_{fp}. At this completion of rotation, and essentially of the factor analysis, the thorough investigator will wish to check that his V_{fp} translates back correctly into the R_v values—for there has been a long series of steps since he left it. The formula for restoring the correlation matrix from the V_{fp} is

$$R_v = V_{fp} R_f V_{fp}^T \tag{8.6}$$

Comparing this with the orthogonal case where $V_x = V_{rs} = V_{fp}$ (where V_x is any orthogonal rotation of V_0) we see that the orthogonal equivalent is

$$R_v = V_0 V_0^T \tag{8.7}$$

i.e., the difference of Eq. (8.6) and Eq. (8.7) is the omission of correlation among factors R_f. Actually, for mathematical parallelism one can put an R_f in Eq. (8.6) but it will be an *identity* matrix, of unities in the diagonal, and so be a superfluity, without effect. The R_v thus obtained by Eq. (8.6) or (8.7) is a *reduced* matrix, i.e., with communalities not unities in its diagonal.

8.10. An Overview of SS Alongside Other Rotational Principles

Since this is the third and last chapter on the rotation problem it is appropriate, even though its main concern is simple structure, to remind the reader of the full armamentarium available to him and of the particular utilities of each method.

In hypothesis-seeking factor analytic research there is only one rival to simple structure, namely, *confactor rotation*, though PROCRUSTES and confirmatory maximum likelihood are available for hypothesis testing use. At present, because of unsolved difficulties in adapting to the oblique case, confactor is little used as a practical method, though sounder in its basic principle. It remains to be seen also what magnitude of differences in factor variance is required between the two samples essential to the design, and how frequently such differences can be found in practice. A trial of confactor on oblique plasmodes and real data has recently been made by Cattell and Brennan (1977) showing that a solution is regularly obtainable when the factor correlations are the same in the two experiments. (See Appendix 3).

A third possible objective in rotation is to bring two studies with the same variables into maximum congruence of their factors. This has been worked out by Tucker (1951). The objective there, however, is not the same as in confactor rotation, for one could ask for maximum congruence in the sense of achieving a least-squares fit of loadings on the two studies without the goal of maximum proportionality of loadings. In fact, Tucker views the procedure as one simply to bring two studies into maximum congruence so that they can be further rotated in lock step together, to some chosen target, e.g., simple structures. The resemblance to confirmatory maximum likelihood analyses (p. 400) should be noted. The equations for this "factor synthesis" of two studies are given by Tucker (1951) and in an independent approach, applicable to several studies, by Meredith (1964 a, b). Nesselroade has programmed the latter.

A further principle of rotation, described by Thurstone (1947), but to which we have given little space because it is more of applied than basic research interest, is that of rotating so that in n variables each of the k factors aligns itself exactly with each of a chosen k of the n variables— chosen, of course, for some special significance. Thus, in a k factor V_{rs} matrix, k variable rows are chosen, usually with a check back into the R_v matrix (correlations among variables) to see that they are tolerably orthogonal to each other and not likely to produce a singular matrix. The values in each row of the V_{rs} matrix for a chosen variable are then divided by the root of the variable's communality (h from h^2) so that the new values when squared will sum to unity, and those "inflated" rows are erected as columns in an L matrix. Multiplication of V_0 by this L will put factors collinear with the chosen vectors. Actually the vector positions can either be those of concrete tests chosen for theoretical significance or some "artificial" composite vector positions thought up by the investigator in relation to tests in the given V_0 and expressed by him as a new numerical row in the V_{rs} and a column for the L matrix.

This approaches an hypothesis-testing use of factor analysis but the pure use of an hypothesis-testing approach is more completely achieved by the PROCRUSTES program or the confirmatory maximum likelihood method. The latter bring the result of the rotation, as close as possible to a set-out hypothesis—a "target V_{rs}." We have suggested in this book that PROCRUSTES be used to test both the fit of the factor patterns *and* of the factor correlations to a prescribed hypothesis. This might be done by expressing the target in a Schmid–Leiman matrix (see p. 213, below), wherein higher-order factors like lower (primaries) can also have their correlations with the variables set out, since the higher orders define the primary factor correlations.

There are also hybrids of these five methods, and also specialized developments, e.g., canonical analyses maximally predicting one set of variables as a whole from another set, which may be read in Harman (1976), Horst (1965), and Van de Geer (1971). The reference immediately above to the greater soundness of the confactor criterion, in principle, to simple structure, rests on the undue dependence of the latter upon a suitable sample of variables and on proven estimate of hyperplane widths, etc. Sampling of variables in an exploratory factoring is guesswork, and only after several coordinated studies can one achieve an efficient design. This and the technical problems above imply that disparities of factor results are likely to be encountered except where there are experienced, skilled, and thorough use of simple structure criteria.

8.11. Summary

1. In successive shifts toward simple structure whether by ROTOPLOT or some automatic program design using successive shifts, certain tactical rules arise from experience, such as (i) preserving approximate orthogonality in the early stages when large shifts are more frequent, (ii) directing the hyperplane at first by trying to get in far points (remote from the origin), which loses few points around the origin. Later one allows the narrow, definitely forming total hyperplane, if necessary, to lose such far points (since high h^2 is no guarantee of factor purity), (iii) "under-shifting," thus making allowance for total shifts to be combined, and (iv) in the last stages allowing precision of simple structure to take precedence over moderate, "undisturbing" angles (correlations) among factors.

2. Since at all stages of the process and at the final resolution it is necessary to keep a check on correlations among reference vectors and factors, the rationale and the formulas for calculation of R_{rv} and R_f are finally given here.

3. In describing these calculations opportunity has been taken to define the transpose and the inverse of a matrix. It is pointed out that normalizing the columns of the transformation matrix brings about unit length reference vectors, and that after calculating R_{rv}^{-1} to get factor correlations the columns must be renormalized, by placing a D matrix before and after the R_{rv}^{-1} matrix, multiplying to get unit length factors.

4. Problems in rotating studies with many factors—say nearer 35 than 15—include (a) the loss of factors in the rotation process, through mutual collapse or loss of significant variance, and (b) the appearance in the obtained factor correlation matrix of degenerative properties introduced in elementary description here as non-Gramian properties and singularity. In general, the inversion of an R_{rv} with moderate intercor-

relations tends to produce much higher correlations in the resulting R_f, and when these get very high the matrix can easily become singular and uninvertible, lacking positive definite roots.

5. The most common cause of high r among factors or complete singularity is the rotating of one reference vector close to, or onto, the plane formed by two or more others. One RV (or factor) can thus become expressible as a linear combination of two (or more) others, as seen in either the L or the V_{rs} matrix. This superfluity is commonly due to having lost the hyperplane for that factor, but possibly to the space (number of factors chosen) proving to be in excess of the true number of hyperplanes. The reduction of values in the R_{rv} (by going more orthogonal) in order to reduce values in R_f is a reasonable direction of movement; but experience shows that the guaranteed solution is prolonged rotation to a really good simple structure which, whatever the angles, is compatible with inversion to a meaningful R_{fv}.

6. The last statement assumes there has not in fact been a mistake in the number of factors taken out originally. But if there has been, we find that rotation offers a second "court of appeal" (checking the scree and other initial tests) regarding the number of factors, and it should be used whenever a sound decision on number of factors vitally affects theory. For, (i) if one or two extra factors are taken beyond the true figure, they will rotate out either (a) by collapsing angularly or (b) by proving to be factors with quite negligible variance when in their correct rotated positions. When excess space is demonstrated by either case the last factor or factors must be struck off the V_0 before completing rotation. And (ii) some four indices derivable from rotation properties together give reliable evidence not only on overfactoring, as above, but also on underfactoring. They are (a) appearance of a wide hyperplane, in underfactoring, dropping to a narrow band when more are extracted, (b) the appearance, in overshooting the true number, of degenerative factor fission as shown by two factors appearing on a hyperplane previously belonging to one and now blurred and extended, (c) a comparatively sudden increase in R_f values at the point where *factor fission* occurs in overfactoring, and (d) achievement of a maximum value, at least over ranges near the true number, of a fraction–sigma of the loading/communality—for loadings in the factor column when the SS position is reached.

7. The theory is stated here that the *number of hyperplanes existing in the hyperspace* (when geometers' hyperplanes are set aside) *is the same as the true number of factors*. Geometers' hyperplanes are defined as hyperplanes of $(k-1)$ variables among n variables in k space. They are exact hyperplanes, i.e., have all variables at zero, but may include $> k$ variables and then blur slightly. They can be very numerous, their

number being that of the combinations of $(k - 1)$ in n variables. Alleged instances of there being more hyperplanes than k can thus be explained either by broad geometers' hyperplanes (enlarged by chance overlap) or reticular factor model (see below) effects and higher-order factor hyperplanes intruding into the primary space. However, good design permits separation of true from geometers' hyperplanes.

8. It is possible that some more refined approaches to simple structure could be made from considering the hyperplanes in the plots of *factor loadings* rather than in the reference vector system correlations (i.e., correlations of variables with RVs) since it is evident that simple structure inherently belongs in the factor pattern matrix. In discussing this and related questions it became appropriate to give an explicit, formal treatment of the differences between factor and reference vector matrices, and of loadings and correlations, the nature of the former as beta weights being brought out. The equations for transformations among V_{rs}, V_{fp}, V_{rp}, and V_{fs} are set out. Although the loadings in the factor pattern V_{fp} are simply proportional (by a D matrix) to the reference vector correlations in V_{rs}, the rotational calculations and shifts on drawings are more complicated and less practical in the V_{fp} than the customary V_{rs}, and retention of the latter is advocated as most practicable except in refined studies involving more expense.

9. After the simple structure maximization has been reached (by whatever process), it is important to test the *statistical significance* of that structure. ["Statistical" here can apply to sampling variables (n) (as in Bargmann's test) or people (n) too, as in the Cattell–Vaughan Monte Carlo tables.] A valuable introduction to this problem is given also by Mosier (1927). By the Bargmann standards, an appreciable fraction of findings among existing published (alleged) simple structure resolutions would be canceled. But the more recent Monte Carlo construction of significance tables in process by Cattell and Vaughan promises a more lenient evaluation. Nevertheless, on whichever postulates one leans for evaluation, it is evident that researchers are going to have to search more thoroughly for maximum simple structure than has often been customary in the past if objective levels of conclusion are to be claimed and cross-research matching achieved.

10. The next steps after reaching a significant simple structure are (i) a brief check that the result conforms with the original R_v and (ii) examination of agreement of factor pattern with that from other researches (sometimes called the "invariance" problem). The latter is deferred to more extensive consideration later; the former is checked by the formula $R_v = V_{fp} R_f V_{fp}^T$, recognizing that (a) $R_f = I$ (an "identity matrix") in the special orthogonal case, and (b) R_v is a reduced matrix, with communalities in the diagonal.

11. In retrospect, on concluding the third and last chapter on rotational issues, one should remind the reader that although ROTOPLOT and automatic rotations to simple structure deserve every bit of the detailed discussion of principles and tactics here given, yet two or three other approaches discussed earlier now call for attention and development. These are (a) in exploratory factor analysis, confactor and (b) in hypothesis-testing analysis, (i) PROCRUSTES, (ii) confirmatory maximum likelihood, (iii) Tucker's synthesis method converging two studies, i.e., rotation for simple mutual matching, followed by simple structure rotation, and (iv) rotation to put factors collinear with particular variables.

CHAPTER 9

Higher-Order Factors: Models and Formulas

9.1. The Need to Recognize the Existence of Higher-Order Factor Influences

The researcher has now achieved a unique solution as regards primary factors. He possesses a factor pattern matrix of oblique factors at a position of demonstrated maximum simple structure, or confactor resolution, and hopefully of sufficient statistical significance in loadings, hyperplanes, etc. He should preserve in file, as the hallmarks of his solutions, the R_v matrix, the V_0 (with a column of communalities), the L matrix, which transformed from the V_0 to the V_{rs}, and the correlations among the oblique factors R_f. For all of these may be needed; for further steps or for examination later, by others or by himself, of the adequacy of his solution; for searches for the possibility of alternative solutions; and for theory checking and development in the next stage of experiment.

At this point two main further possible directions of investigation—both normally undertaken—open up: (1) to enquire into *higher-order* factor structures, and (2) to interpret the factors, aided perhaps by use of the factor scores for individuals in new types of experiment and criterion relation. Since what we learn about interpretation will have to apply both to first- and higher-order factors this second development is best postponed (Chapter 10) until higher-order factors themselves have been described.

The nature of influences in the real world being what they are—a chain of influences acting upon influences—we should expect the primary factors themselves to be contributed to by further "higher-order" factors. This is one more argument against orthogonal factor resolutions—as in VARIMAX, for example—for such a procedure completely locks the door which leads to any higher-order relations. However, it should be clear that the most potent objection to orthogonal primary solutions is not this enforced ignorance of possible higher-order influences on primary factors. It is simply the empirical fact that the best simple structure, and the ultimate tidying of a confactor resolution, almost invariably reveal primaries standing at positions where they have some real and consistent mutual correlation.

9.2. Extracting Factors from Rotated Primaries and from Correlations of Primary Scales and Batteries

The practical procedures and the governing principles in going to second-order factors from primaries are precisely the same as in going from variables to primaries. The only difference is that we now have two possible correlation matrices from which to start: (1) the correlations obtained from the *simple structure resolution* of primaries, and (2) the correlations *among batteries or scales* built up to measure the primaries. In either case the psychologist has a correlation matrix among factors R_f and puts it through the usual processes: a principal components, with unities in the diagonal to determine, by scree, etc., the number of factors; an iterative factoring to get conforming communalities; unique rotation to an SS, confactor, or PROCRUSTES position, and so on. The factors one gets from factoring primary factors are called second-order factors, or, briefly, *secondaries*, and one proceeds to the same collection of defining matrices for them as in a factoring of primaries. It is convenient to write a Roman II as a subscript to these; thus V_{oII}, V_{fpII}, L_{II}, R_{fII} are the principal statements of findings on secondaries. Incidentally, and if we are very formally setting out relations between the two orders, we would for completeness and orderliness write a I subscript to the primaries—V_{oI}, L_I, V_{fpI}, R_{fI}—though generally the I subscript is best left "understood" and omitted in first-order equations.

From the R_{fII} matrix one can factor again and get third-order factors, or *tertiaries*, by repeating the usual steps. Obviously, in principle, this can go on indefinitely, but third-order results are uncommon in research literature and fourth orders are positively rare. Tertiaries have been replicated in the personality field both in questionnaire (Q data) (Cattell, 1973) and in objective test (T data), (Cattell & Schuerger, 1977) as well as in the ability field (see Knapp, 1961; Cattell & Scheier, 1961; Pawlik & Cattell, 1964; Cattell, 1971a) and fourth orders in questionnaire data (Cattell, 1973). In the case of abilities, the first orders are the well-known primaries (spatial, verbal, numerical, etc., abilities), while the secondaries are fluid and crystallized intelligence, fluency, speed, etc. The general model and sequence of calculations is shown in Fig. 9.1.

To get stability of results at higher orders requires decidedly more skill, patience, and precision than are needed in most researches stopping at the primary level. The reason is, as already noted, that to determine the primary cosine angles correctly the simple structure pursuit needs to be carried to higher accuracy—possibly to as many as 20 overall ROTOPLOT rotations. Since scales and batteries are often constructed for the primaries, e.g., the 16 P.F. and HSOA in personality and Thurstone's, Horn's, and Hakstian and Cattell's (1977) batteries in primary abilities, an alternative correlation matrix R_f for second-order factors is available from correlating scores on such scales as the basis for factoring. What is actually a second-order study can thus take off from correlations among directly scored experimental variables provided these variables are *scales or batteries measuring primary factors* really well. And, though batteries for second-order factors are less common, it is sometimes possible similarly to attack third-order structure from second-order factor batteries and their correlations in this same way.

Whenever scales or batteries are thus available for lower-order factors, the useful possibility exists of using this approach as a check on the above-described factoring which proceeds from the cosines reached at the rotation positions. The kind of agreement one may expect is shown for two studies (one on men, the other on women) in Table 9.1. It will be seen that smaller correlations may flip signs, but the more significant stand firm in their similarities. For instance, the major expected correlations for the second-order, e.g., those from the exvia second-order defined by correlations corresponding to positive loadings on A, F, and H, and

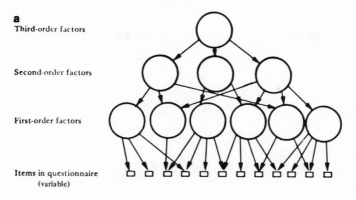

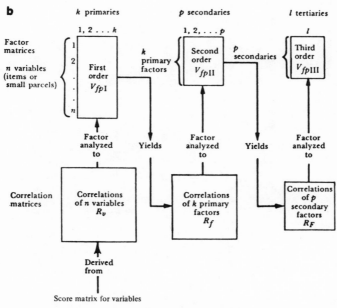

Fig. 9.1. Model and mode of calculation of higher-order factors. (a) Model of action of influences at different strata; (b) mode of derivation of factors. (From R. B. Cattell, *Personality and Mood by Questionnaire*, Jossey-Bass, San Francisco, 1973, p. 105.)

negative with Q_2, can be seen here to be significant and essentially the same, but the small correlations, e.g., A and B, may flip sign. However, if we go beyond inspecting R_{fI} correlations and look at the second-order structure (see Handbook of 16 P.F.) (Cattell, Eber, & Tatsuoka, 1971) we see that it is closely similar for the two sources. Incidentally, it should be said in passing that fairly noticeable differences can sometimes arise be-

Table 9.1. Agreement of Factor Correlations from Rotations and from Concrete Scales[a]

(a) Correlations for male sample

Pure factors: correlations among the rotated source traits of the 16 P.F. as pure factors[b,c]

Source trait	A (1)	B (2)	C (3)	E (4)	F (5)	G (6)	H (7)	I (8)	L (9)	M (10)	N (11)	O (12)	Q1 (13)	Q2 (14)	Q3 (15)	Q4 (16)
A		-11	15	-06	35	26	26	14	-05	14	07	-06	-15	-45	12	12
B			27	37	16	-11	19	02	00	28	-34	-16	27	13	-01	09
C				-07	07	16	38	05	-58	44	-08	-69	-07	-05	28	-27
E					47	-42	48	22	37	18	-65	14	52	07	-38	53
F						12	65	-05	02	-38	-10	10	-60	00	47	53
G							12	-33	-38	-12	38	-29	-56	-63	77	-13
H								08	-16	30	-37	-40	03	-35	03	25
I									03	44	-30	17	13	23	-35	-10
L										-18	-16	58	40	26	-38	49
M											-39	-29	01	05	-06	-15
N												08	-50	-10	34	-26
O													12	18	-40	27
Q1														37	-46	39
Q2															-32	-24
Q3																-29
Q4																

Concrete scales: correlations among the measured source traits of the 16 P.F. as scale scores[b,c]

Source trait	A (1)	B (2)	C (3)	E (4)	F (5)	G (6)	H (7)	I (8)	L (9)	M (10)	N (11)	O (12)	Q1 (13)	Q2 (14)	Q3 (15)	Q4 (16)
A		-01	09	24	39	09	44	24	06	04	-04	-07	10	-43	02	01
B			07	14	09	-09	00	04	02	21	-08	-04	19	08	-03	03
C				-02	09	20	31	-10	-51	06	05	-69	03	-14	50	-69
E					53	-27	54	-01	38	40	-40	-09	48	-08	-23	14
F						-19	66	16	13	13	-37	-12	25	-52	-18	11
G							00	-03	-22	-27	26	-15	-31	-13	56	-17
H								00	-04	18	-25	-37	23	-45	10	-16
I									01	00	39	-03	19	11	08	-18
L										07	-16	40	21	11	-31	55
M											-26	-08	46	16	-24	-03
N												-03	-25	10	25	-16
O													-11	15	-48	75
Q1														13	-15	-02
Q2															-04	03
Q3																-50
Q4																

(b) Correlations for female sample

Source trait	A	B	C	E	F	G	H	I	L	M	N	O	Q1	Q2	Q3	Q4
A		07	-03	06	33	32	29	34	-18	-23	-03	02	10	-47	07	-03
B	-05		-22	07	-07	17	-15	30	10	13	-03	35	06	19	-15	17
C	05	-03		-27	11	07	34	-03	-42	-09	39	-56	-19	-22	60	-50
E	15	03	-09		23	-41	14	00	32	55	-39	-14	65	34	-27	36
F	36	01	08	48		-11	66	-18	03	-06	-37	-31	20	-27	-15	20
G	11	-07	18	-26	-18		03	29	-34	-18	25	24	-14	-35	23	-03
H	32	-04	37	44	57	04		-05	-14	04	-28	-54	18	-41	22	-04
I	12	00	-07	-18	-13	09	-06		-30	18	16	25	-10	-05	10	-09
L	04	-03	-44	45	24	-20	02	-13		14	-43	19	14	33	-42	49
M	-11	11	02	33	03	-30	08	16	04		-32	-15	43	46	-11	19
N	03	00	12	-53	-34	24	-24	15	-27	-30		01	-24	-03	27	-26
O	-12	04	-70	-08	-15	-11	-49	09	31	-08	-07		-16	10	-45	26
Q1	-04	11	-09	46	15	-25	12	-08	19	41	-32	-08		26	-15	37
Q2	-44	04	-17	-02	-38	-21	-40	-02	06	25	01	16	19		-24	38
Q3	02	-11	49	-24	-25	55	15	-01	-32	-21	27	-44	-18	-10		-57
Q4	05	10	-71	20	12	-25	-25	07	50	-20	67	06	06	06	-59	

[a] Decimal points have been omitted.

[b] Based on data from 423 males and 535 females, Forms A + B, 1968–69 edition of the 16 P.F. All subjects were college students. The original data for this analysis were taken from Russell (1969).

[c] Correlation matrices (and all subsequent analyses) were obtained separately for males and females. Data for males are presented above the main diagonal, data for females, below the main diagonal.

tween one R_v and another where the V_{fp}s are actually highly similar, the difference being due to change in variance "size" of one second-order factor relative to another without any real change in pattern.

Each of these two approaches has its advantages and disadvantages. The *pure factor* primaries (as we may call those unit-length primaries from rotation) have only the disadvantage that simple structure can never give as exact a rotation as one would wish, though with, say, more than 80 variables and 10 factors, and a hyperplane narrowed to ±.05 by using about 600 subjects, a patient and skillful rotator can probably define angles to within ±3 degrees. The correlating of scales, on the other hand, suffers from two disadvantages: (1) that the scales may not be pure, and thus add to their correlations with others through shared impurities, and (2) that they are not fully reliable or valid having random error and specifics contributing to their scores. The latter will work systematically to underestimate correlations relative to the pure factor method and to yield lower communalities, therefore, for the second-order factors. An estimate, by correcting correlations for attenuation, can be made that will wipe out (2), i.e., underestimation as such, but final results from the two approaches will remain different through the difference between the nature of (1) and (2) and the sources of random error, namely, imperfect rotation in (1), and contaminated scales in the latter. If one wants to do only a single factoring, the mean of the two essentially congruent R_{fI} matrices (if n and p are similar; otherwise the mean of the covariance matrices) is probably the best beginning. However, most that is discussed here applies most consistently to the first basis—primary factors defined as rotated factors—and the reader must recognize the few situations where this does not apply to primary scale or battery factoring. As a practical matter it should be noted that the simple structure rotation where there are only two markers (the bare minimum) for each primary will sometimes yield second-orders, going directly to them as an alternative structure.

9.3. Factor Interrelations: The Possibilities of Strata, Reticular, and Other Scientific Models

Because of the fact that communalities cannot be uniquely estimated (which is different from accurately known for the *population*) where the number of factors exceeds half the number of variables used, most experimental designs will be planned to yield decidedly fewer factors than variables. If we keep to the notation n = number of variables, k = number of primaries, p = number of secondaries, and t = number of tertiaries, then we can state that $n > k$; $k > p$; and $p > t$ (where $>$ could in the best designs be translated "greater than twice"). That is to say, there will inevitably be a pyramidal hierarchy as we go up the orders, which may well end in a single factor. It seems to the present writer that several personality and ability theorists have allowed themselves to be charmed into near mysticism by this spurious hierarchy. (One theorist speaks with theological respect of the top factor—in private at least—as "Mr. G!") Two principles need to be observed as a corrective here.

1. The fact that a factor is of a higher order does not necessarily mean that as a concept or influence it is more important. Statistically, at least, as we shall see later, systematically better predictions (estimates) of criteria (and variables generally) can be made from primaries alone than from secondaries alone, and so on. In the ability domain this means that rather better predictions of particular school performances can be made from primary abilities than from general intelligence (Cattell & Butcher, 1968), and in the personality realm from 16 such factors as affectia, surgency, ego strength, self-sentiment strength, etc., than from 8 second orders such as exvia, anxiety, cortertia, independence, etc.

2. That if one starts from a base of n_1 variables the factor at the top of that hierarchy of higher orders will not usually be the same as that at the top of the hierarchy from starting with a larger but overlapping set of n_2 variables. Indeed, neither of the apex factors will be an apex factor if one puts n_1 and n_2 together, as shown in Fig. 9.2.

But this is not all we need to be wary about when approaching concepts of factor relations from the familiar operations in factor order calculations. If the factors are determiners then we should loosen up our minds to the possibility that factors will have *many other forms of interrelation than that of strata* and certainly than that of a hierarchy cut out of strata.

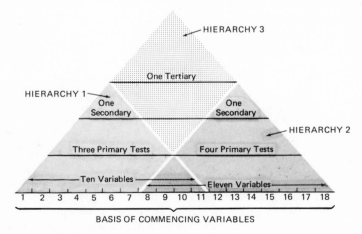

Fig. 9.2. The relativity of hierarchies to the base of chosen variables: strata definite, pyramids uncertain.

Quite a number of terms have been developed by psychologists for particular factor interaction models by which they prefer to try to explain relations. Over the years we have Spearman's *two-factor* (general and specific only) "monarchical" (1927) design; Guttman's (1954) *simplex* and *circumplex*; Holzinger's bifactor; Foa's (1965) *facet* designs; and the present writer's *cooperative factors* (Cattell, 1952b), *stratoplex, feedback*, and *spiral action* models (Cattell, 1973); and so forth.

It should be noted that most of these models require definition in terms of (a) the nature of interactions among the factors themselves and (b) the patterns of their common action on a set of variables. For example, the concept of cooperative factors supposes that the factors will be independent, but that there will be a peculiar similarity in their loading patterns on variables. An instance would be the innervations from the sympathetic and parasympathetic nervous systems on several bodily functions. For, although the signs of their loadings are mainly opposite, the physiological targets—among a random list of physiological expressions on which influences could operate—on which they have marked loadings, are very similar, and those they leave unaffected are very similar.

However, the more important relations, and those with which we are concerned in the present primary, secondary, etc., discussion, are among factors, and here the ultimate extremes of possibility are between (1) orderly strata, as discussed, with one-way action and (2) completely free interaction in any kind of network, which we shall call the *general reticular model*. In that model there can be direct action among factors along with positive and negative feedback in all directions. The model (as number V) is shown in Fig. 9.3, which sets out to represent adequately, despite mini-

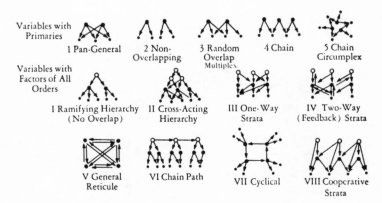

Fig. 9.3. Principal possible models for operations of factors as influences. (a) Patterns of factors on variables; (b) patterns of mutual interaction of factors. (From R. B. Cattell, *Handbook of Multivariate Experimental Psychology*, Rand McNally, Chicago, 1966, p. 214.)

mum of space for discussion, the principal models one may need to consider. Those numbered (1) through (3) are concerned with special patterns of loadings on variables. Those with Roman numerals I through VIII are concerned with mutual interrelations of factors themselves.

There is at present no grand method of procedure—no single system of analysis for testing experimental data relations—by which we can immediately tell which of the above models best represents what is happening in the given data. The researcher has to proceed by the usual trial-and-error application of hunches until he finds out which best fits. Certainly there is quite a gap—a fascinating gap into which factor analysts will hopefully soon enter—between factor analysis as now practiced and represented and the ultimate complexity of a method joined to a computer program that would be capable of handling the general reticular model. In Chapter 1, where experimental design was discussed, it was pointed out that factor analysis can be combined with manipulation, and it is in this direction that we must probably move for discovering feedback patterns of the reticular kind. Except for some excursions in this direction in Chapter 13, however, the scope of the present book asks us to concentrate on the presently practicable *stratum* model.

The commonly assumed model in taking out higher-order factors is the multiplex [number 3 in Fig. 9.3(a)], in which the loadings of the higher order on the lower order are free to assume any pattern of overlap (unlike for example, the pan-general, number 1, where every variable is loaded by every factor, or every first order by every second order) or the Holzinger bifactor type of nonoverlapping loadings, number 2. With an additional stratum this most common random overlap multiplex becomes the stratoplex, number III. Contrasted with the pan-general and stratoplex models are a variety of special condition models unlikely to fit most data. A general model that presents the greatest contrast to the general stratoplex model is the general reticule model, V. Therein, as illustrated here by four factors, causal action—independent contribution to the variance—can run in all directions. As yet no analytic method can directly discover such a structure, though path coefficient analysis, discussed below, approaches the problem. At best a series of manipulative multivariate coordinate experiments is required.

An effective application of even the plain stratum model is not easy, but since it seems to fit well a lot of psychological analyses and can be positively handled, this chapter will be largely concerned with it. To begin with we must make a distinction between *orders* and *strata*. Let us suppose that in factoring scales for *primary* factors an investigator unwittingly includes two scales each of which is really already measuring a second-stratum factor and which shares with others a third order. Then alongside the second-stratum factors which emerge there will be likely to emerge one third-stratum factor—if these two second-strata scales show a third order. The apparent second-order set will then actually have both second-order and the given third-stratum factor in it. If we define *order* operationally, simply as what emerges in a given sequence of factorings, i.e., factoring "variables," factoring the primaries therefrom, and so on, then such mixtures are sometimes likely to occur. Consequently, we may define *order* as what we operationally reach immediately and *strata* as the layers inherently found in the data corresponding to scientifically required concepts of higher and lower action. (They are correct concepts so long as the strata model actually holds in the given domain.) Consequently, the stratum to which a factor belongs may need to be explored and checked by several different experiments operationally yielding only orders. The reader who needs to analyze some specific case where an apparent discrepancy of order and stratum rank is suspected may find Fig. 9.4 and its associated reading (Banks & Broadhurst, 1965) helpful.

The concept of stratum ultimately takes its authority from the concept of *sampling of variables* as the above example of mixing first-order scales with mere variable items most clearly shows. The definition of first-order factors—those in the primary stratum—can be given only by defining a population of what are truly variables. And the stratum after that—the secondaries—are defined as subtending in their loadings two or more primaries and being derived from a sample of primaries. In the domain of personality the most trusted and tried basis for sampling variables was set up in the *personality sphere* (Cattell, 1946b; Goldberg, 1975; Norman, 1967) the operational basis for which could be all the personality words in the dictionary, but a more sophisticated derivation of which is more fully discussed under sampling in factor analysis (p. 496). Sampling can fail by omitting some area of behavior, in which case the worst that will happen is that gaps will follow in primaries and the higher orders built on them. A well-known instance is Eysenck's development of personality theory on a basis of three second-order factors instead of the eight or nine of Barton and others, or Thurstone's PMA being built on five primaries whereas in the fullness of time Hakstian and Cattell's IPAT battery and Horn's wide searches (1965c) came to cover twenty. The effect of such omissions on higher strata findings can quickly be seen by a look at Fig. 9.2.

A second way in which variable samplings can differ, and affect our concepts of higher strata is with respect to *density of representation*— in fact by overstocking an area rather than omitting it. In terms of the personality sphere, we can imagine the surface covered by about 100 variables—the number commonly found to cover most discoverable primaries in objective test (T data) factorings—or by 1000, which is the number of items in Q data, e.g., by the four forms of the 16 P.F., plus the CAQ (*clinical analysis questionnaire*) or MMPI. (The same total would be

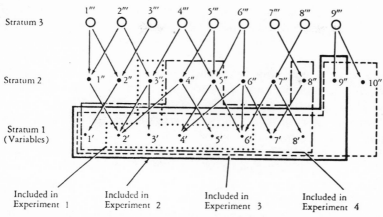

Fig. 9.4. Effects of choice of variables in causing confusions of strata and orders. The figure has set out variables (= •), primary factors (= •), and secondaries (= ○) in what would eventually be recognized as their true strata level (shown also by primes ′, ″, and ‴.) Some common possibilities of false stratum conclusions are shown. In each case the variables included in factoring are encompassed in a "fence." *Case 1.* A primary factor scale properly belonging in stratum 2 (no 3″) is accidentally included with five ordinary variables (2′, 3′, 4′, 5′, and 6′). No problem arises since it has representation on 2 and 3 (see arrows) (light interrupted line) and its second stratum (primary factor) character is recognized by its loading the new factor virtually unity. *Case 2.* A primary factor scale from stratum 2 is again included, but in this case it happens to have no representation in variables. It is missed as a primary and set aside as a specific. *Case 3.* More than one primary factor scales (nos. 9″ and 10″) are included (heavier even interrupted line) with eight stratum 1 variables. The result is a secondary, 9″, appearing as if it were at the primary order. *Case 4.* Second stratum factors (primaries) three in number (4″, 5″ and 8″). Nos. 4 and 5 will appear in true rank as primaries because they have variable representation; no. 8 will be lost as a specific. There is likelihood that a third stratum factor (secondary 5″, loading on 4″ and 5″) will be generated and mistaken for a second stratum (variables taken in unevenly interrupted line). (From R. B. Cattell, *Handbook of Multivariate Experimental Psychology*, Rand McNally, Chicago, 1966, p. 217.)

reached in T data if we went down to the item level.) If one factors, in the Q data case, a set of variables constituted only by every 30th item, present results suggest the density will be so low that the necessary two representatives for each primary may not appear. The first order of factoring might then go directly to the second-stratum factors.

Some embarrassing examples exist in the psychological field of the cost of confusing order and strata. One is the theoretical chaos that has descended on some writers in the personality structure world. In the dense representation of items in the questionnaire medium, now covering over 1000 items of examined loading, the first-order factors turn out not to be alignable with the first orders in objective tests. But, more closely examined by strata observations, the data show that first orders in T data (objective, performance tests) actually correspond to second-stratum factors in the questionnaire medium, and second orders in T data to third-stratum factors in Q data.

The strata levels hold tolerably across both domains of modality—abilities and personality—and possibly the third—that of dynamic traits. The primaries affectia A, surgency F, superego strength G, etc., in the questionnaires seem to be on the same stratum level as the ability primaries spatial ability, numerical ability, reasoning, etc. The secondaries—exvia (primaries A, F, H, and Q_2-), anxiety (primaries C-, H-, L, O, Q_3-, and Q_4), etc., come out aligning with the first orders in T data, which include fluid and crystallized intelligence, fluency, and other second-stratum factors in ability and personality domains.

It is thus incumbent upon an investigator when he proceeds to second and third orders to (a) note at what stratum level of measured data his first R matrix takes off; (b) to have at least a majority of his variables at a uniform stratum level—lest he become entangled later in the uncertain inferences illustrated in Figure (9.4); (c) to make a clear distinction in his report between the orders, reached by his particular procedures and the strata levels at which he believes he has established his factors; and (d) to seek ancillary evidence to factor analysis itself in checking on those levels.

Of course, there is an appreciable possibility that as research advances the structure will be found not to fit exactly the regular *stratoplex* model, but to require the greater freedom and complexity of the *reticular* model, with split-level strata and feedback from below upward. This is a future challenge, but in the fields of psychology and sociology so far pursued the stratoplex model is a promising resolution of the findings and one to be pursued until it breaks down.

9.4. The Theory of Higher-Strata Factors: New Influences or Spiral Feedback Emergents?

The question naturally arises in psychology or any other science as to what we consider second- (or higher-) stratum factors to be. The answer, in most general terms, would naturally be, consistently, that they are of the same nature as primaries, namely, determiners or influences, but acting on lower-strata factors instead of the final variables. That is to say they spread over and contribute to the variance of several primaries as primaries do to the variance of specific variables. For this reason we should expect them to show the usual properties of (a) simple structure and (b) obliqueness. As mentioned under simple structure, because of the fewer "variables" (primaries in this case) involved in most practical instances of factoring primaries, relative to factoring a lot of variables, the determination of simple structure is more difficult, through the small sample presenting possible alternatives. This has happened in the fourth, fifth, and sixth secondaries indexed in Q data in personality (Q IV, Q V, and Q VI) which sometimes "flop over" between two different positions (Cattell, 1973). As in all such quandaries, the only solution is to get the most consistent verdict across several studies, on different samples of people and with new sets of associated primary variables. Since there is by now ample evidence to satisfy the theoretical expectation of simple structure, and of significant factor correlation among the secondaries at the SS positions, there is no excuse for the illogical procedure of rotating secondaries to orthogonality. This inconsistent procedure would artificially terminate the pursuit of higher orders at that point. (The case is different, of course, if one only wants a statistical solution, uninterested in replicable scientific concepts of structure.)

The concept of a second-stratum factor as a contributing influence can probably be illustrated in its scientific setting best by the relatively long-established care of the primary abilities, where we theorize that fluid intelligence, one of the secondaries, is instrumental in improving rate of learning simultaneously in the distinct school learning areas represented by verbal, numerical, and logical reasoning primaries. The unitary character of the primaries in this field is presumably the result of an "agency" (Cattell, 1971a) or reinforcement of a pattern of skills brought about simultaneously in all learned elements by some unitary social institution. At a level which happens to be bare of psychological content it can be illustrated in the ball problem (Cattell & Dickman, 1962) plasmode, where the primaries of weight and size, positively correlated, tell of a secondary which is "the decision to manufacture a larger or smaller ball." Whenever the manufacturer sets out to make a larger ball he will tend to increase both volume and weight (and all the ball "behavior" variables which Dickman's analysis prove to depend on such primaries). Despite this common contributor—"size"—volume and weight (in beach balls, golf balls, and so on) will operate separately as primaries in influencing the bounce, speed of rolling, etc., of the balls.

However, psychology has already run into examples in which secondaries may require an alternative explanation. A classical example is the Q data secondary anxiety (Q_{II}) which loads the primaries ego weakness (C-), threctia (H-), guilt proneness (O), poor self-sentiment development (Q_3-), and high ergic tension (Q_4). It is at once evident that at least in terms of established psychoanalytic theory causal action is required to be precisely the reverse of that which the factor model requires. There is excellent agreement of factor analysis and psychoanalysis over the agencies involved—ego weakness, guilt (from superego), and ergic tension ("libido" to the Freudian). But whereas clinical thinking makes these the causes of the anxiety, the view of a factor as an influence which we have so far followed makes anxiety the cause of ego weakness (C-), guilt proneness (O), etc. Psychoanalysis could be wrong, and the alternative theory that a rise in anxiety reduces the ego strength, undermines the self-sentiment, seeks an outlet in guilt, and increases the general tension of frustrated ergs (Q_4) should be tried.

However, even without any further special checking experiment there are evidences that offer difficulties to this return to the primary concept of a *factor as a determiner*. For example, one component in the second-stratum pattern—the tendency to threctia (susceptibility to threat, i.e., timidity), H(-)—seems to have substantial genetic components which would more readily make it a cause than a consequence. On existing psychological evidence, the best hypothesis is one that moves from a "determiner" concept to what we may call a *spiral feedback* theory (Cattell, 1973) which supposes that a higher level on one of these primaries tends to produce through interaction of personality with environment, feedback which leads to a higher level on another primary in the second-order pattern. For example, an ego-weak (C-) individual is likely to be able to achieve less ergic expression and thus end at a higher level on ergic tension (Q_4), and a threctic person is more likely to transform that frustrated drive into anxiety. No one of the primaries, when deviant, is a sufficient cause of anxiety, but they are (by this theory) permissive conditions for its appearance. A quite specific causal proposition within the spiral-feedback theory has been proposed in an interesting data analysis by Lawlis and Chatfield (1974).

Spiral feedback is not the only model that can be proposed within the general concept that a set of primaries, interacting with a life situation, will tend to create such mutual augmentations that correlations will be created among them leading to the emergence of a second-order structure. Thus the emergence of mutual correlations of C-, H-, O, Q_3-, and Q_4, can be alternatively viewed as due to an interaction more general than a spiral triggering. But in any case the general theory supposes a "permission" or "mutual encouragement" within that subgroup rather than a quite new external determiner or influence operating on all of them. The notion of simple structure is so tied up with that of a determiner that at first one may wonder how and why spiral feedback could lead to a second-order simple structure. But the requirement of simple structure could be met at a first level of analysis by the restriction of such mutual stimulation to an "in-group" of primaries having the requisite properties. However, as factor analytic techniques advance in precision it should be possible to determine, e.g., by the nature of the hyperplane, the orthogonality to other second orders (since "emergent" second-stratum factors should have no significant higher-order determination), and the relation of loading patterns to population situations and age, whether the "outside influence" or the "emergent" model better explains the phenomenon of a second-stratum factor.

Meanwhile, until these techniques develop, and for the sake of brevity here, we shall primarily discuss the higher-stratum factor as if it were what it probably most generally is: an additional determining influence. In regard to such influences we have noted above in passing a somewhat naive tendency to equate higher-stratum position with higher importance. (Just as there has been a tendency to overrate the importance and prevalence of grand "monarchic" factors.) No such systematic bias is defensible, statistically or psychologically. Sometimes, indeed, the higher-stratum factor may depend on relatively fortuitous influences, e.g. situational and physiological influences outside psychology. For example, social status selection (granted a wide status range) causes some common increment upon and, therefore, correlations in the population among, such factors as intelligence, ego strength, surgency, self-sentiment development, etc. Although a single factor corresponding to social status has been demonstrated (Cattell, 1942), its primary loadings are somewhat uncertain, but there is little doubt it can appear as a secondary among the primaries. Is social status a personality factor? Surely it is not in the ordinary sense. Rather it is a second-stratum organizer of personality factors, exercising, as it happens, a relatively faint contribution to common variance. In short, higher-stratum factors may fairly often be expected to lie outside psychology, in sociology and economics for example, or, on the other hand, in some factor important in physiology, such as metabolic rate, which contributes some small variance simultaneously across several personality primaries.

An illustration of a well-replicated second-stratum factor of this general kind in psychology is found in Q VIII, which loads desurgency (F-), superego strength (G), self-sentiment development (Q₃), and, very slightly, ego strength (C). The current theory is that it represents the main effects of good upbringing (which Q VIII is accordingly presently labeled, though sometimes with less interpretation, as control, descriptively) probably in the home but possibly in the social status sense as well. Individuals high in Q VIII tend to be quiet and restrained in manner (F-), conscientious and altruistic (G), controlled and definite in self-concept values (Q₃), which include prudence and self-respect. Here, almost certainly, is a second stratum appearing from a broad external influence. In the above attempt to put higher orders in perspective it is, of course, recognized that when an influence outside the personality domain makes its impress on personality, as a die press shapes the automobile of which it is not part, the resulting pattern is in personality, despite the original influence taking its form from environment. Indeed, the primary sentiment structures in objective attitude data, which can be seen to be due to the pattern of impress (reinforcement schedules) of definite social institutions, e.g., the church, the sports team, come from outside; but they become finally actual response structures within the individual learned with respect to naturally occurring sets of stimuli.

However, the higher the stratum order of a factor is in personality data (and much the same argument applies to other factor analytic areas) the less relevant it may get, in two senses: (1) the scientific sense of having close substantive connection with behavior (with variables) and (2) the statistical sense of contributing adequately to the prediction of variance on specific individual behavior variables. Of course, the last mentioned is a matter for empirical decision and instances of substantial contribution may exist, as in the contribution of Q_f (fluid intelligence) to primary abilities (McNemar, 1964; Vernon, 1969). But in terms of potency and relevance among psychological concepts per se, we must reverse the naive view of "higher strata" meaning higher importance and insist that in magnitude of contribution to predictions (see p. 219), the higher strata factors may commonly be expected to be less rather than more relevant, theoretically and practically, to the given domain of variables.

9.5. Calculating Loadings of Higher-Strata Factors Directly on Variables: The C–W Formula

With this perspective on strata structure and the way in which it is revealed by successive factorings, we may now turn to models for making calculations among strata. Second-strata factors have their loadings on primaries, but one of the first questions psychologists asked was "how do we calculate the influence of such factors directly upon basic variables?" By the latter, they commonly refer to various criterion performances in applied psychology. Actually, we are concerned here, as anywhere else, ultimately with a two-way regression estimation. We need to handle (1) the prediction of concrete behaviors from higher-strata factor scores and (2) the estimation of higher-order factor scores from scores on variables. The latter is important because frequently one has no well-tried primary factor scales and scores from which to work to secondaries. In that case one must refer to *the actual ground of measured items or other variables* for the estimation of secondaries and tertiaries as individual scores. However, in this chapter we shall deal with the estimation of variables from factors, as the first and more necessary required solution, and more illuminative of essentials. The converse procedure—getting scores on second orders from variables—is best left to the next chapter, which handles the general problem of scoring factors.

If we know the factor pattern of primaries on variables $V_{fp\mathrm{I}}$ (note we have not hitherto, in earlier contents, found it necessary to add ı to this V_{fp}) and of secondaries on primaries $V_{fp\mathrm{II}}$, then it can be shown that the loadings of secondaries directly on variables, which we will designate $V_{fp\mathrm{II}_v}$ can be obtained by

$$V_{fp\mathrm{II}_v} = V_{fp\mathrm{I}} \cdot V_{fp\mathrm{II}} \qquad (9.1)$$

$$(n \times p) = (n \times k)(k \times p)$$

This formula is sometimes called the Cattell–White (C–W) formula to distinguish it from the Schmid–Leiman (S–L) formula, the two being superficially similar but vitally different. [The parentheses under the Eq. (9.1) matrices are to remind us of the orders of the matrices and what they are doing. The numbers are n variables, k first-order factors, and p second-order factors.] The rationale for deriving the Eq. (9.1) C–W formula can be seen if one looks at the steps in this matrix multiplication:

$$r_{vF_1} = r_{vf_1} r_{f_1 F_1} + r_{vf_2} r_{f_2 F_1} + r_{vf_3} r_{f_3 F_1} \cdots \qquad (9.2)$$

where r_{vF_1} is from $V_{fp\mathrm{II}_v}$, the correlation of any variable, v, with second-order F_1; r_{vf_1} is the correlation of first-order f_1 (note lower case for primary) with v; and $r_{f_1 F_1}$ is the correlation of primary f_1 with secondary F_1 (in $V_{fp\mathrm{II}}$). Obviously we are using the same principle as way back in Chapter 3, p. 20, namely that $r_{ab} = r_{ag} r_{bg}$. Note that we have used here f for a primary factor and capital F for a secondary. This may not be ideal, since a capital sometimes means a matrix, and since nothing remains for third order; but it gives the clearest statement in this limited context, at any rate. Regardless of whether we are dealing with a C–W (Cattell–White) or S–L (Schmid-Leiman) transformation it is convenient to designate these matrices which leap over one or two intermediate strata to load on variables "vaulted matrices," e.g., a vaulted factor pattern matrix.

Once one has the $V_{fp\mathrm{II}_v}$—the matrix of loadings of the second orders directly on the variables—one can make all the usual transformations (see pp. 120 and 244), provided one has the correlations of the secondaries among themselves, $R_{f\mathrm{II}}$, which of course are known from the second-order transformation matrix, L_{II}, Eq. (8.1), obtained in rotating the second orders to whatever unique resolution one reached. The matrix a psychologist probably most desires after that in Eq. (9.1) for predicting variables is the $V_{fe\mathrm{II}_v}$ (fe stands for factor estimation) since the second-order factor scores will have to be derived directly from the individual's variable scores. However, as just stated, the derivation of the V_{fe} is deferred to Chapter 11, explaining the main issues of factor scores (specifically, p. 274).

If the analysis has proceeded to still higher orders, the calculation of the V_{fpIII_v}, and so on, proceeds in just the same simple fashion by the extended C–W formula (extended here just to 3 strata):

$$V_{fpIII_v} = V_{fpI} V_{fpII} V_{fpIII} \tag{9.3}$$

$$(p \times c) = (n \times k)\,(k \times p)\,(p \times c)$$

Parenthetically, for some completeness of overview here, let the reader note that one can restore the original (reduced) correlation matrix from this vaulted factor matrix V_{fpIII_v} by just the same principles as for a first-order matrix, thus:

$$R_v = V_{fpIII_v} R_{fIII} V_{fpIII_v}^T \tag{9.4}$$

$$= V_{fpI} V_{fpII} V_{fpIII} R_{fIII} V_{fpIII}^T V_{fpII}^T V_{fpI}^T$$

Various other familiar operations, e.g., proceeding from the V_{fp} to a V_{fe} or V_{rs}, can be performed with these "downward-projected" vaulted matrices, provided that for the $V_{fp_x \cdot v}$ one has the $R_{f(x)}$ correlation matrix (x being the factor order). Frequently the V_{fp_x}, at the top of x strata, will represent a culmination on a single factor, so that V_{fpIII} will be, in the case of Eq. (9.3) above, a $p \times 1$ column vector, c being 1.

The result of such calculations of projected factors is usually a series of comparatively small loadings by the typical xth stratum factor on variables. This lesser loading from higher-order factors may mildly surprise the psychologist, unless he has already gained the perspective above, that a higher stratum does not imply higher importance. However, since a higher-stratum factor spreads its influence broadly over more variables, there is a sense in which it has importance, while the estimation of scores on such a factor is not deficient in validity, i.e., in the magnitude of the multiple R with the factor achieved by the weighted score on variables.

Since this is the first time the reader has encountered in this book relatively complex matrix algebra, he may wish to "brush up" in the compact treatment by Hohn (1958) or Horst's (1963) excellent adaptation of matrix practice to the social scientist's needs.

9.6. The Ultimate Factor Theory: The Stratified Uncorrelated Determiner Model

It is important that what one does in mathematicostatistical calculation should parallel and express what one believes to be the most probable scientific model. Our view of the latter remains openmindedly that factors can interact both by reticular and by strata models, though we keep to the latter because it is more amenable. As far as one can see at present the means of unraveling a reticular structure must be in the supplementation of factor analysis by what is called *path coefficient* analysis (Wright, 1934, 1954, 1960; Blalock, 1971; Rao, Morton, & Yee, 1974). A brief introduction to this is given on p. 426, Chapter 13, but here we shall pass on to the main stratified models.

In the strata model the initial result of factoring is a set of factors which, at each and every stratum, are oblique. The above section has shown how, by the C–W formula, we can calculate the pattern of loading influences of each oblique higher-stratum factor directly on variables. This is helpful both in interpreting the nature of a factor from its effects on variables and also in scoring the higher-stratum factor from the only concrete data we have: the scores on the variables. However, let us be clear that this calculation does not suppose that the obtained loading values represent any direct action of the higher-strata factor causally on the variables. The model says that third-order factors actually operate on second orders, and secondaries in turn on primaries, so the C–W calculation, though it undoubtedly speaks of an effect refers to an ultimate, displaced effect, at one or two removes.

Furthermore, we cannot set out one single specification equation, via the C–W formula, to show all orders of factors in their separate equations, brought together to account for the variances of the variables. For part of the loading variance on a variable written down as due to a primary factor is brought in again in the loading worked out for a secondary (since the secondary accounts for part of the variance of the oblique primary itself). This overlap and redundancy comes in again at the third order. The result would be—if we strung along in a single equation the loadings on a variable by the C–W formula from first-, second-, third-order, etc., factors—that the communality of variables would be spuriously large and would usually well exceed unity.

However, a variable specification equation of this comprehensive kind, i.e., with all strata represented, can be written if we first take out of the primaries that part of their variance which they share in common, i.e., that which comes from higher-order factors and is responsible for their being oblique and correlated rather than orthogonal. The orthogonal residues after this removal—the parts of the primaries independent of and orthogonal to the higher order—we shall call "primary stub factors." Such stub factors can be successively separated at each stratum all the way up the strata. The result will be a set of uncorrelated primary stubs, all of them also uncorrelated with a set of mutually uncorrelated secondary stubs, and so on up to tertiary stubs, etc. [See Eq. (9.7).]

If we set aside the "emergent" model of higher-strata structure as something unusual, and stay with the more likely and simpler "broader influence" model, then we suppose that the initially observed correlation

of a set of primaries is due to some other influence simultaneously affecting all of them. It thus contributes some variance to each that is common
to all. Up to a point, the calculation of "stubs" will also fit the emergent
model, but we are primarily deriving the convincingness of this form of
calculation from the strata influence model.

To get to the heart of the new theoretical model which goes parallel
with the mode of calculation, let us note that the basic assumption is that
factors would never become correlated unless some other influence—that
which we locate at a stratum above, in the strata model—acted upon them
in common. This indeed is nothing more or less than a particular corollary
from general scientific determinism, though it is determinism in one of
several possible forms. Its postulate is that two things never get correlated
by chance. (Random numbers on small samples, which do show some correlation, seem an exception, but they are not truly random, being set up
by human action.) A zero correlation of A and B must exist unless the
usual logically required conditions upset it, namely, some of A in B, some
of B in A or some of X in both. Note that the two first possibilities fit
the usual factor model just as well as the last, for in them we are dealing
with a correlation of a variable and the factor within it. Consequently,
when two primaries correlate, the model tells us that there exist two primary "stubs," as we may call the uncorrelated specific parts ("unique
variances") in each common primary factor and some higher-order factor
influencing both and contributing to their variance. For example, measures of a primary of intelligence and another of stature, in subjects all
measured at 10 years of age may be correlated practically zero. But, if
we take a range of 10–17-year-olds, a second-order factor of age comes
in, which extends the variance of both mental age and stature to larger
values. It produces, by introducing common-age variance, a substantial
correlation between the primaries. One sees then that the true nature of
intelligence, for example, is better represented by the "stub," i.e., test
performances with age partialled out, than by the oblique factor as its
pattern appears initially across the wide age range. Thus, in the latter
case the intelligence factor will correlate in boys with the number of
teeth, the amount of beard, and the cumulative number of summer
holidays each has had. Note, however, that it will not load these, in
the V_{fp}, if rotation to simple structure or confactor uniqueness has been
good—it will only correlate with them through correlating with the other
factors which produce them. This, incidentally, reminds us of the importance of interpreting factors through the V_{fp} rather than the V_{fs}.

The present model of quite independent factors at the primary level,
which residual factors we have called, statistically, specifics or stubs, appearing on first analysis as primaries of larger variance, and correlated,
due to the common contribution from one or more second orders, we

shall call the *stratified uncorrelated determiner* (or SUD) model. The title is apt because it indicates that the numerous determiners in the physical and biological world are by their independent natures necessarily uncorrelated, that they become correlated by one acting on the magnitudes of two or more others, i.e., contributing common variance to them, and that this acting most commonly occurs in a hierarchy of strata (not to be confused with a pyramidal hierarchy). It will be seen that what we are calling the stub in a primary (or a secondary) factor is logically and statistically exactly analogous to the specific factor in a variable, and is similarly orthogonal to other specifics and to any broad factors.

9.7. The Calculations for the SUD Model, Beginning with the Schmid–Leiman Formula

We have above essentially two models of action, which may be called the stratified uncorrelated determiner (SUD) model and the serial overlap analysis (SOA) representation, though the latter remains to be set out here. Granted that the SUD is a viable scientific model, then we should be explicit that the SOA is not a scientific model, but only a mathematical *mode of analysis and representation* of the same model. It takes an "as if" position, presenting the values as if the second and higher orders act directly on the variables, but recognizing that in fact the "vaulted factor" exerts the given influences only through intermediate factors. Moreover, as we have seen, if it permitted the calculated effects to overlap, accounting for parts of the variance of variables twice over, it would lead to a summed result which overestimates the variance of variables and thus would not correspond with statistical fact. Paralleling and participating in the calculations on the two models we have two formulas: the C–W and the S–L. The C–W formula, already explained in Eqs. 9.1 through 9.3, and which corresponds to a model of serial overlapping action, we relegate to the status of a particular analytical tool. It is not a likely model but only a means of calculation when second- or higher-order factors are to be interpreted and scored directly from variables.

The formula which best operates in calculations in the SUD model is that of Schmid and Leiman (1959) (henceforth to be abbreviated to the S–L formula). At first, the matrix operations associated with the S–L formula may look more complex than those for the alternate resolution in the S–W, but if followed with care they will become clear. And they deserve to be so followed because our stratified determiner model, which they serve, is by far the most likely in scientific terms. To state our postulate more precisely, the notion of intrinsically uncorrelated stub factors brought to correlation by the influence of other, outside factors is likely to have universal validity; but the restriction to strata—instead of acceptance of the unrestricted reticular model—is an assumption of less universal validity.

The calculation for the stratified uncorrelated determiners (SUD) model, employing the S–L formula, has to begin with the usual procedure advocated as standard for most factoring, namely, rotating the first order to simple structure, the obtained second orders therefrom to simple structure, and so on. These rotations will, of course, be oblique. Real confusion of concepts sometimes arises from confounding the issue of initial obliqueness or orthogonality in rotation which has to be decided in favor of obliqueness, with an ultimate model of orthogonality. It cannot be too strongly emphasized that the ultimate orthogonal solution in the SUD model is conceptually and algebraically entirely different from "*rotating (the primaries) for orthogonality*." SUD is not a simple "orthogonal resolution." One must first get the V_{fpI}, the V_{fpII}, etc., involving of course the factor correlations in R_{fI}, R_{fII}, etc., if ever one is to find the solution that will yield the stub factors in the orthogonal, stratified SUD resolution.

What happens in the SUD resolution can best be seen by considering the second-order factor pattern, V_{fpII}, but including the commonly omitted k uniqueness part of the matrix, as shown in Table 9.2. (A first-order factoring will have preceded this.) The new, "stub" first orders, as we are going to know them, are only the u—the uniqueness, as in variable specifics before—in this table. For we have agreed that all the variance taken out of the original primaries by the secondaries [the broad factors to the left of Table 9.2(a)] belongs to the latter factors and never was, originally, and in the scientific model sense, a part of the primaries. Consequently, to get the true loadings on the variables, belonging to these now truncated "stub" primaries we have to reduce the original variable loadings on the first-order V_{fp} by multiplying the "whole" factor loading column originally obtained in the first-order factoring at the bottom left of Table 9.2 by the fraction defined for us by the corresponding u value (top right of Table 9.2, now moved to middle of bottom row). If, for example, u_1 is .5, the loadings of the stub factor on variables will be those of the old f_1 reduced to a half. The way to do this by matrix multiplication is shown on the lower half of Table 9.2, namely by taking the already provided diagonal U matrix from the second order and multiplying the earlier (first order) V_{fp}, i.e., the V_{fpI}, thereby. Do not forget that Table 9.2 upper row concerns second stratum and the lower row the first-stratum (primary factor) matrix.

For the SUD model as such the same procedure is repeated up the strata, in each case taking the V_{fp} from the oblique solution and multiply-

Table 9.2. Derivation of Matrices in the Stratified Uncorrelated
Determiners (SUD) Model (Derived from the S–L Formula)

(a)

	Broad secondaries				Uniqueness secondaries, to be changed to stub primaries					
Primaries	I	II	$\cdots$	P	1	2	3	4	$\cdots$	k
1	$a_{1\mathrm{I}}$	$a_{1\mathrm{II}}$		a_{1p}	u_1					
2	$a_{2\mathrm{I}}$					u_2				
3	$a_{3\mathrm{I}}$						u_3			
4								u_4		
.										
.		$-V_{fp\cdot\mathrm{II}}-$								
k	$a_{k\mathrm{I}}$	$a_{k\mathrm{II}}$		a_{kp}						u_k

(b)

Original first-stratum factor matrix	Stubs from second order (above)	Primary stub factor matrix

$$V_{fp\mathrm{I}} \quad \times \quad U_{\mathrm{II}} \quad = \quad \underline{V}_{fp\mathrm{I}}$$

[a]Note that the stub factor $\underline{V}_{fp\mathrm{I}}$ is written with an underline to distinguish it from the usual, original $V_{fp\mathrm{I}}$.

ing it by the unique matrix next above, which converts it to loading values appropriate for the stubs, i.e., it continues what is shown in Table 9.2 through the steps shown in Table 9.3.

At this point we introduce a variant of the SUD, which may be treated either as a new model or as a calculation or analysis convenient for certain purposes. (Our ultimate conclusion is that it has not too probable a status as a real scientific model.) In this—which we shall call

the independent direct influence model or calculation, because it supposes the SUD factors now act *directly on the variables* instead of on
the factors next below—the loadings for the SUD factors are projected
for each stratum as they would work out directly on the variables. In this
case, the concept of the factors being at different strata requires ulterior
evidence, since in regard to variables they behave like primaries. The
model employs what has long been known as the Schmid–Leiman transformation. It involves the use, however, also of the Cattell–White transformation. The steps which express the IDI (independent direct influence)
model involve the essence of the S–L transformation, which is

$$\underline{V}_{fpI v} = V_{fpI} \times U_{II} \qquad\qquad (9.5)$$

and combines it next with the C–W transformation

$$(V_{fpII v} = V_{fpI} \cdot V_{fpII}) \qquad\qquad (9.6a)$$

in a formula which converts the truncated orthogonal stub loadings in
9–5 into their loadings directly on the variables. Illustrated for a second-
stratum factor this is

$$\underline{V}_{fpII_v} = V_{fpI_v} V_{fpII} U_{III} \qquad\qquad (9.6b)$$

$$(n \times p) = (n \times k)\,(k \times p)\,(p \times p)$$

The notation may look a little complex, but the retention of all implied subscripts is necessary if the reader is to keep his head in an intrinsically rather complicated operation. Thus the first $\underline{V}$ in (9.6b) has an
underline to distinguish it from the V_{fpII_v} in (9.1) and (9.6a) which arises
directly from the C–W formula and represents the second order at its "full
length" projected on the variables. (The small v in V_{fpI_v} is admittedly usually superfluous, and not used elsewhere, but it helps the alert reader in
actual use of these formulas by reminding him that V_{fpI_v} begins right on
the variables, i.e., as primary factor loadings on variables.)

The upshot of repeating this S–L calculation is to produce as many
separate matrices as there are orders. Thus, for four orders as shown in
the SUD [Table 9.3(a)] we should have four distinct parts to the matrix,
each part with a symbol and a matrix calculation as shown in Table 9.3(b).
The procedure will end when the analysis reaches a single factor, assumed
to happen in this case at the fourth order. As pointed out elsewhere (Cattell, 1965c), the outcome in Table 9.3(b) can be formally set out most
conveniently for those operating in matrix terms as a *super-matrix* (Horst,
1965; Gorsuch, 1973), but we need not pursue that device here (see
Cattell, 1965c, for a brief statement).

To help the reader's grasp of the SOA model concrete instances of
this and the IDI calculations are set out in Table 9.4. At the top, in

Table 9.3. Comparison of the Stratified Uncorrelated Determiners (SUD) Model and the Independent Direct Influence (IDI) Model[a]

	Part I of Matrix	Part II of Matrix	Part III of Matrix	Part IV of Matrix
(a) SUD calculation	$\underline{V}_{fpI} = V_{fpI}U_2$ $(n \times k)$	$\underline{V}_{fpII} = V_{fpII}U_3$ $(k \times p)$	$\underline{V}_{fpIII} = V_{fpIII}U_4$ $(p \times t)$	$\underline{V}_{fpIV}$ $(t \times l)$
(b) Independent direct influence calculation	$V_{fpI_v} = V_{fpI}U_2$ $(n \times k)$	$V_{fpII_v} =$ $V_{fpI}V_{fpII}U_3$ $(n \times p)$	$V_{fpIII_v} =$ $V_{fpI}V_{fpII}V_{fpIII}U_4$ $(n \times t)$	$V_{fpIV_v} =$ $V_{fpI}V_{fpII}V_{fpIII}V_{fpIV}$

[a]An ordinary oblique analysis is here supposed to have proceeded to four factor strata. The four V_{fp}s are written V_{fpI}, V_{fpII}, V_{fpIII}, and V_{fpIV}, and the corresponding unique matrices as U_1, U_2, etc.

Table 9.4(a) for contrast, is set out the alternative analysis of the same correlation matrix—indeed the same first-order V_{fp}—by what we have called above the serial overlap analysis or SOA. In the latter, which emerges from the usual analysis and rotation procedures followed up through the middle of this chapter for each and every order, one gets, initially, as shown in Table 9.5, an $(n \times k)$ V_{fpI} of primaries loading variables, next a $(k \times p)$ second-stratum matrix, V_{fpII}, showing secondaries loading primaries, and so on to third or higher orders. By the Cattell–White formula one converts these at each stratum (beyond the first, which is correct as it stands) to loadings directly on variables. The three-part matrices at this point will consist, respectively, of an $(n \times k)$ primary [10×4 in Table 9.4(a)], an $(n \times p)$ *vaulted second order* (concretely 10×2), and an $(n \times c)$ [10×1 in Fig. 9.4(a)] vaulted third order.

Each of these matrices in the overlapping action analysis correctly describes literally what the factor at a given stratum contributes to the variables, standing in its normal oblique position. But in this analysis the matrices are not independent: they overlap and contribute redundant variance to the variables. That is to say, part of the variance contributed to variables by the primaries as they stand revealed at simple structure is restated again in defining the action of the secondaries, and part of the variance due to the secondaries is restated in the tertiaries. Actually the original oblique primaries alone tell you all the common variance that can be accounted for among the variables. That common variance is, of course, given not only in the oblique V_{fpI} but also in the original unrotated V_0. Incidentally, that common variance—the communalities, h^2— cannot be calculated simply by squaring and adding the loadings on the oblique factors as one does with the V_0 (see explanation, p. 240). The whole higher-order structure simply takes a part of the total variable

Table 9.4. The Serial Overlapping Action and Independent Influence Models and Matrices Compared

(a) Serial overlapping action analysis: higher orders vaulted on to variables (C–W formula):

	First order					Second order			Third order	h^2
	$1'$	$2'$	$3'$	$4'$	h^2	$1''$	$2''$	h^2		
1	.00	.00	.00	.50	.25	.00	.30	.09	.21	.04
2	.00	.50	.00	.00	.25	.00	.00	.00	.00	.00
3	.00	-.50	.60	.00	.61	.36	-.30	.09	.09	.01
4	.50	.60	.00	.00	.61	.25	.00	.06	.21	.04
5	-.50	.60	.00	.00	.61	-.25	.00	.06	-.21	.04
6	.60	.00	.50	.00	.70	.60	-.25	.24	.33	.11
7	-.60	.00	-.50	.00	.70	-.60	.25	.24	-.33	.11
8	.00	.00	.60	-.60	.79	.36	-.66	.28	-.16	.03
9	.00	.00	.00	.50	.25	.00	.30	.09	.21	.04
10	.00	.00	.00	.60	.36	.00	.36	.13	.26	.07

Intercorrelations of first-order factors (Cattell–White)

	$1'$	$2'$	$3'$	$4'$
$1'$	100	00	15	18
$2'$		100	00	00
$3'$			100	-08
$4'$				100

Intercorrelations of second-order factors (Cattell–White)

	$1''$	$2''$
$1''$	100	60
$2''$		100

(b) Independent influence matrix with loadings vaulted on to the variables (S–L formula) a derivation of stratified uncorrelated determiners

	First order					Second order			Third order $1'''$	h^2	Σh^2
	$1'$	$2'$	$3'$	$4'$	h^2	$1''$	$2''$	h^2			
1	.00	.00	.00	.40	.16	.00	.21	.04	.21	.04	.25
2	.00	.50	.00	.00	.25	.00	.00	.00	.00	.00	.25
3	.00	-.50	.52	.00	.52	.19	-.21	.08	.09	.01	.61
4	.43	.60	.00	.00	.55	.13	.00	.02	.21	.04	.61
5	-.43	.60	.00	.00	.55	-.13	.00	.02	-.21	.04	.61
6	.52	.00	.43	.00	.46	.32	-.18	.13	.33	.11	.70
7	-.52	.00	-.43	.00	.46	-.32	.18	.13	-.33	.11	.70
8	.00	.00	.52	-.48	.50	.19	-.47	.26	-.16	.03	.79
9	.00	.00	.00	.40	.16	.00	.21	.04	.21	.04	.25
10	.00	.00	.00	.48	.23	.00	.25	.06	.26	.07	.36

variance, tied up in and defined by the correlations among primaries, and spreads it out over successive higher orders. It is meaningful to do this, because the higher-order structure is actually there, as revealed by the simple structure. However, we shall have the alternative of stating the higher structure by a set of overlapping statements, differently parcelling . the same communality, as in Table 9.4(a) or of orthogonal "stub" factors as in Table 9.4 (b), stating the truly independent contributions to the variance of the variables.

The statistical expression of this difference is seen at once in Table 9.4 in that one cannot square the loadings along a row and add them, in the overlap model (a) to get the communality h^2 for the given variable, whereas in the independent influence model (b) one can. The reader may test any row in Table 9.4(b) to check this. It will be noticed that one comes out over the whole seven columns with the figure given just for the four first orders in Table 9.4(a) in accordance with what is stated above. [Parenthetically, the reader will have realized from Chapter 8, as just stated, that one cannot get the true h^2 values for oblique factors simply by summing the squared loadings in the rows. This fact is expanded upon below (p. 240), but it is not the only reason for the squared loadings over three h^2 columns in Table 9.4(a) not adding up correctly: the reason there is the overlap.]

In conclusion, it will be noted that though both models require that one first get the correlations among the factors, by the usual rotation procedures, the complete statement of the serial overlap analysis of vaulted factors requires that the factor correlations be produced, additionally to the V_{fp}, as in Table 9.4(a). By contrast the independent determiner model tells all that we need to know by the stub matrix as in Table 9.4(b), since the factors are orthogonal. However, the SS obliquities have had to be found before Table 9.4(b) can be worked out. One advantage of the independent influence model (the SUD projected on variables), as will be evident from what has been said about communalities, is that the estimation of a variable, e.g., a criterion, can be made by a specification equation which *simultaneously includes factors of all strata levels*, thus (for three strata):

$$a_{ij} = b_{\mathrm{I}1j}F_{\mathrm{I}1i} + \cdots + b_{\mathrm{I}kj}F_{\mathrm{I}ki} + b_{\mathrm{II}1j}\underline{F}_{\mathrm{II}1i} + \cdots + b_{\mathrm{II}pj}\underline{F}_{\mathrm{II}pi} \qquad (9.7)$$

$$+ b_{\mathrm{III}1j}\underline{F}_{\mathrm{III}1i} + \cdots + b_{\mathrm{III}tj}\underline{F}_{\mathrm{III}ti} + b_{bj}F_{ji}$$

(Here as usual we have used k primaries, p secondaries, and t tertiaries— the last orthogonal.) There is one factor of unique variable variance at the end—the incalculable specific and error which makes the prediction only an estismate. This is a specific to the variable, not any higher-order specific.

In the realm of psychological theory, Eq. (9.7) is a highly important end product, because it states finally how we predict behavior by what we consider to be the most likely scientific model, using the most realistic psychological concepts (or concepts in any sciences) that we have. From the matrix of which it is a row, Table 9.4(b), we can derive equations for estimating all of these factor scores. In practice, it has a blemish, in that the estimation of factors from the same variables may produce some dependence of the factor scores where none really exists for the true factors (since they are orthogonal). However, though usually to a slightly lesser degree, this distortion exists even in estimating first-order orthogonal factor systems, as will be seen when we come to factor score estimation (p. 274).

9.8. Some Bases of Decision among Alternative Models of Factor Action

One must be alert, however, to the fact that the independent influence matrix, Table 9.4(b) above, though derived from the SUD, may deceive us, *in the theoretical domain*, into thinking that as equations they represent what "really happens," rather than offer a basis for computing estimates. If we can be granted that, in some fields of action at least, a hierarchical strata structure exists and can perhaps also rule out the emergent theory of higher orders, then it would still remain incorrect in one way. That is, that Eq. (9.7) shows the higher-strata determiners as if acting directly on the variables rather than, in each case, on factors in the stratum of factors immediately below them. Thus we reject IDI as a model, but accept the S–L independent influence matrix, and Eq. (9.7) as conveniences of statistical statement.

Almost certainly, in most instances in science believed to represent strata, the determiners do not vault over strata to affect variables directly. Our best guess at the present time is that factors act on factors, in a "falling row of dominoes" fashion, as far as strata are concerned, down to the primary factors, which alone act on the variables. For example, rising temperature may (a) cause improved plant metabolism (primary 1) and (b) the snow on ski slopes to melt (primary 2). An increase in primary 1 may cause pines to grow and a deeper green in certain foliage; and an increase in primary 2 may cause a rise in accidents, more departures of visitors, and more time by the remainder in the bar; but the rising outdoor temperature does not itself cause people to gather in the bar. The statements in Eqs. (9.3) and (9.7) are therefore correct as to the ultimate outcome of each factor's contribution to the variance of any variable, but unless we keep in mind what has just been said, these useful equations may lure us into the possibly, or even probably, incorrect assumption that the actual causal action of higher strata is directly on the variables. This throws away other sources of scientific knowledge and calculation, e.g., what is known about the "natural history," over time and changing conditions, of some intermediate factor.

In summary of this discussion, it should be realized that in fields where higher-order factors can be taken out we really have three ways of representing them:

1. The *stratified uncorrelated determiners* (SUD), in which factors exist, orthogonally, at different strata, acting at each stratum directly on factors (not variables, except at the primary level) immediately below them.

2. What for lack of a better name we have called the *serial overlapping action*, in which factors remain oblique and the matrix represents their action directly on variables, but with the overlap and redundancy of lower-strata factors including already the higher-strata variance contributions. This is a form of analytical representation, useful for some purposes, but cannot be a scientific model.

3. The *independent influence matrix*, in which the basic structure is that of uncorrelated "stub" factors as in the SUD, but where it is supposed that the concept of strata loses some of its meaning, because the factors all act directly on variables. This could be a model, but is unlikely and we are inclined to consider it here as only a mode of statistical representation, useful for such purposes as estimating higher strata factor scores directly from variables.

After the above excursion into the matrix basis of calculation, therefore, we finish, as we began, with some metatheoretical discussion. Not only here, but in other areas of factor analytic resolution, e.g., alternative simple structures, we meet a proposition rather widely encountered in scientific theory: that two alternative theories sometimes explain phenomena with such virtually equal potency that an immediate decision between them is not possible. In our present case the alternatives are: (a) in the uncorrelated (orthogonal) determiner model (1) to have each stratum act directly on that below (2) to have all levels act directly on the variables and (b) in the orthogonal vs. oblique overlap to suppose that the oblique factors at least at the first stratum have become full functional unities and are no longer stubs plus a common variance added from the higher order, though they were so originally. The latter choice (b) involves subtleties for which we have no space to digress into here, and is relegated to the Appendix.

The former (a) can briefly be discussed. In the general domain of science, alternatives such as the earlier instanced wave and corpuscle theories of light, or the expanding versus stable universe, which remain long in suspense, do so because no crucial test can be found or, in the latter, no theory which subsumes both contending theories. A possible crucial test here between serial action of adjacent strata, a factor on factor as in the SUD model; and a real common action of all strata directly on variables is offered by observing whether both higher order and lower order alike have simple structure in their loading patterns on the variables. For it has been shown with tolerable certainty that the existence of simple structure of secondaries on primaries does not, by inherent mathematical necessity, bring about simple structure of secondaries directly on variables (Cattell, 1965c). Therefore, if the latter is shown by experiment to exist, it would argue for a direct action model. The empirical evidence as to whether higher-strata factors like F_{IIp} and F_{IIIc} in Eq. (9.7) produce well-defined hyperplanes (other than on the stratum immediately below) is at present ambiguous. Some instances of second-order hyperplanes in variables seem to exist. Thus, with 32 and 64 short scales marking the 16 factors in the 16 P.F., the present writer has shown that simple structure programs will sometimes rotate to a second-order SS on variables at the same time yielding SS on the "stubs" of primaries. Parenthetically, if this is so, our generalization that there are only as many hyperplanes as factors in the variable space would have to be modified to say that there are as many as there are primaries and higher orders, together.

One objection to accepting the shared direct action on variables model is that the tests for number of factors might then be expected to yield a number ($k + p + t$, etc.) equal to primaries and higher strata together, whereas in fact these tests seem reliably to yield only k. The implication of expecting a total of ($k + p + t$, etc.) is that the higher orders now become nothing more than shallower primaries, acting alongside original primaries (as stubs) and with no distinctive features. Hence they would count as further primaries in the number of factors test, which they do not seem to do. Let us consider further our concrete example of a study confined to a set of cognitive ability and physical size measures in an age-scattered high school population. In the serial model we should expect two primaries—intelligence and physical size—and a second-stratum factor—age—loading both, because both increase with age. If we included age as a variable it would be loaded on both primaries (having no other time or age variables with which to form another factor) and when we calculated the projection of the second order directly on variables it would load the second-order age 1.0 (within limits of error). However, we should scarcely expect age to show simple structure on the variables, because virtually all of them, physical and mental, would change with age.

Nevertheless, because of indubitable instances of simple structure sometimes slipping the resolution into second-order positions, the question must be at present left open. It seems, incidentally, that the second orders, when they do have simple structure, only do so if enough factors are taken out ($k + p + c$, etc.) to permit the "stubs" of the lower orders also to appear, i.e., the alternative structure to stopping at oblique primaries is the full stratified uncorrelated determiners model. A difficulty with the latter model we have not so far raised is that ($k + p + t$, etc.) in Eq. 9.7 could conceivably exceed n. This does not present a logical but only a statistical problem, since we have agreed earlier that with few variables (small n) in a complex area it is quite possible for more factors to be operating than there are variables. Recognition of this fact is primarily a call for foresight in planning experiments. If at each higher level the number of factors is less than half of the number of the level below and we begin with $n/2$ the total of this series $n/2$, $n/4$, $n/8$, etc., cannot exceed n. However, even in an infinite series the problem remains that the number of factors test typically gives only the dimensionality of the primaries in Eq. (9.7).

Leaving the specific issue of higher orders as such to comment on the general problem of alternative simple structures across or within strata one may suggest that the concept of factor efficacy is useful. Efficacy means the extent to which the given factor structure is an effective predictive hypothesis across a variety of experimental and life situations. As will be more apparent when we come to comparing R-, Q-, P-, and other experimental designs, some factors stand up as recognizable patterns across several situations, whereas others have lesser generality of conceptual and experimental and predictive reference. In the present context the question one ultimately hopes to have settled is the relative efficacy of primary oblique (or sometimes higher-order oblique) factors as defined by the independent influence analysis [Table 9.4(b)] and by the separate stratified orthogonal determiners (SUD) model. There is no question that statistically we can use the former—and must do so whenever we want to estimate higher-strata factor scores from scores on variables. But in scientific explanation it could be that primary oblique factors only or the SUD model, or the independent influence acting directly on variables lead to quite different predictions. Thus in our example of intelligence, physical size, and age, knowing that intelligence is at an intermediate stratum between age and certain cognitive performances would enable us to understand anomalous changes in those performances in relation to age through other influences on intelligence, e.g., neural disease or injury.

9.9. Summary

1. Empirical studies show that there is a natural tendency of simple structure (and presumably oblique confactor) resolution to yield oblique

factors, both at first and higher orders. This permits factoring of factors bringing representation in a series of ordered strata.

2. The nature of the correlations among factors is much the same whether they are obtained as angles (cosines) reached in SS rotation or from experiment with scales constructed to measure the factors. Consequently the higher-strata factors come out essentially the same. However, the correlation of primary scales, except when they have very high validity, is probably less accurate and the loadings need correcting for attenuation (dividing by the square root of the validity, subsuming reliability) to be comparable with the former values.

3. The strata (or "stratoplex") model is not necessarily the only one making a good fit to known interactions among factors. A hierarchical series of strata need not, in any case, end in a pyramidal hierarchy. The monarch factor in a hierarchy is always only relative to a breadth of base of variables, and convergence in a pyramid is an artifact of the factor analytic process. The "top factor" becomes one of a peer group of factors in a stratum when a domain is more extensively sampled across variables. Other models are the duplex, the simplex, the chain simplex and circumplex, the cooperative model, and most general of all, the *reticular* model. The last supposes a general and varied network of relations among factors as such, and admits circular feedback influences, positive and negative. In the theoretical construction of models two main kinds of specification can vary: (i) the patterns of relations among factors, and (ii) the overlap and other relations of the patterns of the actions of factors on variables. The "parallel loadings" of *cooperative factors* offer an experimentally recognized instance of the latter.

4. The difference of stratum and order is defined. The circumstances leading to their confusion, e.g., to first-order extractions sometimes necessarily yielding some second-stratum factors interspersed with primaries are discussed. One cause is a change in density of representation of variables and its elimination requires that the population of the *personality sphere* of variables, from which a stratified sampling is made, be defined on a basis independent of that given by the magnitude of correlations between "adjacent" variables. A sample of variables of much lower density may factor out, in part or as a whole, directly to second-order factors, i.e., second order relative to the primaries resulting from factoring a more dense representation of the behavioral domain.

5. The most likely meaning of higher-strata factors is that they are influences affecting lower-strata influences, and contributing to their variance, just as primaries do to variables. However, there are already empirical indications that in some cases quite a different cause may be at work, namely, that they may be "spiral-feedback" emergents from interaction of certain members of the set of primaries. These are envisaged as an ef-

fect of mutual interaction or successive stimulation of primary factors, perhaps conditional upon the presence of some environmental situation favorable to such development.

6. The psychologist should guard against the "sympathetic magic" thinking that higher-strata factors are higher in importance. This is true neither statistically nor conceptually, for they generally predict less of a criterion than do scores on the primaries, and in meaning they may be conceptual entities falling right outside psychology or whatever the original field of the variables is, and difficult to interpret. Thus several personality primaries, in a widely ranging population sample, may be affected by a second-stratum factor of age or one of social status. But these factors are of less interest to psychology as such, being properly pursued as influences, of more relevance, respectively, for biology and sociology. Empirical evidence suggests that second-strata factors frequently diminish (in magnitude, as secondary sources of variance) when outside influences, e.g., age, social status, culture are more controlled and held to narrow limits. (Both Garrett and Tryon concluded that Spearman's g reduced and changed pattern with controlled reduction of age and social status ranges.) What really characterizes them and gives them some importance in both theoretical and quantitative terms, is a greater breadth of the variables that are to some degree affected, i.e., a smaller effect and a broader influence than characterizes primaries.

7. The loadings of higher-strata factors are naturally initially worked out as loadings on the stratum of factors immediately below them. However, means exist in the C–W and S–L formulas to calculate their loadings projected directly on the variables, vaulting the strata between. These loadings on variables may, to ensure clarity, be called "vaulted" patterns. Such patterns present the most practicable way of proceeding to estimation of higher-order factor scores since good scale scores for primaries and, still more for secondaries, as required if scores are to be estimated from the stratum next below, are not always available.

8. A distinction must be made, however, between using a formula concerned with *assessment of statistical effects* of higher orders on variables, as in the serial overlap analysis (SOA), using the C–W formula, or the direct influence model (IDI), using the Schmid–Leiman (S–L) formula, both vaulting the projections on to the variables, and accepting those analyses as representing actual models, i.e., postulating what *determining action actually occurs*. A model corresponding to the C–W formula of serial overlapping action, cut off at the line between primaries and secondaries, and called the organic oblique primary could be entertained since it proceeds as if the primaries, at oblique positions, are real, organic entities. However, it would have no way of accounting for the correlations among primaries, and, if merely descriptively continued to higher orders

would result in combined loadings from the various strata that would fit no reality because they would account for greater variance in the variables than the latter actually possess.

9. The preferred model has been called the *stratified uncorrelated determiner* (SUD) model. This supposes that the world of determiners is made up of uncorrelated factors, but lying at different strata levels, each acting on the factors in the level below. The observed correlation of lower orders is due to added, shared variance from a higher-stratum factor (in primaries, for example, from a secondary). These originally uncorrelated factors may be aptly designated the "stubs" of the observed correlated factors. The SUD model therefore is one of factor strata in which the determiner factors are uncorrelated both within and between strata, and it is not to be confused with what could be called a Schmid–Leiman model in which the higher orders are simply broader factors, acting directly on variables just like primaries, and which we have here called the direct (IDI) influence model. The S–L procedure is thus a method of analysis, not a model.

10. The stubs are the unique part of each factor in higher-order factoring, corresponding to the unique factors in variables in primary factoring. The main form of the calculation which separates out these stub factors is embodied in the Schmid–Leiman (S–L) analysis formula. An example is worked out above comparing the projections of higher orders (as oblique factors) on the variables by the C–W formula, and the projections as orthogonal stubs by the S–L formula. However, the composite matrix output for the SUD model is different from the Schmid–Leiman matrices for it is a series of stub factor loading matrices in which the loadings of each stub are on the factors immediately below, not vaulted all the way to the variables as in the serial overlap and independent direct influence models. This SUD model can be illustrated, in contrast to (a), the serial overlapping, and (b), the independent direct influence, matrices in Table 9.4 above by Table 9.5.

11. There are difficulties about accepting the Schmid–Leiman transformation to independent direct influence matrices as a model, rather than the SUD model, namely (a) it would theoretically require the projected (vaulted) loadings of the "higher orders" on the variables to show simple structure, and (b) it would require the number of factors to be greater than the number of primaries k found by the number of factors tests. It is agreed *a priori* that SS as reached for a second order on primaries does not necessarily or normally lead to its having SS also in its projections on the variables. However, since hyperplanes corresponding to the latter seem sometimes to be found, the question of which is the truer model—direct or indirect action—is still open. The notion of *degree of efficacy* of factor structures and concepts is introduced as one which, here

Table 9.5. Matrix Representation of Resolution according to the SUD Model

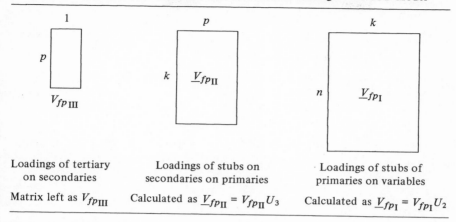

Loadings of tertiary on secondaries	Loadings of stubs on secondaries on primaries	Loadings of stubs of primaries on variables
Matrix left as $V_{fp_{III}}$	Calculated as $\underline{V}_{fp_{II}} = V_{fp_{II}}U_3$	Calculated as $\underline{V}_{fp_{I}} = V_{fp_{I}}U_2$

and elsewhere, can be applied to help choose among these models which must still be considered alternatives pending further research. Thus, it is possible to talk of a *family* of people doing this and that or of actions as emanating from *separate members*, the efficacy of the alternative conceptual structurings differing according to domain of action.

12. As a practical technical matter one must recognize that higher-strata factors are usually not as reliably rotatable to simple structure as are lower orders. Consequently, they are not as reliably definable in their structure meaning and scoring. Regarding the last, their scores are inevitably more contaminated by accumulated inclusion of specifics, from lower-strata factors and variables. A theoretically and practically interesting conclusion from strata investigation is that a single specification equation can be set out for any piece of behavior which fits the SUD model, using the S–L formula projection of higher-order "stub" loadings on the variables, as in the independent direct influences (IDI) model.

13. The most condensed statement of alternatives in a given higher-strata structure is in matrices representing the four ways of considering the action discussed above, illustrated for three strata as follows:

a. The basic oblique strata series, derived in the ordinary oblique resolution:

$$V_{fp_{I}} \quad : \quad V_{fp_{II}} \quad : \quad V_{fp_{III}} \tag{9.8a}$$
$$(n \times k) \quad (k \times p) \quad (p \times l)$$

b. Serial overlapping action calculation, stratified joint influence on variables, employing the C–W transformation:

$$V_{fp_{I}} \quad : V_{fp_{I}}V_{fp_{II}} : V_{fp_{I}}V_{fp_{II}}V_{fp_{III}} \tag{9.8b}$$
$$(n \times k) \quad (n \times p) \quad (n \times l)$$

c. Independent direct influence on variables model or calculation, employing the S–L transformation:

$$V_{fp_{\mathrm{I}}}U_{\mathrm{II}} : V_{fp_{\mathrm{I}}}V_{fp_{\mathrm{II}}}U_{\mathrm{III}} : V_{fp_{\mathrm{I}}}V_{fp_{\mathrm{II}}}V_{fp_{\mathrm{III}}} \tag{9.8c}$$
$$(n \times k) \qquad (n \times p) \qquad\qquad (n \times l)$$

d. The SUD, stratified unrelated (uncorrelated) determiners:

$$V_{fp_{\mathrm{I}}}U_{\mathrm{II}} : V_{fp_{\mathrm{II}}}U_{\mathrm{III}} : V_{fp_{\mathrm{III}}} \tag{9.8d}$$
$$(n \times k) \quad (k \times p) \quad (p \times l)$$

It will be noted that b, c, and d derive from a factoring and a unique
oblique rotation which first yield the solution (i) Eq. (9.8a).

CHAPTER 10

The Identification and Interpretation of Factors

10.1. What Are the Bases of Identification and Interpretation?

Behind us lies exploration and definition of the varieties of scientific factor models, and the grasp of practical procedures for extracting and uniquely rotating factors in conformity with those models. The trunk of the tree of factor analytic procedures is completed and we stand at a juncture where we may branch out in a variety of directions. Among these directions are included: (1) the identification and interpretation of factors; (2) the calculation of individual factor scores; (3) the calculation of factor battery validities; (4) the application in this area of psychometric concepts that go with scaling; (5) the examination of the outcome of

229

using various covariation indices in the correlation matrix; and (6) further pursuit of the model, from traits, into the domain of states and processes, etc.

At this point it is surely most desirable to move closer to consolidation of statistical and psychometric practices and ideas around the main model before extending the model to new uses. So the present chapter and Chapter 11, on scoring, psychometric scaling, and validity problems will be devoted to the first four above. After that the wider developments of models and relation indices will be considered in Chapter 12. These next few chapters are likely to be tougher in demands on statistical know-how than heretofore, and readers who feel the need are advised to get additional elaboration of the concepts and formulas, here stated in base essentials, from such books as those of Gorsuch (1974), Rummel (1970), Mulaik (1972), and Harman (1967), or from a class instructor at points indicated.

When a factor analysis has delivered the unique resolution, in such loading patterns as the V_{fp} and such correlations as in R_f the basic researcher is eager to unwrap the patterns to see how they fit some theory, or to see what discoveries he has developed as novel concepts in his field. But the practitioner may be more keen to calculate factor scores, so that he may proceed to diagnosis, prediction of criteria, etc., with individuals in clinical, educational, and industrial fields. If the scores are urgently needed in his present plans then he should perhaps go directly to the next chapter, for these branches can be taken in different orders. However, interpretation of the factors, and ascertaining what other research findings they match is of primary importance, even for practical use of scores, which is why it is tackled first here.

It may seem as though we have implied that interpretation hinges on the V_{fp} factor loading patterns. But at the outset a broad approach will recognize that interpretation uses two other approaches besides inspection of the loading pattern. These are the behavior of the factor measured, duly scored, in various experimental situations, as well as the correlation of the factor with other factors. These approaches require that we give here at least an introductory statement about getting factor scores, though the complexities in some forms thereof are deferred until Chapter 11.

However, the main interpretative approach considered in this chapter looks to the *factor pattern* output and draws inferences from the nature of the positive and negative loadings thereon, as well as from the nature of the variables not loaded at all on the factor. Although high and low, etc., commonly refer to values on the factor pattern matrix V_{fp} some psychologists have used other VD (*variable-dimension-relation*) matrices, and since these others *do* have a role we shall—after the next section—stop to clarify the meanings and relations of the various VD matrices. Most express, of course, the regression of variables on factors (or reference vectors) and the converse relations.

10.2. Identification and Interpretation from Loading Patterns

Frequently the psychologist seems to approach his interpretation by looking at the high loaded variables only, in the given factor pattern or the highest correlating variables in the factor structure matrix. (The former, V_{fp}, it will be remembered, consists of loadings, the latter, V_{fs}, of correlations.) This simple concentration on *salients* will often lead one astray. For example, both the general intelligence factor (fluid, g_f) and the spatial primary ability factor are loaded highly on spatial ability measures. They differ in what else is loaded on fluid general intelligence and in the fact that verbal and numerical ability are not loaded on the spatial factor. This searching of the domain to decide what is not loaded is a vital principle somewhat neglected in arguments on the nature of factors. It expresses, however, the methodological principle that to understand A we need to look also at what is designated as not-A. (Or in the poet's terms: "What should they know they of England, who only England know?") Consequently, one looks, after a glance at the positive loadings, first at the negative loadings and then turns around to examine the nature of those variables which *do not load that factor at all*. Both procedures require that one keep freshly in mind also the question of what variables were in fact never entered into the experiment.[1]

Probably the chief specific pitfall of misidentification of factors from salients arises from a lack of alertness to the phenomenon of *cooperative factors* (Cattell, 1946a, 1952b). In the broadest sense this simply means that the same set of variables can be fairly saliently loaded by two or more factors. This possibility exists because no fewer than three distinct factors can each load the same variable by .58, and even up to four by .50 with-

---

[1] Hopefully to reply to the criticism that "one only gets out of factor analysis what one puts in" will seem to the modern reader a mere prevention of the beating of a dead horse. Nevertheless, since it was a question appropriate in the early days of factor analysis and still gets raised occasionally, let us point out: (1) In no scientific investigatory method does one resort to the magic of finding relations between variables that have not been put into the experiment. One gets relations, including factors, only among variables used. (2) What one gets out of a well-designed— especially a representatively designed—factor experiment is a whole set of factor-variable relations commonly unsuspected and unknown to the investigator. The relations are new, the variables perhaps old. Quite often these unitary factors and relations are in fact so surprisingly new and in conflict with traditional hypothetical ideas of how variables are organized, that "what one gets out" is too startling to be easily accepted by these same critics.

out overshooting a communality of 1.0. (Even higher repetitive loading can occur with obliquity.) In the highest degree of cooperativeness, as when a particular correlation cluster ("surface trait" in personality) lies between two factors, the similarity of the whole loading pattern of each factor to the other is very high—as shown in the sympathetic and parasympathetic nervous system factors (Cattell, 1950a). Tests of mechanical aptitude, for example, may load not only on the mechanical aptitude factor, but also appreciably load on intelligence and spatial ability, while tests of verbal fluency will load also on V (verbal-grammatical ability), on retrieval (fluency g_r—Cattell, 1971a), and on speed g_s. Only by locating in the same study the slightly contrasting patterns of these other factors, or reasoning from the sample of variables used that they are likely to be absent, can one hope to make a correct identification. The history of personality factorings and namings is strewn with misidentifications. One sees confusions of F, surgency, H, parmia, and A, affectia, with one another and with exvia [the second-stratum extroversion factor loading A, F, H, and Q_2 (–)]. Exvia is different from any sum of A, F, and H in that it loads low self-sufficiency in decision making [$Q_2(-)$] while A, F, and H do not. In every study it is particularly important to refer back to the investigator's list of variables used, in order to ask whether the absence of an expected element in the pattern is simply due to that relevant variable not having been included in the research. One of the surest ways to avoid confounding factor X with factor Y is to have enough of the right variables to have the factors appear together in the same matrix, rather than decide while one is in absentia. This is one more argument for designing important factor experiments to carry a relatively large number of well-chosen variables.

Incidentally, much trouble down the years in efforts to integrate factor studies has arisen from rather rashly giving premature interpretative labels,[2] especially when derived from small studies, as mentioned above. Caution indicates that we should prefer descriptive labels at stage one *if single words existed in the dictionary for describing a factor rather than describing a behavioral variable.* But since they rarely do—factors being

---

[2] It is not the naivety or over enthusiasm of investigators alone which leads to so many interpretive labels being prematurely fixed to factors. A methodologically sound factor analytic finding in any substantive area—yielding a new concept, simply as a definite pattern of numbers, i.e., loadings—is the most valuable kind of contribution in itself. It carries our knowledge one step toward the well-replicated convergence on a final pattern, so badly needed by psychologists for development of dependable concepts. But many editors in substantive areas fail to realize this need for long-term programmatic replication of a pattern *per se* and actually demand that immediate relevance to the substantive area be given by interpretations based on a simple factoring report. It is as if, in a chemical journal, a report on a new determination of the freezing point of helium be required by the editor to "entertain" the reader by the whole textbook history and concept of helium.

new concepts[3]—and since there is always a continuum from description to interpretation, some new, partly interpretative term has often to be given early. The insurance against a snowballing of misinterpretations in such cases can only reside in a factor index system, such as Thurstone (1938), the present writer (1947, 1957a, 1973), and others have proposed and consistently used (see Appendix). Finally, it would hardly need to be said—except for some blatant historical instances—that one should not cause confusion by failing to relate current research findings to previous research findings through employing a new term where a highly suitable technical term has already been given—unless a radical new interpretation can be proved. In chemistry, Lavoisier was allowed his oxygen, Wöhler his uric acid, and Curie her polonium, each through his or her personal discoveries and interpretations, but no good chemist suddenly decided to call nitrogen or electrolysis something else. So such labels for new personality factors as surgency, parmia, and premsia are necessary, but if a factor confirms the ego or the superego structure previously well recognized in psychoanalysis, surely there should be respect for this precedence.

Provided the researcher maintains regard for these logical principles and social obligations, the game of factor interpretation can offer to the theoretician an intellectual exercise at least as entertaining as a crossword puzzle, and as demanding as abstract philosophy. He has to ask: "What is likely to be positively acting on variables a, b, c, etc., inhibiting to variables p, q, r, etc., and unable to act at all on variables u, v, w, etc.?" Thus, Spearman reached "Capacity to educe relations and correlates" as the quintessence of g, from studying the loadings of variables. And this ap-

---

[3] The leaders of programmatic research in substantive areas have usually attempted to be precise, and to do so also by breaking away from sloppy, popular terminology. Thurstone, looking at his primary abilities, coined such two-word labels—keeping to a good descriptive level—as "spatial reasoning", etc., which were distinctive. Spearman, perceiving the danger of popular terms, and the innumerable definitions of intelligence, stuck firmly with his symbol g, for the general factor he had discovered. Speaking of Spearman's g factor, Burt points out that the word intelligence is only two or three hundred years old, though "wit" may have sufficed for it before. But now that we have g_f and g_c, new terms such as fluid intelligence and crystallized intelligence, become necessary. A totally new concept such as the F factor in personality was given the new name, surgency vs. desurgency, by a Latin derivation from *surgere*, which describes this ready "rising up" of verbal expression in the surgent person. This term, interestingly enough, was soon picked up by scientist-novelist Lord Snow in describing characters in his novels. There is no objection to popular writing using psychology's technical terms, especially when correctly used as in this instance, any more than to correctly used concepts from physics in science fiction or journalism. What causes confusion is the opposite, as when some factorists use popular terms like sociability and extraversion—with their trailing misconceptions and journalistic value judgments—in referring to precise concepts like surgency, affectia, exvia, etc.

proach to ability structure through factor analysis was able to integrate and give new insight into the current definitions of intelligence emanating from other approaches ("Capacity to acquire capacity" from education, "Capacity to think abstractly" from clinical study of the defective, and "Adaptability of response to new situations" from animal experiment). Incidentally, since "capacity to make abstractions" is a central expression of intelligence it follows that interpreting factors, by seeking a common underlying determiner abstracted in relation to a given set of behaviors, is a very intelligence-demanding sport!

10.3. The Nature of the Five Most Common Variable Dimension Relation Matrices with an Introduction to Factor Scores

Some tolerably good contingent interpretations have been made from factor patterns alone from time to time; enough to yield satisfactory initial descriptive interpretations appropriate at stage one. But in the end, as indicated above, the other properties exhibited by a factor: (1) its higher-order relations (correlations with other primaries), (2) its relation as a score to various life criteria, and (3) its response to manipulation, e.g., its age change curve, its genetic determination, (and, in the case of a state factor, the provocative environmental stimuli), are necessary to the potent development of a long-lived and scientifically profitable hypothesis. However, before pursuing interpretation through the wide domains of (1), (2), and (3), it is generally best to find out more about the factor in terms of its loading pattern alone. Indeed, since initial loadings are apt to be low, and factor scores consequently not too valid, an extension of the factor pattern and discovery of higher loaded variables, by further experiment, is generally the first requirement. To pursue this, the insightful experimenter forms the clearest possible hypothesis and proceeds (4) to enter a series of fresh factorings with the variables from his hypotheses, in the spirit of the IHD spiral described in Chapter 1.

All four of the further directions of investigation cited above require, however, that we get a more precise understanding of the meaning of loadings as such than we have yet acquired, and this means more precisely understanding the relations among the variable-dimension matrices introduced in the last two chapters. Mostly we have dealt with reference vector structure and factor patterns but now Table 10.1 lists the most important ten VD matrices, i.e., matrices relating variables to dimensions, in one sense or another. The first four are by now known to the reader at least by name, and the fifth V_{fe} has been mentioned in a general sense as that from which factor scores are estimated. To the remainder we will give attention later in this section.

Table 10.1. The Most Important VD (Variable-Dimension Relations)
Matrices (Four Common, Six Less Common)
(All of Order $n \times k$)

1.[a] V_{rs}, *Reference vector structure matrix.*
Giving correlations of variables with reference vectors, as reached by the original rotational resolution.

2. V_{rp}, *Reference vector pattern matrix.*
Loadings of reference vectors on variables. Rarely used.

3.[a] V_{fs}, *Factor structure matrix.*
Correlations of factors and variables.

4.[a] V_{fp}, *Factor pattern matrix.*
Loadings of factors on variables, for variable estimation.

5.[a] V_{fe}, *Factor estimation matrix.*
Weights of variables to get factors, in factor score estimation.

6. V_{dfc}, *Dissociated factor contribution matrix.*

7. V_{afc}, *Associated factor contribution matrix.*

8. $\underline{V}_{dfc}$, *Stub dissociated contribution matrix.*

9. $\underline{V}_{afc}$, *Stub associated contribution matrix.*

10. V_{fm}, *Factor mandate matrix.*

---

[a]Most commonly used in ordinary practice.

Although a thorough treatment of factor scores is deferred to the next chapter, we must, for completeness of view over Table 10.1, deal briefly here with the nature of the important V_{fe} matrix in Table 10.1. From general multiple regression principles it will be recognized that the estimation of a factor score requires that we know (a) the correlations of the variables with the factor and (b) the correlations of the variables among themselves. The factor estimation matrix V_{fe} is nothing other than the weights, from a multiple regression calculation, to be assigned to each variable (in the company of the other variables) to get the best possible estimate of the factor score. Each column carries the weights on all variables appropriate for estimating one factor. Incidentally, it is convenient for ease of clear discussion to speak of "weights" for variables in getting factors and "loadings" for factors in getting variables. If the usual formula for allowing for variable intercorrelations in getting weights on a criterion is organized for calculation in matrix form we have ($\underline{R}_v$ being unreduced)

$$V_{fe} = \underline{R}_v^{-1} V_{fs} \qquad (10.1)$$

$$(n \times k) = (n \times n)(n \times k)$$

That is to say, the direct correlations of the variables with any one factor, given in any one column of V_{fs}, are not themselves the weights to use, but must be corrected for the correlations among the variables by premultiplying by the inverse of the (unreduced) correlation matrix among variables. There is an alternative to this, generally called the "ideal variable" method, which is discussed in the next chapter.

For completeness of view we should notice here, as we have briefly before, that the reciprocal direction of correlation to Eq. (10.1) is the estimation of variable scores from factor scores (regression on variables) and that this is given by the loadings of factors on variables, which we have all along called the factor pattern matrix V_{fp}. In this reciprocal relation it is natural that the correlations of the factors should now enter the estimation, by using the inverse of the R_f, as follows:

$$V_{fp} = V_{fs}R_f^{-1} \qquad (10.2)$$

Actually, in the course of factoring we generally get the V_{fp} output first. It is the reference vector structure which we have used all along in the rotation process multiplied by the diagonal "stretching" matrix of tensors D^{-1}. The D, it will be remembered, is obtained from the $L'L$ matrix by taking the square root of the inverted diagonal values in that matrix, i.e., of $(L^T L)^{-1}$. Accordingly, since we now want V_{fs} for Eq. (10.1) it can be obtained from Eq. (10.2) by

$$V_{fs} = V_{fp}R_f \qquad (10.3a)$$

or, if we go all the way from the reference vector structure

$$V_{fs} = V_{rs}D^{-1}R_f \qquad (10.3b)$$

which means that V_{fe} [from Eq. (10.1)] may be written to summarize its derivation as

$$V_{fe} = \underline{R_v^{-1}}V_{fp}R_f \qquad (10.4)$$

Still keeping in view the reciprocity of the two-way regression between factors and variables, we note that the estimation of a variable score from V_{fp} is

$$\hat{Z}_v = Z_f V_{fp}^T \qquad (10.5)$$

$$(N \times n) = (N \times k)(k \times n)$$

In this, Z_f is a factor score matrix of N people by k factors (note that specific factors are not included) with factors in standard scores. By the rule of matrix multiplication it will be seen that the first score in $\underline{Z}_v$ — the matrix of variable scores is

$$\hat{Z}_{1i} = b_{11}F_{1i} + b_{12}F_{2i} + \cdots + b_{1k}F_{ki} \qquad (10.6)$$

and others similarly, the b coming from the V_{fp}.

That is to say, it is our old friend the specification equation, except that it lacks the specific factor, a score for which we do not have. The specific factor score and loading could be entered in Eq. (10.6) for theoretical completeness, but it would be fanciful and do us no good because in practice we have no acceptable estimate of the specific factor scores other than from the variable itself which we are trying to estimate. For that reason, the $\hat{Z}_{1i}$ score in Eq. (10.6) is only an estimate (as shown by the "hat"), erratic in proportion to the size of the loading (known) for the specific factor. We can suppose that the latter loading is implicitly applied to the unknown individual specific factor score, which is given a mean value of zero. The result is, though the factor scores are here entered as standard scores with a sigma of one, the variable scores will be semistandardized only, falling short of a sigma of one. They will have a mean of zero and a normal distribution of the factor scores as usual, but we must write them as $\hat{Z}$ not Z, because they are only estimates and the sigmas are short of one.

In most practical situations, moreover, we do not even start with fully valid factor scores, such that the true sigmas of the factors are unities, i.e., the broad factors, F_1, F_2, etc., will also be estimates, and in most cases also with sigma <1.0 so that the variable sigmas will be still farther short of unity. For such a practical situation we may more truly write $\underline{\hat{Z}}_v$ as $\underline{\underline{\hat{Z}}}_v$, and Z_f and $\hat{Z}_f$ as follows:

$$\underline{\hat{Z}}_v = \underline{\hat{Z}}_f V_{fp}^T \tag{10.7}$$

In Eq. (10.5) we have supposed the factor estimates from batteries, etc., are in practice restandardized to standard scores, but, one must repeat that $\hat{Z}_v$ will not be in standard score, and the columns of $\underline{\hat{Z}}_v$ [Eq. (10.5)] will have variance less than unity. It is useful as a reminder to keep the symbols $\hat{Z}_v$ and $\underline{\hat{Z}}_v$ instead of $\hat{Z}_v$ for these matrices, the former being a "contracted standardization" of semistandardized scores, as we may call them in addition, of course, to the estimate sign which implies some more random inaccuracies in relation to the true value. The $\hat{Z}_v$ we would use for estimated scores stretched back (by dividing by sigma) to a standard sigma of 1.0.

Now the converse estimate—that of factors from variables—with which we are concerned in this "new" V_{fe} addition to VD matrices, will take a matrix of people's standard scores on the variables Z_v and multiply it by this factor estimation matrix V_{fe}, thus

$$\hat{Z}_f = Z_v V_{fe} \tag{10.8}$$

$$(N \times k) = (N \times n)(n \times k)$$

Here again $\hat{Z}_f$ is both an estimate and short of a sigma of 1.0 in its columns. Or, if we wish to keep in view the developmental steps to V_{fe}, then

$$\underline{\hat{Z}}_f = Z_v R_v^{-1} V_{rs} D^{-1} R_f \tag{10.9}$$

$$(N \times k) = (N \times n)(n \times n)(n \times k)(k \times k)(k \times k)$$

In Eqs. (10.8) and (10.9) using variable scores, we do not have the problem that the predictors cannot be obtained in correct standard scores, as we do when starting with factors, estimated from other sources, such as partly invalid batteries, so no $\hat{\underline{Z}}$ appears. The variables are a real basis in real standard scores. Nevertheless, the resulting factor scores will be short of unity according to the imperfection of estimate, so again we write $\hat{\underline{Z}}_f$ rather than Z_f. Such approaches to standard scores, which have a zero mean but a sigma short of one, we have just agreed conveniently to call "contracted standardized" or semistandardized. Contracted standard scores can quickly be converted to standard scores as mentioned by multiplying $\underline{Z}_f$ by a diagonal matrix D_c, c standing for corrected for underestimation thus (and similarly for $\hat{\underline{Z}}_f$):

$$\hat{Z}_f = \hat{\underline{Z}}_f D_c \qquad (10.10)$$

The values in this D_c matrix would generally be obtained simply by looking at the empirically obtained factor sigmas in $\hat{Z}_f$ and taking their reciprocals. But they can also be worked out beforehand by calculating the multiple R from the correlations along with weights of variables in estimating the factor, and dividing the given sigma thereby as in Eq. (10.11) in the next section.

10.4. Five More Esoteric Variable Dimension Relation Matrices

So much for the precise nature and mutual transformations of the five VDs with which the reader has been in some degree familiar from the beginning of oblique resolution. Four of them, he realizes are in constant use—the V_{rs}, V_{fs}, V_{fp}, and V_{fe}—while the fifth, V_{rp}, though in the same family of transformations, is seldom used.

The above relations between variables and dimensions express either correlations between them or the weight given to a score on one in estimating a score on the other. As one seeks to interpret factors, and envisage further properties, however, a slightly different kind of statistical information concerning how much one contributes to the variance of the other, can throw some additional light. From the calculation of the multiple correlation coefficient R which leads to the fraction of variance $\sigma^2 = R^2$, that a set of predictors contributes to a criterion, the reader knows that one has to multiply the weight by correlation for each predictor and sum to get the desired value, thus:

$$R^2_{x \cdot 1,2,\cdots,n} = r_{1x}w_1 + r_{2x}w_2 + \cdots + r_{nx}w_n \qquad (10.11)$$

where the w's are weights, as in V_{fp} and V_{fe} above, and the r's are correlations, as in the V_{fs}, for each variable involved.

If we are concerned with how well each of the k factors is being estimated from the n variables, then we can obtain the R's by matrix procedures by taking the square roots of the diagonal in the R_x^2 matrix obtained
as follows:

$$V_{fe}^T V_{fs} = R_x^2 \qquad (10.12)$$

$$(k \times n)(n \times k) = (k \times k)$$

The R's are essentially *validities* (concept validities) for the n tests
used as a "battery" for each factor. When squared they also tell us, as
shown above and below, the variance of its estimated factor scores, if
variables themselves all have unit variance, i.e., are in standard scores. The
essential steps in getting an individual's factor score are, therefore, to take
the V_{fp} obtained, drop the specific part—the $(n \times n)$ diagonal matrix of
unique loadings, as a useless appendage—and calculate by Eq. (10.4) the
V_{fe}. One then puts the subjects' scores in standardized form in an $(N \times n)$
matrix and calculates by Eq. (10.8) the $\hat{Z}_f$ factor scores. These can then
be "stretched" to give standard scores on factors, by putting into the computer a diagonal matrix D_c, as in Eq. (10.10), obtained by either method
above. If, conversely [to Eq. (10.12)] we want the extent of prediction
of a variable from the oblique factors we take the root of the diagonal of
R_v^2 where

$$V_{fs} V_{fp}^T = R_v^2 \qquad (10.13)$$

$$(n \times k)(k \times n) = (n \times n)$$

By algebraic transformations of Eq. (10.13) it can be shown that
these R^2's are the same as the communalities h_v^2 in the V_0 matrix before
transformation to an oblique resolution. The communalities of course remain the same under rotation but they can no longer be calculated from
the oblique matrix V_{fs} simply by squaring and adding the row of correlations or from V_{fp} simply by using the loadings.

The formulas and concepts in Eqs. (10.12) and (10.13) have to do
with the total variance contribution to a factor or to a variable. However,
it is useful to have variance contributions broken down by single variables
and individual factors, in VDs that are analogs to the weights and correlations in the variable oriented V_{fp} on the one hand and the factor oriented V_{fe} on the other. (The V_{fs} is oriented both ways.) Such a pair of
variance-contribution VDs, however, can be diversified into three by considering whether the factor contribution to variables is to be evaluated
from the factor in and of itself, i.e., with other factors partialled out, or
as it acts in the company of the set of k factors concerned. The logical
meaning here seems best preserved by calling these alternatives the *dissociated* and the *associated* contribution matrices, respectively, while the

matrix of contribution of variables to the factor can be called the variable contribution matrix, V_{vc}. Let us now define these in more detail.

The *dissociated factor contribution matrix*, V_{dfc}, takes the V_{fp}, which, of course, gives the loadings of each factor in dissociation from (with partialing out of) every other factor. T_c multiplies each term in the V_{fp} by the corresponding correlation of that factor with the variable, using the same reasoning as for variables in Eq. (10.11). The multiplication of two matrices cell by corresponding cell with no succeeding summation must not be represented by the ordinary matrix multiplication symbol, and we shall represent it by a (:) between the two, thus:

$$V_{dfc} = V_{fp} : V_{fs} \qquad (10.14)$$

The sum by rows in V_{dfc} are the communalities and the sum by columns are the magnitude of general effect of the factors. The latter is novel to us, but is the basis for evaluating the "size" of a factor (see Chapter 13). It has sometimes been urged as a basis for factor interpretation, but our further discussion suggests keeping to V_{fp} patterns, though the difference will generally be numerically small. As a peculiarity it should be noted that since a loading usually has the same algebraic sign as the correlation from which it is derived there will normally be no negative values. However, in multiple correlation work and here, where large correlations happen to arise among factors, trivial to small negative variances can appear and the meaning of a negative contribution toward a variance has to be entertained. If the values in the V_{dfc} are to be comparable in size with those in the V_{fp}, V_{fs}, V_{afc}, etc., they must be entered as the square roots of the products above, which is impossible where there are negatives, and so generally one will leave them as variances. It is noticeable that the V_{dfc} (see Table 10.2) has more near-zero values than the others, but this cannot be taken as SS in the ordinary sense. The use of the V_{dfc} (which, because of lack of significant negative values lacks immediate resemblance of pattern to the others) lies principally in reminding us of *the real contribution* of a given factor, as such, to variables. Calculating communalities with oblique factors is done by Eq. (10.14) or its equivalent.

The *associated factor contribution matrix*, V_{afc}, has just the opposite intention from the V_{dfc} in that it seeks (to quote the original articles, Cattell, 1962b, 1966a) "to state how much of the variance of each variable is accounted for by both the variance from the factor under consideration and the covariance which that factor shares with its usual entourage of correlated factors (in that study, population and culture)." Inasmuch as we speak of a particular "entourage," decided by the associated variables and therefore factors in a given experiment, the V_{afc} values must be at the mercy of circumstances, and have no constant or "permanent" value—unless a total entourage can be defined. But they are not more company dependent than the weight values in the V_{fe} matrix applied to particular variables for estimating a factor, similarly depending on the entourage of variables. At least such values are likely to approach constancy as the sampling of the personality sphere of variables approaches that of a comprehensive stratified sample. The correlation of a factor with a variable in the V_{fs} represents what the factor does also in virtue of its associate factors. For example, in Fig. 8.1, if factor 2 approached factor 1 and vice versa, so that they became positively instead of negatively correlated, it is easy to see that variable a_j, which is loaded on both, would have higher correlations with both. The higher correlation would be due to both sharing a second-order factor, due to their positive correlation and thus contributing more to the variance of a_j. The V_{fs} correlations, for a given factor, depend, in short, on its correlations with other factors. The magnitude of the factor's contribution to the variable is the square of that correlation, and the complete V_{afc} is the cell by cell product:

$$V_{afc} = V_{fs} : V_{fs} \qquad (10.15)$$

In both Eqs. (10.14) and (10.15) the value in the matrix (V_{afc} or V_{dfc}) shows how much the variance of the variable would drop with the removal of the factor (reduction of its variance to zero) but the amounts differ because in one case it drops out alone and in the other it is assumed to carry away with it the covariance with other factors. Parenthetically, since the values in the V_{dfc} and V_{afc} are variance contributions not rs or ws—indeed the squares of the V_{fs} values in the V_{afc} and approximately the squares of those in V_{fp}—one must remember in comparing the shape of the patterns that the patterns will have decidedly smaller values in the variance contribution matrices, V_{dfc} and V_{afc}, as can be seen in Table 10.2 comparing (b) with (a).

One can play one more variation on the theme of the above VD matrices, namely that of substituting the "stub" matrices from the Schmid–Leiman (Chapter 9, p. 218) for the "full length" primary factors throughout three matrices. Thereupon all nine of the above VDs will alter, in the direction of lower values, proportional to the unique variances in the $(k \times k)$ diagonal unique factor part. Deriving these, which we can designate by underlines for $\underline{V_{fp}}$, $\underline{V_{fs}}$, $\underline{V_{afc}}$, etc., is a comparatively straightforward exercise for the student. If we are correct in putting the greater confidence in the SUD scientific model it is in these VDs that we should look for the refined views on factor interpretation. For example, the general arguments above on the width of the hyperplane suggest that, when that width is appropriately fixed by the standard error of a loading, it is in the V_{dfc} that we should evaluate simple structure.

Finally we come to the *factor mandate* (or *factor influence*) matrix, V_{fm}. In the domain of bivariate experiment, with manipulation intro-

Table 10.2. Personality Research Illustration of V_{rs}, V_{fp}, V_{fs}, V_{fe}, V_{afc}, and V_{dfp} Matrices

(a) Degree of resemblance of factor pattern, factor structure, factor estimation weights, and reference vector structure on intelligence, extraversion-introversion, and anxiety

Reference number in larger factor matrix	Variable	Master Index Number	U.I. 1 (Intelligence)				U.I. 32 (Invia-exvia)				U.I. 24 (Anxiety)			
			V_{rs}	V_{fp}	V_{fs}	V_{fe}	V_{rs}	V_{fp}	V_{fs}	V_{fe}	V_{rs}	V_{fp}	V_{fs}	V_{fe}
20	Numerical performance	199	+.50	+.505	+.491	+.352	.00	.000	−.015	−.014	−.11	−.112	−.047	−.030
31	Classification	51	+.52	+.525	+.507	+.376	+.05	+.051	+.032	+.034	−.14	−.143	−.069	−.048
32	Short interval memory	249	+.36	+.363	+.370	+.231	−.08	−.081	−.073	−.066	+.05	+.051	+.086	+.057
4	16 P.F. desurgency (F−) scale	F	+.03	+.030	+.009	−.002	+.44	+.444	+.420	+.357	−.17	−.173	−.107	−.098
6	16 P.F. threctia (H−) scale	H	+.02	+.020	+.064	+.041	+.42	+.424	+.471	+.420	+.33	+.336	+.398	+.271
44	Fluency; dreams /total	271	.00	.000	+.010	+.002	+.03	+.030	+.041	+.025	+.07	+.071	+.080	+.042
99	Annoyability	211	+.02	+.020	+.088	+.073	−.01	−.010	+.064	+.001	+.52	+.529	+.531	+.381
16	Readiness to confess frailties	219	+.05	+.050	+.105	+.061	−.03	−.030	+.029	−.017	+.42	+.428	+.430	+.276
55	Tendency to agree	152	−.05	−.050	−.019	−.010	+.03	+.030	+.064	+.027	+.24	+.241	+.212	+.129

(b) Unique characteristics of factors (in DV profiles), shown in associated and dissociated factor pattern matrices

Reference number in larger factor matrix	Variable	U.I. 1 (Intelligence)		U.I. 32 (Introversion)		U.I. 24 (Anxiety)	
		V_{afc}[a]	V_{dfc}	V_{afc}	V_{dfc}	V_{afc}	V_{dfc}
20	Numerical performance	+.248	+.500	.000	.000	+.005	-.072
31	Classification	+.266	+.520	+.002	+.050	+.010	-.092
32	Short time memory	+.134	+.360	+.006	-.080	+.004	+.033
4	16 P.F. desurgency (F-) scale	.000	+.030	+.186	+.440	+.019	-.111
6	16 P.F. threctia (H-) scale	+.001	+.021	+.200	+.023	+.134	+.229
44	Fluency; dreams/total	.000	.000	+.001	+.030	+.006	+.046
99	Annoyability	+.002	+.021	-.001	-.011	+.281	+.373
16	Readiness to confess frailties	+.005	+.053	-.001	-.032	+.184	+.291
55	Tendency to agree	+.001	-.050	+.002	+.030	+.059	+.157

[a] As usual in multiple correlations in extreme values it is possible for variance contribution products to be negative as in the instances above where loading and correlations are opposite in sign. It will be remembered that these figures are combinations of variance and covariance, the latter being sometimes negative. From Cattell, *Educational and Psychological Measurement*, 1962b.

duced to hold all else constant, as typical in some readily controlled "brass instrument" experiments, the psychologist frequently sees linear scatter plots of dependent variable and independent manipulated factor which are so narrow as to amount to a correlation of 1.0 between the factor influence and the dependent variable. (In some cases to thoroughly randomize all else over large numbers of variables will produce the same effect.) By comparison, in a multivariate experimental design one is apt to forget that the loadings of .2, .4, .5, etc., typical in a factor pattern matrix really mean, if they are significant, that the correlation *would be 1.0 if all other influences were held constant.* In other words, the essential conclusion is that a factor either influences a variable or it does not and that variations in loading are due to the particular "company" of other uncontrolled factors systematically active in influence on the same variable. In terms of psychological concepts and theory often this stark truth, rather than any precise loading value, is the important discovery. One must not thereby belittle the findings of the actual V_{fp}, V_{fs}, etc., values since they are, of course, not only statements about the other factors in the situational setting but also useful numbers for a practical statistical estimate or prediction in that kind of situation. However, the main property of the factor is given by the variables it significantly influences. To a chemist it is conceptually important to know that phosphorus is capable of interacting with potassium, with iron, with many organic compounds, and not, say, with fluorine or with argon. And it is of general theoretical importance regarding fluorine that, say, it reacts with many chemicals whereas helium has dealings with very few.

Much of what we want to know about the nature of a certain factor or factors can be stated sufficiently therefore by a matrix of 0's where no effect exists, and 1.0's or -1.0's where there is definite interaction, positive or negative, augmenting or reducing the score in any significant degree on the other variable. *Factor mandate* is somewhat preferable to "factor influence," as a term to designate such a "statement" matrix, since influence also exists in all quantitatively graded matrices, whereas here we have a straight statement of whether, in terms of natural properties, the factor does or does not have the capability or power invariably to affect the given variables. All that is needed to write a factor mandate matrix is a statistical decision, preferably gathered over a good share of experimental matrices, whether a V_{fp}, or better, a V_{dfc}, entry is or is not significantly different from zero.

10.5. Discussion and Illustration of Relations of VD Matrices

There are yet some other possible "hybrid" VD matrices from combinations of values on the primary four or five VDs set out at the beginning of Table 10.1 that could be discussed if one wished to pursue issues further. But their relevance to practical work is relatively remote and we shall stop with the above (see, however, for some other considered entities, Cattell, 1962b). It will help the reader's grasp of the above, however, to see them in concrete terms, and Table 10.2 therefore shows the values worked out, by the formulas above, for three second-order factors, intelligence, exvia–invia, and anxiety, in relation to nine variables cut out of a larger matrix. [It has also the psychological interest of showing a second-order questionnaire factor as a first-order T data (see Chapter 9, p. 204) and is on a sufficient sample, 500, for relatively small loadings to be significant.]

It will be noted that though the shape of all is quite similar for any one factor the V_{fp}, V_{rs}, and V_{dfc} form a family. The V_{fp} has the same hyperplane structure as the V_{rs} and virtually the same as the V_{dfc}, but in the rest the essentially zero values are somewhat otherwise distributed. We see also that the weight assigned a variable in the V_{fe} can differ appreciably from the factor loading. For example, the Annoyability Test (indexed in Cattell & Warburton, 1967) MI 211 loads the anxiety factor about twice as highly as does tendency to agree (MI 152), but in measuring anxiety by a battery (the V_{fe}) it becomes three times as important. One may also notice [Table 10.2(b)] that the loadings on any primary variable that should be truly independent are always less in the V_{dfc} than in the V_{fp}. It will be noted that the V_{afc}, as would be expected, gives a less clear picture of the factor itself than the V_{fp} and V_{dfc} though at a practical level it gives a truer picture of what the factor, in interactive covariance with its entourage of associated factors, contributes to the variance of variables. It tells how much a higher score on a factor will finally affect the individual's behavior, and thus has practical value, whereas the V_{fp}, and, better, the V_{dfc}, give more abstracted "clean" patterns which help us grasp the theoretical meaning of the factor. Although the task of interpretation, as stated at the beginning, eventually requires more than attention to patterns, yet it is important that the right patterns be intelligently employed and compared. Our position here has been that the V_{fp} (and ultimately the V_{dfc} if variance contribution is the criterion) is the essential matrix to be considered, both in assessing simple structure (along with the V_{rs}) and in interpretation. The objection is sometimes offered that since the V_{fp} consists of multiple regression weights, certain objections normally applied to beta weights apply here, e.g., that as obliqueness becomes great it is possible to have a variable weighted opposite in sign to that of its correlation with the factor (in the V_{fs}), and that suppressor action could bring variables out of the hyperplane that should be in it. These objections suppose the obliqueness, however, to be, as it were, accidental instead of the intrinsic part of the (second order) factor structure that theory requires it to be. Rather than the loading it is the *correlation* of the variable with the factor (in the V_{fs}) that is unstable, depending on the correlation of factors produced by second-order influences. But the *loading* (in the V_{fp}) should be constant across variations in factor correlation and changes of second-strata factor variance (Thurstone, 1947). However, the above discussion of the eight matrices in Table 10.2 surely strongly indicates to the reader that a really shrewd interpretation of a factor requires a look at what differs and changes in a comparative study of at least the V_{fp}, V_{fs}, V_{dfc}, and V_{fm} matrices.

10.6. Planning for Factor Identification by Matching: The Four Experimental Possibilities

Granted that we understand what we mean by the pattern of a factor, the next step after producing a pattern uniquely related to a certified significance[4] of simple structure or confactor relationship is to

---

[4] Objections are sometimes raised to the procedure of (a) attaining significant SS independently in the two studies, and (b) then applying matching indices, in favor of dispensing with SS and rotating the two V_0s immediately toward some best agreement (Korth, 1978). There are two arguments often run together here, and a comment on each is needed separately.

(1) First, one reviewer expresses a not uncommon view that the test of significance is not necessary, for "unless the researcher is a bumbling idiot in the use of graphical ROTOPLOT, or uses a seriously defective analytical rotation program he cannot help finding a position with a significantly better count than a random spin, causing the null hypothesis to be rejected." The answer is that something better than "better than a random spin" is needed, namely, proof of a position unimprovable by prolonged trial. Simple structures satisfying only a low but "statistically significant p value can be diverse, and differ appreciably in position and still more in the second-order structure position. One simply has to demand stiffer limits (as in the Bargmann test) and, above all, proof of maximization, as in the *history of hyperplane* plot. (2) A more confused objection, mistaking the logic of scientific inference, is that implied in the notion of rotating to maximum mutual agreement followed by demonstration that excellent agreement is obtained. This can capitalize upon all kinds of chance. Moreover, in this procedure there is confusion between the objectives of *confactor rotation*, which calls for deliberately designing two experiments to facilitate a judgment on whether a *parallel proportional profiles* (Cattell, 1944d; Cattell & Cattell, 1955) relation can be obtained, and, if so, allowing it to fix the rotation and the objective simply of a least-squares type of fit of two rotations. The latter was proposed by Tucker (1951) as a basis for beginning a rotation of the joint studies to maximum simple structure, and by Mulaik (1972) and others as rotating V_{fpA} to an unrestricted best fit to V_{fpB} (A and B being two experiments and the V_{fp} any kind of initial rotational positions). Such a complete match by such a transformation shows the two systems are convertible, which is not surprising with the same variables and equivalent samples, but it begs the question of what the desirable criteria are for the unique rotation itself. Meredith's (1964) method brings several studies to maximum conformity, but he recognizes that this is not confactor rotation.

check the pattern against those in other, independent researches to see if some identification of the factor can be sustained. To do this effectively presupposes in the first place that the researcher has chosen enough *marker variables* from relevant studies to permit patterns to be compared on identical (not roughly similar!) variables. Incidentally, it also supposes that hyperplane stuff variables have been included. For example, when somewhat similar (partly cooperative) factors like surgency (F) and parmia (H) in personality, or fluid and crystallized intelligence among abilities, are involved, it is important to include variables hypothesized to be the hyperplane of, for example, fluid intelligence but not in that of crystallized, and vice versa.

In spite of injunctions of the above nature, and general appeals to the strategy and polity of science, published factor analyses over the past forty years have all too frequently indulged in a sterile isolationism of unconnectable studies, without marker variables. The result of such narcistic planning in "private worlds" has been to scatter one-story shanties over the scenery where an architectonic growth of cathedrals of integrated knowledge might have soared. One-shot experiments in a tightly controlled area of precision in, say, physics can be meaningful, but because of the intrusiveness of unavoidable sources of error, confidence is justified in factor analysis only in results well-replicated at least once. And when some adequate factorings using representative design, e.g., the personality sphere, have broken ground in a new area and revealed markers in a dozen or more dimensions, there is no excuse any longer for studies in that area to be encapsulated, in some subjective choice of variables, against scientific integrative advances.

Fortunately, examples of good planning for subsequent matching and integration procedures are available for the student in a number of excellent studies in both the ability and the personality field, in Horn and Anderson's work on fluid and crystallized intelligence and the work on primary abilities from Thurstone to Hakstian. Unfortunately, such planning occurs less widely in the field of personality, motivation, and emotional states, where researches abortive for this reason abound. Gradually, however, the practice of factor analysis has gained sophistication in this matter, and, in the latter field, one can point to the studies by Baggaley, Baltes, Barton, Bolz, Burdsal, Cartwright, Curran, Dielman, Damarin, Dermen, DeYoung, Digman, Eysenck, French, Goldberg, Gorsuch, Hundleby, Krug, Knapp, Nesselroade, Norman, Pierson, Royce, Saville, Schaie, Schuerger, Sells, Sweney, Wardell, and Watterson, among others in Britain and the USA, and Adcock, Delhees, Fahrenberg, Gibb, Meschieri, Pawlik, Pichot, Schmidt, Schneewind, Schumacher, and Tsujioka, in other countries, carefully carrying markers, of an exact kind, to make research architectonic and to preserve the international character of scientific concepts.

Owing to the statistical requirements of the pattern similarity matching indices we are about to discuss, it is desirable not only to have a minimum of two or three good markers per factor but also to have a total variable sample of twenty and preferably forty variables in common to two analyses for psychometric soundness in using matching indices. Some tolerably definite conclusions from the congruence coefficient have nevertheless been obtained with as few as ten. This common-marker requirement demands farsighted and programmatic planning in research because, on the one hand, testing the developing new hypotheses requires that a sufficiency of new, theory-crucial variables be added at each experiment, while retaining enough markers firmly to hold the framework of recognized structures so far known. Thus a conflict arises between keeping sufficient length of time to make each test measure reliable and the difficulties of getting ample subject time for, say, as many as 50–100 different subtests such as are needed to encompass 10–30 factors.

Granted the right choice of variable to elucidate patterns we have two kinds of matching situations [(1) and (2) below]. But if we consider the whole problem of trying to match research results, then we logically have more, namely, four possibilities, as follows (Cattell, 1966a, 1969, 1973; Buss, 1975; Royce, 1973):

1. Same variables, same subjects
2. Same variables, different subjects
3. Different variables, same subjects
4. Different variables, different subjects

The first situation is not commonly presented in the usual comparison of independent researches, since it hinges only on different analyses and rotations of identical data. For example, there may be two different technical methods of resolving the same data, the outcomes of which we may need to compare. In this special case, in addition to the methods of pattern comparison available in (2) and discussed below, the two solutions can be compared by correlating the factors deemed to be the same, either as factor score estimates (since the individuals are the same) or by calculating $L_1^T L_2$, where L_1 and L_2 are the two discovered transformation matrices, applied to the same V_0, whereupon the diagonal will give us the correlations within the pairs of factors judged to be the same (if previously arranged in order according to theory). Obviously the latter is better than going through estimated scores, if the same R_v and V_0 are the starting points. If the researcher is not starting with the identical V_0 and R_v as when the same subjects are measured on two different occasions, the case falls under (2).

Case (2) merits our main concern, so we shall discuss (3) and (4) briefly and return to settle down to (2). Case (3) requires that the two sets of variables be correlated in the same R matrix and factored to the same V_0. Then one may wish to rotate the upper (set 1) and lower (set 2) parts separately to whatever simple structure, PROCRUSTES, or confactor unique resolution one intends. Since the rotations start from the same V_0, it is easy again to use $L_1^T L_2$, as before, to give the correlations of the factors. Alternately one can rotate together, but this seems less of a check on "factors being the same." Nevertheless, the present writer used this latter approach (1955) to test the hypotheses that the second orders Q I, exvia, and Q II, anxiety, in Q data are the same as first orders U.I. 32 and U.I. 24 in T data. Actually, in this case, a single common simple structure was found. The resemblances of set 1 and set 2 patterns, respectively, to their earlier known expressions as Q I and U.I. 32, etc., patterns were determined by methods for case (2). Tucker's interbattery method (1958), which operates only on one quadrant of the R matrix just considered—namely a matrix of r's between set 1 and set 2 variables—has sometimes been brought to bear on this problem. A good account of procedure will be found in Gorsuch (1974, pp. 284–286). It treats the "common matrix" $R_{v_1 v_2}$ as analyzable, thus:

$$R_{v_1 v_2} = V_{f p_1} R_f V_{f p_2}^T \qquad (10.16)$$

The pursuit of simple structure can be made separately in the two V_{fp}'s or together. A slight objection to Tucker's method (in the present writer's opinion) is that one is operating only in the space common to the two sets of variables. If certain factors in what would have been the V_{fp_1} on its own (as in our first solution) are poorly represented in the second set of variables, in V_{fp}, they will barely appear in the Eq. (10.16) solution, and can be distorted as not to be easily recognizable (see Hakstian and Cattell, 1976). The method has some special uses, but as a matching method its shortcomings outweigh the gain of factoring a more limited matrix ($v_1 \times v_2$) and generally one does better to factor the larger matrix ($v_1 + v_2$) $\times$ ($v_1 + v_2$) in the regular method (3). A third possibility in (3) is to separate v_1 and v_2 factorings, follow up by estimation of factor scores and proceed to direct correlation of the two sets of SS factors on the same subjects. This is weaker because of the less than perfect estimation of the factor scores, but its directness appeals to some and it is a useful check.

Case (4) is "logically" impossible of direct solution, and yields only to conjuring tricks involving additional data. In short, one can handle it only by going beyond the data system, by the following alternatives.

(a) By taking primary correlations and proceeding to second orders among the primaries providing one has reason to believe the two sets of variables are in the same domain one then asks if a sorting of the order of primaries can be obtained for set 2 which yields the same R_f and $V_{fp\text{II}}$ matrices as for set 1. The likelihood of really good second-stratum similarity among what are really quite different primary factors is remote and if found suggests a basis for matching. However, one must end on the practical note that this requires superb accuracy in rotation.

(b) By correlating set 1 and set 2 with a common new set 3 of variables on both subject samples. One possibility then without doing two new factorings, i.e., of $(1 + 2)$ and $(2 + 3)$, is to project the factors existing in set 1 and in set 2 on set 3, through using simply the R matrices $(1 + 3)$ and $(2 + 3)$ using the Dwyer extension (explained on p. 501). One then examines the similarity of patterns on set 3. Another, longer way, is actually to factor each set again, matched up with set 3, rotate the set 1 and set 2 parts of the V_0 by PROCRUSTES to the accepted resolutions obtained earlier, apply L_1 and L_2 so obtained to set 3, and examine similarity of patterns. Of course, any comparison of correlations of factors X and Y, respectively, in set 1 and set 2 with other "outside" variables (without any further factor analyses), e.g., with life criteria, age change, etc., helps serve to check their identity or difference, in a rough way, but that just proposed—their similarity on the whole of set 3 correlations—is more systematic and exact.

Reciprocal to the solution of the different subjects–different variables case by bringing in such common "third party" variables is that of bringing in common people. That is to say, one finds a set of people to whom both sets 1 and 2 can simultaneously be given. Though this is a solution to the matching question asked, it is, of course, no longer a type 4 case but a reversion to 3. However, it is at least a way out of the type 4 conundrum that has often been followed. For instance, in cross-cultural research where, wishing to compare the structure of tests which cannot be exactly translated and must logically be considered as possibly different in the two translations, a group of people is tested which speaks both languages and does both tests. The test factors can be brought to simple structure separately and compared by method 3.

It will have been noted that where the situation involves the same subjects, namely, in (1) and (3), the comparison is obtainable either by comparing rotations (i.e., by comparing L_1 and L_2 either directly or by the correlations given in $L_1^T L_2$) or by correlating scores. But when it involves different subjects—which is the type 2 case, with same variables—we encounter the commonest form of the factor matching problem, deserving the fullest study. Here the method has to be that of comparing the V_{fp}, V_{dfc}, and higher-order structures. Let it be emphasized that comparing the primary V_{rs} matrices which are closest to hand is but a beginning. One should go at least to comparing the V_{fp}s where the size as well as the shape of the pattern can be taken into account. Attention needs also increasingly to be given (see Cattell & Vogelmann, 1976) to the forms of the associated higher-order structure, to absolute magnitude of factor contribution to the main salients (highly loaded variables) as shown in the V_{dfc} and several less commonly considered features (see Cattell, 1962b).

10.7. Indices for Matching: The Congruence Coefficient r_c

Accepting the type 2 case as most appropriate to more intensive study, we propose to ask just how the comparisons of the patterns on the variables are to be conducted. The comparison of V_{fp}, V_{dfc}, V_{fpII}, etc., patterns based on two distinct groups of people has to take place, of course, *after the unique factor resolution, by one method or another, has already taken place* in each. The definitions of the unique position are essentially three: SS, confactor, and an arbitrary statement of an hypothesis, usually to be compared with the best possible approach to it by PROCRUSTES. The above reminder is in italics because the problem and the research purpose here unfortunately often gets confused with that of trying to bring two studies to maximum congruence. But actually the sole purpose is to examine the degree of congruence of independent resolutions. Thus if the resolution of two studies has been carried out from the beginning in a joint planned *confactor rotation* (p. 550) or if they have already been brought to maximum congruence by Tucker's synthesis method (1951), Mulaik (1972), or if one has been rotated as closely to the position of the other by PROCRUSTES, the principles of evaluation of goodness of mutual fit is more complex and difficult than when comparing unquestionably experimentally independent solutions.

Starting with two factor pattern matrices V_{fpA} and V_{fpB} independently reached (or two V_{dfc}'s) the investigator will find five main matching approaches open to him:

1. Burt's (and Tucker's) congruence coefficient r_c
2. The salient variable similarity index s
3. The configurative transformation method
4. The pattern similarity coefficient r_p or intraclass r
5. Confirmatory (proofing) maximum likelihood

All, of course, require the same variables in the two studies, but if there are variables they do not share, these are simply cut out of the V_{fp} of each before the comparison begins.

The first index has a general form analogous to the correlation coefficient, and is close in principle in that it evaluates shared variance. The second is nonparametric in that it distinguishes categorically between variables that are loaded and variables that are defined as not loaded, i.e., by hyperplane variables, and examines the probability of overlap of two studies on the variables of these categories. The third attempts to bring test vectors from both studies into a common space and determine the angles between the factors in that space. The fourth brings some refinement on the first method. The fifth is complex and left to Chapter 13.

The *congruence coefficient* (originally due to Burt, 1941, but developed by Tucker, 1951, and Wrigley & Newhaus, 1955) aims to avoid the defects of the ordinary correlation coefficient when finding the similarity of two series of loadings. They are: (1) with positive and negative loadings the ordinary *r* takes its deviations from the mean of each series. Thus with large positive loadings predominating, smaller positives become negative deviations. In judging similarity of factors the question of whether a loading is positive or negative, as it stands at an absolute value, is often more important than its degree of magnitude, so it is important in a good index to give credit to similar signs in the loading series from the two experiments. The congruence coefficient, which is

$$r_c = \frac{\sum\limits_{j=1}^{j=n} b_{j_1} b_{j_2}}{\left(\sum\limits^{j=n} b_{j_1}^2 \sum\limits^{j=n} b_{j_2}^2 \right)^{\frac{1}{2}}} \tag{10.17}$$

where b_{j_1} and b_{j_2} are the loadings of variable a_j on the compared factors, F_1 and F_2. It will be seen the r_c value gets augmented only when the loadings b_1 and b_2 are of the same sign.

The second defect of ordinary *r* is (2) that if the loadings rank in the same order on the two factors but one factor has a much larger variance or mean size of loadings than the other, *r* will show no difference from the situation where the factors are the same size. If the absolute size of the factor is part of its identifying characteristic, as argued above, then *r* is inadequate. But so, too, by this standard is r_c, which is why it needs supplementing, at least in many instances, by the nonparametric *s*. On the other hand as seen in discussing confactor rotation, the general size of loadings of one and the same factor is likely to vary from one population to another, so some insensitiveness to this is desirable in a coefficient, within moderate limits. Thus if the researcher assures himself that the V_{fp} on inspection have nothing more than a moderate population sample discrepancy of loading sizes, r_c is an excellent index. The converse of this, however, as Pinneau and Newhaus (1964) have shown, is that if two factors both load a set of, say, six variables highly, even though the pattern among them is different, a large r_c will result. This, however, is not necessarily entirely misleading. At any rate the *s* index next to be described would do the same.

The practical problem with the use of r_c is that statisticians have not given us a clear test of its significance derived from first principles. Accordingly, Schneewind and Cattell (1970) generated distributions of congruence coefficients from typical loading patterns (V_{fp}) over many actual psychological experiments by Monte Carlo procedures. As with virtually all parameters in factor analysis the distributions of r_c values tend to vary with the number of variables and number of factors, though the variation in this case is not great. Although considerable numbers were taken, and reasonably smooth distributions obtained, the table offered for practical convenience in Appendix 7, p. 568 (see also Appendix 8 for s values) has been a stopgap. Starting from random normal deviates as loadings and with an interesting difference concerning the null hypothesis being tested, Korth (1978), from simulated data worked out with Tucker (Korth & Tucker, 1975), has provided the values in Table 10.3. Due to the difference in assumptions, and possibly to the use of random normal deviates rather than actual empirical loadings in the Monte Carlo progress, this table demands noticeably higher values for significance than those on p. 568, although the random normal deviates basis being a theoretically clearer assumption we must accept them.

Table 10.3. **Significance Values for Congruence Coefficients by the Korth and Tucker Method**[a]

Number of variables in common[b]	Number of factors in each study	Critical $p < .05$ value of r_c	Number of variables in common	Number of factors in each study	Critical $p < .05$ value of r_c
10	4	.93	50	10	.41
30	4	.46	60	10	.38
50	4	.34	70	10	.36
70	4	.32			
28	8	.58	28	11	.61
30	8	.52	30	11	.55
32	8	.53	32	11	.55
28	9	.55	28	12	.58
30	9	.54	30	12	.58
32	9	.55	32	12	.56
			30	15	.56
28	10	.60	50	15	.42
30	10	.55	60	15	.43
32	10	.51	70	15	.38
			54	23	.46
			50	25	.58

[a]From Bruce Korth, "A significance test for congruence coefficients for Cattell's factors matched by scanning" *Multivariate Behavioral Research*, 1978.

[b]In the past the number of published researches that could count even as many as 10 variables measured in a precisely similar way in two researches was very few indeed! The more programmatic research of recent years has made reliable comparison, with up to 40 variables in common, attainable. The range of 10–40 variables in common has been taken in Appendix 7 as best representing the needs of psychologists in the foreseeable future (Schneewind & Cattell, 1970).

The Schneewind–Cattell table in Appendix 7, p. 568 is more complete in p values. It also contains reference to a second form of the congruence coefficient which we have called the intracongruence coefficient and written r'_c. It is obtained by inserting a reflection of every variable in the V_{fp} matrix immediately below the unreflected, so that there are twice as many variables $(2n)$ as before. These loadings will now have a mean of zero (since every positive has an equal negative) so that r_c now has the same form as the correlation coefficient and could have its significance tested by the usual values for correlation coefficients, at least, if loadings followed the distribution of random normal deviates. Since simple structure says they do not it has seemed best to apply Monte Carlo methods here too (fourth column, Appendix 7) but the ordinary r evaluation is not far out. For example, a significance at the .01 (two-tailed) level with 10 cases needs to be about .71 for r and is .77 for r'_c here, and with 50 cases, .35 for r and .38 for r'_c here. The appearance of a negative r_c, r, or r'_c, of course, simply means that one factor should be reversed in its direction and, as such, evaluated for a positive match.

In using r_c it will be found that rather the more r_c's between factors that one definitely *knows to be different* reach indices of *moderate significance* than one would expect to find. This argues that real influences are rather more frequently cooperative in pattern than sheer chance, i.e., random loading patterns, would indicate, and is an interesting and valuable commentary on the character of natural influences when detected as factors. It is as if target variables for any one influence tend to have some property in common which then makes them simultaneously, as a group, also susceptible to some other influence. The classic case of such variables as skin resistance, pupil diameter, heart rate, etc., being a special group of variables susceptible to both of two autonomic factors—the sympathetic and parasympathetic—has been mentioned before, while ball behaviors responsive to both weight and size (p. 137), and human ills responsive to both cold and damp are other instances of genesis of cooperative patterns. Parenthetically, "cluster chasing" rotation programs are particularly susceptible to missing the two distinct factors in such cases, for it collapses them into one.

The tendency we are discussing—for the appearance of significant r_c where we know factors to be different—thus has interest for general methodology and theory, but is merely a disturbance when an r_c table (bounded by factors from one study by rows and the other by columns) is being examined. For one hopes to be able to find an ordering of factors that will bring high matching values down the diagonal and nothing much elsewhere. Occasionally, inspection of loadings will show that an abnormally high value is clue to a single unusually high variable producing an otherwise anomalous similarity, and such a spurious match may perhaps be dropped. The clearest way to examine the matching and remove doubts from this source is to attempt a rearrangement of the order of factors on the second (perhaps "new") study such that the $k \times k$ congruence matrix will give, if possible, *the highest coefficients falling consistently down the diagonal*. If this can be achieved—and in most studies we know where careful SS has justified expectation of matches it can—then a one-to-one identification of the factors in the new and old studies can be considered to be demonstrated. For the chances of reaching highest values straight down the diagonal in, say, a 16×16 matrix, by random data are very remote. This susceptibility to "diagonalization" means in the case of any putatively matched factors, X and Y in, say, 16 factors, that X's congruence with Y is higher than with any of the 15 remaining factors (in the row) in study B, and that Y's congruence with X is higher than with any of the remaining 15 factors (in the column) in study A. Some perspective on the probability of encountering lower but still statistically significant (Table 10.3) off diagonals in the given domain, in relation to the percentage actually found, can be reached by examining the mutual congruences in square matrices within study A factors alone, and study B factors alone. For it seems likely that the cooperativeness, as shown by an undue number of moderately high r_c or s indices in some substantive domains, e.g., in physiological and social psychology, may be characteristically higher than in others, which will show within any one of the studies. If the matrix of congruences between studies A and B can be "diagonalized," and if all diagonals are significant at the $p < .01$ level—a condition not infrequently attained in well-planned studies—then a reasonably firm conclusion can be drawn. We conclude in that case that the two unique resolutions reached indeed bring mutual support. And in more general terms we obtain support for the hypothesis that such simple structure or confactor rotations have the quality of being invariant.

10.8. Indices for Matching: The Salient Variable Similarity Index s

This mention of the statistician's term "invariant," however, evokes the comment that the factor patterns, as such, should not, in terms of theory, be strictly and exactly invariant across samples and populations when the factor influences are the same! It is inevitable that from sample to sample and still more from population to population the variances of first- and second-stratum factors (and therefore the correlations of primaries) should alter. Consider, for example, a set of correlations resolving into just two common factors X and Y which from one sample to another change in relative size in this way—the total h^2 remaining constant. It is evident that a drop from, say, .8 to .7 in the loading of factor X on a_j must bring about a relatively large rise on factor Y's loading on a_j, namely, from .4 to .58 (h^2 being held at .8), even if Y does *not really change in its own variance* as such. It changes only to accommodate to the change in absolute variance size of X. A little arithmetic with changes in variables in such a case, beginning, say, by producing a $\frac{7}{8}$ sigma change in X and none in Y will show the loadings of X do not uniformly decrease by $\frac{7}{8}$, and that Y's loadings do not increase responsively in a uniform fashion over its column of loadings.

In short, any change in the variance of a factor as a single influence or determiner will not bring about a proportional change simultaneously in its loadings over all variables. Invariance in a factor does not ensure constancy of loading shape and we are required to pursue the more sophisticated game of discovering what modification of shape of loadings is proof of invariance of factor influence *per se*. Although invariance of factor pattern is not the desideratum when one wishes to argue that the factors themselves are invariant, the difference of pattern invariance and factor influence action invariance is usually not great. What the nature of the manifest patterns should be to indicate true identity is a complex question to be discussed under real base factor analysis below (Chapter 13, Section 13.8, p. 410). Meanwhile, one has to take the position that an index responsive to a high degree of shape invariance is desirable, but one which does not penalize the matching value for departures originating as indicated above.

An index designed to have these properties of resistance to lesser pattern change was proposed by the present writer (1949c) as the "salient variable similarity index." It is listed as the second main choice of an index above, for its record of dependability is good especially since it was improved (Cattell & Baggaley, 1960; Cattell, Balcar, Horn, & Nesselroade, 1969; Cohen, 1969). In theoretical basis it is consistent with the position already described in the *factor mandate matrix* (p. 235) above, that a factor either loads a variable or it does not. In that case the presence of low but significant loadings, not far from the true hyperplane, is simply an appearance due to several other factors happening to operate simultaneously on the same variable, reducing its loading on this one, as pointed out above. In that case the moderate loadings should be given as much weight in recognizing a factor as the substantial loadings. The unique character of a factor is nothing more nor less than the unique pattern of its actions in the mandate matrix.

It follows logically from this that we should make decided attempts to recognize a loading level which marks the edge of the hyperplane and distinguishes loaded from unloaded, allowing for measurement error. Below this value we could be confident that most loadings would really be zero loadings, varying therefrom only because of the effect of error of measurement operating through and interacting with the sample size. Above this all would be real, salient loadings. Earlier we suggested cutting at about ±.15; but with Finkbeiner's evidence that ±.07 is nearer the limit in adequate studies—adequately rotated, on sufficient samples and with tests of reasonable reliability—we would advocate a cut at a lower value. The error of calling significant variables nonsignificant is, if anything worse than the converse, since significants are normally decidedly less frequent. Obviously, the best estimate of the cutting line must vary with sample size, etc., according to Jennrich's formula (1973, 1974). But, in general, a cut at ±.10, in a good study (despite current opinion being inclined to consider this as too low) is probably the shrewdest.

The s index—*the salient variable similarity index*—therefore differs from r_c, in proceeding *nonparametrically*, in principle in accordance with the mandate matrix. In procedure one first classifies all variables by loadings on a given factor (by, say, the ±.10 cut) as *salient* or *nonsalient* variables (the former grouped again as positive or negative). At this point it has been suggested that chi square or even an ANOVA method (see Pawlik in Hundleby, Pawlik, & Cattell, 1965) be used and, indeed, s (and Cohen's development of it) can be considered a special development of chi square but, except possibly for Cohen's (1969) development, s seems to have more convenient properties.

The s index begins, as in all comparative methods, by segregating out from any larger sets of variables in the two studies the variables they have in common. Operating with these it asks for any given factor "If we take the salients on factor X in study A, and find how many of them are also salients on factor Y in study B, what is the probability of X and Y being the same factor?" Salients can, of course, be positive or negative, and to match they should agree in sign, so the first procedure, is to find and enter the frequencies—the f and n—in a table as shown in Table 10.4, with consideration of sign. Some complicatedness and uncertainty of interpretation in

Table 10.4. Tabulation Required for Calculating s Index

	Factor 2			
Factor 1	PS[a]	H[b]	NS[c]	
PS[a]	f_{11}[d]	f_{12}	f_{13}	n_1
H[b]	f_{21}	f_{22}	f_{23}	n_2
NS[c]	f_{31}	f_{32}	f_{33}	n_3
	$n_{.1}$	$n_{.2}$	$n_{.3}$	n

[a]Positive salient variable.
[b]Hyperplane variable.
[c]Negative salient variable.
[d]Joint frequency. The same n variables in both studies.

Table 10.5. Significance Tables for s: The Salient Variable Similarity Index

n		p values for hyperplane count, 60%											
10	s	.76	.51	.26	.01	.00							
	p	.001	.020	.138	.364	.500							
20	s	.63	.51	.26	.13	.01	.00						
	p	.000	.004	.086	.207	.393	.500						
30	s	.59	.51	.42	.34	.26	.17	.09	.01	.00			
	p	.000	.001	.005	.019	.054	.131	.256	.411	.500			
40	s	.51	.44	.38	.32	.26	.19	.13	.07	.01	.00		
	p	.000	.001	.005	.016	.041	.090	.169	.282	.426	.500		
50	s	.46	.41	.36	.31	.26	.21	.16	.11	.06	.01	.00	
	p	.000	.001	.004	.011	.026	.061	.115	.200	.301	.428	.500	
60	s	.42	.38	.34	.30	.26	.21	.17	.13	.09	.05	.01	.00
	p	.00	.001	.003	.008	.018	.039	.078	.132	.215	.318	.440	.500
80	s	.35	.32	.29	.26	.22	.19	.16	.13	.10	.07	.04	.01
	p	.000	1.001	.003	.008	.018	.034	.063	.107	.170	.243	.338	.447
	s	.00											
	p	.500											
100	s	.36	.31	.28	.26	.23	.21	.18	.16	.13	.11	.08	.06
	p	.000	.001	.003	.006	.011	.019	.034	.058	.091	.136	.196	.271
	s	.03	.01	.00									
	p	.353	.449	.500									

n		p values for hyperplane count, 70%											
10	s	.67	.34	.01	.00								
	p	.002	.052	.316	.500								
20	s	.67	.51	.34	.17	.01	.00						
	p	.000	.002	.027	.135	.357	.500						
30	s	.56	.45	.34	.23	.12	.01	.00					
	p	.000	.002	.016	.061	.190	.383	.500					
40	s	.51	.42	.34	.26	.17	.09	.01	.00				
	p	.000	.002	.007	.034	.098	.222	.403	.500				
50	s	.47	.41	.34	.27	.21	.14	.07	.01	.00			
	p	.000	.001	.004	.018	.052	.123	.247	.407	.500			
60	s	.39	.34	.28	.23	.17	.12	.06	.01	.00			
	p	.000	.002	.007	.025	.066	.142	.262	.415	.500			
80	s	.38	.34	.30	.26	.21	.17	.13	.09	.05	.01	.00	
	p	.000	.001	.002	.007	.019	.046	.092	.174	.283	.425	.500	
100	s	.34	.31	.27	.24	.21	.17	.14	.11	.07	.04	.01	.00
	p	.000	.001	.002	.006	.016	.037	.069	.125	.207	.318	.438	.500

n		p values for hyperplane count, 80%				
10	s	.51	.01	.00		
	p	.012	.187	.500		
20	s	.76	.51	.26	.01	.00
	p	.000	.003	.041	.279	.500

Table 10.5 (Continued)

p values for hyperplane count, 80%

30	s	.67	.51	.34	.17	.01	.00	
	p	.000	.001	.011	.083	.316	.500	
40	s	.51	.38	.26	.13	.01	.00	
	p	.000	.003	.024	.115	.347	.500	
50	s	.51	.41	.31	.21	.11	.01	.00
	p	.000	.001	.006	.036	.142	.361	.500
60	s	.42	.34	.26	.17	.09	.01	.00
	p	.000	.002	.012	.052	.164	.369	.500
80	s	.38	.32	.26	.19	.13	.07	.01	.00
	p	.000	.001	.004	.022	.076	.199	.391	.500
100	s	.31	.26	.21	.16	.11	.06	.01	.00
	p	.000	.002	.010	.036	.105	.220	.402	.500

p values for hyperplane count, 90%

10	s	.01	.00			
	p	.052	.500			
20	s	.51	.01	.00		
	p	.003	.099	.500		
30	s	.67	.34	.01	.00	
	p	.000	.007	.133	.500	
40	s	.51	.26	.01	.00	
	p	.000	.012	.167	.500	
50	s	.41	.21	.01	.00	
	p	.000	.018	.198	.500	
60	s	.51	.34	.17	.01	.00
	p	.000	.002	.029	.217	.500
80	s	.38	.26	.13	.01	.00
	p	.000	.004	.045	.251	.500
100	s	.31	.21	.11	.01	.00
	p	.000	.007	.061	.286	.500

the *s* formula (see Cattell, Balcar, Horn, and Nesselroade, 1969, for details) ensues unless certain conditions are met. They are (a) that the same number of variable be counted as lying in the hyperplane in the two researches. This may require widening or narrowing the hyperplane a little in one relative to the other. (This is often justified in any case by different sizes of errors and samples.) And (b) that the number of positively and negatively loaded salients be equalized in any factor by reflecting the sign of the original set of variables until this is achieved. This may happen differently on the two factors to be compared, but, of course, count of agreement and disagreement in the f_{12}, f_{13}, etc., cells proceeds on the result after reflection. The signs of variables within the hyperplane are ignored, since all are postulated as actually 0.

When Table 10.4 is so completed the formula for s is

$$s = \frac{f_{11} + f_{33} - f_{13} - f_{31}}{f_{11} + f_{33} + f_{13} + f_{31} + \frac{1}{2}(f_{12} + f_{21} + f_{23} + f_{32})} \tag{10.18}$$

where the f's are the frequencies of variables in the cells as numbered in Table 10.4.

It can be seen that this will equal 1.0 with perfect agreement, 0 for chance agreement, and -1.0 when one factor is the complete reflection of the other. (Note that in final assessment this -1.0 is also an agreement, since in general we are at liberty to reflect a factor 180°, naming it in the verbally opposite direction when we do so.) The formula is readily programmed.

To proceed to a significance test for s, we have to note that its distribution will alter with the percent in the hyperplane and the number of variables. Thus it can be shown that a value of .32 with 50 common variables and 60% in the hyperplane is significant at the $p < .01$ level, whereas with 100 common variables and a 90% hyperplane an s of .20 is significant at the same level. The following tables have been worked out for 10–100 common variables and 60%–90% in the hyperplane. To enter them take the obtained s value and number of common variables n and run along the row in Table 10.5 appropriate for hyperplane percentage until the given s value is reached.

One can, of course, interpolate to get (approximately) intermediate values in relation to the table. There seems little point in programming Table 10.5 since relatively few values are usually to be evaluated though a 20 × 20 factor matching table with 190 values calculated can be incorporated with the above program for calculating the s. Cohen's (1969) K statistic, which performs essentially the same function, and s statistic, might well be programmed together for a check.

10.9. Indices for Matching: The Configurative Method

The third matching procedure listed above—configurative matching—seeks (still in the present category of same variables on different subjects) to determine resemblance geometrically by bringing the two studies into the same space and finding the cosines (correlations) between supposedly matched factor axes. It arose from discussions of Kaiser, Schönemann, and the present writer in 1960–1965 in relation to confactor and PROCRUSTES formulations, the resemblance to which will be apparent. An initial statement was given by the present writer (1965, with credit to Kaiser), and a fuller development by its main author appeared in 1971 (Kaiser, Hunka, &

Bianchini, 1971). The presentation here will keep to the statement of essentials in the former, but those pursuing the method intensively are advised to read the latter.

To understand the principle it is necessary first to recognize the difference between ordinary factor correlations in ordinary Euclidean space and what we shall call *field factor correlations*, because they get their correlation secondarily from the projections upon them of the field of variables around the factors. When discussing correlations of factors to obtain higher-strata factors (Chapter 9, p. 198) it was pointed out that "correlation of factors" has an ambiguity of meaning, to which we would return at this later point for further clarification. The meaning of factor correlation so far has been that the pure factors, as axes, are oblique, their correlations, with unit-length factors, being the cosines of the angles between them. Within that meaning we have pointed out that the correlation of estimated factor scores, though the factor loadings are based on the rotational position, only approaches and characteristically underestimates the true direction cosine correlation. To these two ways of calculating factor correlation we now add a third. This meaning of "correlated," frequently used (alas, often without mentioning the distinction), is that the columns of loadings of the two factors are significantly correlated. The first two calculations can be regarded, respectively, as an exact and a rough way of getting at the same concept—correlations of factor axes—but this third calculation applies to a quite different conception. The present chapter has already allowed us to get familiar with this idea, in the notion of *cooperative factors*, and in the use of the congruence coefficients and s to test the degree of similarity of factors by their loading patterns.

With this distinction in mind we should, incidentally, note that the statement that principal components are uncorrelated needs qualification. In this case they happen to be uncorrelated in both senses—as axes and as loading patterns—and essentially the same is true of successive centroid factors.

The moment the principal axis solution is rotated, even orthogonally, however, the condition of being uncorrelated ceases to be true in the second sense; the factors are orthogonal but the columns of their correlations (or loadings) on the variables no longer mutually correlate zero. Consequently, since the majority of final psychological results are rotated it is true that even if they remain geometrically orthogonal they will correlate (with varying significances according to sample) in this sense of having some positive or negative relation of profiles on the loading pattern columns. (This, if significant, represents a cooperative factor phenomenon, but the generally more likely, smaller magnitudes can be seen in the off diagonals of any of the r_c matrices discussed in Cattell, 1975, p. 114.)

Let us compare the three calculations in terms of formulas. As pointed out above, the correlations of estimated factor scores will usually be relatively close to the correlations of the factor axes in the test space. The former [see Section 10.3 above, from Eq. (10.8)] will be

$$R_{\hat{f}} = V_{fe}^T R_v V_{fe} \qquad (10.19)$$

and the latter (the ordinary correlations as defined by direction cosines) will be (bringing in R_v and V_{fe} for comparability)

$$R_f = V_{fp}^{-1} R_v V_{fe} \qquad (10.20a)$$

[from Eq. (10.4) above] or since V_{fp} is not directly invertible

$$R_f = (V_{fp}^T V_{fp})^{-1} V_{fp}^T R_v V_{fe} \qquad (10.20b)$$

Finally, the correlation of the factors by similarity of the V_{fp} columns will be more radically different,

$$\underline{R}_f = D \underline{V}_{fp}^T \underline{V}_{fp} D \qquad (10.21)$$

where $\underline{V}_{fp}$ has deviations from the mean loading and D is the inverse of the root of the diagonal of $\underline{V}_{fp}^T \underline{V}_{fp}$.

It will be seen that these three views of the factor correlations are distinct though likely to be roughly similar. One reason that Eq. (10.21) is more different is from the effect of the cooperative factor phenomenon, for it is the existence of clusters in the configuration of the variables falling in this quadrant or that which produces "cooperative" correlations even when all factors are exactly orthogonal in the ordinary sense. However, the usual slight obliquities also make some contribution to cooperativeness of patterns. Indeed, if variables were completely randomly distributed in space, i.e., with no such clusterings as produce significant cooperativeness, factor patterns would correlate only to the extent that the factors are oblique and the "field correlation" would be essentially the same as the "cosine correlations." This difference in the two kinds of "factor correlation" is shown in Fig. 10.1. Since correlations of the third type, Eq. (10.21) from similarity of columns, depend on features of the field or configuration of variables, we shall continue to distinguish them as "field correlations."

The above arguments for the existence of clusters being responsible for cooperative factor patterns is easy to see in Fig. 10.1(a). For variables a, b, c, and d will simultaneously appear with substantial and similar signed loadings on F_1 and F_2. In Fig. 10.1(b) the simplest representation of an equalized (random with large numbers) distribution is given by placing a, b, c, and d each loading equally and oppositely on the orthogonal factors F_1 and F_2. Their loadings—the V_{fp} will be two columns of four numbers in each—will correlate zero on the orthogonal factors. But on the oblique projections [Fig. 10.1(b), on the right] drawn, of course, for F_1 parallel to the F_2 axis, and vice versa, the sum of cross products will no longer balance to zero, and a negative correlation will appear.

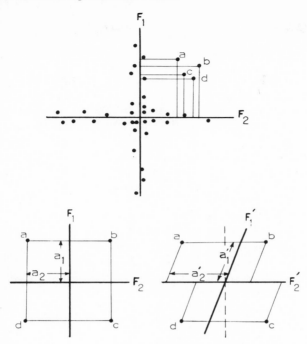

Fig. 10.1. The two kinds and sources of correlations among factors. (a) Cooperative-ness due to surface traits, illustrated by the cluster a, b, c, and d of commonly involved salients in orthogonal factors F_1 and F_2. Note that though the formula for the field correlation will make F_1 and F_2 positively correlated, they stand in space orthogonal. (b) True factor (direction cosine) correlation (obliqueness). Note the effect on the loadings, on moving from an orthogonal to an oblique position, is to make the loading patterns (V_{fp} columns) for variable a, b, c, and d more similar for F_1 and F_2 (a and c are high; b and d become reduced on both) though on the left they were uncorrelated (actually all orthogonal loadings are here equal).

Now what the third device for calculating factor matching—the configurative method—aims to do is create a space from the combined configurations of the two R_v matrices and put the axes in each in such positions spatially that they will yield the known loadings on the config-uration of variables. This means somehow getting the configuration of variables on the first matrix R_{vA} and the second R_{vB} into a compromise common space with the best possible mutual fit, as when one man fits the spread fingers of his right hand as closely as possible to those of another right hand. And when we have that new space we shall aim to put the factor axes as found in the rotation in A into that space and similarly for B, and then find the correlations of the "corresponding" A and B vectors (if there are any so corresponding), i.e., see what mutual angles they have.

To suit later manipulations, the original expression of field correlations of factors according to loading, as in Eq. (10.21), is here taken in terms of the factor structure matrix, thus:

$$\underline{R}_f = \underline{V}_{fs}^T \underline{V}_{fs} \tag{10.22}$$

The D's in Eq. (10.21) are omitted on the assumption that these $\underline{V}_{fs}$ columns have already been divided by the root of the summed squares in each.[5] ($\underline{R}_f$, the matrix of field factor correlations, is so written to distinguish it from $R_{\hat{f}}$ and R_f above, the ordinary estimated and actual factor correlation matrices.) The diagonals of this factor field correlation matrix will be unity, just as in R_f, and the off diagonals will be the correlations of one factor with another. If we now take the V_{fs} from the two different matrices (each treated as in 10.22), however, we have

$$\underline{R}_{f_A f_B} = \underline{V}_{fs \cdot A}^T \cdot \underline{V}_{fs \cdot B} \tag{10.23}$$

and here the correlation will be those among the two sets of factors. Like an r_c matrix, the matrix will have the largest values along the diagonal if the order of the truly matching factors is the same. In this formula we are also assuming that the configuration of the variables themselves in the two cases can be considered approximately the same.

There are approximations in Eq. (10.23)—in the interests of making the general nature of the argument clear—which render it something of a caricature of the best configurative approaches.

The more accurate calculation can be briefly stated as:

1. Calculate

$$K = V_{0A}^T V_{0B} \tag{10.24}$$

where V_{0A} and V_{0B} are the unrotated matrices in experiments A and B.

2. Calculate KK' and $K'K$ and take the latent vectors. This means stopping the ordinary factor program one stage short in the factor extraction, thus[6]:

$$KK^T = L_x D_x^2 L_x^T \tag{10.25a}$$

---

[5] It will be noted that whereas two rows in the V_0 matrix give at once an inner product which is the correlation of the variables (in $R_v = V_0 V_0^T$), the reciprocal product of two columns in $V_0^T V_0$ does not immediately give R_f. The reason is that a variable's row is normalized (granted inclusion of the unique factor loading) whereas a factor's column is not. Consequently in any evaluation of "field correlations" among factors by correlating their columns, such as occurs in Eq. (10.22) and Eq. (10.23) the V is not the ordinary one, but is written $\underline{V}$ to show that its columns have been normalized, i.e., divided by the square root of the sum of its squared values. This is just the ordinary conformity required with the formula for the correlation coefficient in standard scores.

[6] A matter which the mathematician, looking at the mathematical origin of factor analysis, would regard as the Chapter of Genesis is that factors come from the *latent roots and latent vectors* concept. The latent vectors and roots usages will be explained later, though we have mentioned them briefly as a step on the way to a principal axis

$$K^T K = L_y D_y^2 L_y^T \tag{10.25b}$$

3. Take L_x and L_y out of this result and calculate

$$L_{AB} = L_x L_y \tag{10.26}$$

4. If L_A and L_B are the existing transformation matrices from V_{0A} and V_{0B} to the two unique resolutions to be compared, then

$$R_{f \cdot AB} = L_A^T L_{AB}^T L_B \tag{10.27}$$

where $R_{f \cdot AB}$ is the correlation of the factors in A with those in B, in the composite "pseudo" space. If all factors match it should be possible to rearrange the order in one of them so that all high values are down the diagonal, as in the other matching matrices above, and then factors in both will be in the same order. For more detailed justification of the above steps see Schönemann (1966b) and Kaiser, Hunka, and Bianchini (1971). Because this is a "compromise" using "pseudo" space and for other reasons, the exact standards by which one could calculate the significance of the correlations obtained has not been worked out, but a rough answer can be gained by the usual r significance using n as the number of variables involved.

10.10. Comparative Properties of Various Approaches to Matching, Including r_p

Some comment has been made on the properties of the above three in describing them, but more explicit mutual comparison, bringing in the r_p intraclass, yet to be described, is appropriate. Since no one of the above three methods—congruence r_c; salient variable similarity s; and configuration correlation—is without defects from some assumption that one would prefer not to have to entertain—the present writer advocates and practices taking in important cases (the Cattell and Vaughan studies on 16 P.F. structure were such)—at least two of the above, and accepting their joint contribution to the matching verdict. The r_c and s values in the cases tried out show appreciable agreement, though there are also instructive differ-

---

and components analysis, and the reader may encounter the terms in collateral recommended reading. The main point is that latent vectors are not our factors of unit length, but have to be multiplied by a diagonal matrix of square roots of the latent roots, each latent vector being appropriately "stretched." What comes out of the computer program is the parts on the right of Eq. (10.24). The psychologist will recognize it if we write: $ED^2 E^T = (ED)(DE^T) = V_0 V_0^T$, where E is the matrix of latent vectors and D^2 of latent roots. The historical origin of factor analysis in psychology (Spearman, 1904–1924) followed quite a different course, but the independent origin in mathematics came by the latent roots and vectors theorem. The summed squared values in all E columns are the same, unlike V_0 columns.

ences, as one would expect. For example, in the factors in the 16 P.F. for which high loaded variables have not yet been found (M, N, and Q_1) the r_c coefficient agrees less across different 16 P.F. factorings than s does. This reflects the fact that low salients change their rank order rather easily, which upsets r_c whereas all stand categorically as salients to s. Moreover, r_c gives some weight to difference of signs of loadings when those loadings are really only of small loadings in the hyperplane—a defect which s does not show. For these reasons when loading patterns are distorted from experiment A to B by big changes in size of other factors one may advocate, with Kaiser, the configurative matching, or, more simply s. At any rate, experience from personality data shows that r_c and s give tolerably equivalent results, though occasionally one reaches significance when the other does not.

It will be noted that none of these three methods gives "credit" for the two factors matched being more close in general size, as distinct from being of the same pattern or rank order of loadings. In this respect, a further index, the pattern similarity coefficient r_p is theoretically superior to all of them, though it has apparently not been given a tryout. A minor practical difficulty is that the differences of loading of any given variable (of the n variables) on two factors $A \times B$—which is a d value in the r_p formula

$$r_p = \frac{2n - \Sigma^n d^2_{(A-B)}}{2n + \Sigma^n d^2_{(A-B)}}$$
(10.28)

has to be expressed in terms of the standard deviation of loadings on the given variable (Cattell, Coulter, & Tsujioka, 1966). To get that accurately would require a whole series of factor analyses! In this situation an acceptable approximation for the sigma would be to estimate it over this relatively small sample of $2n$ variables and $2k$ factors by pooling all loadings. The significance test for r_p given by Horn's (1966, Chap. 18) table based on n, the number of variables in the factor comparison, could be used, though again the capacity to "diagonalize" the k by k matrix of pattern similarity coefficients between the two studies will be finally the important evaluator. Since the intraclass correlation has fairly close resemblance to r_p, in the crucial matter of taking difference of level into account, it will have here the same general appeal as r_p. One may guess that experience will show r_p to have greatest validity among the above three methods, provided the estimation of the standard deviation of loadings is accurate enough.

As we approach conclusion, it becomes desirable to remind the reader of our statement above that the ordinary factor analytic factor pattern, V_{fp}, essentially used in all above, is ultimately not the proper expression of factor effects in which to expect constancy and to test invariance. Its defect, as pointed out above, is that a size change in one factor interferes with the constancy of pattern of another even when the latter does not intrinsically change at all. Only in *real-base factor analysis*, proceeding through *factoring of covariances*, would we expect the same factor to retain its pattern from experiment to experiment with a high degree of invariance, regardless of what happens to the others. Only real-base analysis can give proper regard to absolute size of a factor in decisions on matching. Real-base factoring cannot be treated until Chapter 13, but the reader should have no difficulty in transferring the application of the above matching and identification procedures to real-base factor patterns, since no essential change is necessary in applying the above formulas.

Finally, in a consideration of all methods that might be thought of as testing matches of independently reached rotations one must take a side glance at PROCRUSTES rotations and at confirmatory maximum likelihood methods. These ask if there exists a rotation of A that will fit an independently reached rotation in B, and are thus independent on one side only, but in certain research strategies they may ultimately function on a matching evaluation procedure.

Illustrative instances of results of interpretive and matching procedures in psychology have happily become more plentiful just in the last decade. Hakstian, Horn, and others have applied them in identifying and separating old and new primary ability factors. Sweney and Cattell (1964) used them in matching factors of dynamic conflict across three experiments. There have been matching also of personality factor primaries and secondaries across age groups (Cattell, 1973; Cattell & Nichols, 1972) and across cultures (Tsujioka & Cattell, 1965; Schröder, Cattell, & Wagner, 1969; Cattell, Schmidt, & Pawlik, 1973) and between Cattell's, and Rummel's national culture pattern studies, in which two or more of the above have been tried. In the field of states the history of U.I. 24, anxiety, and U.I. 32, the exvia factor, are of interest. U.I. 24 was first well replicated (matched), but as an essentially uninterpreted factor. By situation 2—different variables–same people—methods it was shown that the replicated T data factor U.I. 24 was the same as the replicated Q data factor Q II. Examination of positive, negative, and zero loadings then led to the interpretive hypothesis that it was anxiety. Decision against Eysenck's competing hypothesis that it was "neuroticism" was made by the disparity of the profile of loadings on primaries on Q II (anxiety) from the total pattern distinguishing normals and neurotics—the pattern obtained by criterion rotation (Cattell & Scheier, 1961). The decision on the basis of these pattern resemblances was in this case extensively checked—with support for that matching—by the nature of higher- and lower-order relations, by rating and clinical criterion relations other than to neuroticism, notably to anxiety levels and to responses to manipulation, e.g., by threat situations and by tranquilizers (Rickels, Cattell, Weise *et al.*, 1966). These supported in the highest degree the ultimate interpretation of U.I. 24 (Q II) as anxiety, and illustrated methodologically the combination of identification procedures we mentioned at the beginning.

The subject of factor matching should not be left without a brief aside on proofing (confirmatory) maximum likelihood, described in Chapter 13. The reason it is not given a cognate role with other methods here is that it does not compare two independently reached positions, but uses a given position in one as a target for the other, as in PROCRUSTES. It then delivers a verdict on the goodness of the match when one is brought to maximum congruity with the other. This verdict is statistically more sophisticated than the means available for testing a PROCRUSTES match, but is more cumbersome than the latter. However, although emphasis here has been on matching two empirically obtained results, the methods apply in pattern statement which is why proofing maximum likelihood is listed as a fifth possibility above.

Much of the topic here studied has been handled in factor texts under the rubric of "factor invariance." However, as we have seen elsewhere, and under the topic of confactor rotation, if the same underlying real influence is at work in different experiments the factor pattern should generally not be invariant but slightly or moderately different in response to circumstances. One seeks replication rather than invariance. A very clear treatment of the effect of selection effects upon "invariance" will be found in Mulaik (1972), Chapter 14.

10.11. Summary

1. After unique resolution of a factor analysis at all strata, *interpretation* and factor *scoring* are the next steps, but interpretation takes precedence, since the sooner a factor score has contingent meaningful interpretation the better can ensuing criterion studies be chosen.

2. Indexing of recognized and replicated patterns as such, before interpretation proceeds, is desirable since premature naming confuses subsequent interpretive researches. Naming should initially be more descriptive, though there is a continuum from description to interpretation.

3. The first interpretation has to be from the pattern of (a) high positive and negative loadings, (b) the nonloading, and (c) the absent variables. However, this alone is never enough and regard must be paid to (i) second-order loadings (primary correlations), (ii) the criterion correlations, (iii) manipulative responses of factor scores, and (iv) the IHD spiral of programmatic work in which the hypotheses produce designs for new variables whose loadings would be crucial to the hypothesis chosen.

4. While concentrating on the first stage of interpretation by abstracting a common principle among high loadings one encounters the question of which VD (variable dimension) matrix is best used for this, as well as for matching. This prompts an examination of ultimately some ten principal VD matrices, as a topic in itself, also bringing out the important statistical relations among them.

5. An important understanding needed here is a grasp of the *reciprocal* regression in VD matrices—that of factors on tests and vice versa—presented in exact form by the V_{fp} and V_{fe} matrices. The four most frequently used VD matrices are the V_{rs}, V_{fp}, V_{fs}, and V_{fe}, which are statements of regression coefficients (V_{rs}, V_{fs}) and weights (V_{fp}, V_{fe}). But there also are matrices of magnitudes of variance contributions and these, in the direction of contribution from factor to variable, are called the *dissociated* and *associated contribution* matrices, V_{dfc} and V_{afc}. The former derives from the factor pattern, which partials out other factors, and the latter from the factor structure, which allows effects to be included from associated factors. Additionally there are both regression and contribution matrices corresponding to the stub factors, the latter written $\underline{V}_{dfc}$ and $\underline{V}_{afc}$. In this equivalent stub set to the ten here there are, of course, also the stub $\underline{V}_{fs}$ and $\underline{V}_{fp}$ set out in the Schmid–Leiman formulation. All of these—$\underline{V}_{fs}$, $\underline{V}_{fp}$, $\underline{V}_{dfc}$, and $\underline{V}_{afc}$—are in the system of the SUD (stratified uncorrelated determiners) model. Both sets are statistical statements of VD relations and change values with population, etc., but the *factor* mandate matrix V_{fm} is the ultimate unchanging statement of the nature of a factor, simply in terms of what it does or does not affect at all.

6. Factor matching is an adjunct to factor interpretation, since there is little point in interpreting a pattern until it is shown to be invariant. However, it is by no means the only basis of factor identification for criterion relations to factor scores and general evidence on natural history play their part. Effective matching studies presuppose good experimental planning, usually attainable only in programmatic work, in which marker variables are exactly carried forward, with particular regard to discriminating among "cooperative" factors.

7. The matching task faces four situations: (i) same variables, same people, (ii) same variables, different people, (iii) different variables, same people, and (iv) different variables, different people. Devices of calculation are presented for each.

8. Situation (ii) is most common matching when the same variables are present and is given extended consideration under the four chief available methods: (i) congruence coefficients r_c, (ii) salient variable similarity indices s, (iii) configurational correlations, and (iv) use of the pattern similarity coefficient r_p or the intraclass r. (Only quite rarely are these precise methods unnecessary, namely when a factor in each of

the compared experiments loads one or two variables so highly—.8 or more—that no other factor could load those variables more highly or indeed than quite slightly.) Significance testing tables for (i) and (ii) are presented, and a reference for that for (iv) given.

9. A comparison of the methods shows some weaknesses in the special assumptions in each. The pattern similarity coefficient r_p has the advantage of avoiding the assumption in the other three that difference of size (level of loadings) is irrelevant, but it has not been given much trial as yet, especially concerning any effects from its rough estimate of the sigma of loadings. Empirical results show tolerable agreement of r_c and s. Since these two emphasize different principles it is suggested that, in the absence of r_p, they be used jointly to give a final verdict in important cases.

10. Evaluation of matching between two studies should be based (since matching indices of moderate significance level can occur between truly different factors due to "cooperative" factor action), additionally on the *capacity to order the factors* in the $k \times k$ matching matrix so that the highest values in each row and column fall exclusively down the diagonal. Illustrative instances of use of matching indices are given.

11. Granted the usual variation of variance size of factors from one population sample to another an invariant factor would not be expected to retain its pattern across studies in the ordinary (unit length) factor pattern matrix. It would do so only in real-base factoring; so in truly precise work the above methods are better applied to real-base (covariance-derived) factor patterns (see Chapter 13).

12. Indexing of factors by code is initially essential, and naming less so. A systematic indexing for traits and states has been in use by an increasing circle of psychologists for thirty years, in which patterns are tied to A, B, C, etc., symbols for L and Q data, and to U.I. numbers for T data (see the Appendix and Cattell, 1944a, 1946a, 1957a, 1966a, 1973; and Hundleby, Pawlik, & Cattell, 1965). Names need at first to be more descriptive, settling as soon as research permits on interpretive meanings, and in either case avoiding popular terms.

PART II

CHAPTER 11

Factor Measures: Their Construction, Scoring, Psychometric Validity, and Consistency

SYNOPSIS OF SECTIONS

11.1. Orientation of Factor Analysis to Psychometric Practice

Once having reached unique resolution of factors in a domain and—by matching and interpretation—established them as replicated and meaningful concepts, the psychologist can enter many useful directions of pure and applied research. Indeed, it is here that multivariate and bivariate experimentalists can most profitably come together, for the latter can now avail themselves, in their ANOVA designs, of meaningful factor scores, as independent and dependent variables. To take these steps the next requirement is to set up measuring batteries and scales appropriately scored and of definable reliability and validity. To serve that need we turn in this chapter to the principles of scoring, to the salient issues of battery construction in the light of those principles, and to the evaluation and maximizing of psychometric properties of factor scales, as reflected in validity and consistency coefficients.

The scoring problems *might* be treated here as in most factor analytic tests—namely, as an abstract game of matrices between factors and "variables." But while describing those matrices we propose here also to give attention to some criticism of the nature of variables in the setting of tests important to the practicing psychologist. Indeed, we plan to follow the factor analytic penetrations into general psychometric concepts more fully than is usually done in either factor analytic or psychometric texts.

11.2. The Initial, Common Form of Estimation of Scores for Primaries and Higher-Strata Factors

So far we have briefly introduced [Eq. (10.8)] the most common method—the ordinary multiple regression method from a factor estimation matrix—for getting factor scores from a matrix of variables whose scores are known. Later in this chapter other approaches will be considered; but for the moment we shall go a little more fully into this familiar regression method, as carried out from primaries to higher-order factors. Repeating here Eq. (10.8) from above as Eq. (11.1) we have

$$\hat{\underline{Z}}_f = Z_v \cdot V_{fe} \qquad (11.1)$$

$$(N \times k) = (N \times n)(n \times k)$$

where the factor scores in a column of $\hat{\underline{Z}}_f$ will be in contracted or semi-standardized standard scores (see definition, p. 238). As we saw, they are quickly transformable by a linear stretching (D matrix) into ordinary standard scores. [Note the derivation of V_{fe} is given in Eq. (10.1).]

In discussing various aspects of the estimation of common broad factors it helps perspective to take occasional side glances at the reciprocal process of estimating variables from factors, in what we have called the specification equation and which for any given variable is constituted by a row from the V_{fp} matrix. In this connection it sometimes puzzles the student following the theoretical specification equation to find that in practice the specific and error factors, which are part of the full V_{fp} matrix, are dropped from the variable estimation equation and procedure without a word of apology or explanation! By factor estimation the practitioner means *using common, broad factors* and these alone enter in his use of "factor scores" to estimate variables.

If one also *wants* estimates of the specifics it suffices simply to restore to the V_{fe} the $n \times n$ square diagonal matrix of correlations of variables with specifics in Eq. (10.8) which has been tacitly dropped. (Since the specifics are orthogonal to everything the V_{fp}, V_{rs}, and V_{fe} values remain the same, see Table 11.1.) However, nothing is gained, practically, by bringing into the V_{fp} or V_{fe} estimation matrices the specific factor loadings or weights, because we have no specific factor scores. Theoretically we do not know (in 999 cases out of 1000) what the specific is, so we cannot bring ulterior psychological knowledge to help the specification prediction, and practically it can be obtained only as a fraction, by a regression coefficient, of the very variable we are trying to estimate. What has just been said about estimating primary factors from variables applies, mutatis mutandis, to second-order factors with primaries as scores, and to tertiaries from secondaries, etc. But the special issues of higher strata are taken up in Section 11.5.

11.3. Specifics and the Practical-versus-Complete Theoretical Specification Equation

Practically and psychometrically then, the attempt to assign specific factor scores to individuals amounts to nothing more than giving to each person the same fraction of the variable one is trying to predict. If one did, in a purely circular exercise, the addition would merely bring the estimates of the variable into a standard score rather than a contracted standard score form. Consequently, it is usual—amounting to a tradition—not to bother with the square $(n \times n)$ diagonal specific factor matrix that could be tacked on to V_{fp} or V_{fe} either in estimating broad factors from variables or in estimating variables from factors. Table 11.1 shows what the matrices should "theoretically" be like, but the practicable factor estimate and variable estimate specification equation has the last, unique (specific and error) terms and matrices chopped off from the theoretically complete (p. 67) equations.

In both directions of estimate—factors to variables and vice versa—

Table 11.1. Matrix Calculation of Scores when V_{fe} and V_{fp} Are Theoretical-Complete with Unique Factor Terms

(a) Calculation of factor scores from variable scores

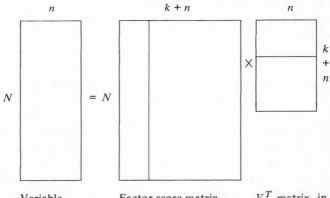

Factor score matrix: Z_f, expanded to include unique factors

Variable score matrix: Z_v

V_{fe} matrix in complete form (not usually presented) with weights for unique factors

(b) Calculation of variable scores from factor scores

Variable score matrix

Factor score matrix

V_{fp}^T matrix, in complete form (not usually employed) with loadings on unique factors

N = number of people
n = number of variables
k = number of broad factors

the estimates reached will have contracted standard score properties, i.e., a mean of zero and a sigma less than one (in what we just called "contracted standardized" scores), and if one wishes to operate next with estimates in the standard scores—as often one does—then it suffices to put into the computer calculation a diagonal matrix of inverses of the obtained sigmas as in Eq (10.10) to stretch the sigmas to 1.0. The D values can either be obtained by waiting for the output of estimates or can be calculated beforehand from Eq. (10.12), since R (the multiple correlation of variables with the factor) defines the sigma. Parenthetically, in looking again at this two-way estimation we may note that the V_{fp} is likely to be a more constant matrix from study to study than the reciprocal V_{fe}. For both the company of factors and their correlations in a given domain and population will tend to constancy, whereas the company of the set of variables by which the same factor is estimated will alter from research to research and battery to battery even in the same domain. It suffices only to introduce a new variable y correlating substantially with an existing variable x (and both correlating with the factor) to produce a marked drop in the V_{fe} weight of x. The weight for a variable in estimating a factor is thus not a fixed property of the variable but a function also of its company.

11.4. Is Estimation Necessarily a Circular Process?

Another incidental source of perplexity needing clarification here arises in the exclamation that "Estimating factor scores from variables and variables from factor scores is a merely circular process." It certainly would be if the investigator were confined to the matrices of a single research. However, even there he could estimate the factors for predicting variable j from the $(n - 1)$ variables left when the row for that variable has been omitted, i.e., j would not enter into its own estimation. Thus even with one matrix the process *need not* be circular and lose independence. However, in most *practical* uses of the factor analytic specification equation this special little problem does not arise since factors such as intelligence, anxiety, surgency, etc., are scored (estimated) from batteries that have nothing to do, in construction origin, with the criterion variables that are being predicted. In theoretical structure it is true that we *initially* get the meaning of factors by abstraction from variables, and explain why a variable behaves the way it does by reference to the factors that influence it. But as a factor matures from an empirical construct to a theoretical concept, it brings in all sorts of meaning beyond that of the original variables, as pointed out in discussing interpretation in the last chapter, and its value in the specification equation is experimentally independent of the predicted criterion variable. In any case, this mutual relation of empirical data and theoretical concepts, in which each relation enlightens in both directions, is part of the normal and necessary logical structure of scientific understanding.

11.5. Estimation Problems Peculiar to Higher-Strata Factors

The above instructions for obtaining factor scores, in standard score form, for primaries can also be used, as mentioned in Section 11.3, for higher-strata factors. It will be remembered that higher-strata factors can be conceived and measured either as overlapping factors or as independent "stub" factors. In either case, one has the alternative of estimating the given stratum either from the scores on factors in the immediate structure below it, or, in the "vaulted" models (p. 241), from the more remotely placed variables themselves. And the factor stratum below that from which one is estimating the higher strata may itself be estimated directly from variables or consist of, say, primary *scale* or *battery* scores, commonly an unweighted sum of variables in such scales.

The alternative procedures here are rather numerous because the possible alternatives of the model are several. Taking the two main models— the SUD (which uses stub factors) and the SOA (or serial overlap analysis)—we have in each the alternative of estimating the higher strata from the next immediately below or of estimating directly from the "floor" of variables. The former (SUD) will use the Schmid–Leiman breakdown (p. 209) and the latter (SOA) the Cattell–White. Therefore, in either, knowing the V_{fs} or V_{fp} and the R_f we can use the V_{fp} on variables to estimate the factors from variables, without concern that we are operating with a vaulting matrix. In the overlapping (SOA) we correct the V_{fp} on variables to the V_{fs} on the variables and, knowing the correlations among the variables, R_v and the correlations among the second order factors R_{fII}, we can then calculate the V_{feII_v}, and so estimate S_{fII}. In the nonoverlapping case (SUD), one will work out the V_{fpII_v} as in Table 9.5 and proceed similarly. As pointed out above, for the primary factors the V_{fp} loadings for the "full length" and "stub" form are simply proportion and except for difference in the intermediate contracted standardization $\hat{Z}_{fII}$ will result in the same $\hat{Z}_{fII}$, i.e., the scores corrected to unit sigma for both. But in higher orders the simple relation no longer holds and they will be different.

The reasons for primary factor estimates initially falling short of unit sigma and of full validity are (apart from error variations) the insufficiency of high loaded variables and dragging in of the numerous specific factors in each variable that contributes to the broad factor score. When a V_{fe} from one study is used to estimate factor scores from the same variables in another experiment we have, additionally, the effect of sampling error or even of real differences of population on the weights. Cross validation—applying the V_{fe} on the new sample and comparing the validities (R values) obtained—often shows appreciable drops in validity. Thus whenever possible one should calculate and use the V_{fe} as calculated for the given sample. In contrast to this sophisticated weighting common psychometric practice often simply adds the items in a scale or the subtests in a battery not only with equal weight among themselves, but consequently, with unchanging weight from population to population, and this can cause appreciable inaccuracy, avoidable by the use of V_{fe} as in Eq. (11.1).

Practically all these sources of error have typically been greater in higher-order than lower-order factor score estimation. Not only are values for higher strata generally less stable and accurate, but, if one is estimating directly from variables one meets the situation that weights (like the loadings) in higher-order factors directly on the variables tend to be systematically lower than those for lower orders. On the other hand, compensation comes from the fact that one is using decidedly more variables. Particularly in little explored areas where not many variables have yet been accumulated to estimate primaries, the secondary, standing on its broader base, may be more validly and reliably estimated. Thus a general intelligence test of one hour and 200 items is likely to have a validity equal to or better than a spatial or numerical ability primary, of, say, ten minutes and 40 items. But if the primaries were each measured on 200 items the principles above tell us that the primaries would typically have higher validity than the intelligence test.

11.6. Basic Concepts Regarding Types of Validity

Since the term validity has inevitably entered at this point, as we consider scoring, it should be pointed out, even to the reader matured in more recent test psychometric concepts, that we use validity and consistency here in definite systematic meanings belonging to quite recent and putatively more elegant systems than some in use in older textbooks. As this is a book on factor analysis our digression into validity *per se* must be brief but the reader may study it at leisure as developed in various articles and books (Cattell & Warburton, 1967; Cattell, 1964a, 1964b, 1973). Meanwhile the essentials are summarized in Fig. 11.1.

Figure 11.1 will remind the reader of the three dimensions of validity.

1. From *concrete* (relating to a single particular criterion variable) to *conceptual* (relating to a concept, e.g., intelligence, anxiety) along a dimension of abstractness.

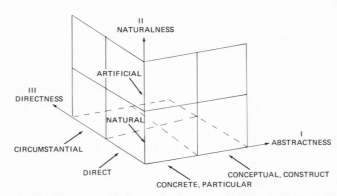

Fig. 11.1. The three dimensions of validity arranged in a cube of bipolar dimensions.

2. From *natural* (a variable naturally in the environment) to *artificial* (e.g., another intelligence test, or a laboratory performance) along a dimension of *naturalness*.

3. *Direct* (evaluated directly against the criterion) to *indirect* (evaluated by similarity of relations to many variables to those possessed by the criterion) along a dimension of *directness of referral*.

These dimensions being dichotomous there are $2^3 = 8$ possible varieties of validity coefficient. However, the only ones we need discuss closely in validating estimates of factor scores from research variables or practitioner's batteries are *direct conceptual* validity and *indirect conceptual* validity, assuming both natural and artificial validity may exist in either. (Parenthetically, concrete validity *as* validity is of no interest to psychological theory. It tells us that pegboard X predicts with such and such validity an industrial output performance Y at a certain workbench.) Such relations to specific criteria, e.g., that the fluid intelligence factor measured by culture fair tests predicts rate of learning Chinese in China (Cattell, 1971a), or English in America, or rank order in a chess championship are statements of *relevance*, not validity, since no one knows initially whether the test should predict these particular performances. Relevance coefficients extend our knowledge of what a factor means and tells us how valid, say, the rate of learning Chinese is as a measure of intelligence, not the converse, since, no one knows as just indicated whether learning Chinese should be a saturated measure of the intelligence factor, although someone may, of course, entertain that view as a personal *a priori* hypothesis.

The direct conceptual or construct validity of a factor battery score is its correlation with the pure factor, which defines uniquely the concept concerned. Its indirect concept validity is the extent to which its correlations with other factors and variables are similar to those which the pure factor itself has with these surrounding variables. Initial experiment shows (Cattell, Eber, & Tatsuoka, 1971) that when a series of batteries are ranked as to their direct validities the order tends to be approximately the same as the ranking on indirect validity coefficients. This can be seen in illustration by comparison of the direct and indirect validities of the various 16 P.F. factors (Cattell, Eber, & Tatsuoka, 1971). The mode of calculation of the latter in that case was to determine the correlations of the given factor X with 15 others, as a pure factor, and compare the series with the series of correlations of the scale for factor X with the 15 others. The two columns of correlations are then correlated to give the indirect validity. To obtain the correlation of scale X with the 15 other pure factors requires that it be placed as a variable in the pure factor space, i.e., in the simple structure resolution.

Everything of importance in the present discussion can be handled in terms of the more familiar notion of direct validity, and it remains only to explain why in that domain *conceptual* validity is in most situations a more correct expression than the presently somewhat popular "*construct* validity." A construct—strictly an "empirical construct"—is restricted to properties arising solely out of the immediate empirical data. A factor appearing only in one experimental matrix, and thus discovered for the first time, is indeed only a construct, and a test's correlation with it is a construct validity. But serious validation deals with established, replicated factors, experimentally related also to many criteria outside any factor analysis, as in the instances of fluid intelligence, anxiety, perceptual speed, surgency, etc. These are scientific concepts, enriched by and definable in a complex network of operational relations. Though most readily and uniquely defined operationally by the factor, it is no longer a construct but a concept, with added meaning and reference from many sources.

In general, the researcher and the practitioner are alike properly most concerned with concept validity. They want to know whether a given factor battery or estimated factor score measures something of wider theoretical meaning, such as intelligence or exvia. But provided the factor is defined on a sufficient basis of variables to give reliable unique rotation that concept validity can be adequately determined by correlation with the factor on one appropriate population sample. It can be obtained (a) by estimating the factor score by the given scale or battery [Eq. (11.1)] and correlating it with the pure factor, or (b) by calculating the multiple R of the battery subtests, as in Eq. (10.12). The correlating of a battery score as a literal battery score, with the pure factor can be achieved by (and only by) putting the battery score into a factor analysis with enough other variables to measure several other factors and fix a multidimensional hyperplane; for the *pure factor* is the unit length factor vector in the configuration of variables, and, of course, must not be confounded with the score on a factor battery the validity of which has to be established.

11.7. Definitions of Homogeneity, Reliability, Factor Trueness, and Unitractic and Univocal Batteries

Before pursuing factor estimation further, in terms of asking about special practical problems, it is necessary to clarify the central concepts of *test consistency*. Again the reader must be referred for wider supportive discussion elsewhere (Cattell & Warburton, 1967; Cattell, 1973). What is best referred to in the most general concept as test consistency has three aspects, logically derivable from the basic data relation matrix (p. 322):

1. Homogeneity: the mutual agreement of different parts of the same test, e.g., agreement among items or subtests. It will be recognized as we come to the data box (Chapter 12) that we are here comparing along the stimulus–responses id dimension.

2. Reliability: the agreement of the test with itself across occasions in time. There are at least six important varieties of the reliability coefficient, e.g., those examining agreement with repetition across time and administrative circumstances. These varieties can be sufficiently represented here in the Dependability Coefficient (Cattell, 1973), which is the *immediate* or time-partialled test–retest correlation. Reliability is comparison along the *occasion* and *observer* coordinates of the data box.

3. Transferability: the agreement of the test with itself across varieties of people and subcultures. This concerns sections along the person coordinate of the data box.

The last, requiring fuller familiarity with the data box, carries us' into other domains not yet discussed (see Chapter 12). But the two preceding forms of consistency can be quickly dealt with. Therein one runs into the unfortunately continuing confusion of reliability coefficients with homogeneity coefficients. This can perhaps only be entirely dispelled by reference to fuller discussion (Cattell & Warburton, 1967; Cattell, 1973). However, it may help in separating them to consider the illustration that a very heterogeneous, low homogeneity "test battery" (like the sum of a man's height, weight, and visual acuity) can have high reliability over time as shown by the dependability (test–retest) coefficient. Conversely, a test of very high homogeneity, e.g., people's abilities to sing correctly two high-pitched notes, which could correlate highly, could yet have quite low dependability, i.e., from occasion to occasion. A comprehensive overview of consistency coefficients—though not all of it concerns us here—is set out in Table 11.2.

Table 11.2. Diagram of Varieties of Consistency Coefficients

Property of test examined	Reliability				Homogeneity		Transferability
Name of coefficient	Dependability reliability	Conspect reliability	Intrinsic reliability	Stability	Scale homogeneity	Form equivalence	Transferability over standard populations
Data coordinate over which there is variation	Occasions (ambient stimulus)	Observers (administrators and scorers)	Superficial form (not content) of items	State of subjects	Items or variables within a scale	Two forms of same scale	People (type of subject)
Mode of examining variation in testing operations	Correlation of two occasions with same administrator and scoring	Correlation with different scorers, observers, etc. (administrative dependability when administrators change)	Correlation with same items but different writing style, form of response system, etc.	Long-term retest permitting change of trait and state	Correlation among items, subtests, forms	Correlation of two forms	Pattern-similarity coefficient applied to behavioral index vectors in factor specification equations
Symbol for specific coefficient	r_{rd}	r_{rc}	r_{ri}	r_{rs} or r_s	r_h	r_e	r_{pt}
Symbol for corresponding index	d_{rd}	d_{rc}	d_{ri}	d_{rs}	d_h	d_e	d_t

A sorting out of other concepts—such as factor-true, univocal, uni-tractic, and factor-homogeneous—can perhaps be most briefly achieved by Fig. 11.2 (a) and (b). If tests or items are plotted as points in space (two-factor space necessarily in Fig. 11.2) then their homogeneity is low if they have to be encompassed by a large circle and high if they can be contained in a small one. The factors in Fig. 11.2 have their axes drawn for simplic-ity in (a) as orthogonal. Measures lying on or close to either axis, as in the *a* and *c* circles on the vertical axis or scales that could be similarly drawn on the horizontal will have summed scores that are *factor-pure* (or "factor-true"), and if the circle is small, as at *c*, they will also be very *factor-homogeneous*. But an equally small circle at *d* will be *test-homo-geneous* yet not *factor-homogeneous*. That is to say, the items will inter-correlate highly but all of them will be mixtures of two or more factors. It

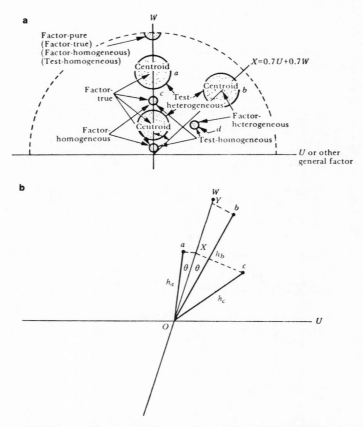

Fig. 11.2. Definition of homogeneity, trueness, and validity in factor space. From R. B. Cattell, *Personality and Mood by Questionnaire*, Jossey-Bass, San Francisco, 1973, p. 380.)

will be noted that, as in the large circle at a, on the vertical axis it is possible for a test to have low homogeneity, yet for its total score (the center of the circle) to be factor-homogeneous and true. We shall return to this in connection with "suppressor action" and "optimum homogeneity."

A clear recognition of the difference between the factor *trueness* and factor *validity* of a measure can be gained through Fig. 11.2(b). If W is the wanted factor and U represents one of any number of unwanted factors in the test items, roughly orthogonal to W, then the purity of a test is given by the cosine of its angle, i.e., by its angular closeness to the factor. Thus tests a and b (angle θ) are truer than c, which has a larger angle. The correlation of a test with the factor, however, which is its validity, is

$$\text{Validity} = r_{aF} = h \cos \theta \qquad (11.2)$$

where h is the root of the communality and θ is the angle of a to W. Thus b, being longer, i.e., of higher communality, is more valid than a (as OY is to OX) though their trueness is the same since both are at an angle θ. (And c which is less true than a is, however, just as valid, namely, having the projection OX.) It is assumed that the rest of the length of a short-test vector lies outside the common factor space, and trueness, therefore, is assessed purely with reference to contamination with other common factors.

The attachment of psychometric desirability to trueness, on the grounds that regardless of validity, trueness means relative purity of measure, in the sense of noncontamination by other factors has sense in some circumstances. But one has to recognize, as just stated, that there is contamination by a possibly large specific. Relative to contamination by some broad, psychologically systematically important but unwanted factor, as occurs in c, this may not be dangerous for use of the test as a predictor, for if the specific is very narrow this contamination has the same status as random error, reducing but not biasing prediction. (Though we must always remember that today's specific factor may, with the advance of scientific exploration, become tomorrow's common, broad factor!)

The definitions of *univocal* and *unitractic* also need clarification and distinction, since Guilford and Michael (1948) and some other psychologists have used univocal in a different sense. Whatever attachment of terms we use the fact is that there are two concepts. The concept which we call *univocal* is that the items of a test have only a *single* broad factor in them, apart from specifics, as in Spearman's definition of a good battery for intelligence. To the statistician the definition is that we have a *rank one*, i.e., single broad factor, correlation matrix among the subtests. However, the proof—as Spearman and others require for a univocal battery—of either (a) a hierarchical correlation matrix among the parts, and/or satisfaction of the tetrad difference criterion (Spearman, 1927), or (b) the factorial proof of a rank—one matrix *does not guarantee that the single com-*

mon dimension is a pure source trait factor. The above checks may show that it is mathematically one factor but in a psychological or scientific sense, where rotation has gone in some random direction, one factor can be a proportional combination in each subtest, of two quite different factors, from a psychological point of view. For example, in the ball problem (p. 95) it is obviously possible to draw an axis designating a general, broad factor X which is a combination of, say, 0.8 of weight W, and 0.6 of volume S (size), and to find a set of variables which will contain nothing but specifics after this factor is taken out. (This last condition is important in getting a univocal scale, as Spearman's careful selection of variables showed.) Incidentally, there is little experience of how far selection of variables in any domain, undertaken to meet the conditions of a rank-one univocal scale, is or is not likely to yield a mixture of common factors, i.e., a mixture as far as meaningful SS rotated common factors are concerned.

By contrast, a *unitractic* scale (from Latin "tractus" from which our word trait is derived) is here defined as having its axis along a meaningful, uniquely rotated factor trait. It is a univocal scale with an additional condition; for in the mathematical sense, both a univocal and unitractic scale have items or subtests which form a *rank-one correlation matrix*, but except in the instance of Spearman's "g" (intelligence being a much analyzed and tested concept, hard to miss in test choice) the present writer would guess that construction of a univocal set of items in a scale rarely hits also a unitractic set—unless separate conditions are introduced from the beginning. For that matter, even with the intention to make a unitractic scale it is highly unlikely that in practice it will be achieved, i.e., that we get, say, twenty items with nothing but loadings on the required source trait and twenty specifics. A scale can be constructed, the final total score on which is factor-pure, but the items will, short of a miracle, have loadings on other unwanted common factors too (as well as on specifics) the elimination of which from the final score is obtained by mutual *suppressor action* studied in Sections 11.8 and 11.9 which now follow.

11.8. Relations among Homogeneity, Reliability, Transferability, and Validity

In considering interrelations of validity and consistency coefficients in the field of factored tests the dependability reliability coefficient (test–retest) can quickly be dismissed as having no necessary systematic relations to homogeneity or transferability. The size of the reliability (dependability) coefficient derives largely from the style, answer choice distribution, item frequency, and degree of ambiguity in items. Its relation to validity is only the familiar one that reduction of length or increase of random error will reduce validity pari passu with the direct reduction of reliability. In the opposite direction of action the effect of differing factor combinations in a scale and different final validity is not upon the *dependability coefficient* (immediate retest) but upon the *stability coefficient* (with which the former is rather frequently confused). The stability coefficient (see Table 11.1) is derived from a long-term retest, and its contrast with the immediate test–retest value, the dependability coefficient, permits us to separate the *function fluctuation* in the trait or state itself, from the unreliability of the test itself, evaluated by the immediate retest or dependability coefficient r_d. Considering the correlation coefficient as defined by the predictable fraction of the variance, and dividing the variance of any given measure into (a) true trait variance σ_t^2, (b) variance from fluctuation of this trait, σ_f^2, and (c) variance from error of measurement, σ_e^2, we can write the dependability (immediate test–retest reliability coefficient, r_d, as in Eq. (11.3a). This equation to r_d rather than r_d^2 holds only for a symmetrical relationship of test components (Cattell, 1973):

$$r_d = \frac{\sigma_t^2 + \sigma_f^2}{\sigma_t^2 + \sigma_f^2 + \sigma_e^2} \tag{11.3a}$$

Note $r_d = 1 - \sigma_e^2$ when the total observed variance—$\sigma_t^2 + \sigma_f^2 + \sigma_e^2$ —is unity, i.e., a standard score, which contrasts with the long-term stability coefficient r_s, constituted thus:

$$r_s = \frac{\sigma_t^2}{\sigma_t^2 + \sigma_f^2 + \sigma_e^2} \tag{11.3b}$$

It follows that a coefficient r_{tc} for the constancy of the trait as such can be written and obtained by calculation as in (11.3c), thus:

$$r_{tc} = \frac{\sigma_t^2}{\sigma_t^2 + \sigma_f^2} = \frac{r_s}{r_d} \tag{11.3c}$$

The analysis of the fluctuation of a trait measure as such is a special problem in factored measurements that will be given perspective in Chapter 12. But at least an important warning can be given here against considering a high r_{tc} for a psychological measure, and therefore a high r_s for a test, as invariably desirable. As far as validity is concerned, one can see that if a test ought to measure say, a fluctuating state, such as anxiety, then the higher its validity, e.g., the more it excludes contaminating trait variance, the lower will be the stability coefficient of the test, since anxiety as a state should not be constant.

In this connection the reader is reminded that the practice some years ago of contrasting the "concurrent" and "predictive" validities of a test shows basic conceptual confusion. If a test predicts less well "predictively" (ambiguously used for *future* prediction, better called *foretelling*) than currently, this is a property of the trait, not of the test. It is properly expressed by the above r_{tc} coefficient. Predictive vs. concurrent is inapt.

Thus reliability–dependability and validity relations can be relatively briefly stated in this factor analytic context; but among the homogeneity, transferability, and validity coefficients of factor measures more systematic factorial and general relations exist. Transferability is defined (Cattell & Warburton, 1967; Cattell, 1973) as the degree to which a factor battery can be applied to different populations and still measure the same factors. If a battery j has loadings (in the specification equation) of $b_{xj1}, b_{xj2}, \ldots,$ b_{xjk} on subculture x and $b_{yj1}, b_{yj2}, \ldots, b_{yjk}$ on population subculture y, then r_t the transferability coefficient can be calculated:

$$r_t = r_p \cdot (b_x b_y) \tag{11.4}$$

where $r_p \cdot (b_x b_y)$ is the pattern similarity coefficient here calculated between the two series of b values (loadings) (Cattell, 1973). Alternatively, if there is doubt about the estimate of the standard deviation of the loadings in the given domain, one could use r or intraclass r. The rationale of this is, of course, that what the test is measuring factorially is defined by the specification equation. The test's concept validity is the pattern of loadings on the factors, and commonly the best design will call for its being high on one factor only. But regardless of what pattern is required, the constancy of that pattern across the various groups tested is a measure of the exactness of its transferability. The r_t value, for example, is emphatically higher on a culture-fair than a traditional intelligence test.

It is argued in what follows that this transferability will be better in batteries (a) where the homogeneity is low and (b) where a single prominent specific plays no part. The reason for (a) is that the behaviors which most purely express a factor in a given culture, and have very high loadings, are more likely to be peculiar to a culture than the general run of moderately loaded behaviors. For example, in a society of mathematicians the high loadings of crystallized intelligence are likely to be in mathematics and in a subculture of footballers in football tactics, but otherwise they will agree in wider fields of moderate loadings of intelligence behavior that will hold reliably for both groups. As to (b) above the argument is that specific factors, as "bloated specifics" (below), are also likely to be more peculiar to a culture. For example, suppose we aim to measure a supposed broad factor of "spatial and mechanical thinking" believed to be largely present before training. There is a strong probability that a couple of items having to do with, say, the mechanisms of pneumatic drills, not used in one population but much used in the other, will carry their essentially specific variance as a bloated specific (described further below) into the general mechanical aptitude factor estimate, yet be quite irrelevant to the second culture. This could happen to a mechanical aptitude test applied in, say, New York and Fiji. The transferability of tests in which some specific variance is confounded with broad factor variance, as happens in tests of high homogeneity, is therefore likely to be poor.

11.9. The Arguments for Low Item or Subtest Homogeneity in Constructing Tests for Factor Estimation

A fashion in test design which took on the status of an authoritative generalization arose from some superficial thinking on homogeneity in the first half of this century. A side glance at this bit of social history is necessary, because the unsuspecting student is still as of 1978 being exposed to textbooks confusing homogeneity consistency with reliability consistency, and given to naively quoting reviews commenting critically on low homogeneity, e.g., in the latest issue of Buros's (1972) otherwise valuable evaluations of personality tests. Any text that offers psychometric test constructors an unqualified injunction to maximize item homogeneity should be regarded with doubt. The issue has been further confused semantically by calling homogeneity "reliability." For tests aimed to predict *highly specific performances* or attitudes a high homogeneity, as expressed for instance by a high Cronbach alpha or Kuder–Richardson coefficient, is desirable. But in measuring naturally occurring broad common factors a test boasting high homogeneity is at once suspect. There are two reasons for this, one simple, the other requiring some discussion.

First, one recognizes that what is, in terms of the personality sphere or other bases for stratified sampling of a population of variables, essentially a very *narrow specific* can be blown up to the apparent status of a common factor in any given matrix by entering the experiment with several items that are close variants on the specific variable. For example, in a test of attitude to army discipline an item which deals with attitude to the mess sergeant can be developed into ten similar items, all involving the sergeant, whereupon a broad factor—which we shall call a *bloated specific*—is almost certain to be found loading across the ten variables. As shown in Figure 11.3, the variable b which would have stood at a with respect to the true common factor space stands out also on a specific S orthogonal to the common factor F. Suppose it is now joined on S_1 by a set of variables c, d, e. These share the extra variance due to the "common specific" and become a quite highly correlated cluster, pulling b out onto this new broad factor. In fact, they have created a new pseudo-broad factor. It is not unlikely that a necessary extra dimension will not be taken out for this extra factor and it will be taken to be one with F, the original broad factor. Thus finally a "bloated" specific sits on top, as it were, of the real, meaningful common factor and by the extra com-

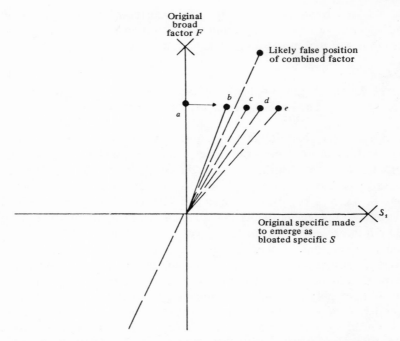

Fig. 11.3. Raising item homogeneity by confounding with a bloated specific. (From R. B. Cattell, *Personality and Mood by Questionnaire*, Jossey-Bass, San Francisco, 1973, p. 361.)

mon variance it has added, it raises the variables *b*, *c*, *d*, *e*, etc., to a higher degree of homogeneity than they would otherwise have had due to the true common factor. The persistent selection of items for high homogeneity as such in a scale intended for a broad common factor undoubtedly often leads to a test with this contaminated factor structure. However, for lack of any factor analysis, or through a careless factor analysis underestimating the number of factors, the situation remains unexposed. Naturally, the estimation of the score for the supposed unitary factor *F* due to the spuriously high common variance has a spuriously high apparent validity and encourages the notion that high homogeneity is desirable for validity.

The second reason for being critical of or at least suspicious of high homogeneity on a test to estimate the score of a broad factor is that except in, say, primary abilities, the single behaviors that are really highly loaded (.7–.9) on a factor *F* and free of other broad factors, are as hard to find as large diamonds. Personality and motivational behavior, for example, is complexly determined and it is extremely rare to find instances of "pure" dynamic trait behavior into which other broad factors than that desired do not intrude. The craft of test construction teaches us that these intrusive "unwanted factors" can be kept out of the final factor battery or scale score only by arranging for suppressor action. That is to say, on obtaining a reasonably good item for F_w that turns out to be positively loaded on unwanted factor F_u, one seeks to balance it by finding another sharing a loading on the wanted factor F_w but having a negative loading on the same unwanted factor F_u. This suppressor action is illustrated in the simple case of just two factors in Fig. 11.4, by items a and b, and, at lower loadings, by a' and b'.

Even if single items were discovered defining behavior which emanates almost purely from a particular personality factor, one would hesitate, certainly in the questionnaire medium and generally in others, to rest the factor measure on a very small but "psychometrically adequate" (for validity) sample of such highly valid items. For such items, as we have argued, are probably more likely than items of middling loading to be unstable from subculture to subculture.

One must also remember in connection with validity that the factor loading pattern of a *broad trait* describes a *common trait*, i.e., one defined, in the personality structure (not the statistical "common" sense) as similar in pattern across many persons. The usual factor across people (defined below as R-technique) cannot yield a *unique trait* pattern within any in-

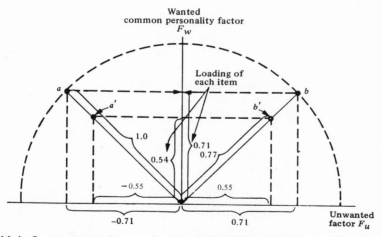

Fig. 11.4. Suppressor action producing high validity with low homogeneity. W = wanted factor; U = unwanted. $r_{ab} = 0$, $r_{a'b'} = 0$. Demonstration of high validity with zero homogeneity. Loadings are as shown. Validity of two-item scale is 1.0 if only two broad factors are involved. If a specific is involved, taking 40% of variance and reducing length of item vectors as shown at a' and b', then their joint validity is reduced to 0.77, but their correlation remains zero. (From R. B. Cattell, *Personality and Mood by Questionnaire*, Jossey-Bass, San Francisco, 1973, p. 358.)

dividual (see discussion on fit of the model, p. 311). The expression of a trait, e.g., dominance, among several people in whom it happens to stand at the same level, will nevertheless stand at different levels in any particular area of expression for each. Granted that this is the nature of a statistical measuring of a factor—an "ideal" pattern attenuated by the peculiarities of individuals—then the best scientific approaches to validity and comparability lie in sampling behaviors over a broad area. Such behaviors, even though significantly loaded on the wanted factor, are likely because of their diversity to show small or trivial mutual correlation, i.e., very low homogeneity. (Two items each loaded 0.3 on the factor will, even *without* suppressor action, correlate mutually less than 0.1.) However, a moment's reflection reminds us that in terms of getting a substantial multiple correlation validity, such discrepancy between item validity (high) and item homogeneity (low) is precisely what we should be aiming for. Further discussion of the actual formulas relating mean factor correlation, number of items, and optimum item homogeneity belongs to general psychometry and may be read elsewhere (Cattell, 1973, Chap. 9). Here a typical actual experimental instance of truly low homogeneity, required by the above considerations, producing substantial validity of the whole scale (of only 13 items) is shown in Table 11.3. From these theoretical and experimental contributions we may conclude that, when limited testing time (the usual condition) commands one to

Table 11.3. High Factor Score Validity Achieved by Taking Items of Optimal[a] Low Homogeneity

Personality factor H: parmia vs. threctia from 16 P.F.												Item	Correlation with factor
												1	45
29												2	53
26	10											3	26
09	16	00										4	15
17	26	10	15									5	37
16	21	12	06	22								6	25
18	11	14	19	19	17							7	23
34	28	30	-01	13	10	07						8	52
22	23	18	05	14	03	03	28					9	42
20	15	29	08	15	10	10	31	21				10	36
10	16	10	08	02	-10	00	17	18	14			11	29
13	26	17	06	23	19	08	23	14	22	08		12	42
10	10	15	-09	05	06	01	18	20	14	11	10	13	23

[a] As in multiple correlations generally, the optimal homogeneity is a value which is low relative to the correlation of the items with the criterion (in this case the pure factor). Just how low rests on principles of calculation elsewhere (Cattell, 1973). The above calculations are on a sample of 780 men and women. It will be noted that some homogeneities (correlations of items) are actually negative; but other samples, averaged, show all to be positive. R scale validity by weighting = 0.79. Mean item validity = 0.34. Mean homogeneity = 0.14. The validity of *un*weighted (equally weighted) items is 0.75.

construct a test to get a factor score from a limited subset of the array of variables, one should examine not only the factor structure matrix but also inspect the R_v matrix along with this V_{fs} to pick items with the highest possible ratio of mean validity (correlation in the V_{fs}) to mean homogeneity (mean correlation in R_v). The factor concept validity of the resulting score, if items are unweighted, can be assessed from

$$r = \frac{n \, \bar{r}_{jw}}{\{n[1 + (n-1)\bar{r}_{jm}]\}^{1/2}} \qquad (11.5)$$

where $\bar{r}_{jw}$ is the mean r of items with the wanted factor and $\bar{r}_{jm}$ is the mean interitem r (homogeneity). All that is said here about n items applies of course also to subtests or other elements in a factor battery.

11.10. More Theoretical Issues in Factor Score Estimations

As the identification and agreed interpretation of traits and states in various fields of psychology mature in this decade, there is likely to be a switch of maximum effort from both exploratory research and hypothesis-testing research on structure to relating factor scores to criteria in everyday life and to manipulative studies on factor scores in relation to physiological, genetic, and social influences. One is compelled to predict that a certain amount of surprise and dismay will accompany this shift to factor score measurement. For the straightforward application of the regression method which we have so far explained will often, in practice, yield rather discouragingly low correlations between factor scores for the same factor concept obtained from different sets of variables scores. Furthermore, turning to different theoretical aspects of the estimation process will reveal to the statistically sophisticated some perplexing uncertainties in underlying psychometric theory.

The present section, dealing with these problems, must inevitably exceed the level of mathematicostatistical sophistication invoked up to this point, and many readers are advised to be content to get the general sense rather than follow particular equations. Such a grasp of the general logic of conclusions should suffice in following subsequent sections, but here we wish to penetrate the full complexities in the interests of those who wish to do so. However, initially the problems are relatively straightforward. Correlations of other variables with factors, including the validity correlations above, will rest on one of two bases: (1) estimation of factor scores and correlation of them with criteria, or (2) incorporating the criterion variables in the factor analysis and getting their correlations and loadings with respect to the pure factor. Before proceeding here to the estimation problems, let us note the relations of values from these two sources. The latter procedure will give correlations with the pure factors, whereas the former will be attenuated relative to these by the unavoidable degree of invalidity in the factor.

First one must look at the classical paper of Guttman (1955) and the further clarifications of Heerman (1963, 1964) and Schönemann and Wong (1972) to the effect that there is an "indeterminacy" in factor scores derived from variables. That is to say, granted a unique rotation, there is still normally an indefinite series of possible factor scores (most substantially mutually correlated, however) that will account for the given variable scores of individuals. The basic argument here is that an orthogonal relation matrix C can be found such that

$$C \cdot V_{fp}^T = V_{fp}^T \tag{11.6}$$

i.e., the multiplication ends by restoring the same values. V_{fp} here is the full factor pattern with specifics, i.e., $n \times (n+k)$, and C is $(n+k) \times (n+k)$. Since C is orthogonal

$$CC^T = I \qquad Z_v = Z_f V_{fp}^T = Z_f IV_{fp}^T \tag{11.7}$$

(Note Z_v here has no estimation sign because the uniquenesses are in V_{fp}.)

$$Z_v = Z_f C^T CV_{fp}^T$$

and substituting Eq. (11.6) we have

$$Z_v = (Z_f C^T) \, V_{fp}^T \tag{11.8}$$

the expression in brackets meaning a transformed set of factor scores.

Disturbing though this diversity is, one has to remember that the transformation is limited. It can best be envisaged geometrically by Fig. 11.5. The infinite series of alternatives to the Z_f are all contained on the circle at the top of the cone. Swain's (1975) evaluation of estimation procedures is helpful here.

Note that the equation above deals with *full* restoration of variable scores, and vice versa, using both broad and specific factors. But since in real life one will actually deal with estimates only, from the broad factors, we have translated the consequence of Eq. (11.8) into an estimation situation, with communalities for the batteries of h^2. These estimates of the factor score will all have the same validity, namely, $h \cos \theta$, and be equally good at yielding predictions of the variables but (a) they may not show very high mutual correlations. Thus if a and b (Fig. 11.5) are two estimates at opposite sides of the circle, $r_{F_a F_b} = h^2 \cos \theta^2$. For example, a validity of .8 will lead to a mutual correlation of estimates of only .64, and (b) the correlations among factor scores for different factors will not be as accurate as if they were actually taken from cosines at the simple structure "pure factor" rotation position.

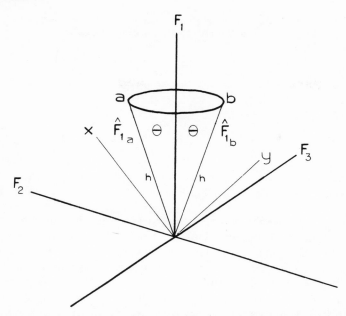

Fig. 11.5. Inherent indeterminacy of factor scores. Shown in three space.

The practical inference from the Guttman–Heerman–Schönemann proposition is that the ratio of variables to factors in a V_{fe} matrix should be large. At various points in this book we have encountered arguments that tend this way—or, at least, toward designing factor experiments with *many variables*, say, from 40 to 200. But note that we make an exception to getting this effect simply by taking a large number of short items, as in questionnaires, since one then gains numbers at the expense of reliability. To retain stability in questionnaire or other "short-variable" analyses we argue later for *parcelled factoring*. In short, it is no answer to Schöne- mann's request for a high variable-factor ratio to take many small variables of borderline reliability. With a sufficiency of variables communality values become better (pp. 30–31), simple structure is more accurate (p. 112), the risk of confusing cooperative factors is reduced (p. 525), and now, as implied by the above, the role of specifics relative to broad factors in the score estimate is reduced. Gorsuch gives results as in Table 11.4 to guide us in this matter.

The above problem concerns the situation with identical variables, but something representable in the framework of Fig. 11.5 by the factor estimate vectors x and y occurs when factor estimates are made from different sets of variables. No uniform cone is involved because different sets will give different validities (lengths and angles of vectors relative to F) but here the best vector for the given battery will lean to one side or

Table 11.4. Agreement of Factor Estimates (Same Variables) at Various Variable-to-Factor Proportions[a]

Ratio factors to variables	Minimum average correlation of estimates
1/20	.90
1/10	.82
1/5	.67
1/4	.60
1/3	.50
1/2	.33

[a] As stated in text, these assume the same V_{fp}. Reproduced by courtesy of Dr. Richard Gorsuch.

the other of the true factor according to the zones in which the variables used in estimation happen to cluster. Thus we have already pointed out that if the factors themselves have been brought to, or left at, orthogonal positions the factor score estimates of such orthogonals will not remain mutually orthogonal. This applies to the ordinary V_{fe} method above, but if for some relatively specialized reason the experimenter is going to stay with orthogonal factors and wants score estimates for them that will behave as if orthogonal, it is not beyond the ingenuity of the mathematical statistician to provide him with a transformation. The use of the Anderson and Rubin method (1956) is advocated by Harris (1967) and McDonald and Burr (1967) for this purpose. The calculation of the V_{fe} is

$$V_{fe} = U_{fp}^{-2} V_{fp} (V_{fp}^{T} U_{fp}^{-2} R_v U_{fp}^{-2} V_{fp})^{-1/2} \qquad (11.9)$$

Or if the V_{fp} is written in the "complete" form

$$V_{fe} = U_{fp}^{-2} V_{fp} (V_{fp}^{T} U_{fp}^{-2} V_{fp})^{-1} \qquad (11.10)$$

where U_{fp} is the square, $n \times n$, specific plus error factor part of the full V_{fp}. Theoretically, this seems a sound and attractive proposition, but in practice it has been extremely little used, so that at present one can make no report from the field of experience. Doubtless such experience will now develop as factor scores become more used. It yields estimates which, in the orthogonal case still stay orthogonal but in the general oblique case are nonorthogonal though *uncorrelated with any pure factor in the system other than that which they are estimating*. Unfortunately, this does not represent, any more than the Anderson–Rubin solution, what the SS oblique solution requires, namely, estimates which correlate as closely as possible to the way the pure obliques correlate. The latter is discussed below, but the above approaches deserve mention as statistical model possibilities, i.e., if one adopts for some reason the orthogonal model and also because they become relevant if one goes over to complete adoption of the SUD model.

Another type of factor estimation, discussed aptly by Tucker (1971), has more scientific interest. It is of interest because it reverses the direction of regression, which *could* imply a reversal of causal action if we make more than statistical assumptions. The regression of F on v, which is what we have in V_{fe}, can most naturally be interpreted, if one connects regression with causality, as that v operated on F. But precisely the reverse position has been adopted here, in the scientific model, namely, that the factors are the origins of variance on the variables (as expressed in V_{fp}). Tucker's penetrating logic is that the two-way regression is better treated *as if* in the second case (V_{fe}) the important regression is still in conformity with the first, namely a regression of variables on factors. His method is sometimes called the "least-squares method," which is too generic for identification and for which we would substitute "Tucker factor estimate." It determines the weights to be applied to the variables so that a least-squares fit of the observed variable score to the score estimated by regression on the factors is obtained. This requires weights to be found for the variables to give factor scores that will have this property. They are found by the equation

$$V_{fe}^T = (V_{fp}^T V_{fp})^{-1} V_{fp}^T \qquad (11.11)$$

In discussing this solution Tucker (1971) also gives some valuable comments on the general problem of working with factor score estimates *appropriate to intended uses.*

11.11. Approximate Estimation Procedures

To be realistic one must recognize that most practitioners in this area are perhaps rightly less interested in refined but elaborate methods meeting special conditions and assumptions, than in saving time even on the standard V_{fe} procedure by some acceptable approximations. While one should not pander to appeals for bypassing the precise rotational and other procedures for defining the factors themselves it must be admitted that there are practical clinical, school, and industrial situations where catering to acceptable approximations in scores is realistic. For example, in most schools and clinics the practice of adding intelligence test subtests on the raw scores as they come is unquestioned. It would be better if its results were at least understood.

Approximations may be considered in two senses: (1) using computing formulas more suited to the bulk of practitioners with limited computing facilities, and (2) using parameters from smaller samples or less numerous or reliable variables. As to the first, in the days when electronic computers were unavailable or small, the calculation of the inverse of $R_v(n \times n)$ in Eq. (10.1) $V_{fe} = R_v^{-1} V_{fs}$ was a substantial obstacle, suggesting a need either for some "short cut" of calculating V_{fe} in a new way or using an approximation. Ledermann (1939) and Thomson (1949) suggested short cuts which inverted the smaller R_f matrix instead, and this was tested by Baggaley and Cattell (1956). However, the procedure is irrelevant to most people in the present computer culture and is relegated here to a footnote.[1] Baggaley and Cattell (1956) also estimated the error in an approximation then more prevalent than now, though often used in rough work still. This consists in using the correlations of the variables with the factor (the V_{rs} values) or even the loadings (the V_{fp} values) instead of the V_{fe}. Actually, this is done almost unconsciously in the ability field, where the origin of the choice of subtests in the WAIS and WISC will be found to be based on correlations found with factors rather than on a correctly calculated V_{fe}. Of course, except where correlations among factors or among variables get high there is appreciable similarity of V_{fs} and V_{fp} with V_{fe}. Really high values in V_{fp} and V_{rs} tend to remain high in the V_{fe}. Cattell and Baggaley ended with weightings which agreed by congruence coefficients (V_{rs} with V_{fe} columns) from .67 up to .94. This may seem encouraging to the sloven or the overbusy, but an r of, say, .8, leaves room for much error in conclusions from experiment. The biggest contributor to discrepancy of the V_{fp} or V_{fs} with the V_{fe} is correlation among the variables, and large correlations occur often enough to cause real trouble in taking V_{fs} as the basis of weighting. Since certain approximations in factor analysis are unavoidable the less one succumbs to avoidable approximations the better.

What may seem a grosser approximation (though actually less systematically biased than the above substitution of V_{fs}—or V_{fp}—for V_{fe}) is to reduce the number of variables from which estimation is being made, keeping to just the really high-weighted salients in V_{fe}. In this case, taking, say, the top 13 out of 100 variables in a given V_{fe}, one should recalculate the $V_{fe} = R_v^{-1} V_{fs}$ using the chopped-off V_{fs} and the inverse of that part of the big R_v which is a square now cut to 13×13. This procedure was used for the values in Table 11.3. With a judicious choice the agreement of this approximation with the full V_{fe} result is very good, as Nesselroade, Horn, Humphrey, and others have shown. Indeed, the top 8 to

---

[1] Calculating V_{fe} takes the form (Thomson, 1938, 1949):

$$V_{fe}^T = (1 + J)\, [V_0 (L^T)^{-1} D]^T U_f^{-2} \tag{11.12}$$

where

$$J = [V_0 (L^T)^{-1} D]^T U_f^{-2}\, [V_{rs} D^{-1}]$$

where D is as usual from $(L^T L)^{-1}$ diagonals.

12 variables in a 100 variable study can usually be taken with confidence as a substitute to weighting all 100. The general relation of estimation procedures to purposes is well discussed by Tucker (1971).

One can carry this a step further by taking the same subgroup of significantly or highly loaded variables and merely giving them equal weight, namely, 1.0, compared to 0 for other variables in the analysis. Such all-or-nothing weighting is, after all, what we do without a blush in most current applied psychology, e.g., with the Wechsler and WAIS and other intelligence tests and in most questionnaires. In fact, we do something even less rationally justifiable, for we add the subtests in an intelligence test where the subtest scores are simply the values *as raw scores*. Among these subtests the raw score ranges often have decidedly different sigmas (and sigmas generally unknown to the practitioner!). What validity attaches to the score in such procedures is often an unknown quantity. The user has abandoned true weights for equal weights, as he imagines, but actually he is not even getting the rough justice of the latter, for two reasons: (a) that just cited, namely, that we are not dealing, in these raw scores, with standard scores, and (b) the further reason that the varying correlations among the subtests convert our weighting from 1.0 in the V_{fs}, by multiple correlation rules, to inequalities in the V_{fe}.

In this domain of scoring it is useful to distinguish and apply three concepts, all of which rest on the standard score system. First, we have *nominal weights* which are the weights we actually apply to the scores on their unit sigma basis, either deliberately, explicitly aiming at some best values, or accidentally, by using the raw sigmas as they stand, which are not unities. But the variance contributed by two variables a and b that are correlated r_{ab} is

$$\alpha_{(a+b)}^2 = \alpha_a^2 + \alpha_b^2 + 2r_{ab}\alpha_a\alpha_b \qquad (11.13)$$

If a third variable c is added uncorrelated with a and b, the variance of the sum of the three is

$$\alpha_{(a+b+c)}^2 = \alpha_a^2 + \alpha_b^2 + \alpha_c^2 + 2r_{ab}\alpha_a\alpha_b \qquad (11.14)$$

from which it will be seen that because of their positive correlation a and b are individually contributing more than c, though the experimenter has given them equal nominal weights, i.e., added them with equal sigmas.

If we want to add factor subtests with correct and maximum prediction of the pure factor by its estimate there is no alternative to using the "beta weights" in the V_{fe} applied to standard scores on the subtests. [Incidentally, one should note, as pointed out above, that if one asks what each variable actually contributes to the factor score it is $w_F r_{vF}$, already mentioned in the transformation from the V_{fe} to V_{dfc}, the dissociated variable contribution matrix ($V_{dfc} = V_{fe} : V_{fs}$) among the variable-dimension matrices: p. 235].

Our arguments for greater respect for the factor mandate matrix might seem to support giving a real weight of unity to every variable that is unquestionably (significantly) affected by the factor at all, and 0 where not. But this is fallacious, for quite apart from falling back on an all-or-nothing weighting in the mandate matrix we are implicitly speaking of regression of variables on factors and here it is the regression of factors on variables that we are probably concerned with. Since the term "salients" has been specialized to mean variables saliently loaded on factors confusion would be avoided if we used "contributors" for the best variables from which to estimate a factor. To get contributors, the best column to inspect is that of the V_{fe}, in which one will be evaluating the size of b *weights* (w in the factor matrix world). The next best is columns in the V_{fs}, and there we are choosing by what would most easily be chosen as nominal weights. In the end the factor battery constructor is compelled to set up tentative batteries culled on the basis of the V_{fs}, and then to convert them to V_{fe} to see what the correlation in the particular company he has chosen will do to the weights. After dropping some that turn out of low weight he will add new variables from the V_{fs} and thus proceed iteratively to the best V_{fe} collection.

Seemingly in spite of the above theoretical arguments for graduated weighting, the present writer and several other investigators (Guilford & Michael, 1948; Horn, 1965a; Moseley & Klett, 1964; Trites & Sells, 1955; Schmidt, 1971; Wackwitz & Horn, 1971), giving systematic trials to estimation methods, have found surprising little gain of exactly weighting over all-or-none weighting of variables. Thus in our own work the particular finding in Table 11.3 of a fall only from 0.79 with the full V_{fe} grading to 0.75 when salients are given equal weight is a bit unusual but not really exceptional. [Other such comparisons of exact and nominally equal weights on items in questionnaire source traits are .84 to .67; .76 to .63; .96 to .41; .56 to .55; and .61 to .61 (Cattell, 1973, p. 363).] Other investigators report correlations between scores on the exact and rough methods of about 0.9, but this should not hide our finding of the former being generally the better, and of occasional gross declines as in .96 to .41 in the third case above, just mentioned.

Wackwitz and Horn (1971), however, add the interesting and important empirical finding that the drop in cross validations from sample to sample is less in the ungraded than the finely graded (exact) weighting, and this evidence of greater toughness in the unweighted leads them even to argue that the ungraded is better. Two comments on these findings are justified: (1) the rough has less distance to fall anyway, and (2) it holds best where differences of weights among salients are small; but where the weight range is wide, the story may be different. For lower loadings are probably likely to retain their values better in cross validation across groups. It should be noted that in the above instance of dropping only from .79 to .75 we were dealing with 13 items of reasonably equal loading out of 186 possible items. Had the cut been taken at the top 40, with an appreciable range, the rough ungraded loading would not have done so well. Also the empirical results depreciating the argument for exact loadings have mainly come from questionnaire items, not parcels or subtests, and the intrinsic instability of the former may justify not taking their exact weights too seriously.

The answers regarding methods of approximation in factor analysis scoring procedures therefore cannot be given in categorical generalizations, but require calculations alert to several influences. Some of these calculations either have not yet been brought to satisfactory form, or would take us on too remote journeys into esoteric psychometry, so the answers here will therefore be presented as partly experimental. Thus it is safe to say that use of the V_{fe} (or Tucker's reverse weighting) is best, that more variables give better results than few; that nevertheless if one extends the set until it includes variables falling to a low common factor and a large specific variance the estimate will deteriorate again, especially in cross validation; that salients should first be chosen by taking a relatively substantial fraction from the V_{fs} and then putting them to the test in the V_{fe} (inverting the R_v only for the subset) and dropping those poorer on the V_{fe} that had seemed relatively likely from the V_{fs}; and that low homogeneity of items should be preserved for reasons of wide behavior sampling, good transferability, etc.

11.12. Some Special Issues of Factor Estimation among Oblique Factors

It is not unusual for discussions of factor score estimation, and design of high validity scales, to get fixated in somewhat academic isolation on evaluation of the immediate score validity as such. But both in research and in applied psychology the construction and use of factor batteries the evaluation of validity and the improvement of validity alike have to pay regard to wider demands. The nature of a good validity, and the designs to improve validity must take account also of effects on other properties of batteries and scales, notably of consistency (reliability, homogeneity, and transferability), factor-trueness, correlations among factors, relations to higher- and lower-strata factors, and the use of the factor scores in behavior specification equations for criteria, etc., that are to be employed under varying conditions. In relation to use in some total test installation one may also want to consider effects on the *efficiency index, which is the validity per minute of testing.*

Let us consider first factor trueness and higher-strata relations. It has been pointed out that Bartlett's formula for diminishing the contribution of specifics in scoring a factor will give factor-trueness to estimates in the orthogonal case, and Anderson and Rubin's (1956) method will bring factor estimates of oblique factors to orthogonality. However, what we really want, if factor score estimates are to use the equations obtained for the true factors, is that they should be *factor-true* [as illustrated in Fig. 11.2(b)]. That is to say, orthogonality of estimates is not desirable and the vectors for the estimated scores of those factors should lie in the common factor space *collinear with the true factor vector positions* (as reached by rotation) and these will normally be oblique.

The reasons for wanting them factor-true are important, as follows: If various scales and batteries estimate personality and ability factor scores differently [though perhaps with equal validity, as in (a) and (b) in Fig. 11.5], the specification equation weights, and the diagnostic profile deviations for each form of estimation will be different. Since the aim of psychology as a science is to build up universal generalizations, and in practice to accumulate criterion specification equations of universal value, science is frustrated by these local effects in its aim of converging by architectonic research on generally applicable values. What we would like is a means of transforming each local result to true factor values, for general theory, and, reciprocally, to be able to take the general theoretical values for specification equations, second-order factor weights, etc., and convert them to values for each particular population and each alternate measuring instrument in the field for a given factor. For example, in most multifactor batteries, such as the personality questionnaire series (16 P.F., G-Z, HSPQ, CPQ, etc.) or the Hakstian and Cattell (1977) and Thurstone and Thurstone (1948) primary ability batteries, one normally proceeds also to get second-stratum factor scores from primaries. The structure of second-stratum factors, as expressed in weights derived from correlating scale scores will depart somewhat from weights appropriate to the true primary factors, except in the rare case where the scales are completely factor-true.

As an ideal response to this need for having concrete scales correlate as the true factors do (except for the specifics in the former) one may set up the test construction goal of introducing suppressor action to reproduce these true correlations as completely as possible. As the reader is reminded by Fig. 11.4 this involves taking the V_{fe} matrix and choosing variables as high as possible in contributing to the wanted factor and at the same time mutually balancing with positive and negative weights on the unwanted factors. The aim, of course, is not as simple as in Fig. 11.4, which is to produce scores on primaries that are mutually orthogonal. Instead, the aim here is, by suppressor action, to produce scales that, in the factor space, have the *same cosines to one another as the truly placed primaries have.* To see how far one is achieving this an index can be worked out for the goodness of the suppressor action by summing the columns of the unwanted factor columns in the V_{fe} for that set of variables chosen as equal weighted salients for scoring the wanted factor. If they are not to be equally weighted then the rows must be multiplied across by the weights chosen before the vertical addition is made.

The desired relation of weights for n items or subtests on two factors, X and Y is

$$r_{xy} = \frac{\sum\limits_{j=1}^{j=n} (w_{xj} - m_x)(w_{yj} - m_y)}{\left[\sum\limits_{j=1}^{j=n} (w_{xj} - m_x)^2 \, (w_{yj} - m_y)^2\right]^{1/2}} \tag{11.15}$$

where r_{xy} is the required correlation between the factors, m_x is the mean of loadings on X, m_y on Y; w_{xj} and w_{yj} are weights for variable j on X and Y, respectively.

One may note here a rather important practical difference between the ability (and sometimes motivation) modality and personality modality. In the former, as Spearman (1927) noted long ago, *most* variables have positive loadings on the factors, and, as Thurstone noted (1947), the primary ability factors also definitely form "a positive manifold." In the personality domain, by contrast, variables are just as frequently negative as positive, in their loadings on factors, and the basis for suppressor action is much better. Thus a positive variable for factor X has perhaps a 50/50 chance of being negative on Y. Consequently, there is no inherent probability that the V_{fe} weights for estimating X will systematically correlate with those for estimating Y, or any other unwanted factor, and, if they initially do so, some inspection and alteration of choice of the items included (employing suppressor action technique) will remove it. (Indeed, it is through this combing procedure that we can bring the X and Y scales to the desired true correlation between them.) With ability primaries, however, the almost uniformly positive correlations among primaries tend to bring about a positive correlation between two higher-order factor score estimates, e.g., between fluid and crystallized general intelligence tests, that is greater than that existing between the true second orders.

In the personality and other domains the distortion in factor scale (estimated factor scores) correlations, relative to those of true factors, may not be in the direction of greater positiveness, but it still exists. In the 16 P.F., despite the improvement of suppressor action among items, which brings scale correlations, when each side is scored only on its own items, reasonably close to true factor correlations (see Cattell, Eber, & Tatsuoka, 1971, pp. 113–114), a problem arises when one shifts to *computer synthesis scoring* (see below). The latter uses all items for each factor, with different weighting, and though it increases all scale validities, it yet yields somewhat distorted mutual scale correlations, not as good for further calculations, e.g., second-order estimates from standard primary weights, as one would like.

Fortunately, transformations are possible either of initial primary factor scale estimate scores, or of item score weights, which will give estimates having the particular desired relations, namely an approach to the correlations among true factors. We have seen that Bartlett has an expression for obtaining factor estimates orthogonal to other true factors, and that Anderson and Rubin (1956) and McDonald and Burr (1967) have formulas that yield actual orthogonal estimated (scale scores). Respecting the fact that most scientific concepts require scores, as argued above, that will correlate as indicated by the true factor obliquities (from simple structure or confactor) rotation the present writer and Hakstian, encouraged by the demonstrations of transformations to orthogonality, independently reached transformation (different in application) for score estimates *simulating the oblique true factor correlation.*

These two methods will now be described as variants in presenting the second of the following two main purposes in transforming raw scores for addition to a factor score.

1. *Computer synthesis* (or variance reallocation) aims to get more valid factor score estimates for each factor than are obtainable from adding only the items in the given factor scale itself (whether the latter be unweighted as in most practice, or V_{fe} weighted). This can only be done in an "omnibus" test or battery, i.e., one containing scales for other factors than the one being considered. Such computer synthesis can be applied (a) directly to every item score in the whole omnibus test, or (b) to a combination of the score of the scale itself and the scores of all other scales in the test, generally each such contributing score being obtained by simple unweighted addition of the items in the given scale.

The relative merits of these two alternative procedures, i.e., of applying multiple regression weights for scoring factor x (a) to the n items in the k scales of an omnibus test such as the 16 P.F. or Guilford–Zimmerman, or (b) to the k scale scores from these n items (including the initial score on x itself), could be discussed at length. But in brief we note that it can at least be said for the latter that the ratio of specific to common variance is higher in the average item score than in a scale score, which makes scale scores attractive, also the sheer reliability of scales is higher, and likely to yield more constant weights for application across samples (cross validation of weights). Further it is generally computationally easier to apply a V_{fe} to k scales (which would be scored for other purposes anyway) than to n items. For the former it can be argued on the other hand that whatever success is achieved in item suppressor action loses in the scale scores some potential contribution to the variance of other factors (see below).

The ability to "squeeze" more evidence on the score of factor x from the $(k - 1)$ remaining scales (or from all items) arises from the fact that even if they were pure measures they would correlate with x, permitting some prediction of it, but that in any case they are not pure but contain items which contribute in part to other factors and therefore contain variance usable for other factors, i.e., it could be usefully "reallocated" to where it belongs. Thus, the reader will realize that to the extent that suppressor action has been perfectly realized the prediction of x from the $(k - 1)$ scales becomes inferior to prediction from the n variables, for the latter, individually, retain their ulterior factor portions. Whether computer synthesis procedures are better applied to the scales or the items on an omnibus test therefore depends on the perfection of the test scales, the computing facilities, and whether scales are to be scored in any case. What can be said for certain, however, is that in any existing omnibus test a gain in validity is possible by "computer synthesis" on either basis. The essential formulation below applies to either kind of element (scale or item), though we will state it in scale symbols as follows:

$$\hat{S}_{F_x} = w_1 S_1 + w_2 S_2 + \cdots + w_x S_x + \cdots + w_k S_k \qquad (11.16)$$

Here $\hat{S}_{F_x}$ is the estimated score on factor x and S_x is the given scale score on x, the weighting on that particular scale, w_x naturally being the largest w. It should be noted that this equation has the same form as the estimation of a second-order factor score (Section 11.5).

The calculation of the above weights is

$$V_{Ffe} = R_f^{-1} V_{Ffs} \qquad (11.17)$$

$$(k \times k) = (k \times k)(k \times k)$$

where F reminds one that it is a weighting of factors in estimating factors. Here although the V's are square and deal with factors as rows as well as columns, we have retained the V symbol, because the factor scales (here written as f and defining rows) are being treated as variables on the true factor space. V_{Ffs} is the correlation of the scales with the true factors we are seeking to estimate, and derives, of course, from factoring the scales (in which A and B and preferably also C forms should be used, to provide 2 or 3 times as many variables (rows) as factors, though we are supposing for simplicity in the 11.17 square matrix that k rows only—say, the A form only—are cut out to give only k rows). R_f^{-1} is the inverse of the correlations among the k scales, and V_{Ffe} is the desired factor estimation matrix of weights for computer synthesis. As Table 11.5 shows, by comparisons

Table 11.5. Increments in Validity from Computer Synthesis Scoring of Multifactor Questionnaire Scales

Forms		Source trait																
		A	B	C	E	F	G	H	I	L	M	N	O	Q_1	Q_2	Q_3	Q_4
Complete 16 P.F. (A + B + C + D) $N = 606$	Using computer synthesis scoring	93	84	92	90	93	90	96	93	91	78	77	79	89	89	91	95
	Using only the four equivalent scale scores	91	78	87	83	90	86	94	90	80	49	47	54	82	81	82	89
Longer forms (A + B) only $N = 958$	Using computer synthesis scoring	90	72	82	90	93	89	92	88	85	85	84	94	84	88	90	87
	Using only the two equivalent scale scores	86	53	77	71	88	77	94	80	67	71	64	86	68	80	80	63

of validities from the pooling of 2 and 4 scale scores, with and without computer synthesis from the whole omnibus, there is relatively little gain from the 4 vs. 2 scores but a substantial gain from computer synthesis.

Scale computer synthesis obviously makes virtue out of vice, and some of its power vanishes if the suppressor action has been too perfectly achieved. Consequently, as scales improve we may see synthesis applied more to items. In either form, since the same scales and the same items are brought into the estimated score of each and every factor (though with different weights), some spurious correlation is produced among the latter. If one uses only the item set belonging to the given factor, the item synthesis approach is not liable to this production of systematically spurious factor score intercorrelations.

2. *Scale score estimations collinear with the true factors.* Regardless of which of the methods of factor score estimation so far discussed—unweighted, weighted for items of the factor only, or "synthesized" across a whole omnibus test of items or subtests—is used the correlations of the factor estimates with one another or a criterion will not be quite those of the true factors. The true outside criterion correlations can be matched, approximately, by a correction for attenuation, but to make transformations on the estimated factor scores that will bring their intercorrelations to the same values as those of the true factors is a more complex undertaking.

Yet there are advantages in aspiring to such goals. The storage of information for applied psychological work—notably the specification equations for criteria and the weights by which higher-strata factor scores may be derived from primaries—is best made in terms of the pure factor measures. The practitioner can then adaptively modify these according to the particular battery—e.g., in anxiety factor measurement the IPAT, or the Eysenck or the Manifest scale—he is using, knowing its validity, consistency or other psychometric properties. This is analogous to the chemist's definition of melting point, atomic weight, etc., for the pure element, leaving the practitioner to calculate therefrom what is to be expected in thousands of impure mixtures and compounds.

However, when we say that the aim is to get transformed values from concrete scale estimates that will correlate correctly, there are conceptually still several possibilities. They can, for example, be brought to correlate correctly consistently with the attenuation due to their degrees of invalidity; or one can aspire to transform them to scores that will literally correlate as do the true factors. And when we say collinear with the true factors we must recognize that it would be possible to get factor estimates which mutually correlate like the true factors, but are still not collinear with them.

The issue here is whether direction cosines or correlations shall equal the desired true values. The former is easier but leaves an extra calculation in applying known specification and second-order weights. The main problem in all this work is that the new factor estimate scores contain the specifics as well as the common factors, even if perfect suppressor action could get rid from each scale of the undesired common (broad) factor components. When each factor is estimated only from its own items this specific variance is peculiar to it, and correction for attenuation will handle any corrections of weights needed in, say, the specification equation for a criterion, or the estimation of a higher-order factor. But when computer synthesis has been used, all the specifics in the omnibus are shared, in various degrees, by all broad factor estimates. The latter are no longer vectors in k space, but in $(k + n)$ space, and since the true factors are in k space, we can talk of the estimates being collinear only as projections on k space, or of having the same mutual correlations in some region of a $(k + n)$ space.

Let us consider first the simpler case where each factor is estimated by its own block of variables, and, unlike "synthesis," no specifics from variables have become part of the common space of the estimates. In that case, having the estimates and the true factors representable in the same k space, it is only a question of using an L (transformation) matrix to bring the estimates collinear with the true factors and then extending them to unit-factor length. We shall use $\underline{R}_s$ for the correlations among the extended (unit length) scales and R_s for those literally among the factor scales as estimated scores. We will begin the statistical argument, for the moment, from these extended scales $\underline{R}_s$ values, reverting to R_s if certain purposes require it later. The extended scales $\underline{R}_s$ is obtained from the concrete scale correlations, R_s, by

$$\underline{R}_s = D_h^{-1} R_s D_h^{-1} \tag{11.18}$$

where D_h is a diagonal of square roots of factor communalities. Since $V_0 L_s = V_s$ and $V_0 L_f = V_f$ (V_s and V_f being loadings of the scale and factors, respectively, on their factors at their existing positions in k space—the latter 1's and 0's):

$$R_f = (L_f^T L_f)^{-1} \tag{11.19}$$

and

$$\underline{R}_s = (L_s^T L_s)^{-1} \tag{11.20}$$

Further, the $k \times k$ correlations of the f with the s (R_{fs}) can be determined, if we solve for L_s and L_f above by factoring R_f and $\underline{R}_s$ by

$$R_{fs} = (L_s^T L_f)^{-1} \tag{11.21}$$

We now wish to estimate (calculate not estimate if scales are unit length) the true factors from the scale factors as follows. The "estimation" matrix V_{fe} is calculated:

$$V_{fe} = \underline{R}_s^{-1} R_{fs} \qquad (11.22)$$

Introducing true (collinear) factor scores as S_f and standardized scores on scales as S_s we now have

$$S_f = S_s V_{fe}^T = S_s(\underline{R}_s^{-1} R_{fs})^T \qquad (11.23)$$

If we consider our purpose requires scale scores not declared of full validity (unit length in the k space) then

$$\underline{S}_f = S_s(R_s R_{fs})^T \qquad (11.24)$$

The alternative of deriving estimated scale scores from all items in the omnibus test that will correlate as do the true factors has been solved by Hakstian (1976) as follows.

Two conditions are to be met: (1) that $\underline{R}_s$ shall equal R_f, and (2) that the factor score estimates shall have maximum validity within the constraint in (1).

Let $U = (V_0^T R_v^{-1} V_0)^{-1/2} V_0^T R_v^{-1} V_{fs} L_f$ [R_v being matrix of variables, $n \times n$, and L_f a $k \times k$ arbitrary component set, as above, from factor $R_f = (L_f^T L_f)^{-1}$].

Employing the canonical decomposition

$$UU^T = WD^2W^T$$

$$UU = VD^2V^T$$

and

$$U = WDV^1$$

we can find the maximum validity of estimates at the point $T = \underline{W}V^T$.

Hakstian illustrates this by the "nine test problem" in Harman (1967), using the oblique solution on pp. 243–246.

$$R_f = \begin{matrix} 1.00 & .65 & .46 \\ .65 & 1.00 & .53 \\ .46 & .53 & 1.00 \end{matrix}$$

It is interesting to compare the obtained correlations and validities with the above constraint with those from other estimations.

(a) Usual regression method:

Validities	.96	.98	.93
Correlations	.68	.52	.56

(b) Least squares (reverse regression, see p. 403):

Validities	.95	.97	.91
Correlations	.61	.39	.49

(c) Bartlett's method, adapted to oblique factors:

| Validities | .96 | .98 | .92 |
| Correlations | .61 | .40 | .50 |

and (d) Hakstian's method for true factor collinear estimates:

| Validities | .96 | .98 | .92 |
| Correlations | .65 | .46 | .53 |

It is encouragingly evident that the loss of validity through the constraint of true factor collinearity is trivial in Hakstian's method, while the collinearity sought is not achieved at all so well by the other methods. Preliminary use of the present writer's method above (collinearity from separately estimated scale scores) shows similar but not quite so accurate representation.

The question arises when using the results of such *directional corrections* (as we may call them) of estimates to collinearity with true factors whether the weights given to the new factor scores in specification and second-order estimation equations should be corrected for differences in validities of the estimates. If the validity of a scale x is only, say, .9, then only 81% of its variance is true factor variance. If a weight properly applied to the pure factor is W, then a value $W = wr_x$ or in general $W = wr_v$ (r_v being the validity coefficient) if applied to the new scale scores would bring each factor closer to its true representation in the total. However, it has the obvious danger of increasing random error by giving larger weight to less valid measures.

Reverting from this special true factor collinearity of estimation problem to that of general regression estimates and the special regression in computer synthesis, in which one works from scales instead of items, certain general conclusions may be noted. First, as illustrated in Hakstian's data the usual regression (V_{fe}) method is typically slightly more accurate than the reversed regression, "least squares," "Tucker's method" (p. 297). Second, the distortion of correlation among factor scales by the sharing of item specifies in different factor scores derived by V_{fe} from the pool of items is typically not great—certainly not as great as was feared before the tryout. The upshot is that there is much to be said in accurate work for the collinearity correction. In principle this makes possible the practical use of a bank of specification equations for various criteria, and second-order estimates, etc., when using different actual published scales for the same conceptual factors. For the weights can be derived from cumulative research leading to standard weights corrected to the pure factor positions and these standard weights can be repeatedly used, with different actual scales, as the scores on the latter are direction corrected. Indeed such corrections could be of value in all sorts of calculations, e.g., on age curves, nature-nurture ratios, profiles for clinical diagnosis, as is illustrated today for example in the handbook of supplements for the 16 P.F. test (Cattell, Eber, & Tatsuoka, 1971).

11.13. Estimation of Factor Scores Required for Comparisons across Different Populations: The Equipotent and Isopodic Concepts

In the course of psychological research there is nowadays a frequently occurring need to compare scores across populations, e.g., in groups at different ages, or in different social classes or national cultures. As Humphreys, Gorsuch, Child, and others have regretfully had to point out, the bulk of cross-cultural research flourishing in the sixties and even the seventies has amateurly ignored the essentials of this score comparison problem. Concepts such as intelligence, anxiety, and exvia have to be operationally defined as unique factor measurements if comparisons are to be made. A cross-culture, or cross-subculture, or cross-age comparison consequently involves the conscientious researcher in the problems we handle here of (a) proving the identity of the factor obtained in two settings, and (b) solving some new problems of factor measurement.

The first task—that of demonstrating that one is dealing with the same factor—has been handled in Chapter 10, with respect to the four empirical possibilities: same variables, different people; different variables, same people, etc. It should now be added to that discussion (a) that factors of substantially different pattern at different stages in age development can be considered cross identified, if *continuity of the pattern, with gradual change* can be demonstrated in the intervening years. Thus, the pattern of crystallized intelligence, g_c, is substantially different (compared upon the same variables, of course) at, say, 6 and 60 years of age, but its continuity is easily to be seen from matching across short—say, 3-year—intervals. In such recognition of identity we are (i) using the various objective matching methods—r_c, s, etc.—described above, and (ii) allowing for the purely statistical effects of changing population variances and covariances upon the pattern of an identical influence operating in different circumstances and conditions. To this approach to the problem we now need to add (b) that a conception of identity can be set up quite different from the above, (a), namely that the two factors belong to the same class, as a categorical type. That is to say, all that can now be claimed is that they are of the same conceptual species. For example, the sentiment factor in the MAT (motivation analysis test) called "attachment to spouse" is obviously going to have different qualities not only in different cultures but even in different spouses. The latter degree of uniqueness has to be handled by unique personal trait patterns, P-technique, but the former involves truly different, yet sufficiently similar, common-trait factors. The differences in factors that are of this kind, in psychology, sociology, and physiology, are of epoch, as well as culture. An important instance is that of change in the pattern of super ego strength (G factor). The 16 P.F. gives evidence of G having changed substantially, in the values it respects and statistically loads, between 1949 (when the first edition of the 16 P.F. was published) and 1968 (when the latest edition was made available). Cross identification in extreme cases of this kind can be established only by continuity research (over epochs now, not personal ages) as just mentioned, or by finding samples of people who live in both cultures and speak both languages. Most fundamentally and objectively they need to be established by a typological approach (taxonome program) in which categorical allotment of similar factor patterns to within a type is accepted on both factor analytic pattern and functional grounds.

Granted evidence for fundamental conceptual identity one still faces the problem that the two factor scores to be compared are derived from the same variables, in the case of literal continuity of identity, by substantially different weights, and in the case of only "species identity" even by patterns that include quite a number of categorically different variables. Comparison of scores in this latter case (it is essentially matching problem 4 above—different variables, different population) is possible only by getting subjects who live in two cultures and speak two languages—as in the extremes in continuity research just referred to. Here our concentration will be on the first type, where continuity is demonstrable. In this domain there are two problems, especially if the factors lie at substantially different absolute score levels. There is first the above problem of (a) difference in weights, and secondly of (b) the possible differences in size in true scale *units on the variables* to which the weights will have to be applied, i.e., at different levels on the variable scale. Opening discussion on the latter is a terrifying prospect since it could open the floodgates to detailed overview of the almost boundless literature on scaling of ordinary variables and tests which psychometry has developed over half a century. Rather than digress so close to the end of this chapter to justify a definitive theoretical position in so complex a realm it seems best to refer the reader ahead to the conclusions on scaling *per se* reached in Chapter 14. From that source we shall take as a working principle here that the relation of raw score units to true scale units may alter as one moves over the range of raw scores, but that correction for this alteration is possible. To take difficulties one at a time we shall first assume here that the transformation to an equal interval scale over the range of raw score comparisons to be studied has been made, i.e., that we can trust the units.

If the data available permit all kinds of analyses, four ways of proceeding to score comparisons exist.

1. Two (or more) groups may be thrown into a single factor analysis. Although one may suspect there really are systematic differences between the two groups in mean and pattern one nevertheless gets a single V_{fe} from the scores from which both individual scores (even of people in different groups) and group means can be compared. If differences in means exist this process throws the intergroup sources of variance–covariance in with that of the intragroup source. Some factorists tend to object to this procedure if the differences of means and sigmas of the groups are large. However, if, as frequently is assumable, the within and between variance–covariance spring from the same factor influences, there can logically be no objection on this basis; but there is some brutality to possibly different inherent patterns in forcing estimation on both groups through the same V_{fe}.

2. If one balks at including the between and the within variance–covariance together, it is easy to bring both groups from raw to standard scores before bringing them into a common factoring. There will be a common V_{fe} for comparing individuals, but no longer any possibility of comparing groups since they have been brought to this same mean. Some factorists are lukewarm on this approach unless the variances are reasonably homogeneous, since it can be argued that relations over a large span could be significantly different from a small. Like (1) it handles both groups as if they had one and the same factor pattern.

3. If there is a large number of groups one can take the means of these groups on the various variables and correlate and factor the means, using groups as entries in the score matrix. Just as (2) ignores the between groups variance–covariance so this ignores the within group variance–covariance. Its greatness aptness is in comparing the factor scores of groups; but, unlike (2), it is practically possible to use it also for the second purpose, too, in this case the comparison of individuals. Its disadvantage is that one rarely has enough groups. It has been called the transcultural or transgroup method.

As a general tactic in the above three methods one should examine the means and sigmas for significance of difference. If both are homogeneous the pooling methods above are undoubtedly sound. If means differ, but variances do not, then (2) above, using covariances or standard scores before pooling, is satisfactory. If both are significantly different, then separate factoring, or (1) above or (4) below are tentatively to be followed for a pooled factoring.

4. If we wish truly to respect the effects of differences of factor pattern expression for the same influence operating in different situations and groups none of the above three is satisfactory and it is necessary to concern ourselves with a radically different approach which we have called the *equipotent* and *isopodic* methods (Cattell, 1970). These cannot be

fully understood until we deal with real-base factor analysis, but a general statement of their action and an approximate procedure for most calculations will be given here.

Two (or more) population groups of subjects will typically differ in means, raw score standard deviations, and factor loading patterns. Our aim with the equipotent or isopodic methods is to respect the different factor patterns but nevertheless to reach factor scores for the groups that will recognize and express the differences in means and sigmas. When one considers differences of factor patterns one is also talking of ensuing differences in the other VD matrices between the two groups—the V_{fs} and the V_{fe}, etc.—and in this case of seeking factor scores the change in V_{fe} due to change in V_{fp} is the practical issue.

Adequate discussion of the proper calculation procedures requires that one go back beyond mere statistical principles to the model itself. Why do factor patterns and variable variances alter significantly from situation to situation and group to group? By our model the total variances of variables derive from the summed variances of all factors involved and from error variance. (The variances of factors derive from variances of variables only in a statistical regression analysis.) The statistical derivation in this direction is nevertheless by regression and we are accustomed to the correlation coefficients in the V_{fs}, and their corresponding weights in the V_{fe} and V_{fp}, altering, for one and the same set of variables, with group and situation. For example, the correlation of formboard performance with Spearman's g is high in five-year-olds and low in twenty-five-year-olds, and correlation of pegboard proficiency with the manual dexterity factor alters with the weeks of practice people have had. It seems reasonable to speak of the underlying correlation value as an inherent correlation, to be fixed on some definite population sample referrent, and then to separate out any real changes in its size from psychological changes (or other conditions in the scientific model affecting the population) from the sample changes due to changing sigma of the sample group. Beyond this effect of the group sigma directly on the factor–variable relations there lies, in the purely statistical field also, the change of loading due to changes in variance in accompanying factors.

The latter "intrusive" effect, already discussed as part of the standard factor model but not existing in the real-base (covariance) calculation and model, causes us to ask such questions as: Do formboard correlations with intelligence drop with older groups for intrinsic reasons, namely, because dexterity rather than intelligence largely enters, or through statistical changes of variance? And do pegboard correlations with a dexterity factor increase with practice because differences in motivation factors decrease, i.e., is it an intrusion–withdrawal effect from other factors? Answers will be forthcoming only when more real-base factoring is done. Meanwhile, we take a position on the factor mandate matrix which suggests that a factor either operates fully upon, or is irrelevant to, a given variable and that differences in loading for factor X on variable j merely express statistically the changes in relative size of the various factors acting on j. Though we do not propose at this stage to proceed to a complete handling of the problem, to bring a qualitative treatment of the above problems to the reader's notice is desirable. Our theoretical position is that there is on the one hand what one is tempted to call an *intrinsic relation* of factor and variable and on the other a *manifest correlation* as seen in any given study. Actually the intrinsic relation is not a correlation or regression but a statement of the rate of change of variable x with increase in factor A (to be fixed in suitable units) *when no other factors are operating*. In standard scores, if there is any relation at all, this tangent is 1.0. Otherwise it is 0.0 and these are the values that we put in the factor-mandate matrix. But in the ordinary V_{fp} it will not be 1.0 and if we keep to raw scores and a real-base factor system it will still not be 1.0, because of scaling effects and, in the former case, of the intrusion of other factors. For example, the amount of iron oxide produced by a $100°C$ rise in the factor of temperature and a 100-mm rise in the factor of oxygen pressure could be very different, and at some agreed scale of factor size in real-base terms and of variable scaling the tangents for increase under the two factors would still be realistically very different. These values representing concrete curves in agreed scale units we will call the *intrinsic relations*. They are the factor potencies (p. 413) for specific variables and could be set in an "improved" (more information-packed) form of the factor-mandate matrix instead of mere unities. The manifest correlations, in our usual standardized factor system will be a function both of this intrinsic relation index and of the relative raw score variances of variables and factors relative to the position at which the intrinsic index is calculated. One must note also the hidden assumption about linearity; for the relation of the intrinsic and the manifest values will depend on the linearity of the former, and the subsequent (correlational) linearity of the latter.

A condensed presentation of the main influences that can *statistically* affect the manifest regression value, in connection with intrinsic effects, is shown in Fig. 11.6. In each of these drawings, two different experiments A and B are represented, in which the horizontal axis is the change on the factor and the vertical on a variable. The conditions that change between A and B are the intrinsic one of real steepness (tangent) of the relation and the statistical one of level and scatter. In graph (1) both conditions remain the same and the manifest regression stays at r_1. In graph (2) the intrinsic effect has altered as it reaches a new level of the two variables (implying a curvilinear relation of factor and variable), but the sample

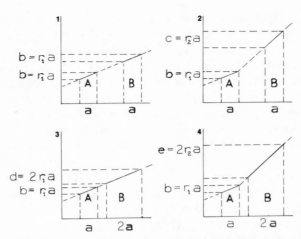

Fig. 11.6. Effects of changing (a) intrinsic relation and (b) variable variance. A and B refer to two population samples factored on the same variable. These r's are not ordinary factor regressions or manifest correlations which would not simply double unless real-base factoring (p. 407) were used with raw units for variables and factors.

sigma on the factor remains the same. The value r_1 then changed to r_2, and the increment in the variable is c not b. In graph (3) the intrinsic relation stays steady, but a doubling of the *real* range of the factor doubles the variable change, from b to d. In graph (4) the intrinsic relation and statistical effect combine,[2] constituted by a shift of regression from r_1 to r_2 and a doubling of the factor range, resulting in a change in variable represented by the range e rather than the original b.

With this glance at some fundamental problems in factor estimation, arising from effects on factor-variable correlations, let us return to the difficult problem of comparing factor scores across essentially different populations, but where evidence of true conceptual factor identity across the two exists. As the reader was warned earlier, this problem requires presentation of an intrinsically complex idea and he may prefer to skip it at first reading, or, better, invite some instructional help. The methods of com-

---

[2] Until the general model of real-base factor analysis is clarified below, some questions may arise in the reader's mind about the above relations that cannot be briefly answered. There is such a reciprocal relation between scores on variables and scores on factors that the two cannot be in fact quite so neatly separated as in Fig. 11.6. Thus, if we suppose that in a case like (4) the relation to other variables than b remains the same then the increase in contributions to the factor score by b from both causes will raise the correlation a little less than the gains in factor contribution immediately suggest. For example, if b accounts for 16% of the variance of the original factor (an r of 0.4) and its contribution increases by 50%, it will not contribute 24% but only $24/108 = 22.2\%$, because the factor has also increased in size. Its correlation would thus go from 0.4 to 0.47.

parison are those of the *equipotent* and *isopodic* models (Cattell, 1970), the latter being an extension of the former. The calculation approach to the equipotent method is first to throw the two groups into a single distribution and work out a common mean and sigma for the variables. Essentially it then applies to the resulting standard scores not a common V_{fe} but the V_{fe} weights appropriate to the nature (variance–covariance structure) of each group. It can be argued that these V_{fe}'s should be corrected for the difference of variance on each variable between the A and B groups on the one hand and the final pooled group, since this is not part of the inherent correlation of factor and variable in each case. And when that correction is made (Thurstone, 1947) a second correction—which expresses the meaning of equi- in equipotent—needs to be made. For even if the very same variance existed in groups A and B for all the variables, group A would have the larger factor-score range if the correlations in that group were such as to produce a V_{fe} with larger values, leading to a higher multiple R in estimating the factor scores.

If the greater efficiency of estimate in A were due to truly higher average correlations—intrinsic correlations—of factors with variables in that group it might be questionable to bring the two V_{fe}'s to the same efficiency and therefore to the same factor score sigmas when variable sigmas are the same. For it could be argued that in some sense we have evidence of larger factor sigmas. But probably much of the difference is due to error, i.e., to relative reliability of tests, and there is no reason to expect the real, broad factors to differ in the way the efficiency of estimate does. Accordingly, after calculating the multiple R yielded by the V_{fe} for the A and for the B groups we bring the V_{fe} to equipotence by multiplying the column in one by a value which brings its R equal to the others. The simplest way to do this, and get other advantages, is to multiply each by $1/R$ to bring it to a factor estimate that will have a sigma of unity (in k factors by a D matrix of $1/R$ values, as described on pp. 238, 274).

Since in the common pool of scores the variables for A and B can still retain unequal sigmas, the resulting factor scores by the equipotent method are still free to differ in sigma, and in mean between groups A and B, and a common metric thus exists by which to test significances. In a cross group comparison of individuals, of group means, and of group sigmas by the equipotent design, the *peculiarities of loading pattern characteristic of the nature of each original group or culture* have been respected, while reducing extraneous accidental differences.

The isopodic ("equal footing") method proceeds one step beyond the equipotent by substituting for raw scores used above certain rescaled scores having some claim to genuine equal interval units. Its discussion is postponed until scaling (Chapter 14), and real-base factoring (Chapter 13) have been handled. (See p. 406.)

Meanwhile one must recognize that many differences of population, e.g., those of sex, age, and social class produce substantial differences in the factor pattern for a given trait. While psychometrists, with reservations, have been willing to handle certain comparisons between groups by method 1 (p. 313) above, i.e., pooling the groups and doing a single factor analysis, it is not always ideal, because it ignores the differences of pattern which the equipotent and isopodic methods respect. However, it was appropriately used, perhaps, in the plotting of factor scores across age by Horn and Cattell (1966) and in studies by Tucker, Humphreys, and others. Nevertheless, the objections to it are (1) that a refined comparison is not achieved, and (2) that by the relatively poor fit of this working procedure to a model that could recognize *two distinct factor pattern systems* more variance of variables is thrown into specific and error factors, and estimation is therefore not as good.

The equipotent method—and the isopodic method yet to be fully described—are therefore preferable. In summary of the equipotent method:

1. The groups are factored separately and the separate V_{fs}'s and V_{fp}'s calculated.

2. The subjects are pooled and, for each variable, the sigma and mean found. If the original sigmas on any variable were s_1 and s_2 and the new is s_{12} then the inherent correlations in the joint situation are calculated by applying the fraction s_{12}/s_1 to each row in V_{fs1}, and s_{12}/s_2 to each V_{fs2}.

3. The V_{fe} are calculated ($V_{fe} = R_v^{-1} V_{fs}$) for the separate groups.

4. The multiple R's are calculated (from $V_{fe}^T V_{fs}$, taking the diagonal values. These are, of course, the validities of the factor score estimates against the true factors). For each factor the column in the V_{fe} is multiplied by $1/R$, so that the V_{fe} gives factor scores of unit variances.

5. The resulting application to the common score distribution ($S_f = S_v V_{fe}$) gives comparable individual scores for members of the distinct groups and preserves differences of means and sigmas between groups.

Examples of comparisons of American and Australian groups on anxiety and exvia scores by this method are given in Cattell (1970).

11.14. Summary

1. The estimation of factor scores involves one in discussion of the usual psychometric parameters of consistency and validity of scores, and of designing scales and batteries to maximize desirable psychometric properties.

2. A brief revision is therefore given here of the definition of *con-*

crete-versus-conceptual (construct); *natural-versus-artificial*; and *direct-versus-indirect validities*; of the three forms of consistency—*homogeneity, reliability, and transferability*—and of the difference between *factor trueness* and *factor validity*.

3. Some confusions are pointed out in the past use of *univocal*, and it is suggested that the two distinct concepts be given distinct labels: univocal for a measure whose elements form a rank one matrix, and *unitractic* for a measure which, in its final score, measures only one meaningful source trait (rotated to meaning). (A univocal scale could be a proportional mixture throughout its items of two—or more—real source trait factors.)

4. Necessary distinctions are drawn among factor-homogeneous and test-homogeneous, factor-pure, factor-true, and factor-valid.

5. A warning is given against the obsolete concept (repeated, however, in quite recent factor test reviews) that a good factor scale or battery should have high homogeneity (these reviews additionally miscall homogeneity reliability). Both on statistical and on psychological grounds (transferability and fair measure of persons from diverse backgrounds) the elements (items and subtests) in a good measure of a broad, common factor should have low mutual intercorrelations relative to their correlations with the factor, which should be as high as possible. Artificially high intercorrelations and spurious claims to high broad factor loadings are often due to making up highly similar items which introduce an undetected "bloated specific," i.e., a specific blown up to a broad common trait superimposed on the true common factor.

6. Different factor estimation procedures are discussed, and it is pointed out that some indeterminacy in estimated scores is reduced by a higher ratio of variables to factors. In weighting procedures the difference between nominal and real weights is clarified. In multiple scales the factor synthesis (or "computer synthesis") method is shown to give decided improvement in validity. On the other hand, similar empirical studies show that unweighted scores are sometimes practically as good as weighted and that on new samples the equal weights may transfer better than the more sophisticated weightings. The estimation of scores of higher-order factors can be either from the scores on the lower stratum or directly from variables; but the former may be better.

7. Estimates of uncorrelated (orthogonal) factors normally give correlated scores though corrections to orthogonality can be applied. There are advantages in transforming factor scale estimates to give correlations (obliquenesses) among the scales the same as those among the true factors, even though some slight loss of direct validity occurs thereby (since the straight V_{fe} gives the maximum validity). The increase in factor trueness at the expense of factor validity gives gains in using standard specification

equation, standard second-order weighting, etc., values which research can build up for the true factors. Although approach to correct interfactor correlations is achieved, as empirical results show, in the 16 P.F., HSPQ, etc., by careful choice of suppressor items, this method necessarily stops short of absolutely correct obliquities and the additional use of the *directional transformation* (by Hakstian and others) set out here may be desirable.

8. Many psychological investigations require comparison of factor scores across different populations. Four ways of doing this are described. Some of these "force" the two or more groups to the same factor pattern, but the important fourth method respects differences of pattern and uses the *equipotent* and *isopodic* method of transformation, the latter going one step further than the former.

CHAPTER 12

Broader Experimental Designs and Uses: The Data Box and the New Techniques

SYNOPSIS OF SECTIONS

12.1. Perspective on Uses of Factor Analysis in General Experimental Work
12.2. The BDRM Initially Studied as the Covariation Chart
12.3. The Misspent Youth of Q Technique
12.4. Choice and Sampling Principles for Relatives and Referees
12.5. The Complete BDRM
12.6. The Five Signatures of a Psychological Event
12.7. The Factoring of Facets, Faces, Frames, and Grids
12.8. Sampling and Standardization of Measures across the Five Sets
12.9. The Nature of Factors from Each of the Ten Main Techniques
12.10. Differential R Technique (dR Technique)
12.11. The Problem of Difference Scores and Their Scaling
12.12. The Experimental Design Called P Technique
12.13. Trend and Cycle Problems Peculiar to P Technique
12.14. P Technique with Manipulation, Lead and Lag, and Chain Designs
12.15. The Relation of dR- and P-Technique Factors
12.16. Manipulative and Causal-Analysis Designs in Learning and Other Factor Analytic Treatment Experiments
12.17. The Relation of Factors from Various BDRM Facets, Faces, Frames, and Grids: n-Way Factor Analysis
12.18. Comparison of n-Way (Disjunct and Conjoint) and n-Mode Factor Analysis
12.19. Summary

12.1. Perspective on Uses of Factor Analysis in
General Experimental Work

Our choice of direction at this point lies between exploring the rich and diverse roles of factor analysis in psychological, social, and biological experiment, on the one hand, and scrutinizing, on the other, purely psychometric problems and statistical requirements. Both research and applied science must certainly in the end study intelligent application of sophisticated formulations in the latter area. But it has seemed best to us to continue at this juncture the momentum of our expanding general view of factor analysis. Perspective must precede the detailed technical pursuit of ultimate accuracy, for we should not be doing justice to accuracy unless we viewed it also in the context of these wider approaches. And there may be some readers who desire the broad methodological overview without intending to progress to meticulous computing issues.

Perspective could mean either looking broadly across different areas of content—psychology, biology, etc.—or across varieties of research designs to be used on any kind of content. The latter is our purpose here; to open up broader and more diverse designs and uses of factor analysis applicable to such a wide spectrum of substantive research as would be met in the former.

At the beginning of this book (Chapter 1) we began by "placing" factor analysis itself in relation to some general dimensions of experimental design (p. 7). It is from that statement of basic principles that we shall take off here in further directions. Up to this point the reader has probably been assuming in almost all contexts that correlations are between pairs of psychological or biosocial variables *measured over series of people, animals*, etc. Indeed, most of our illustrations have been referred to that setting. But at this point we must introduce the concept of the data box or *basic data relation matrix* (BDRM) which opens up numerous other settings in which correlations and factor analysis can be applied.

12.2. The BDRM Initially Studied as the Covariation Chart

Every theory—except those bogus rhetorical theories which, alas, characterized much early writing in personality and social psychology— has to express itself ultimately in a *set of relationships*. One is reminded again of Mach's dictum that the aim of science is to formulate (in general, quantitatively) the equations expressing the relations among observed elements of data. And it is through examining these specified relations, by ANOVA and CORAN procedures, that any theory can ultimately be tested. At first the number of such possible relations in the universe of science may appear to be infinite. But systematic analysis shows that in fact, by recognizing their categories, they can all be contained in the case of behavioral sciences in the basic data relation matrix—the BDRM. This is a "table" or matrix of scores by considering which the possible relations to be examined by experiment can be systematically viewed.

The BDRM seeks to categorize all the kinds of data that can enter into behavioral relations (or, relations elsewhere, in other sciences) and then to explore mathematically the number and kinds of relationships that can be examined among those data sets. In behavioral science the possible observable and recordable sets are people, stimuli, responses, etc., as set out below. It has turned out that such orderly examination of the possible relations among possible sets, in this purely theoretical way, led to domains of research, and to the creation of concepts, which the ordinary, subjectively directed expansion of research had not encompassed.

The BDRM—later called more briefly and colloquially the *data box*— was first proposed by the present writer (1946a) in the more limited form of the *covariation chart*. On the Herbartian maxim that the best natural teaching order recapitulates history we shall momentarily look at this embryonic conception in Fig. 12.1.

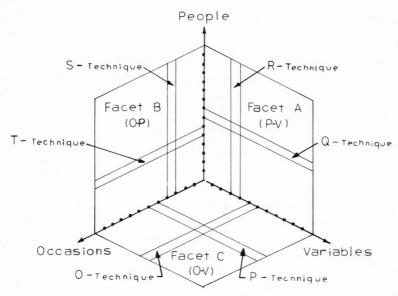

Fig. 12.1. The covariation chart. P-V = person–variable facet; O-P = occasion–person facet; and O-V = occasion–variable facet. Parallel lines in pairs are correlation series used in each technique.

The covariation chart is a parallelepiped of three dimensions. Along one we have people (organisms, groups, or other entities of the class of "behaving organisms"). Along another we have *variables* (tests, responses, behavioral measures, etc.) on the *individuals* of the first coordinate, made initially on one occasion of observation. Along the third we have time, or, more correctly, a series of occasions defined by their characteristic conditions (it being understood that ambient conditions change so that each occasion is unique). The term *dimensions* for these sets could be misleading and the more general expression *Cartesian coordinates* is better. For the coordinates are not quantitative dimensions *per se*. Each is a series of *discrete entities* of one type—people, tests, and settings (or occasions). (The general terms *"ids"*—taken from Latin as Freud did, but, with a meaning easily distinguished from his "id"—is conveniently used to cover individual items in all these types of discrete entities whatever they may be.)

This parallelepiped has three main facets, each of which is a score sheet or matrix. Facet A is the familiar classroom score sheet with people down the left edge and tests listed across the top (or bottom), defining columns. A score entered into any cell therein is a *relation* between an id of the row set and an id of the column set. It says in this case how person *X* performs on test *Y*. Facet B of Fig. 12.1 concerns "people over occasions" and facet C, "occasions by variables." Each facet is a slab of data only one id thick. For example, facet A as seen here is *people* on a series of *tests*, but measured on only one occasion. If instead we preferred to average the values over all occasions it would be like collapsing the whole box along the occasion axis, flattening it into a single face. Such an *averaged* score matrix, across tracts of the third dimension, we shall in fact call a *face* to distinguish it from a *facet* which is marked by only one particular id in the third coordinate, i.e., is only one id thick.

Now the pair of parallel lines drawn in facet A in Fig. 12.1, marked *R technique*, represents the data basis of a correlation of values in two series constituted by all people's scores on two variables, *j* and *k*. This correlation of *j* and *k* over all people is the correlation typically entered into the R_v matrix most commonly referred to in factor analysis. That R_v matrix, of course, takes not two but all series like the pair shown, going over all mathematically possible pairs in that number of variables. In a monograph report in 1917 Sir Cyril Burt proposed factoring across the score matrix, i.e., taking series the other way in the facet, using a pair like that shown in Fig. 12.1. This is *Q technique* (though at that time, however, he did not use the expression Q technique). That is to say, he correlated two people over a series of tests. By regular matrix terminology R and Q are properly called *transposed techniques*, and Q is transposed factor analysis though for a while the new developments were unfortunately popularly called "inverse factor" techniques.

Corresponding to R and Q we have P and O transposes in the occasion–variable facet, and S and T transposes in the occasions–persons facet—on the floor of Fig. 12.1. The publication of the covariation chart (1946) extended the almost accidental stumbling on the R–Q transpose into a systematic, explicit discovery of many new possible relational analyses. These opened up to factor analytic investigation, with surprise to practitioners, vast new substantive areas—some previously thought possible only for the classical bivariate experiment. The results were soon evident in new concepts in personality and clinical psychology, in experiments on emotional states, in coordinations in psychophysiology, in group dynamics analyses, in social psychology, and even in learning theory (Tucker, 1966b, Chap. 16; Fleishman, 1957). Actually, most of the relations were as new to classical, bivariate, ANOVA analyses as to CORAN, for, of course, relations in the two are parallel, difference of means in the former being substituted for correlations in the latter. Nevertheless, due to the incapacity of the former directly to apprehend patterns, doors on which experiment of the classical bivariate type had long beaten in vain, e.g., locating, measuring, and defining the number of drives, readily opened to these more appropriate, if complex, methods. Nevertheless, the pioneer work with these new multivariate experimental designs, like most pioneer work, ran into many inherent problems (not to mention those created by popular misunderstanding) which we must discuss.

12.3. The Misspent Youth of Q Technique

It was perhaps natural that experimenters should tend to begin with the forms nearest to home, namely, Q and R techniques, rather than venture with intellectual courage into the entirely new pairs presented by P- and O-, S-, and T-technique domains. Many became interested solely in the Q-technique transpose (though misunderstood as "inverse" factoring) of the familiar R technique, as can be seen by looking at the publication lists of the fifties. Stephenson's popularization of Q technique and Rogers' use of it in clinical psychology to measure the similarity of real and self-ideal, etc., personality in "Q sorts" involved at least two major misunderstandings. Both could have been avoided if psychologists were more ready to recognize that profound writings 60 years back in the literature can be better than superficial up-to-date current articles. Most of both the pitfall warnings and the essential concepts regarding Q technique are in Burt's 1917 writing, his later principles on sampling of variables, and his 1937 article on relation of transposes.[1]

The two main problems in Q technique, one real, the other the above misunderstanding, concern (a) the need for introducing sampling methods with variables, and (b) the realization that though it correlates people it is not a method for finding types but for finding dimensions. Further, as in all transpose pairs, the dimensions from Q technique are systematically related to, and, with proper conditions, identical with, those from the corresponding R-technique analysis. Discovering types as such on a basis of factor analytic dimensions or other variables is specifically discussed later in this chapter. Here it suffices to point out why Q is not a type search method. The dimensions yielded by Q technique are naturally marked by *high and low loaded* (*scoring*) *people* instead of high and low loaded tests, as in R technique. Thus the intelligence factor in R is marked by high loadings on, say, analogies and classification, and in Q by people, who are known to score high on analogies and classification. However, whereas in true type search, by taxonome (p. 447) and similar approaches, each person belongs only to one type, if we "type" people by Q technique a person will belong to a different type for each dimension found. For example, we shall find that the high and low groups on intelligence will not be the same as the high and low groups on, say, anxiety. And in all dimensions we shall not have discrete types but a normal, continuous distribution of people from high to low.

As regards *dimensions*, Q technique tells us nothing we do not know from R technique, and vice versa (with a small exception discussed below). The choice of R or its transpose (or of any other technique and its transpose) is therefore not a matter of end goal but of convenience and of the ease of meeting statistical requirements. Thus if a factor analysis lacks degrees of freedom, having more variables than people (or, in general terms, more *relatives* than *referees*), and if the experiment unavoidably gives us a score matrix wider than it is high, e.g., with many variables describing symptoms and only a dozen clinical cases, we may well turn to Q technique. In that case we turn the score matrix, as we should with any oblong matrix, so that it is "high" rather than "wide" and the correlate the columns that were formerly the rows, i.e., we use the transpose.

---

[1] Burt's argument depends on being able to obtain a *double-standardized* score matrix, by having standard scores over people and over tests (not to be confused with a *double-centered* matrix in which any row and any column has a mean of zero, which is thus semistandardized; Tucker, 1956). The present writer's empirical studies show that successive standardizing of columns (which throws off rows) and thereupon of rows (which throws off columns) will, after perhaps ten such shuttles back and forth, in general ultimately converge on double standardization, though the final matrix will vary according to whether the first step was by rows or columns (see Ross, 1963).

As for species types—true types—by any adequate definition (below), the *totality* of the organisms surely has to be considered, i.e., the whole profile of factor scores. It makes no sense to describe people who are high or low merely on a single dimension as "types," since the matter is much more easily handled simply by saying they score high on surgency, or intelligence, or whatever. If 100 people go through a Q-technique analysis, the ten (say) recognizable as extremes on the intelligence dimension are separated into (say) five high and five low, and most of the former or the latter could be either high or low on surgency, and so on. There are no stable types revealed: the groupings of people change with every factor. However, a couple of distinctly different uses of Q technique by Bolz and by Gorsuch, which are not strictly Q technique and miss some of its defects, are discussed below under type search *per se*.

Just as the misunderstanding over Q technique and typing leads us, below, to a systematic study of typing, by the taxonome approach, so the second pitfall into which Q-sort fell—that of ignoring sampling of variables—alerts us to a problem running through all uses of the data box. Rogers' index of similarity of two people (or an ideal and a real person) by Q-sort overlooked that the correlation coefficient is blind to differences of level in their profiles, while other "Q typists" have overlooked that any different choice of variables could lead to a radical difference of correlation between two people and of subsequent sorting. Variables in the ability field and in the personality field might, respectively, show two persons to be very different and very similar. Any dependable and meaningful figure for the similarity of two people must be based on a principle of comprehensive sampling of variables. This is relatively new to statisticians, for hitherto they have been concerned only with statistical concepts of *population* and of *samples* confined to the BDRM set we call people and things.

12.4. Choice and Sampling Principles for Relatives and Referees

Concern with guiding concepts for sampling variables began with the writings of Brunswik (1956) on "representative design" in experiment, and with the present writer's development of bases for the concepts of the *personality sphere* of variables, as a defined population for personality and ability research (1946a). The latter, reinforced by Burt (1941), also contained proposals for quantifying the concept of variable distance and variable density, in complete independence of correlational evidence, for correlation as a basis would have led us into circular definition. The concepts received most explicit development in Burt's (1941) and Cattell's (1952c; but see also 1949b, 1952b, and 1958a) critiques of Q-technique usages.

When the issue is considered more widely than in just the person–test facet of the data box it is useful to clarify discussion by two new terms—*relatives* and *referees* just mentioned. In any score matrix the ids that are being related by correlation are called the relatives. Thus tests are relatives in R technique and persons are relatives in Q technique. The ids over which relationship is being assessed and to which the correlations are referred are called the referees. Sampling concepts developed in statistics, both in CORAN and ANOVA, in regard to referees—indeed, specifically, *people.* However, as we shall see, good experiment in *representative* designs calls for attention simultaneously to sampling domains of *relatives.* Previously, referees, and specifically people, have been the subject of most advance in sampling of populations in R technique. In the choice of the relatives two different principles need to be followed, according to whether the factor analysis is hypothesis testing or hypothesis creating. In the latter, and in a special sense in the former, the central arguments in CORAN parallel those of Brunswik in classical experiment with ANOVA. In hypothesis testing there will be a deliberate choice of marker variables—old (to relate to past definitions) and new (to test new deductions about expected factor structure and factor meaning). Furthermore, if the investigator is sophisticated, there will also be concern for sampling another set of variables, chosen as "hyperplane stuff." In exploratory, *hypothesis-creating* factor experiment, on the other hand, the main concern will be with a different goal of sampling, but one that should be no less explicit and determinate than in hypothesis-testing designs. In the human behavior domain two main sources of a defined population have been proposed for sampling the total personality sphere, namely, (1) the dictionary of language about human behavior, covering three to five thousand terms according to Allport and Odbert (1936) and other surveyors, and (2) time sampling of behavior through day or month at regular intervals, in the given culture. The latter requires (Cattell & Warburton, 1967, Chap. 5) additional sampling across modes of scoring, e.g., time, errors, performance ratio. The beginnings of empirical bases for situation sampling have been given us in the work of such as Barker (1963, 1965) and Sells (1963). In introducing and using these personality sphere concepts the present writer found (1946a, 1957c, 1973; Hundleby, Pawlik, & Cattell, 1965) that like Columbian maps of the world they gradually became extended in area, by new explorations in this case by including new variables for factors, old and new, initially found thin in representation by variables in the original personality sphere.

Perspective should remind us that this extension of sampling and population definition from referees to relatives in the R–Q score facet is only the beginning. Principles have yet to be developed for defining populations and stratified samples along the other coordinates of the data box, i.e., for in stimuli, responses, occasions, and observers. These can better be approached as we now move from the introductory conception of the covariation chart to the full BDRM.

12.5. The Complete BDRM

Although the covariation chart and BDRM are discussed in this book in connection with CORAN methods, it needs to be made clear in passing that they are a guide equally to ANOVA designs. The "charts" simply show the totality of *possible relations*—using means, variances, correlations, etc.—among ids. From these relations the investigator may single out those most relevant to his theory, since every theory is embodied in a set of relations. The method of *examining* size and significance of such relations as well as their linear or nonlinear form, belongs to a second stage and remains to be decided upon by the experimenter. Thus, for example, the relation of the variable "intelligence" to the variable "achievement" can be examined by CORAN methods, as through R-technique correlations above, or by ANOVA methods, taking a high- and low-achievement group and testing the significance of their difference of mean on intelligence—with or without allowance on a covariate. The terms that have grown up for the factor analytic techniques, R, Q, O, P, etc., could therefore be paralleled by, say, *R, Q, O,* and *P* in ANOVA, but hopefully if ANOVA proceeds to such ordering it will employ some modification, for example, as by using italics (see Coan's beginning 1961) to distinguish from CORAN usages. Certainly some agreed convention both of symbols and names is needed to avoid confusion. The developmental divisions of CORAN will go through partial and multiple regression to factor analysis and path analysis and of ANOVA investigation through the elementary "higher or lower mean" relations, and interaction evaluations to analysis of covariance. (Incidentally, the question of substituting in a correlation matrix for factoring a relational measure closer than the correlation coefficient to ANOVA statistical evaluations is discussed in Chapter 14, Section 8 below, under "effects of different coefficients.")

More intensive examination of the covariation chart (Cattell, 1966) led to the recognition that behavior analysis, exhaustively pursued, deals with *five* rather than the *three* classes of ids initially set out in Fig. 12.1. Indeed they can become ten if we find it desirable to make a distinction between each id in an average, typical identity state and in a modified state. However, the five basic *sets* as we shall call them are: *person, environmental occasion, response, stimulus,* and *observer* (mnemonic

P, E, R, S, O). The observer set adds to a measurement that astronomers used to call the "personal equation," in their otherwise exact measures. But in behavioral measures and assessments the observer component includes the observer and his instruments—in short, it defines *error of measurement*, and, if we are concerned with inferences to population, also *sampling error*, since these inferences are made by a fallible human observer observing only a subsample of reality. Since the kinds and degrees of misperception of a psychological event are much greater than for an event in astronomy or physics, the observer relations are proportionately more important in psychology. They justify a whole theoretical development, present instances of which are found in trait view theory (Cattell & Digman, 1964), desirability distortions (Edwards, 1957), and so on.

There is no inherent difficulty in settling on definitions clearly to separate these five sets of ids, with the exception perhaps of what we have called the occasion or situation, on the one hand, and the stimulus on the other. The environmental situation is principally in need of distinction and in some contexts we have called it (Cattell, 1963a; Cattell & Child, 1975) the *ambient stimulus* distinguished as such from the *focal stimulus*. The "occasion" or ambient situation is the totality of conditions, as defined by time, place, and psychological nature, in the environmental situation outside the specific stimulus to which the subject reacts. Unless we separate ids and states of ids in the ten coordinate BDRM, this occasion definition will include the state of the organism at that time. The focal stimulus as distinct from this ambient stimulus is definable frequently as that to which he has agreed to react, and with a response of a particular nature, which we measure in a particular way. But in any case the stimulus is the immediate temporal provoker of his response.

It is essential to keep in mind in understanding and exploiting the potentialities of the BDRM that the ids in each set are not values along a coordinate. Each is a pattern *entity* in a given species—people, stimuli, responses, etc.—which has many dimensions and can be measured as an entity as a vector quantity—a profile—on many variables. A subset of those dimensions can be derived from correlating the various measures that can be inserted in the cells of the BDRM. The choice of insertions will decide the dimensions revealed. As psychologists we tend to take it for granted that the entries will be behavioral measures and the discovered dimensions will be traits, states, etc. But one could hold responses constant and insert characteristics of, say, the stimulus, such as its weight, brightness, cost, and thus issue with dimensions of, say, the properties of stimuli which people encounter in their lives rather than the behavioral properties of people.

The question "to which set do the dimensions belong?" will be decided by which set is referee and which relative. When people are referees the factor scores belong to people, but in regard to the nature and definition of the trait both referees and relatives are always involved. A personality trait, for example, has its definition tied both to the stimulus–response relatives and to the population of referees on which it was established. However, we should recognize that so long as our entries in the cells of the matrix are behavioral responses the dimensions emerging will be only psychological dimensions, not physical, economic, or chemical dimensions.

The principles involved in the BDRM concept are universal and its coordinates can be adjusted to different domains, but we must confine ourselves here to those concerned with psychological dimensions. Mostly, but not entirely, we shall enter measures of behavioral response magnitude in the score matrices. However, even here it can be noted that the entries in the score matrix can be broadly one of two *entry modes* as we may call them. The most usual is (1) an *attribute mode*; this is a measure of the relative attached to a particular person. It is the basis of finding attributes of persons in the domain of the given relatives. In more esoteric work, as notably illustrated in the work of Rummel (1967, 1972), one deals with (2) *a relation mode*. Here the scores are on relations between one referee and another, whereas in the attribute mode, in psychology, the entry is simply behavior, as in, say, reaction time. (In Rummel's work, where nations are the referees, an entry might be "trade between Cuba and Honduras.") Special sampling conditions need to be met in pairing referees in the latter mode. The relational mode could alternatively be handled in an attribute design, placing "trade with Honduras," "trade with Mexico," etc., as distinct variables. For in the last resort, relational behavior is an ingredient in defining attributes.

Just as there are three facets and six techniques (from their being two transpose alternatives in each facet) when we deal with three Cartesian coordinates (sets) as in Fig. 12.1 so with five there are $^{5}C_2 = 10$ facets and 20 techniques. These, and the associated notation for distinguishing them systematically can be studied in depth elsewhere (Cattell, 1966a, Chap. 3), and pointed up by the discussions of Coan (1964). The new domains of substantive research which they open up are remarkably varied and have been basically stimulative of new concepts and experiments. Here we can afford space only to refer to some problems which arise in sampling, calculating, and interpreting in particular sets of ids and to discuss the relationships of findings from the different techniques.

12.6. The Five Signatures of a Psychological Event

In the usual R-technique analysis and equations, as so far set down, it has been considered sufficient to designate the behavioral response or performance to be estimated by just *two* subscripts—*i* for the individual concerned and *j* for the particular variable. But we now realize, from the BDRM, that some other subscripts are always implicit, and are omitted only because they are constant throughout the given experiment. However, it is important to remember that they are there.

Any psychological (behavioral) event actually has *five* subscripts, thus:

$$a_{hijko}$$

concerning which we shall systematically keep to the notation which has begun to develop in many substantive writings (see Anderson and others in Cattell, 1966), namely that: *a* is a measure of an act (performance or response); *i* is the individual acting; *h* is the stimulus to the act; *j* is the nature of the particular response. The manner in which *j* is measured, —since various measures of the same response are possible—is given by *a*; *k* is the occasion, which implies a time, place, and set of ambient conditions; and *o* is a particular observer. (Note that alphabetic sequence— *hijk*—is broken, by substituting *o* for *l*, in designating the observer, for the practical reason that the letter *l* is easily confused in print with a one, or even an *e*. However, this break has meaning, too, in that the observer source of variance, as "error," is in a special class in relation to the others.) As stated when describing the data box (which term we shall henceforth use "colloquially" for basic data relation matrix, BDRM), each of these is an entity—an id—which is by nature a pattern expressible as a vector, or even a matrix in itself. A behavioral event requires that all five of these ids come together (including an *o* if the event is to pass into the records of science).

The reader is accustomed by now to calling a sheet of person–test scores (facet A in Fig. 12.1) a matrix. Each of the many facets or faces in the data box is a matrix. But now we ask the reader to recognize that the whole data box is a five-dimensional matrix. Such a matrix is commonly called a *super matrix* in matrix algebra terms and the whole thing can be written out in two dimensions by marking out matrices within matrices as shown in Fig. 12.2. This is a simple instance in which there are only three dimensions and therefore a series of two-dimensional facet matrices entered as shown suffice to describe all entries in the whole three-dimensional box.

It will be seen that we have an $a \times b$ matrix in which each of the entries is itself a matrix of order $x \times y$.

The above could be extended, matrix within matrix, to represent the five-dimensional data box, but so long as the reader has no difficulty in thinking in the solid, and imagining that he sees a five-dimensional paral-

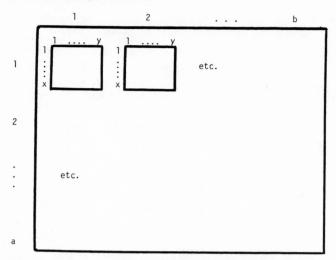

Fig. 12.2. A simple super matrix. A super matrix equivalent to a three-set data box. In this, each submatrix represents a facet in the basic data relation matrix. Yet another coordinate set could be added by making each figure in the small matrices itself a matrix.

lelepiped with numbers entered in every cell we need not, in discussion, unravel it into a conventional super matrix[2] as in Fig. 12.2. Whichever way we view it, each ultimate cell of the super matrix has in it a scalar quantity—a score a on a response j—the other subscripts of which give its latitude and longitude (and two other 'tudes) in the matrix. In any given experiment, for the sake of brevity, it has been the practice to write as subscripts only those ids over which one is considering variation. Thus in R technique there are N persons and n variables and i denotes a given person and j a given variable, so that the typical entry is a_{ij} and the last is a_{Nn}. But this particular person–response variable facet needs also to be located at a particular h in the series of q stimuli; at a particular k among v environmental (occasion) conditions, and with a particular o among x observers. The full five id identification of the psychological event helps us to keep in mind even in a single facet score matrix the conditions that are implicitly there.

[2] If the reader feels that statistics afflicts him with this degree of complexity he may be encouraged to reflect that biology faces a similar problem:

> So naturalists aver, a flea
> Hath smaller fleas that on him prey
> And these have smaller fleas to bite 'em
> And so proceed ad infinitem.

12.7. The Factoring of Facets, Faces, Frames, and Grids

In understanding the range and role of different factor analytic techniques it will help if we recognize that some four kinds of score matrices can be taken out as subsets from the data box. They are *facets*, *faces*, *frames*, and *grids*. A facet (two coordinates) has already been defined, and a face (also two coordinates) is a facet in which the entries have been averaged (or otherwise collapsed) over some or all of the other coordinates. A frame can be defined as a "box" of three or more dimensions, left as a "solid," and the extreme case of a frame is the whole data box. However, any frame, including the total data box, can be "unraveled" and spread out end to end as shown in Fig. 12.3, when it becomes in effect a species of super matrix.

It must be noted that any frame can be unravelled into different grids according to which set we decide to use as the relatives. From a three-dimensional frame three different kinds of grids can in fact be formed—and, of course, five from the full data box. They differ according to which sets of ids are placed along the top and sides of the grid.

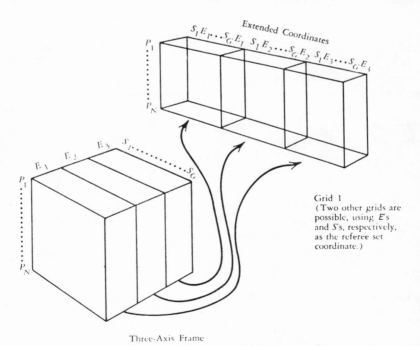

Fig. 12.3. A three-dimensional data box (frame) unraveled to a grid. (From R. B. Cattell, *Handbook of Multivariate Experimental Psychology*, Rand-McNally, Chicago, 1966, p. 103.)

Thus in a persons–test–occasions frame one of the three grids would have tests across the top and person–occasion combinations down the side. In the ensuing calculation step—that of correlation—the relatives are tests as usual, but the referees are both people and occasions, in combination. Thus if there are 20 people and 10 occasions there will be 200 referee entries, each representing one person measured on each of the same several occasions.

The result of factoring such a grid will be essentially an aggregate of the factors obtained from separately factoring a face of tests over persons and another of tests over occasions. It will be remembered that a face collapses a whole series of facets, which means that the first face has scores of persons averaged across all occasions and the second, the scores of occasions averaged across all persons. The first will yield trait factors and the second, state factors. If traits and states have different patterns presenting k traits and p states, then the grid factoring will yield $(k + p)$ factors. But if some are the same, e.g., if anxiety as a trait loads the same variables as anxiety as a state, then the grid factoring will combine the covariance across people and occasions in a single factor pattern.

With this observation we enter the field of multimode and multiway factor analysis, in which we must ultimately proceed to precise statements about the relationships of factors from two, three, four, and five set factorings of various kinds. However, in an introductory sense it can be seen that the psychological and general meaning of patterns found from such extended grid factorings is always inferable from consideration of the nature of the referees. A grid from a four frame would, of course, have referees each of which is a threefold combination of the referees in the separate sets, e.g., over people, occasions, and stimuli (granted we deal with diverse stimuli to which the same response can be measured, e.g., reaction time). Any grid, of course, will also admit of transposed factor analysis, though precautions will be needed to gather enough cases to insure that the transpose is taller than it is wide. Thus one might factor tests-by-observers over people and obtain factors marked by tests as administered by certain experimenters (observers).

It should also be noted that grids can be constructed with combinations along both coordinates of the grid. For example, one could have all combinations of stimuli and responses along the top, as relatives and all combinations of persons and occasions down the side as referees. Discussion of the meanings of various combinations could take us far afield. However, in the example just given the factors would be characterized by high performance on particular stimulus–response performances (which are after all only tests arranged with orderly analysis) which operate as trait and state concepts.

The reader who wishes to read further into the novel mathematicostatistical concepts in the data box and its associated techniques, including a clean-cut notation as shorthand for handling discussion may do so in Chapter 3 of the *Handbook of Multivariate Experimental Psychology*. Here we are simply glancing along the main avenues and passing on. The parallelism to analysis of variance is to be noted, in that each new coordinate is a new "effect" and the role of interaction terms is discussed in the chapter just mentioned. Thus, in the example above, the variance-covariance across people and the variance–covariance across occasions add to give the total variance–covariance when combined people and occasion entries are used for the referees (DeYoung, 1975). By that analogy we shall refer to the grid factoring as a *two-way* factoring when it runs across two kinds of referees and a *three-way* when across three (not to be confused with Tucker's three mode analysis below).

The question will naturally arise in planning the various data box designs—especially when we approach sampling, in a moment—as to how far one is in fact free to vary the choice of ids independently in the different sets, for various sampling and design purposes. The awkward case which usually prompts this question is the attempt at independent choice of sets of stimuli and responses. It then quickly becomes evident that certain responses are scarcely natural—or, indeed even possible—in our culture—or even in animal experiments—to certain stimuli. If the response measure is the rate of eating, and a representative experimental design on stimuli has produced a range of objects from cheese to brass knobs, the entries are going to be very poorly distributed when zero scores are given to the latter. With a small minority of exceptions, which we leave to common sense,[3] it is possible, however, to consider many choices of id sets as independent and leading to many possible combinations.

12.8. Sampling and Standardization of Measures across the Five Sets

The problem of sampling on id sets other than people was encountered here in pointing out that ignoring this problem in Q-sort produced worthless results. What principles do we have for defining a population and a sample in the four sets (outside the first set of organisms, e.g., people, rats, countries) we call stimuli, responses, occasions, and observers? Suggestions for determining population and samples on the environmental variables—focal stimuli and ambient conditions—have been made by the present writer (1946a, 1978), Sells (1963), Norman (1967),

---

[3] Already there are instances in factor analysis, however, as in Osgood's "semantic differential" (Osgood & Miron, 1966) where stimulus–response connections become bizarre and unreal. With what reliability can one, for instance, respond to the stimulus "chair" with "good" or "bad"?

Goldberg (1975), and others. The sampling of responses is similarly a matter to be referred to empirical counts of the totality of behaviors observable in a given culture. The matter has been systematically discussed in Cattell and Warburton (1967) and particularly in relation to the concept of the personality sphere of trait behaviors in Cattell (1946, 1973) and Norman (1967). The former has suggested that in advancing from the original suggestion of the dictionary as a population of behavior descriptions to the more demanding one based on actual time sampling of daily behavior we have to recognize the principle that the quantitative score from any given response can be measured in a wide variety of ways. Consequently, the population of responses is actually one derived both from time sampling of behavior and a system of sampling the possible ways of measurement. The latter can be put on an objective, systematic basis by factoring a large number of possible measures of a particular response, to reduce to a few factors, each of which can then be represented once. If m is a stratified sample of daily response behavior in the culture, and there are p measurement dimensions, on an average, for each response, then the final stratified sample of response variables is $n = mp$.

The set with the most Protean and seemingly uncategorizable varieties of ids is clearly that of occasions (ambient stimuli, ambient situations, experimental conditions). Occasions can obviously be first and most easily fixed in time and space, and in physical terms, e.g., temperature of room, volume of background noise. But, in general psychology, cultural, historical, and personal experience dimensions are, in general, more important and harder to bring into a population sampling scheme than time and physical dimensions. For example, in respecting learning experience of a situation as part of its definition we have to ask immediately "How familiar is the situation?" and, especially in factor analytic learning experiments, "How frequently has the person responded in that situation?" The nature of this set overflows even into defining the situational excitation the subject has been receiving prior to the immediate occasion of measurement. A person grossly annoyed by a previous visitor will react differently to the stimulus presented by the next visitor. The choice is forced upon us between (a) defining the ambient situation k not as what is present at the instant in the external environment, but as the present plus preceding situations, or (b) including, in the instant present, both the external situation and direct evidence of the internal state of the subject. The choice is not easy. For whereas the first loses the character of "an instant event in time," common to the definition of the other ids, and lets in an arbitrary decision on "How far back does the ambient occasion go?" the latter may introduce ambiguity as what we mean by the individual. Seemingly we now mean "an individual in a particular internal condition," but the internal condition should, whenever separately definable, go into the occasion coordinate set.

When one examines all five sets carefully, it becomes evident that all actually share this ambiguity of entity and state of the entity, though in more subtle ways. The observer Smith is not always the same Smith and the stimulus "Halt" cannot be kept (in say two successive experiments with the same people as others) to just the same quality. For this reason the ideal goal has been mentioned here of dealing really with a 10-coordinate data box, each "average entity" coordinate being paired with another of "state within the entity," i.e., the deviation at the moment of experiment from the central characteristics of the entity. Without space for further discussion here we would conclude (see Cattell, 1966a) that the ambient situation ("occasion") should ideally be considered as a strictly external entity, though with 5 rather than 10 coordinates we may collapse internal and external. However, factoring on the 5 set basis is necessary initially to locate the states that would go along the second "occasion" dimension in the 10-set battery.

Along with population and sampling problems in these new sets come questions of standardization. Because of the differences in nature and quality of such standardizations, and the convenience of quick definition in factor analysis, the present writer suggested some years ago that distinct terms be used for standardization on each of the five sets. *Normative* has long been in use for standardizing "over people" and "*ipsative*" has for some time been firmly adopted for "over occasions" ("within one person," as used by Clemans, 1956). *Performative* has been proposed for across *responses* and *abative* for across *stimuli*. *Spective* standardization (from spectare, to look at) might be suitable for across observers. However, except for normative, ipsative, and perhaps performative standardizations, a regular usage for these standardizations is scarcely dependably established. For example, ipsative has sometimes been tied to any series within one person, when sometimes performative (see Table 12.1) is intended. To define both the referee series and the particular position of the facet combined terms might be used, e.g., ipsative performative, for across stimuli and within one person. Convenient terms for precise communication are likely to crystallize in meaning as factor analysis occupies its increasingly broad kingdom of application and discussion.

12.9. The Nature of Factors from Each of the Ten Main Techniques

A brief but more systematic statement is needed regarding the relation of factor meaning to the relatives and referees chosen in a facet or grid factoring, as yet merely touched on above. In the interpretation of use of any facet, face, or grid one must finally clinch the perspective that *the relatives define the natures* of the factors found and the referees define the species of object to which the dimensions apply. However, because of the possibility of recognizing equivalence—indeed identity—in the dimensions obtained from the two alternative transposes from any one facet (Section 12.3) one has to recognize that there are only $^5C_2 = 10$ cases to consider. The factors from these, assuming response scores as entries, would be as shown in Table 12.1.

Table 12.1. Nature of Factors Obtained from the Ten Possible Score Facets[a]

Facet sets	Nature of factors
1. People × Responses (response magnitudes as entries; for one observer, one occasion, one stimulus)	Trait dimensions of people in relation to one stimulus
2. People × Stimuli (contact frequencies as entries)	Person dimensions of stimulus contact
3. People × Occasions (responses to one stimulus as entries)	Person sensitivity to occasion modulation factors
4. People × Observers (response to one stimulus as entry)	Dimensions of effect of people in terms of observers
5. Responses × Stimuli (response magnitude entry for one person or mean person)	Factors of response in terms of stimuli
6. Responses × Occasions (response magnitude entry one stimulus or one mean person)	Dimensions of modulation of response by occasions
7. Responses × Observers (response magnitude for one stimulus, occasion, and person)	Dimensions of response misperception due to observers (state factors)
8. Stimuli × Occasions (coincidence frequencies as entries)	Dimensions of stimuli in terms of occasions in which they are likely to occur
9. Stimuli × Observers (entry: number of persons observed, one response, one occasion)	Dimensions of frequency of action of stimuli in terms of observers
10. Occasions × Observers (entry: frequency of observation of each occasion)	Dimensions of occasions in terms of observers' frequencies

[a]Corresponding to these dimensions are different kinds of standardization, each related to the properties of a particular species of referee. The best known are ipsative, abative, and performative scoring and standardization, over occasions in one person. Nomenclature has, however, not yet settled down as to whether to refer these to different sets of relatives, or to referees (normative "ordinary" standardization), relatives (ipsative), and other sets orthogonal to these (abative, performative) as suggested elsewhere (Cattell, 1966a).

Although it is correct in stating that 10 facets are exhaustive in an initial sense, Table 12.1 is actually illustrative rather than exhaustive of the varieties of factor meanings from factor analyses. For (a), if response magnitude is the standard entry, as in the majority of the instances above there are only 10. But if other entries are considered, e.g., contact frequencies as in (2) or number of persons observed, as in (9), then far more are possible, and (b) since two sets bound the facet, the entries can be relations between them in characteristics of themselves or of any of the remaining three sets. Furthermore, Table 12.1 is only facets and, as seen above, there are also faces and grids of various combinations of sets. The possibilities of general meanings of factors is therefore very great, though finite. It is up to the scientific investigator to reason out each on its own, along the lines above. He should also be aware of the systematic relations of factors (a) between transposes, as discussed above, and (b) between facet, face, and grid factorings using the same relatives over different composites of referees. The latter, as indicated, are generally to be understood in terms of additive combinations of the variance–covariance matrices, as they extend over additional sets.

As a practical observation on the present stage of research experience in these fields the comment should be entered that most of the above avenues are as yet neither theoretically nor substantively explored (see, however, French, 1972). Also unexplored (but see the *Handbook of Multivariate Experimental Psychology*, p. 89) are the results of using attribute and relation score entries in the cells that are not behavioral measures. These are illustrated in (2), (8), (9), and (10) in Table 12.1. Such entries could even be of physical, social, economic, etc., nature. For example, in (8) above the entry would be the frequency of association of a given stimulus with a given ambient situation in the life of a typical member of a given culture and would yield factors important to the cultural anthropologist. However, since space forbids systematic exploration of more detailed conceptual and methodological problems than has yet been given above we shall choose to concentrate on two of the most important techniques—P technique and differential R technique—among the above designs. These are of central interest to the psychologist (once he steps beyond elementary R–Q technique designs) and illustrate the type of statistical and experimental problems encountered in the various further designs.

12.10. Differential R Technique (dR Technique)

So far all the techniques above deal with either single "absolute" scores (as in a facet) or a mean "absolute" score (as in a face) across one or more other sets. The first important innovation to be studied is that of using gain scores (or difference scores generally) in place of absolute scores. This has been called generically *dR technique*, for *differential R technique*.

As we approach the prospects of using factor analysis in learning experiments, in unearthing new developmental concepts, and in clinical psychology—all involving the analysis of change—we perceive two alternatives. (1) We can stay with ordinary R technique, taking two (or more) studies on a cross-sectional R-technique basis. Employing either single or equipotent methods we can then compare factor scores obtained at the different points in time or (2) taking change scores on the variables themselves and converting these by factoring into factor change scores we can deal with factors as they exist in the learning or growth process itself. It will be seen that in some respects—notably in avoiding comparisons of somewhat different factors and in staying close to the data—we can perceive more clearly what is happening if we pursue the latter course, though it has its problems. In passing we may note that dR technique has so far been applied only to response score differences, but that it is potentially applicable to the whole group of "absolute" designs implied by Table 12.1. Among such further uses of obvious importance are the factoring of occasions as relatives, across people as referees, with response increases as entries, to see what dimensions of increased motivation (for people) are attached to various life situations. Or, again, the factoring of responses across occasions (as referees) to discover what the dimensions of occasions are in terms of reinforcement of kinds of responses in the culture generally (score entries being means for the population). This would put descriptions of cultural incentives on an objective and quantitative basis.

Regardless of the sets involved, the use of difference scores in the entries means that the factors are dimensions of change. Let us now keep to the response–person facet (or person–test if one wishes to run stimulus and response ids together as a single "test" variable id) as most familiar to the psychologist and basic to his concepts. In such a design if the retest is after months or years we presumably shall deal with long-term dimensions of development, but if retests are over only a few hours, then we are likely to find our largest factors in fluctuant psychophysiological state ("emotional") change dimensions. The difference score naturally removes any trait variance—if operationally a trait is defined as an entity that does not change in a few hours—and leaves us squarely in the domain of emotional, physiological, and dynamic state change. The substantive findings by dR and P techniques over the first 15 years of their use are not for detailed comment here, except as they reflect on the general practicality of the methods. But to anyone in the least familiar with that field the substantive consistency and richness of the results shows that the model fits the data pretty well. Indeed, in the domain of psychology and especially in its conceptions of emotions, anxiety and depression states, and drive tensions, this design has placed firm concepts of state patterns in place of the conflicting speculations started by Wundt's arbitrary choice of three dimensions and the subsequent speculations on number of drives by Murray, McDougall, and one must add, Freud.

Although the difference score removes the trait pattern variance from dR that is present in R technique, no converse removal of state pattern variance from R technique analysis occurs when we shift from dR to R technique results. If one reflects that at a given moment of testing on any day subjects will stand at different state levels just as they do at different trait levels, then it is clear that both kinds of factors will emerge in company in R technique. Immediate, direct inspection of the patterns in R technique—unless content is extremely different—will not, however, in fact tell us which are states and which traits. Let the reader consider a person unfamiliar with the ocean who is shown a photograph (naturally instantaneous, like R-technique single-occasion pictures). The viewer will surely be likely to confuse mountains and mountainous seas, if small features are eliminated. The mountain at sea is a wave "frozen" by instantaneous view, and the pattern of a state factor such as anxiety will similarly be frozen to look like a trait, in the "one-occasion" R factors. It is only when R and dR are put side by side and matched that the state factors can be recognized and eliminated from the R series. Parenthetically, many R-technique studies have probably mistaken some states for traits, since follow-up work to get factor scores and watch them over time, by dependability coefficients, is rare. Unless one is prepared to compare dR and R results to eliminate states, any study aimed at pure trait patterns requires that one also carry out what may be called a stabilized, averaged R- technique experiment. In this, one takes measures on a whole series of occasions, collapsing these person–test (one-occasion) facets into a person-test face, i.e., a score matrix in which each entry is the mean over many occasions. With means over a sufficiently large number of occasions an R-technique design could virtually eliminate state factors. (In passing, let it be noted that we speak for brevity of state and trait factors. Actually, with more semantic precision, a given state is a point on a state dimension, but it should be clear here when "state" is being used for the dimension itself, on which particular state levels occur.)

The procedures, including rotation, in dR factoring are the same as in R, and the fact that simple structure has been found to work very well implies that single states, like single traits (or trait change factors in the case of developmental studies), each affect only a minority of any well-diversified set of variables. The only difference noted is a tendency to higher correlations among oblique states, probably due to the fact that certain rhythms and certain life stimulus situations tend *simultaneously* to provoke certain states, e.g., fatigue and depression.

12.11. The Problem of Difference Scores and Their Scaling

The problem in dR technique that is really new is that difference scores as such have lower reliability than absolute scores. This obvious weakness has been somewhat too severely pilloried by Cronbach and Furby (1970). The inhibition of growth and learning research which this threatened to produce in some circles was fortunately brought to more rational handling by such papers as those of Nesselroade and Cable (1974) and Overall and Woodward (1975). But let us examine the difficulties. The basic fact to consider is that an observed measure will have a variance equal to the sum of the true and error variances, thus:

$$\sigma_o^2 = \sigma_t^2 + \sigma_e^2 \tag{12.1}$$

In many learning and state fluctuation studies using dR technique the first and second measures will be essentially uncorrelated so the variance of the difference score will be

$$\sigma_{o_2-o_1}^2 = \sigma_{t_1}^2 + \sigma_{t_2}^2 + \sigma_{e_1}^2 + \sigma_{e_2}^2 \tag{12.2}$$

For error is by definition uncorrelated. If we assume true and error variance not different on the two occasions, then

$$\sigma_{o_2-o_1}^2 = 2\sigma_t^2 + 2\sigma_e^2 \tag{12.3}$$

In the use of dR technique in experiments on growth, in a middle range, and in some learning, the first and second true measures may be significantly correlated. If r is this correlation, then

$$\sigma_{o_2-o_1}^2 = \sigma_{t_1}^2 + \sigma_{t_2}^2 - 2r\sigma_{t_1}\sigma_{t_2} + \sigma_{e_1}^2 + \sigma_{e_2}^2 \tag{12.4}$$

In the first case the ratio of error to true score is the same for a difference score as for an absolute score [compare Eq. (12.3) with Eq. 12.1)] though the absolute amount of error is greater. Consequently, in standard scores, as commonly used, it cannot be claimed, in this case, that error is any more disturbing of results in dR than in R technique. In the second case, Eq. (12.4), the ratio of error to true variance will be greater.[4] One

---

[4] The student may read in certain texts of a "paradox" in difference scores, in that the more reliable the absolute measures the less reliable the difference scores. The paradox is self-created by those who fail to distinguish the varieties of reliability coefficient on p. 283 above—especially between the dependability and stability coefficients —and the difference between error and trait fluctuation. Increase in the dependability coefficient (immediate retest) will increase the dependability of the difference score over an appreciable time interval, reducing σ_e^2 in the above equations. Increase in the

cannot determine r directly, because only the observed scores are available. But if one knows the dependability coefficient (immediate retest) of the measure and the observed score stability coefficient (retest after experiment) everything in Eq. (12.4) can be determined and the ratio of error to true variance evaluated.

Whatever the evaluation we are "stuck" with this error. No attempt to squeeze this error out afterwards by statistical means will work, unless quite unlikely assumptions are made. Lord (1958, 1963) in particular has explored these possibilities. One which has appealed to some statisticians is his proposal to partial out the first score (or alternatively the last) from the difference score (Damarin, Tucker, & Messick, 1966). Since σ^2_{e1} is part of the first variance, and e_1 of the first observed score, it is true that partialing out σ_1 from σ_2 gets rid of one error component in the difference score. However, it also produces (except for random error of correlations) a condition in which no correlation of the change score with the original score is possible. Such a "corrected" difference score patently fails to meet the model in most growth researches, where increment and absolute level are commonly significantly correlated for some real reasons. The occasionally advocated reduction of variance of the two absolute scores by an error variance subtraction based on dependability coefficients has some point (see below) if the dependabilities are very different, but as far as individual scores are concerned it will leave the rank order of the difference scores unchanged and the dR correlations unchanged. The fact is that once an error gets into measurements, be they absolute or difference, statistics can do nothing as far as individual scores are concerned, though it can correct various group parameters. But it is reassuring to remember that in state research, as contrasted with developmental research, r in (12.4) is usually zero and error, therefore, no greater than in R technique.

The real difficulties in dR technique need careful study, but they largely lie elsewhere than where the critics (Cronbach & Furby, 1970) have attacked. Our evaluation of a complex and methodologically fascinating field, recently approached in sophisticated fashion by Baltes (1968), Bereiter (1963), Overall and Woodward (1975), Schaie (1958), and especially Nesselroade and Bartsch (1976) and Nesselroade and Cable (1974), mainly in developmental connections, can touch only briefly on essentials. The sources of distortion of difference score correlations one needs to

---

stability coefficient will decrease the variance in the difference score, as will be seen if r increases in Eq. (12.4). (r is not the observed stability, but is a function of it.) If the variance of the true difference increases and σ^2_{e1} and σ^2_{e2} remain the same the reliability of the difference score will be increased and the stability coefficient decreased. There is no paradox; for when we say dependability of difference scores decreases as the stability coefficient increases, we are simply saying there is no important function fluctuation for dR technique to concern itself with. Precisely the same is true, for that matter, of R techniques when true individual differences get small relative to error of measurement.

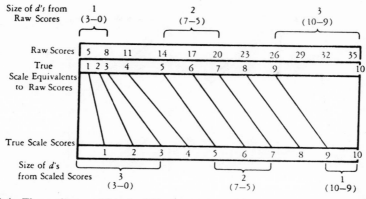

Fig. 12.4. The scaling problem in difference scores. (From R. B. Cattell, *Handbook of Multivariate Experimental Psychology*, Rand-McNally, Chicago, 1966, p. 366.) The rank order of difference scores in row units is here precisely the opposite of that among the differences on properly scaled scores.

watch are (1) different reliabilities of before and after measures, (2) different variances of before and after measures, (3) departure of scaling from equal interval properties. And, of course, because of what Eq. (12.4) above tells, (4) the need for greater efforts for reduction of experimental error (with accompanying emphasis on larger samples) than in R technique. See Sorbom (1975).

As regards scaling effects dR results are liable to dangerous distortions from (a) ceiling and floor compressions in scales and (b) greater departure of the raw score units from true equal interval units in one part of the range than in another—a state of affairs which would affect R technique results little, and not at all, in most analyses, if rank correlations are used. The *raw difference scores* can even assume totally different rank orders from those of the *true difference scores* in extreme instances as shown in Figure 12.4.

Figure 12.4 illustrates that though there may be a monotonic relation between raw scores and true scores in the separate before and after measures, i.e., an increase on one always goes with an increase on the other, leading to the same rank order of true scale and raw scale scores in each, yet this rank agreement no longer necessarily holds with the difference scores.

The pioneer empirical study on dR technique may be said to be that of Woodrow (1945) and, though he encompassed enough ability variables to reveal Spearman's *g*, failed to demonstrate a *g* growth pattern. Some similar oddities are found among the present writer's early studies on school achievement variables and sometimes they can be shown to be related to initial ignorance of the importance of the four points above. As consistencies in results developed it became evident that most trait factors

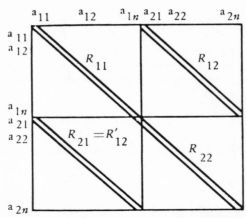

Fig. 12.5. A method of factoring two-occasion measures. (From R. B. Cattell, *Handbook of Multivariate Experimental Psychology*, Rand-McNally, Chicago, 1966, p. 372.) Variables 11 to 1n are taken on the first occasion and 21 to 2n on the second. Three kinds of submatrix, R_{11}, $R_{12(21)}$, and R_{22}, are combined.

can be matched by trait-change (growth or fluctuation) patterns in dR technique. A searching analysis of the statistical bases will show, however, that such patterns could appear as artifacts if (a) the sigmas of many variables are in general larger on one occasion than another and (b) if test reliabilities are systematically different on the two occasions. dR studies should regularly check or rule these out, but in fact when they are controlled the isomorphism of trait and change factors is still confirmed, and must be due to real psychological unity of trait growth.

Both the principles of isopodic (or equipotent) factor scoring and general principles of calculation suggest that in dR work one should see that the before and after scores are (1) free of ceiling and floor effects, (2) then brought separately to standard scores, and (3) combined in a single standard score distribution, within which the difference scores should be calculated. Minor differences in this procedure have been considered in various writings, and Nesselroade and Bartsch (1976) have even suggested a major difference in treatment by a joint factoring of two-occasion measures as shown in Figure 12.5, i.e., an attempt to infer dR factors without using difference scores. The new loadings are here to be found as difference between before and after factors.

It would be a mistake to assume, however, that this is an escape from the alleged criticisms if real or the need to meet the real standards of dR technique as applied directly on difference scores. It is just another avenue to the same result, and one that does not escape the intrusion of error (which now appears in the differences of loadings). Indeed one even encounters fresh problems in the rotation process here. (Ordinary dR yields good simple structure.) Nesselroade's approach is useful, however, in bringing out immediate relations of R and dR, and in providing a check on the analytical procedures.

For more detailed discussion of this and the other issues of the last few pages the reader is referred to the *Handbook of Multivariate Experimental Psychology*, especially pages 365–380. The proof of the pudding being the eating, the real answers to the above doubts appear in the already extensive use of dR technique which has yielded results consistent (a) internally, among different dR experiments, (b) in relation to other psychological and sociological findings, and (c) in cross reference matching to P technique (below).

12.12. The Experimental Design Called P Technique

P technique and its transpose, O technique, are initially defined by reference to a score matrix of tests (or stimuli or responses separately) over occasions (ambient situations in time). This facet is normally anchored to one person—hence P in P technique—but in what has become known as chain P technique a grid of several successive such person facets is factored. Its theoretical properties have now fortunately been explored by thirty years of empirical work. (The first actual P-technique experiments were by Cattell and Cattell and Baldwin in 1945.) By the nature of the measurements, individual differences—which are the basis of trait patterns—are ruled out here as effectively as in dR technique and the factors resulting should be state dimensions or trait change (developmental) factors, depending on the time intervals among the successive retestings. Psychologically its great contribution has been to the taxonomy and measurement of human emotional and dynamic states, but it has also been applied, in the analyzing of socioeconomic data over a century or more, for defining dimensions of sociocultural change.

With the main practical problem in P technique—that an individual must sit typically for as many as 40 tests on every day for 100–150 days—we are not here essentially concerned. However, these practical exigencies must be briefly discussed in methodology in relation to effects of occasion sample size and the intrusion of trend factors from the influence of such things as boredom and learning.

It has been of interest in regard to method cross-fertilization among sciences that when P technique was taken up later in economics there were noticeable additions and subtractions of emphases, and some changes of assumption. For example, economists interested in the causes of business cycles and other changes seemed to decide without hesitation that any steady secular trend should be partialled out of the data (say over a hundred years) before the factoring began. Social psychologists attacking socioeconomic–cultural data on the other hand have not thought it desirable to partial out simple calendar time trends. The arguments against doing so are that the same factors are probably contributory to the trend as to more irregular movements (though they may get more correlated in the trend) and that one loses, in partialling, a true picture of the pattern and time course of these factors because much variance–covariance is thrown away.

12.13. Trend and Cycle Problems Peculiar to P Technique

Although trends do not deserve to be rejected in the crudely categorical way just criticized, a suspicion of which factors are likely especially to be involved in trends will often help design. We may suspect that in psychology the effect of repeated exposure to the measures could produce learning, boredom, and other instrumental–measurement-specific effects we would like to be able to recognize and separate from true trend and state changes *per se*. Such factors could occur rather at the second than the first order and one cannot help noting from actual experiment that second orders (a) are often large in P technique and (b) can sometimes be identified as purely situational influences producing common changes in first-order state responses. In one such instance (Uhr & Miller, 1960, p. 451) the second order seems to represent an approach to grappling with a situation as opposed to evading it, loading stress and adrenergic primaries positively and the anxiety primary negatively.

P technique encounters another peculiar problem differing from that due to trends, but associated with its longitudinal character. This is the above mooted existence of *cycles* and *rhythms*, of various lengths. As Yule (1926) pointed out long ago, a quite peculiar problem arises from the fact that if two variables with the same temporal sine curve rhythm frequency occur in one experiment but stand in phases half a cycle out of step they will, despite this obvious mutual relation, show zero correlation. It may be argued that such a phase difference indicates they are *not* part of the same factor expression; but exceptions in science can be found to this conclusion and the remedy seems to be to introduce *lead-and-lag correlation* as described below.

Whether certain rhythms that may exist in the action of factors, and apparent in complex rhythms in variables, will be caught or missed in P technique depends on the closeness of the repeated measures in time. Repeated measures at long intervals should theoretically catch both the short and long cycles (granted the moment of measurement is short enough), while very short intervals with a short total span may of course miss factors with long undulations. Experiments so far have mainly been diurnal, at one or two sessions per day, though some physiological studies have gone to recordings every few seconds. The detection of rhythms in variables or factors can be achieved by what is called serial autocorrelation: the correlation of the time series on X with itself with various lags, i.e., starting one, two, etc., occasions down in matching occasions on the second with occasions on the first listing. Maxima will appear in correlations when the lag is as long as the wavelength.

In this problem of time interval and in other ways P technique presents an instance of the new sampling problems mentioned above as due to be encountered in factoring new facets of the data box. In each new set of referees or relatives the question will arise as in the old domain of stimuli and responses, whether the plan is to pursue an exploratory or an hypothesis-checking design. In P technique, assuming the emphasis initially on *exploration* of dimensions of states, it is clear that a "state sphere" definition of variables is needed analogous to the personality trait sphere in R technique. It can be based on rating (L) and questionnaire (Q) data, either on the language vocabulary or on cultural time sampling of stimuli and behavior (applicable to L, Q, and T data). The latter in this case, however, must include internal physiological behavior in a systematic way, for P technique has already demonstrated (Bartlett & Cattell, 1971; Van Egeren, 1963) that it is one of the most powerful tools for locating psychophysiological functional connections and unitary response patterns that we have. In other domains, e.g., economics, the same two strategic principles obviously hold equally well for selection of relatives.

It is in the selection and sampling of the P-technique referees, namely, ambient situations combined with same or different focal stimuli, that novel and resourceful transposition of general principles is required. Incidentally, though P technique necessarily involves a time sequence one should not be trapped into accepting time itself as the measured coordinate. The real variation of occasions is that of situation and condition. Only in special cases, where time is simultaneously a measure defining conditions of fatigue or hunger, or serial effects of learning, does it enter the definition of the occasion.

This independence of time in ordinary P technique is brought home to us abruptly when we realize that the same P-technique factors will emerge from a research regardless of whether the occasions are kept in order or shuffled. Of course, the estimates of the factor scores for each occasion will need to be put in calendar order if one is to make further sense of the results. But ordinary P technique is not the whole story and when we get to lead-and-lag and manipulative designs preservation of serial order and even calendar scores become important. Making further sense of the results will often mean relating the ups and downs of factor scores over time to events in the subject's life, as in the clinical studies by Kline and Grindley (1974), Cattell and Cross (1952), and Birkett (1978). In this connection, it should be noted that although most P-technique studies have been concerned to discover, define, and distinguish the states as *response patterns*, and content to leave the provocation of the states to the unknown stimuli and deprivations of everyday life, yet a P-technique design is possible which incorporates experimental manipulation or information on strength of stimulation on each occasion in the correlated data.

12.14. P Technique with Manipulation, Lead and Lag, and Chain Designs

The appropriate use of manipulation in factor analytic design (see general designs, Table 1.1) has hitherto been quite inadequately discussed in factor analytic texts. We propose to introduce it at this point in regard to P technique, and then below (Section 12.16) in more general settings. As the student will realize from earlier perspectives two misperceptions[5] have befogged the true relation of multivariate experiment to classical "brass-instrument" bivariate experiment: (1) that manipulation cannot be applied with multivariate designs and (2) that manipulation alone is a basis for establishing causality, whereas in fact both manipulation, on the one hand, and defined time sequence relations in unmanipulated data in the other suffice (see p. 356).

Either without manipulation of the stimulus or ambient situations as stated at the end of the last section or with manipulation, as now discussed, one can introduce into the relatives in the P-technique matrix measures the magnitudes of stimulus or ambient situation. A pioneer run with such measured stimulus variables was made with electric shock and situational ratings by Cattell, Cattell, and Rhymer (1947) and more explicitly in a clinical experiment with stress interviews (Cattell & Scheier, 1961) and proved practicable and instructive.

If more than one stimulus variable is introduced, and one uses manipulation, the independent variables can be kept mutually independent, as in the controlled distribution in the usual ANOVA design. If one does not use manipulation the probability is that the experimental, psychologically independent variables will not be statistically independent. This may not cause trouble in interpretation, but if it does one can enter with correlations of response with the given stimulus variable from which other stimuli have been partialled out. Alternatively, one can factor stimulus and response variables separately and bring their factors into correlation. The factor analyst will think of a variety of canonical, joint matrix (p. 359), and other designs available for handling relations of two domains in this way, but some relations will have obscure meanings in terms of a causal model, and the present writer would favor determining factors in the re-

---

[5] They illustrate the all too frequent predominance of historical accident over logic in human affairs. Wundtian imitation of the physical sciences, brass instruments, manipulation, ANOVA, and process study grew together. Galton, tests, nonmanipulative recording of natural data, CORAN, and individual difference study formed a second cluster. A broader view of experimental possibilities (see Table 1.1, Fig. 1.2, and Cattell, 1966a) shows the logically possible constructions of experiment to resolve into six dimensions, not two clusters, and illuminates many new designs.

sponse part of the matrix and determining their correlations either simply with individual stimulus variables or with factors among stimulus variables.

In any case, whereas ordinary P technique has been typically an analysis of response into basic dimensions of state response, *stimulus-inclusive P technique*, with or without manipulation, asks the further questions about the relation of the state responses to environmental stimuli and ambient situations. Used in an exploratory way, with environmental sampling, it is an ideal way to locate, for example, the main stimulus situations in a culture which systematically evoke anxiety and various ergic tensions. It is to be expected, and research already verifies this, that simple structure will exist in the response variables, but the question is open as to how far it will exist for factors in the situational measures.

Since the reader has just been reminded of the historical difference of individual difference and process psychology (footnote 5 above) it should be noted that P technique and dR technique are the principle servants capable of determining structure in the process domain (learning, perception, maturation, cultural trends, adolescent development patterns, etc.) as is beginning to be shown by the work of Baltes, Birkett, Nesselroade, Sweney, and others. The economic and meaningful description of any individual process is a factor matrix, covering both situations and responses, over occasions. The mapping of typical, recurring processes in nature is a further step of applying pattern similarity coefficients and taxonome (p. 447) to a collection of such matrices.

To get to grips more thoroughly with the pursuit of causal relations by factor analysis (or, for that matter, with any factorial analysis of data involving time-recording observations) a further development is needed. One has to note that it is possible, and even probable, that in most instances of waxing and waning influences effects will follow not instantaneously but with some time lags. The examination of time-lagged correlations involves recognition of immediate, intermediate, and remote causes in a causal chain. Cause is an ambiguous term, as when one might argue whether X was killed by a car accident or by loss of blood. Certainly on initial sampling of environment and response variables some links are likely to be missed and more refined searches must follow. To search for causal connections lead-and-lag P technique has been proposed (Harris, 1963, p. 185), in which each variable can be "staggered" in time on another. Thus, if one variable is alcohol consumed and the other is severity of hangover headache, the score of the former on the nth day would be put alongside that of the $(n + 1)$ day in the score column for the headache to produce maximum correlation of the two series. Since we do not know—though we may have a hypothesis—in which direction causality is acting, an exploratory study needs to pursue by trial and error—there is no royal road—the mutual lags and leads, i.e., relative shifts of the series in either direction that will most clearly bring out the sequential, causal relations.

Up to the present such "lead-and-lag" correlations have been more widely used in economics than psychology. It will be noted, however, that when operating on variables alone one is in any case not free in a typical R matrix to make any combinations of leads and lags that he might wish. If b is arranged to lag one day on a, and c one day on b, obviously a and c cannot be kept simultaneous. If we force a to meet both relations, then a will become two variables, not one, and we cannot retain a consistent R matrix. In any case, if the "factor-as-an-influence" model is correct lead and lag should be done on *factors in relation to variables,* not among variables. One should complete an ordinary factor analysis and then try, by trial and error leads and lags of variables against *factor scores,* to see where significant increases of correlation can be obtained. In the last phase of such analysis one must also experiment with the length of the experimental measurement intervals as suggested above. Lead and lag becomes an iterative procedure, since at each shift of a variable by one or more occasions the estimate of the factor score on each occasion alters. No adequate empirical data yet exists on the craftsmanship of what we may call *iterative lead and lag factor analysis.* But there seems little doubt that the procedure should result in clearer definition of factors and particularly in the recognition of particular causal stimuli or conditions by their appearing at the top, in correlation magnitude, of the factor loading structure which expresses the factor action. There is no analytic procedure, but only one of trial and error, by which to discover the lead and the interval size which maximizes loadings, but we can be confident that programs will be written capable of making the trial-and-error search conveniently rapid. Finally, iterative lead and lag P technique needs to be enriched with both coefficient methods discussed in Chapter 13, Section 14.

A last special development of P technique needing comment is what Scheier and Cattell (1961) introduced into experiment as *chain P technique.* Because of the onerous demands on subject time, the temptation is great in P technique to shorten the series of occasions. The good rule of thumb that there should be three to five times as many referees as relatives is apt to be transgressed. The point made by statisticians that there is loss of degrees of freedom whenever significant autocorrelation values exist in the individual variables heightens the need for ample referees (occasions). (Autocorrelations has just been defined—as self-correlation, when a variable is lagged on itself. It is not to be confused with serial correlation which is the more general term for any correlations over time series.) Unless the sequence is as completely erratic as a minor stock over some months on the stock exchange, significant autocorrelation values will commonly exist in time-spanning variables, and where there are smooth trends ("warming up," fatigue, learning) and wave forms it will be considerable. The number of degrees of freedom lost is the number of occasions one series has to be moved over from the starting point until the correlation is lost (Quenouille, 1957). Further excellent discussions of this and other statistical problems in factoring over time are given by the distinguished contributors to Harris's survey (1963). The total number of occasions needed for a good P technique is therefore larger than one would expect from experience with corresponding R factorings, and a 10 to 20% increase in sample size over what would be appropriate for R technique would often be indicated.

Since experimenters evidently find difficulty in retaining subjects long enough for such an acceptable occasion sample, Cattell and Scheier (1961), as mentioned above, experimented with the properties of chain P technique. The design is one in which several subjects are used, restricting the length of the series of occasions for each, so that, for example, 10 people sit for 20 occasions each instead of one for 200. In combining these links into a single chain each person's scores are first brought to standard scores (over occasions, which means, to use technical precision, that they are ipsatized). However, if there is reason to believe that larger variations are differently interrelated than smaller ones then *semi-ipsatization* can be used, i.e., different people are brought to the same mean but not the same sigma, before combining into a single-score matrix. Either procedure gets rid of the "standing" differences between individuals, i.e., the trait variance–covariance, and leaves the analysis entirely in the domain of states, though in semi-ipsatization different people will contribute in different degrees to the character of the state pattern.

12.15. The Relation of dR- and P-Technique Factors

It will be noted that chain P technique is an instance of analysis of a grid—a matrix (see Fig. 12.3) for correlating responses simultaneously over persons and occasions. However, it has the special condition applied of maintaining common means for persons, thus eliminating between-person variance–covariances. This grid has a systematic relation to that in dR technique in that both deal with a three set of people, tests, and occasions in which tests are the relatives. The difference is that P technique takes a facet one person thick, whereas dR technique takes a facet one occasion thick. To be exact the dR score matrix is one score thick, but that score is the difference between two contiguous occasions. If we consider this difference score as two, equivalent to deviations around a mean of two occasions, then dR can be considered as the beginning of correlating over a whole series of occasions. Thus, a chain P technique and an extended dR technique in essence handle the same data. The relations of an extended P technique, by taking several people in a chain and of extended dR technique by taking several occasions in a chain can be seen by Fig. 12.6 to be nothing more than a different ordering of the rows when the three-set frame is extended into a grid. The relatives (tests) are the same in both and the referees are merely in different order, which will not prevent the correlation matrix becoming identical in the two cases. Notice, however, that for this to be a true dR technique (right of Fig. 12.6) the scores of each occasion must be a difference of two occasions of the deviations from the mean over all occasions.

From this relationship one would expect that the factors would eventually be essentially the same, though the single person P technique and the single difference dR technique might have extreme boundary properties. Because the typical dR is kept to one change occasion and P technique to one particular person the emphases and inferences will be different. We see that dR would present the common state pattern, across many people, but unique to one occasion, while P would present a state response pattern unique to one person but common to all kinds of occasions of its provocation. There is little reason to believe that a state pattern would show a uniqueness of response pattern to one occasion. Anxiety, with its typical physiological response, is anxiety, whether provoked by an overdraft at the bank or a visit to the dentist, though the particular combinations of anxiety with stress, depression, etc.—which produce clusters

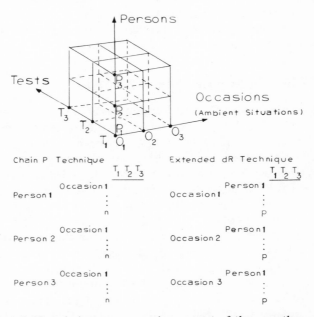

Fig. 12.6. P- and dR-technique score matrices as part of the same three-set grid when P is extended over several people and dR over several occasions.

not factors—will differ with occasions. Nor should we expect, in human species with similar hypothalamic structure and largely similar experience, that there would be much person uniqueness to different P technique patterns. Indeed, we now have empirical evidence, across both general states and motivational states, which bears out, by congruence coefficients, the essential similarity of the dR-, and P-technique (and especially chain P technique) loading patterns. In such comparisons certain statistical rules will need to be followed, having regard to what means define the deviations, and whether covariance factoring or ordinary factoring is involved, but the basis of transformations is provided in Fig. 12.6. However, on statistical, factor analytic grounds, it is clear that whatever uniqueness of person and of occasion exists can be examined by comparisons of results among these techniques. (See Nesselroade and Reese, 1973; Nesselroade, 1972).

A further possibility of uniqueness that has to be considered is that the pattern on a rising emotion may be different from that on its decline (and similarly in economic boom–depression cycles). Obviously the same variables as go up have to come down, but they may do so at different rates. Consequently P-technique studies may need to be extended by selecting occasions first on the up-beat with regard to some specific emotion and analyzing up and down separately for possibly unique patterns and similarly in economics and biological studies.

Another special design which has begun to be tried in P technique is the simultaneous longitudinal analysis of two or more people having special relations, e.g., husband and wife, mother and child, psychotherapist and patient, members of a closely interacting group over the same series of occasions. Howard and Diesenhaus (1965) have approached this in clinical work, Bolz, Cattell, and Barton (in preparation) on a married couple, and Cattell and Adelson (1951), Gibb (1956), and Rummel (1976) on national groups, over the same 100-year periods. The contrast of results from factoring each individual separately, and relating the factor score changes, and, on the other, factoring both in a common matrix, might throw light on group emergents, i.e., social interaction factors not encompassed by factors in the two individuals.

12.16. Manipulative and Causal-Analysis Designs in Learning and Other Factor Analytic Treatment Experiments

As mentioned in Section 12.14 few mistaken stereotypes have persisted so long as that which relegates classical, bivariate, manipulative experiment to the study of process (perception, learning), accompanied by causal inferences, and factor analytic experiment to individual differences, and static structures, supposedly without causal inference. Except for the linearity assumption in factor analyses (and in CORAN generally) and the examination of interaction in ANOVA (the most frequent analysis of bivariate work being by ANOVA) there is complete parallelism of utility. This parallelism of CORAN and ANOVA at the statisticomathematical level has been clearly brought out by Cohen (1968) and Cohen and Cohen (1975) while Cattell and Scheier (1961) and others have illustrated it at the experimental design level.

At several points in this chapter on experimental designs the question of inclusion of stimulus variable scores along with response variable scores has been touched upon. Such coordination of stimulus scores with response scores, whether controlled, manipulated stimuli or stimuli selected from natural distributions can occur with virtually every analysis technique (O, P, Q, R, S, T, etc.) but it is one most naturally encountered in P and dR techniques which have correspondence to repeat designs in ANOVA. Regardless of manipulation of stimuli or absence thereof, the stimuli will normally be the independent variables in the scientific sense, so generally it is desirable to make them so in the statistical sense. When manipulation is possible, and undertaken, one will usually want to randomize stimuli mutually, i.e., have a correlation matrix in which they are uncorrelated, but correlated (if the null hypothesis is rejected) with the variables, while the response variables will generally show correlations with one another. If the stimuli are thus to be orthogonal, then the same balancing arrangement of sets of "high" and "low" subjects on the various stimulus variables is necessary as in a two-, three-, etc., way ANOVA design. In either case it is convenient to arrange the correlation matrix with the stimulus or condition (ambient situation) variables standing in one corner of the matrix, as shown in Fig. 12.7.

While an overall total factoring is possible, and was in fact used without confusion by Cattell and Scheier (1961) in simultaneously revealing state factors (of anxiety, stress, etc.) and the degree of significance of relations of stimuli to them, one may wish to factor responses separately (the lower right corner of Fig. 12.7a) and relate factor scores by ANOVA or CORAN to stimuli or factors in stimulus situations. For in the joint factoring these will normally be correlations between stimuli and responses which, with poor factoring (poor rotation principally) could appear simply as loadings of the stimuli on the response factoring. With adequate design—notably a sufficient attention to expected stimulus factor markers when naturally occurring (nonmanipulated) stimuli are concerned—the stimulus factors should rotate clear by simple structure of the response factors and the relation between them should appear as correlation between the factors. Campbell (see Harris, 1963) has conservatively referred to designs of this kind as "quasi-experiments," but

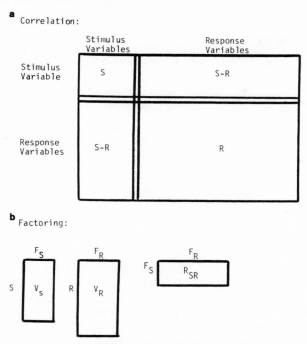

Fig. 12.7. Design for multistimulus–situation–multiresponse ("condition–response design") experiment with or without manipulation.

by our analysis of experiment in Chapter 1 they are true experiments. The advantage of this condition–response design over ANOVA, indeed, is that the relations are revealed not as to significance only (as in ANOVA) but as to *magnitude*, by the size of the correlation coefficient (in CORAN). The disadvantage of CORAN is that if a curvilinear relation exists its true magnitude will not be revealed. But neither will ANOVA tell us anything at once about curvilinearity. As proposed elsewhere here (Chapter 13) the only solution to the curvilinearity problem and the interaction problem in factor analysis is the plotting of curves of b values (loadings, tangents) for successive factorings over short ranges, followed by curve fitting with powered and product factor terms expressing the curvilinearity.

Misunderstandings of the full role of factor analysis in causal connection investigation, with or without manipulation, seem to have arisen partly from a conservative inertia, naturally leading the less enterprising to cling to classical manipulative bivariate experiment, and partly from retaining antique or unexamined concepts of causality not to be found in modern scientific philosophy and method. Granted that we recognize that an event may have several causes, some sufficient, others necessary, and that a cause may produce several consequences, the condition of necessary sequence is the hallmark of a cause (Rickard, 1972). One finds thunder to be preceded by lightning and so it is a cause, but so also is the presence of air where the lightning discharge occurs. It is all a question of what is held constant and taken for granted. Also we usually have no immediate guarantee when A is reliably followed by C, and C reliably preceded by A, that there is no intermediate term B. Only the course of investigation can check on intermediates, but the presence of a necessary intermediate is usually signaled by some irregularity in the A–D sequence relation. When Jenner noticed that milkmaids do not catch smallpox the first established association was obviously a cow. Only later did it become cowpox. Normally the presence of more immediate causes is signaled by the (lead–lag) correlation being less than unity, due to possible contingencies between.

In the use of P technique, or dR technique, or of R technique with before and after measures (relative to stimulus, learning experiences, or other "effects" as ANOVA somewhat obliquely refers to other causes!), another advantage over ANOVA, beyond the handling of many stimuli at once is the exploration of a considerable and more continuous range of levels on each stimulus. [MANOVA has approached the same goal in this respect, but multivariate analysis of variance (see Jones, 1966, Chap. 7) still does not reach the peculiar flexibility of factor analysis of change, or the capacity to *structure* the changes even while observing them.] We need not, in factor analysis, require as in ANOVA that all subjects in the experimental groups be subjected to the influence under study at exactly the same level. The level can vary over individuals, provided we know what each individual is experiencing. If the stimulus is at fixed levels, as in ANOVA, we enter with a point biserial or contingency coefficient, and if, as usual, at many diverse levels, with an ordinary correlation coefficient, for relating response to stimulus levels.

In connection with this structuring of change, a hybrid design between R and dR technique to which we have already briefly referred (p. 340) can also be useful in the present concern of relating stimulus factors to response factors. It may be called the *factor change score* method since its aim is to analyze the effect of learning or other "stimuli" directly in factor terms. It has two forms, which might be called the *person-tied* and *variable-tied* forms. As illustrated in Fig. 12.8, the first form enters each person's score on a variable before and after learning (or other influence) as if it were from two people—the person before and the person after. The *variable-tied* method, on the other hand, enters each person once in the score matrix but treats the variable as different before and after, entering two variables instead of one in the correlation matrix.

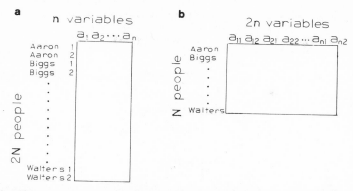

Fig. 12.8. Change score factor analysis: (a) person-tied and (b) variable-tied. The first gives each person two scores (as if for two people) on exactly the same variable pattern. The second uses the same person but two different sets of variables and, therefore, weights.

The person-tied analysis will give a single set of k factors such that there is no difficulty, from the common V_{fe}, in scoring each person, in his duality consistently on the same factor before and after. It assumes, and assures, that the factor form *per se* will not change from the "stimulus." The *variable-tied* method, on the other hand, will yield two different scores on the same factor for the same person, because one applies two different V_{fe}'s (on odd and on even v's—a_{11} and a_{12}) for each of the same k factors. The latter's shortcomings are (a) that there may be difficulty in handling the doublet factors resulting from double entry of each variable (see Tucker, 1966a), and (b) the separate V_{fe}'s are not typical V_{fe}'s if one pursues the usual calculation since they consider intercorrelations of all variables (odd and even) not just those actually in the V_{fe} for odd alone or even alone. For simplicity two occasions of experimental measurement have been considered in Fig. 12.8, but the number could be three or more.

Another form of analysis which can perhaps best be brought into perspective as an instrument of experimental change analysis by factor methods is that of Tucker's (1958) *interbattery* method. In this we have a set of variables applied to a sample of people, where the set of variables can in some meaningful way be divided into two subsets. For example, they might be questionnaire measures (Q data) and objective test (T data) measures of personality, or, as in the recent experiment of Hakstian and Cattell (1976), ability measures and personality measures or, in the present context, a set of stimuli to which people are exposed at individually different levels, i.e., given stimulus strength scores, and a set of reaction levels on the various resulting responses, i.e., response scores. The question is then asked "What factors are common to the personality and

Table 12.2. The Matrix for Interbattery Factoring

Sets of variables

		a	b
Sets of variables	a	R_{aa}	R_{ab}
	b	R_{ab}	R_{bb}

ability space?" or "How much and what kind of variation in the responses is tied up with variation of stimuli?"

The two sets will yield four correlation matrices as shown in Table 12.2. In this interbattery method one concerns oneself only with the common covariance in the north-east (or south-west) corner, namely with R_{ab}. It may be considered constituted by

$$R_{ab} = V_{fpa} R_{fab} V_{fpb}^T \qquad (12.5)$$

The reader is referred to Gorsuch's clear handling (1974, p. 285) for details of computation, and for references, though one should note some uncertainty over notation principles in this complex situation. The outcome of an interbattery factor analysis, i.e., of the matrix R_{ab} is often not easy to interpret in psychological concepts. In the recent study of an ability and personality interbattery by Hakstian and Cattell (1976), for example, the factors seem largely to coincide with already known personality factors with some projections on ability tests. The issue of what one has after essentially partialling out one domain from another and what use to make of that information remains a matter of debate. In rotation, moreover, although the above experimental instance seems to lead to an intelligible projection of factors of the personality modality into the ability modality one wonders what determines the choice of this versus the converse direction of interpretation. The present writer is inclined to view the interbattery method as less definite in its function than the others considered in this section, and perhaps more valuable in applied or social research, where some immediate pragmatic value is sought than in basic personality–learning research. Often a separate factoring in the two domains, learning experiences and resultant response changes, followed by relation of factors would seem more clear; but other investigators have felt that the interbattery analysis may yield information not obtainable by thus taking the whole into separate pieces.

In the brief space of a section, and with the scarcity of experience yet available on manipulative, causal investigations by factor analysis (for instance, the interbattery design seems actually not yet to have been applied to learning and response variables) it is not easy to evaluate with precision the role of these approaches. Their advantages relative to classical ANOVA approaches to experiment may be checked off as (a) simultaneously including a greater range and variety of stimulus situations and response reactions, so as to permit determining the structure of each, simultaneously with the structure of their interactions. (b) Permitting, in circumstances where stimulus situation manipulation is inappropriate, unethical, or impossible, the determination of stimulus–response relations. (c) Allowing a wide range of level of impingement of causal influences ("effects" in ANOVA) instead of fixed effects. This has importance as a more thorough exploration of the relationships than occurs from two or three fixed values, and makes causal investigation possible where manipulation is not possible. (d) It permits separation and partialling out of influence effects on an objective, factor-unity basis of interpretation. (e) It may also yield us a picture of the structure of stimuli in some natural, social setting. The approach still leaves nonlinear and interaction effects open to investigation, as in ANOVA. Incidentally, a possible but little used approach in factor analysis for examining for interaction effects exists where factors can be manipulated, or selected, so that each varies without the other doing so. Comparison of loadings (using real-base values) on separate and combined variation experiments should then reveal interaction effects on dependent variables—among factors, which is what one is interested in, not just among variables *per se*.

It will be noted that virtually all the score matrices to be analyzed in these designs, covering people, responses, and occasions (ambient situation conditions) are typical three-frames "unrolled" to grids for whatever direction of correlation is to be applied.

12.17. The Relation of Factors from Various BDRM Facets, Faces, Frames, and Grids: *n*-Way Factor Analysis

As the systematic survey of possible facets, faces, and frames made earlier has shown, their number is very great. As a result, the discussion of the relations of factors obtained from them could proliferate into an impracticable number for study here. Consequently, nothing exhaustive will be attempted in this section on the relations of such factors, and we shall concentrate on general principles and a few illustrations.

It is necessary at the outset to distinguish between relations of factors that are (a) due to mathematicostatistical circumstances and are precisely expressible in terms of necessary variance–covariance relationships between two analyses, and (b) due to the nature of the psychological or other scientific model concepts and their logical relationships in the real world.

An example of the first would be the relation between the factors from R and Q transposes of the same score matrix, or between the grid analysis of tests over persons and occasions and the factors from each of the test–person facets, in analyses repeated over the occasions. An example of the second would be the relation of P and dR factors from the same variables, presumably dealing with the same real psychological states, or the relation of state factor patterns in R technique to those in dR technique, where exact statistical accountancy regarding derivation of loading patterns from one to the other is scarcely to be expected, but where the model indicates that essentially the same influences are at work.

As an example of necessary mathematical relationship we have already considered the R–Q transpose, with Burt's theorem (1917, 1937) that once a double standardized score matrix is reached the same dimensions though clothed, respectively, in test and person markers, will result. The relationship has been widely discussed in clinical perspective by Broverman (1961, 1962), and brought to a new level of statistical clarity by Ross (1963), Cronbach and Gleser (1953), and Harris (1956).

As far as the matter may be put in simple concepts we can say that the first factor normally obtained from R technique is missing from the corresponding Q technique and that the first factor extracted (from an ordinary, not a double standardized matrix) by Q technique is missing from the R technique solution. This can be seen if we imagine a matrix of bodily measurements, from which R technique would normally first produce a general size factor. The corresponding Q technique correlation will bring each person to the same "size" as another (mean and sigma) so the size factor will not appear. But in R technique the standardization implicit in the correlation coefficient will, however, reciprocally ignore the fact that the length of a nose and an arm are very different, and will throw away the "general human proportions" factor present as the initial factor in Q technique.

Double standardization and covariance factoring, will, except for the possible convergence problem in double standardization mentioned elsewhere (Cattell, 1966a), reach the same result in both because we have thrown away that special information from each (the first factor in each) which makes them unequal. The corresponding comparison of factoring of double-centered (not double-standardized) matrices has been made by Tucker (1956) but the equivalence is even more obscure, and no special use for it has yet been found. (Double centered means zero means simultaneously for all rows and columns of the score matrix.) At a practical level the upshot is that one would usually do best to stay with R technique until data is reached where it is definitely impossible to get more people than tests. When that occurs, the necessary oblong matrix for statistical independence of correlation coefficients[6] then calls for correlating people over many tests. What one finds, however, is still not generalizable to people in general beyond what the size of the small person sample permits.

A face and facet of a person–response set combination differ in that the person's scores in the facet are on a single occasion, or single stimulus, or single observer in the case of the observer facet. In the case of the face the entries are averages over several occasion–conditions and perhaps over several stimuli and observers, depending on how many co-ordinates of the data box we choose to collapse in the face matrix. If the collapse is over occasions, numerous enough to reduce the error from over-occasion variance to triviality, then the factoring of the R × Q face will yield traits only, whereas the factoring of the facet, retaining later occasion variance, will yield both traits and states.

For research design—particularly to get maximum illumination from programmatic research—the comparisons of results from among the various facets, faces, frames, and grids, we are here discussing is particularly im-portant. It is a means of more deeply understanding structure, and of separating factors corresponding to diverse conceptions, i.e., those of traits, states, processes, observer instrument factors, dimensions of en-vironmental situations, etc. Table 12.3 shows an instance of some prac-tical importance to clinicians, concerning the recognition and separation of trait and state anxiety and some related states. The separation cannot be made dependably on content alone either directly or in a contrast of P and R patterns as suggested earlier. For both in P technique and dR technique we may expect possibly to get both a state and a trait-change factor (defined as a slow change having the same pattern as a known trait), and we may not know which is which. The separation of the trait, the state, and the trait-change factor patterns must rest in part on the theo-retical, statistical expectation of the relative variances of these factors in R, dR, and P technique (averaged across enough subjects) on the same variables. Since all factors are brought to a variance of unity in ordinary factor analysis, they have here to be transformed into real-base factors (p. 407) using the raw sigmas on the variables. Table 12.3, recognizing

---

[6] That is to say, the correlation of the $(n - 1)$th with the nth variable is already fixed by the latter's correlation with each of the preceding $(n - 1)$ variables, when we stan-dardize by rows and is not free to assume its psychologically meaningful value. Seen in other terms, as Tucker points out, the common and the specific spaces cease to be separable when $n = N$. Consequently, $n = N$ is to be avoided and when $n > N$ one must seek to obtain the common factors by Q rather than R technique.

Table 12.3. Variance Relations of Trait and State Factors

(a) Assumption: All people have the same state mean and sigma. There is only one
trait and one state involved so that the variance of the variable is wholly accounted
for by them. State factor sigma equals trait sigma equals unity and with either
assumption (a or b below) the basic treatment is by covariance real-base factor
analysis. However, factor size is kept at unity and variable variance change handled
by change in b. The variance of the variable is the same for R and dR, if $b_{jt} = b_{js}$,
but the state change variable is twice as big as the state variable in R. With these
conditions the variance of the actual variable, e.g., a factor scale in raw units,
should be half as big in P as in R and dR.

R technique:
$$\sigma^2_{ij(k)} = b^2_{jt}\sigma^2_{ti} + b^2_{js}\sigma^2_{sk} = b^2_{jt} + b^2_{js}$$
dR technique:
$$\sigma^2_{ij(k2-k1)} = b^2_{js}\sigma^2_{sk} + b^2_{js}\sigma^2_{sk} = 2b^2_{js}$$
P technique:
$$\sigma_{ijk(i)} = b_{js(i)}\,\sigma^2_{sk} = b^2_{js}$$

Note: Subscripts in parentheses indicate what is constant for all measures: i in
P, k in R; j is the specific variable, k the occasion. By contrast a roman letter shows
over what the variance is measured. b_j is loading for variable j. k is roman in R and
dR because, although it is the same occasion in time, the condition is different for
each person in regard to his state provocation. s = state; t = trait.

(b) Assumption: People have different means and sigmas for their states, which are
positively correlated over people.

R technique:
$$\sigma^2_{ij(k)} = b^2_{jt}\sigma^2_t + b^2_{js}\sigma^2_{smi} + b^2_{js}\sigma^2_{ski} + 2r_{mki}\sigma_{smi}\sigma_{ski}$$

Note: σ_{smi} is the variance of state means across people; σ_{ski} is the variance of
individuals from their means where both occasions (ks) and the sigma of individuals
(is) vary. This sigma across two sources is considered approximately the mean of the
sigmas of all individuals.

dR technique:
$$\sigma^2_{ij(k_2-k_1)} = 2bjs^2\,\sigma^2_{ski}$$
P technique:
$$\sigma^2_{jk(i)} = b^2_{js(i)}\sigma^2_{sk(i)} = b^2_{jis}$$

Note: This P technique differs from that in (a) only because σ^2_{sk} is peculiar to
the individual whereas in (a) it was assumed the same for all individuals. Otherwise the
variance of factors and variables is the same for dR and P in (b) as in (a). But in R
the variance of the variable will be greater and the variance of the state factor will be
considerably greater (using raw scores and covariance factors). Such experimental comparisons are the means of deciding between models (a) and (b). The rough generalization is that the state measure will tend to have twice the variance in dR that it has in
the R- and P-technique series.

that states and trait change but not trait factors will appear in P and dR technique, and that traits and states but probably not trait change factors will appear in R technique, summarizes the expected variances.

An important development of recent years is the examination of the formal relationship of factors in a *three-way* and *three-mode* analysis system. They should not be confused and it should be recognized that in principle they are *n-way* and *n-mode* analyses, depending on how many of the dimensions one likes to deal with in the data box.

It is convenient and appropriate to deal first with what the present writer introduced (1966) as three-way analysis since it is simpler than three-mode and keeps closer to psychological concepts now under experiment.

The discussion here is of *n*-way analysis and we shall use three-way as an illustration. According to whether one accepts particular "condition" sets or not the psychological data box is 10-set or 5-set. Taking the 5-set (P, people; E, environmental occasion conditions; R, responses; S, stimuli; and O, observers) with response measures as *attributes*, i.e., the entries representing measured features of any set of ids, let us first consider the data matrices possible from it. In passing let us note the desirability of avoiding confusion from the habit of calling the two sets of ids which bound a matrix, "objects" and "attributes." Both sets of ids are objects (patterns) in themselves and we can best call *relatives* those we choose to correlate and *referees* those ids (objects) we use as the correlatable series, i.e., the entities on which the pairs of measures are made. The ids (objects) have a variety of attributes and it is scores on these, as relations between relatives and referees, that we enter into the score matrix. We are so used to measuring attributes of responses (the relation of a person and a stimulus) as entries in the score matrix that we forget the 5-set data box could be entered with attributes of any one of the sets; though much of the result would be altogether outside psychology. [For example, a stimulus set (S), by an occasion set (E), with frequency of stimuli at each occasion as the attribute is the basis for a purely physical analysis of environment.] Here we shall assume the attribute entries are properties only of responses, e.g., speed, correctness, etc.

The preliminary introduction in Sections 12.6 and 12.7 above has reminded the reader of the nature of facets, faces, grids, and frames. For example, a frame is an *n*-dimensional box, which is a score matrix of which the 5-set box is the ultimate, and a grid is a partial or complete extension—unrolling—of such a frame into a two-dimensional score matrix. From a 3-set frame of P, E, and S, ids, 3 full grids may be unrolled according to the direction of unrolling. The first will have stimuli at the top (relatives) and persons by occasions down the side (referees), the second persons at the top and stimuli-by-occasions down the side, and the third will have occasions at the top. The resulting R matrices will be among stimuli, persons, and occasions, respectively. For less condensed and more diagrammatic treatment of this whole matter of possible score matrices the reader is referred to Cattell (1966a, p. 122 on). The number of possible faces will also be three (combination of 2 from 3 things), namely, stimuli-by-persons (using scores averaged over occasions); stimuli by occasions (using scores averaged over persons); and persons-by-occasions (scores averaged over stimuli). Each of these has a transpose, e.g., R and Q techniques from the first, but since we have learned that transposes yield essentially the same factors our sources of factors are three.

This is the simple geometrical statement of possibilities, but when we turn to extracted scientific meaning our concern in psychology is almost always with explaining the variance–covariance of responses to stimuli. In that case the third face (stimuli by occasions) and the second and third grid are generally (a) irrelevant and/or (b) statistically superfluous, since one grid or two faces already handle the total variance in the given frame, i.e., cover all entries. Others merely break down the same information in new ways. Thus when we have the factors from the grid of stimuli at the top and all possible combinations of people and occasions down the side we have exhausted all that can be said about the contribution of factors in stimuli (tests, if one wishes) from people and occasions.

As to the "irrelevance" this of course depends on breadth of interest. The alternative breakdown of factors among occasions (as relatives), stating dimensions of occasions which account for some variance in person–stimulus response, is certainly psychology. However, it happens that the influence of occasions is already taken into account by occasion loadings in the grid analysis of stimuli (tests) over persons and occasions. It is helpful to think of this alternative as an alternative of $b_{(j+k)}T_i$ and $b_{(j+i)}T_k$. That is to say, in one case the variance on occasions, the ks, goes into the different magnitudes of loadings $[b_{(j+k)}]$ and in the other it goes into different magnitudes of factors (T_k).

One should note what happens to the number of possibilities in relation to the number of sets. If n is 3, as in the example we mainly use, there are 3 grids only one of which is useful and sufficient, and 3 faces, 2 of which are sufficient (tests over persons and tests over occasions) to cover the variance–covariance concerned. If n is 4 there is still one grid (now over combinations of 3 referees) and 3 faces, if 5, then one grid over four referees in combination, and 4 faces.

Let us now consider the relation of the grid and the face results, keeping to the illustration to the one set of relatives about which the others hinge—stimuli or tests. In the face factoring with standard scores the first of the two sufficient faces will use as entries the mean response of each person across all occasions, and the second the mean of each occasion across all persons. Obviously, these will start from the same "grand mean"—that peculiar to each test. Consequently, with independence (no interaction) the deviation of a single entry in the frame a_{hik} (since h is a stimulus or test, i a person, and k an occasion) from the grand mean is the sum of the contributions of the two sets of factors which are

$$a_{hi(k)} = \sum^{x=p} b_{hx} T_{xi} \qquad\qquad (12.6a)$$

$$a_{hk(i)} = \sum^{y=m} b_{hz} T_{yk} \qquad\qquad (12.6b)$$

where the number of factors, p and m, is not, at present, assumed necessarily the same in the two faces, and where the subscript in parentheses means "averaged over that set of k's (or of i's)." Note that the absolute size of the factors over occasions as referees may be larger or smaller than over people (in terms of raw score sigma of variables) so we will suppose this taken care of by covariance, real base, treatment in order that we may add (a_{hik} being from the same mean of grid and faces) to

$$a_{hik} = \sum^{x=p} b_{hx} T_{xi} + \sum^{y=m} b_{hy} T_{yk} \qquad\qquad (12.7)$$

the measure on the left now being from the grand mean. An already psychologically developed instance of this exists in *trait view theory* applied to observer's ratings of subjects. A rating on a_{hio} (i.e., by observer o) is in that case the outcome of a context effect of the real traits of the subject, in the first expression on the right of Eq. (12.7) and of the traits of the observer in the second expression $\Sigma b_{hy} T_{yo}$. Even if p and m are equal, i.e., as many factors in occasions as in people, and of the same loading pattern, we cannot throw them into one set of factors, because the T_y's belong to particular occasions and the T_x's to particular people. A complete condensation in fact becomes possible, in present experience, only in the psychological case of distortion in questionnaires, handled by trait view theory (Cattell, 1973). For there the subject is also observer, the number of factors is the same, they have the same pattern, and i and k are one and the same id.

In the case of the corresponding grid analysis (relatives are stimuli; referees are people by occasions); there is necessarily only one set of factors. Just conceivably they could be rotated to correspond to factors operative over people and others over occasions. Upon getting factor scores one should then discover that the former give values which are constant for individual people over occasions and the latter for the same occasion over all people, and a_{hik} is deviation from the grid mean.

The analysis in Eq. (12.7) has relevance to the special modulation model discussed elsewhere (p. 418). In that case we have psychological evidence that the factors are the same in nature and number for people and occasions, e.g., the pattern of anxiety found in individual differences among people is the same as the pattern shown to change functionally for everyone under the impact of different environmental occasions. In that case, by factoring at different environmental (occasion) mean levels from a central core value (as in real-base factoring) we derive a modulator index s_k such that

$$s_k = b_k/b_o \tag{12.8}$$

where b_o is the covariance loading at the core position and b_k at some situation k. Incidentally a comparison with Eq. (12.7) shows that we could just as correctly have used an additive as a product model, such that if we write the modulation increment to b_k as s_k, then

$$s_k = b_k - b_o \tag{12.9}$$

instead of

$$s_k = b_k/b_o$$

Thus the new value at k would be $(b_{ox} + s_{kx})T_{xi}$, instead of $b_{ox}s_{kx}T_{xi}$ as from the method of Eq. (12.8).

It is one of those instances where two models are mathematically equally good, and the decision between them must rest on the fit as a scientific model to yet other types of experiment employing the alternative mathematical expressions.

To step beyond the three-way to a five-way situation we see that we have the alternatives from factoring faces of

$$a_{hijko} = \Sigma b_{hw} T_{wi} + \Sigma b_{hx} T_{xj} + \Sigma b_{hy} T_{yk} + \Sigma b_{hz} T_{zo} \tag{12.10}$$

Here a_{hijko} is the deviation (standard score form) of the entry in any cell from the grand mean of cell entries in the whole data box.

If the experimental results should prove the factor patterns to be the same (on the variables—the h's), they can be joined together, either additively (in covariance factoring),

$$a_{hijko} = \Sigma(b_{hw} + b_{hx} + b_{hy} + b_{hz})T_{ijko} \tag{12.11}$$

or multiplicatively,

$$a_{hijko} = \Sigma s_w \cdot s_x \cdot s_y \cdot b_z \, T_{ijko} \tag{12.12}$$

Scientific settings which lead to the latter possibility—similar factor patterns on stimuli across all other sets—may be rare, though we have already encountered the convincing instance of modulation of psychological states.

12.18. Comparison of *n*-Way (Disjunct and Conjoint) and *n*-Mode Factor Analysis

With the above description of *n*-way factoring by *faces* and *n*-way factoring by *grids*, let us make more explicit their mutual relations and their relations to an entirely different and novel treatment called three-mode analysis.

It will be recalled that from an *n*-set there are $(n - 1)$ possible face factorings but always only one grid factoring—provided we are concerned in both cases with the structure of variance–covariance for one given set. This latter is typically the stimuli or test response set and in the grid analysis we resolve the scores on those into factors which predict stimulus–response (test) scores in terms of loadings applying to factor scores of combinations of persons, occasions, observers, etc. These we may call *conjoint factors*. In the face approach, which we may call *disjunct n-way analysis*, the factors in people, occasions, observers and varieties of response (in the 5-set case), will be distinct, as a rule, and correspond to distinct meaningful concepts. They will be concepts of the dimensions of people, of occasions, etc., in *terms of their expression* as effects upon responses to stimuli. This is the matter of greatest relevance to the psychologist, though, as we have seen, *n*-way analysis could be made to hinge on other sets than stimuli and other attributes than responses, even in the P, E, R, S, O sets of the psychologist. Thus in an *n* set there could be (without change of the response measure attributes used) *n* disjunct factorings, each covering $(n - 1)$ matrices, i.e., $n(n - 1)$ factorings in all. Secondly, we could make *n* conjoint factorings, each running over a composite set of scores from $(n - 1)$ sets and using *n* different sets of attributes.

The relation of disjunct and conjoint factors from the same set of relatives, e.g., tests, and the same attribute is unlikely to be simple, though we have granted that identities could exist across all disjunct factors, and in that case a relatively simple, summative relation would exist at any rate between the covariance, $V_{fp \cdot c}$, matrices of disjunct and conjoint V's. The former could simply place the latter in one matrix, separating the factors on, say, people and occasions, into two distinct vertical sections of the matrix, the first with loadings significant only on variables contributing to person factor scores and the second on variables contributing only to occasion factor scores. *A priori*—but more from other experience—such clean separation seems unlikely, except with special rotation and we must expect that the "composite" factors of conjoint factor analysis, contributing to "test" variance from ids each of which is a particular person in a particular environment measured by a particular observer, will have no really simple relation to those from disjunct factoring. One may intuit that the factor concepts of disjunct factoring will prove to have more useful scientific meaning and be negotiable in more predictive laws.

In considering these questions the training the reader has received in ANOVA and ANCOVA will probably have been tempting him by now to look for familiar analogies in that system of analysis. It is worthwhile to take a side glance, though any adequate treatment would require a foundation of the not-so-simple relation of ordinary factor analysis to analysis of variance. A number of excellent discussions thereon exist in Burt (1966a,b), Bentler (1976), and several others, while the relation of regression analysis in general to ANOVA and ANCOVA is thoroughly treated by Cohen and Cohen (1975) and Tatsuoka (1971b). In the first place the score entries are one to a cell in the factor treatment whereas there would have to be grouping (at least two cases) at the points of combination of the several effects to permit ANOVA. The perhaps puzzling point is that with ANOVA there would be five sets of effects (five-way ANOVA) each at several levels with the 5-set box, whereas with factor analysis there is always one factor matrix fewer than the number of sets. Thus in Eq. (12.17) we have four sets of loadings. However, the other set is present in the set of factor scores. Both systems account, but in different ways, for the total variance of entries on the dependent variable—the attribute—in the data box; but factor analysis has its uniqueness and its disregard of interaction effects (nonlinear summation) other than those arising through correlation, and it disregards nonlinear relations which ANOVA can recognize. On the other hand the variables in ANOVA are arbitrary and it is blind (until it becomes ANCOVA) to covariance structure.

It needs no psychological Columbus to perceive, relatively easily, that there is a rich domain of substantive, scientific model possibilities awaiting investigation by the above forms of n-way analysis, as instanced by recent developments in trait view theory and state liability modulator action above. However, it must share its lure with the n-mode approach now to be discussed, in terms of beckoning to new worlds of development in psychology. This second approach to handling variation over all sets jointly is statistically more complex and sophisticated. Indeed, most psychologists seem so far to have found the brilliant development of three-mode analysis by Tucker and his students not to be assimilable so easily into analogy with ANOVA modes of thought as does the above three-way (or n-way) factoring. Its first results also present puzzling perplexities in terms of psychological concepts. The steps of development by Tucker (1963, 1966) will be briefly sketched here, but will be comprehensible to other than the mathematically minded only by supplementation with the background reading indicated.

It begins with the Eckhart–Young theorem (Eckhart & Young, 1936)—a central proposition in the history of "decomposing" score matrices to which quite a number of studies in our bibliography refer. The proposition is that a variable score matrix S_v can be broken down into matrices of lower rank as follows:

$$S_v = S_f \quad D \quad V^T + E \qquad (12.13)$$
$$(N \times n) = (N \times k)(k \times k)(k \times n) \quad (N \times n)$$

Although this is more general in meaning than for factors as usually understood, we have used the familiar k as the number in the new component score matrix S_f, and V as the equivalent of a factor matrix $(n \times k)$. However, the approximation of a higher-rank matrix S_v by the given lower-rank matrices is not the same sort of approximation as in the regular factor model, in which the absence of fit is accounted for by *uniqueness* u peculiar to each variable. Here the error matrix E which expresses the approximation of fit, is random and formless. Some psychologists view this difference as inessential except in early formulations, pointing out that the common factor structure in calculations turns out to be the same in this and ordinary factoring. They then suggest that how one likes to conceive the error—as "test" uniqueness plus error or wholly error—is a theoretical matter, subject to appropriately adaptive forms of calculation.

In Eq. (12.13) D is essentially a diagonal of square roots of the eigenvalues of the cross products matrix $S_v^T S_v$. The formulation is thus closely analogous to the familiar $\underline{S}_v^T \underline{S}_v / N = R$ (the $\underline{S}$ denoting standard scores) and $S_v = S_f V_0$, except that the eigenvalues have not been absorbed as usual into the V_0 matrix ($V_0 = VD$ where D^2 is the eigenvalues).

Tucker's approach consists in applying the Eckhart–Young decomposition in a novel development to a three-set (eventually n-set) score matrix instead of the usual two-set facet or face. The m tests are reduced to r factors in an *attribute factor matrix*; the n persons or other ids to q factors in an *object matrix*, and the v occasions to s factors in an *occasion* matrix. In the mathematical process, which must be by-passed here, Tucker had to invent a three-mode matrix multiplication. The ultimate emerging reconstitution of the score matrix is one in which each variable score is expressed as a product of scores on each of these three sets of factors, plus, of course, the random error term from the error matrix E above. However, this is not quite the whole story for the estimation requires also the intervention of a *core matrix*, which Tucker writes as G, and which expresses the relations among the factors in the three systems. This matrix will have the order $q \times r \times s$, i.e., representing the number of factors in each system, which of course need not be the same. If we write the factor scores as a in the person matrix, put b in the stimulus (or test) matrix, and c in the occasion matrix, then the ith person, in the jth test on the kth occasion will have a score of

$$x_{ijk} = \sum_{l=1}^{p} \sum_{h=1}^{q} \sum_{t=1}^{s} a_{il} b_{jh} c_{kt} g_{lht} + e_{ijk} \qquad (12.14)$$

where l, h, and t are particular factors in each series.

In the case of a grid in which correlations are within the a set over all combinations of b and c such as we used in conjoint factoring, then more simply we can write in matrix terms

$$X \quad = \quad A \quad \cdot \quad G \quad \cdot (\ B \quad \otimes \quad C \ ) \quad E \qquad (12.15)$$

$$(i \text{ by } jk) \quad (i \text{ by } p) \quad (p \text{ by } qs) \quad (q \text{ by } j) \quad (s \text{ by } k) \ (i \text{ by } jk)$$

where A, B, C, and G are the matrices referred to above, G is the core matrix, and E is the error matrix. $\otimes$ describes a Kronecker product and the row below refers to the values fixing the orders of the matrices.

Tucker refers to these A, B, and C factors as "idealized" person, stimulus, and response (or occasion, test, etc.) factors, but until more research is done their conceptual status remains rather obscure. On the other hand, the system is an entirely workable one, with internal consistency, mutual transformability of scores and factor values (estimates in one direction), and the possibility of rotating the factors.

There is no simple translation between n-way and n-mode analysis, but the reader may be helped in seeing some relations by considering the steps by which n-mode analysis moves to its ultimate conception. The first matrix, X, i by jk, in Eq. (12.15) is our familiar grid, in this case with persons as relatives and stimuli by occasions as referees. We are used to the breaking down (by several steps) in ordinary factoring, of this score matrix into a factor score matrix and factor pattern matrix.

If we bring this to our usual "tall" matrix form—S_f now being the factor score matrix and S the variable score—[and leaving Eq. (12.15) in transposes as Tucker presents it], and having N persons, n tests, and v occasions, then, considering we begin by factoring the person set, we have

$$S \quad = \quad S_f \quad \cdot \quad A$$

$$(nv \text{ by } N) \quad (nv \text{ by } p) \quad (p \text{ by } N)$$

Three mode now takes the new factor score matrix S_f and rearranges it by a grid transformation to the matrix

$$S_f$$

$$(pv \text{ by } n)$$

i.e., it has tests as referees and person factors by occasions as relatives. This score matrix is again decomposed into the product of a factor score matrix

$$S_f$$
$$(pv \text{ by } q)$$

and the factor matrix for Tests B as in Eq. (12.15). In the three-way case

$$S_f$$
$$(pq \text{ by } v)$$

(after rearranging) is again decomposed leading to the factor matrix for occasions C and the core matrix G:

$$S_f = G \cdot C$$
$$(pq \text{ by } v) \quad (pq \text{ by } s) \quad (s \text{ by } v)$$

It will be seen that factors from one set of ids are, by this process, factored in composites with another set to yield the third set. As stated initially, three-mode analysis is of a high complexity, demanding perceptions of relations among relations.

The reader may pursue the statistical analysis procedures further in the main statistical texts, notably in Gorsuch (1974) and Mulaik (1972) and in the articles by Tucker (1963, pp. 122–137; 1966; 1972), Levin (1974), and others. It has been applied to cognitive, clinical, instrument factor, and social (see Triandis) empirical problems and Tucker has recently worked out the corresponding multidimensional scaling treatment (1972) and begun extension to four-mode analysis. Besides the interpretation problem above, however, there has remained the objection that in the original Eckhart–Young form, it does not fit directly the general factor model with broad primaries, specifics, and higher-strata factors though this objection can perhaps be overcome (Snyder, 1968). It also makes assumptions about variables such as occur in the restricted "fixed effect" ANOVA treatment. However, Snyder's (1968) treatment promises to handle the noncommom variance in terms of uniqueness in each variable. As we have seen, n-way analysis, in either disjunct or conjoint form, completely takes care of the data prediction (if specific scores are known) in the n-way data box. Thus it may, for example, account for all the response variance–covariance on tests by structures in people and occasions. What n-mode analysis does as an alternative is consider prediction also in terms of *additional* structures in occasions in relation to stimuli, etc.; though it goes no further in actually accounting for, i.e., predicting, the scores as they exist in the data box.

The present writer would assert that there is an exciting mathematico-statistical future for this model, which constitutes a more fundamental and brilliant construction in recent years than anything but maximum likelihood (if one may compare two very different services to factor analysis). However, the psychologist as such is likely in practice to find n-way analysis linking up more readily with various facet, face, and grid analyses and permitting relation more closely to substantive concepts resting simultaneously on factor analysis and other experimental approaches.

12.19. Summary

1. Broader and more potent experimental designs are now being developed in factor analysis from broader concepts of the possible available score and analysis matrices, and from novel incorporations of manipulative and sequential arrangements of variables, people, and other sets in the experimental procedures. The familiar two-set score matrix was first extended to the covariation chart, with three sets (persons, tests, occasions) and then to the full five sets possible in psychology—persons, stimuli, responses, occasion–conditions, and observers. The score frame has been called the basic data relation matrix or data box.

2. The various subsets of data matrices that can be chopped out of the BDRM are called *facets* (of which there are 10), and *faces*, *frames*, and *grids* (of which there are definable numbers). In each, the set of ids called the *relatives* constitutes one coordinate and these relatives bound the subsequent correlation matrix for factoring. The correlations of relatives run over the series of *referees* (commonly thought of as the population sample), which may be uniformly of one set (as in a facet or face) or run over up to four sets combined (as in a frame—a box—which unrolls to a grid). A score matrix whose entries relate to several sets of id coordinates can be presented geometrically as a box, or one can represent it algebraically as a super matrix. But in either we are reminded that a psychological measurement has *five signatures* which define its uniqueness as an event.

3. The recognition that CORAN and ANOVA designs can be carried out over many more facets than that of persons against tests, brings the need for broader *sampling* and *standardization* concepts. Recognition of sampling problems beyond those of people (or organisms) first appeared in sampling variables by the personality sphere concept and the time sampling of occasions. This development led, incidentally, to the concepts of ipsative, abative, and performative standardization. Sampling principles on other id sets are still rudimentary.

4. Extension of factoring begins with explanation of transposes of the same face or fact matrix, as in R and Q techniques. It has not been widely appreciated that R and Q factorings yield essentially the same factors which are dimensions not types. In raw-score matrices each loses a factor present in the other; but in double standardized matrices with covariance

factoring the number and nature of factors from transposes is, as Burt showed, the same.

5. Using absolute scores, i.e., not moving into differences, five different techniques of correlation and factor analysis (with corresponding ANOVA breakdowns) are possible, each divisible into two transposes. The most familiar six of this ten are R and Q, P and Q, and S and T techniques, being the three pairs of transposes on person, test, and occasion sets. The natures of the factors derivable from each are discussed.

6. Additional to the absolute score techniques, difference scores, in any direction can be factored, and of the *differential* techniques, dR is the best known. The factors from change over occasions (which, in psychology, are *states*, and *trait change* patterns, the latter due to maturation and learning) should theoretically be essentially the same in dR and P technique, and empirically prove to be so.

7. The main problems in using difference scores arise from (a) their adding the error variance of the two absolute scores, though the true variance also adds simply, unless they are correlated, and (b) poor scaling in absolute scores compounding the distortion in difference scores. Partialling out one absolute score, along with its error, is rarely meaningful in terms of a scientific model. Equal variance and reliability in pre- and post-scores is desirable if artefact factors from trait structure are not to appear in state patterns.

8. P technique introduces new problems of sampling, as well as opportunities of introducing manipulative experimental "independent" variables. It also encounters the need to recognize and handle effects from cycles and trends. Two important developments are *chain P technique*, and *staggered P technique* with lead-and-lag correlations between factors and variables. When chain P technique extends over several people, and dR technique enters with differences from the mean over several occasions, they operate essentially on the same score grid converging on the same factors.

9. What might be viewed historically as sporadic hybrids of classical and factor analytic experiment—but more comprehensively and correctly as normal extensions of factor experiment to include the manipulative and causal sequence effects possible in all designs—are systematically described. The application of time sequence and manipulative factor experiment to process psychology—learning, perception, developmental studies—leads to the multistimulus–multiresponse design, to double-entry factoring, to lead-and-lag P technique, to interbattery factoring, and other novel designs. The advantage of CORAN, and, more specifically, of factor analytic approaches relative to ANOVA have been briefly discussed, in terms of breadth of information gained, significance evaluation, forms of interaction, handling of nonlinearity, etc.

10. A broader experimental design illustrating the combination of manipulative with factor analytic features is that for the structured learning theory model (Cattell, 1978). Therein a factor loading is broken down into four contributory parts by comparing obtained loadings *b*'s under four manipulatively controlled conditions: with perceptual recognition response only; with change in modulating conditions; with evaluation purely of motivation strength; and with full executive performance response combining these influences with executive skill. This is one of several broader designs presently awaiting application.

11. The relations among factor patterns from different facets, faces, frames, and grids of the data box arise in some cases as mathematico-statistical necessities and in others only if the relations in the real data fit a particular conceptual model. The relations of factors between R and Q transposes illustrate the former since the appearance of trait-development factors similarly in dR and P technique mutually, must be considered due to consistencies in the psychological influences themselves. Comparison of the P and dR factor patterns on the one hand, and R factor patterns on the other, should yield definite variance ratios in the same factors, and lead to separation of state, trait and trait-change patterns. The theoretical and statistical examination of these relations and the deliberate use of comparisons of results from different techniques presents an extensive field as yet little explored.

12. The most advanced designs to develop relations among factors within the data box are found (a) in *three-way* analysis and (b) Tucker's *multi-mode* (usually *three-mode*) analysis. Both are potentially extendable to four coordinate systems. The multi-mode design is a greater departure from ordinary factor analysis, and though the procedure is clear the concepts need more relation to scientific models. Three-way analysis breaks down the variance of the dependent variable in terms of factors from mutually orthogonal faces in the data box, and in conception parallels ANOVA. Much remains to be researched regarding the form of the relation (sometimes similarity, sometimes derivative, sometimes supplementary) of factors sets from the various facets and grids.

CHAPTER 13

Varieties of Factor Models in Relation to Scientific Models

13.1. Scientific Aims in Modifying the Factor Model

It is desirable at this point to reiterate the distinctions between a *scientific model*, an *analysis model*, and a *factor extraction procedure*. A scientific model is by no means confined to the factor domain. It can be anything that we can set up in definite, testable form, couched in mechan-

377

ical, or mathematical, or chemical or biological, etc., form. It is some precisely defined set of relations to which we suppose there is an analogy in the working relations of the data.

By an analysis model we mean a purely mathematical set of relations which may or may not claim to be also a unique scientific hypothesis model representing causal relations, etc. For example, principal components is a linear and additive model in which there are as many common determiners as there are variables. As a scientific model it may not recommend itself at all where there is evidence of curvilinearity, of product effects, and of fewer determiners than variables; but it remains a definite analytical model. Once again, regression is an analysis model, but since it works two ways, only one, if that, could be set up as the scientific model. A factor extraction procedure is less basic: it is an algorithm or sequence of calculations of which there may be several appropriate to a given analytical model. Thus the centroid, the principal axes method by Jacobi and the principal axes by Hotelling are all compatible with the components analytical model, or, for that matter, with the factor model. For an extensive study of the varieties and conditions of factor extraction from data matrices, the reader is referred to Horst (1965).

Our main concern in this chapter is with the various analytical models and their suitability for consideration as scientific models, though we shall also have something to say about the efficiency and convenience of extraction procedures in relation to those models. The principal analytical models for the present domain are the components model, the factor model and its various forms, the image model, and the real base factor model. The two last are sufficiently deviant from the factor model to be considered as distinct variants. But what a given factorist will consider a minor or a major variant depends upon his point of view, particularly on whether his emphasis is on mathematicostatistical properties or on the essentials of a scientific model.

Furthermore, though we have argued for the clarity of distinguishing between method of extraction and factoring procedure, on the one hand, and models, on the other, the line between them inevitably gets blurred. For example, we have chosen to consider alpha factoring (which, with enough variables, becomes ordinary factoring) as a deviant only in extraction calculations, whereas real-base factoring, which initially takes the same procedural form as ordinary factoring, we would insist is a distinct scientific model, because it ends with factors of different real sizes. To some recent writers on Jöreskog's and others' work on maximum likelihood and least squares (weighted and unweighted) extraction, it has seemed that the great division is between least squares and likelihood solutions, which, to the present writer, seems a mathematicostatistical fitting issue rather than a difference as fundamental as one of scientific models. Even within the extraction methods there are major and minor differences. Unweighted summation (centroid), weighted summation (principal axes), and image analysis involve differences of a fairly important kind. On the other hand, minres ["minimal residual" (Harman, 1976)] has recently received a lot of attention as if it offered some radical difference of model, whereas in fact it is a minor (but useful) variant on methods of extraction, and, as we shall see later, gives results close to other methods, especially maximum likelihood and least squares.

Naturally, those who experiment with models and methods can always present one with hybrids going beyond the two or three models and the half dozen extraction methods here described. For example, one can depart from the components' model by taking out fewer than n components and from the factor model by starting with unities in the diagonal, as Humphreys has done. If then iterated, the latter becomes factoring, but it is left with the particular communality first reached and is a somewhat messy model scarcely to be recommended. Incidentally, in the model versus extraction-method field, principal components—a model—and principal axes—a method of extraction—are most often confused. (This may be an argument for using Burt's term—weighted summation—for the latter, contrasting clearly with unweighted summation for the centroid procedure.)

It is feasible, and follows a best natural sequence, to pass from the last chapter's discussion of experimental designs—in which any of the present analysis models could be used—to analysis and extraction procedures, leaving more refined issues of statistical inference, scaling, etc., to the succeeding chapter, though we necessarily make contact with them in such approaches as maximum likelihood. In considering different modifications of the factor analysis model (the components model we shall set aside for reasons already given), it is helpful to notice their origins. Some arise from playful explorations of variants of the mathematical model. Like non-Euclidean geometry or Boolian algebra they express the ingenuity and inventiveness of the mathematician rather than any need to adapt to experimental findings, but may, as in both of these examples, find useful application later. The rest arise from attempts to get the analysis model to conform better to the perceived needs of a scientific model. Alpha factoring might be an example of the former and real-base factoring of the latter.

The reader wishing to evaluate the relative aptness of the pure mathematical modelmaking and the more *ad hoc* fitting of perhaps relatively clumsy models to the apparent demands of real data might well gain insight from looking at their roles in the history of science generally, as seen, for example, in Kuhn (1964) and Easley and Tatsuoka (1968). The attempts in factor analysis to adapt to the needs of a purely scientific model, i.e., to what other aspects of psychological data suggest to be necessary, are well instanced by McDonald's nonlinear factor analysis, by the various attempts at more than additive relations of factors, by the "spiral action" theory of second-order factors, and by real-base, true-zero factoring. They are naturally more tentative and generally appear less elegant than the models which have sprung from the brains of mathematicians, unembarrassed by considerations of many awkward natural restraints. But, especially since the mathematicians' products have already received considerable attention elsewhere, we shall place compensating emphasis on the initial fitting of perhaps mathematically odd models to relations directly suggested by the data. This fitting of the factor model to practical requirements includes, incidentally, the bringing of ANOVA and CORAN designs into closer integration, and, at the remoter boundary of the factor model, looking at the promise of integrating factor analysis with path coefficients, as discussed below.

13.2. Departures from Assumptions of Linearity and Additivity

Particularly among clinicians, who encounter the strangest complexities of human nature (and often respond by creating dynamic theories which are in almost equally strange and untestable form as regards logical structure!), factor analysis has sometimes been reproached as an oversimplified model, in two main respects. It assumes linear relations between factors and variables. And it assumes that factors produce their common effects by a simple additive action. One might perhaps expect that the first assumption implies also that relations among variables are treated as linear and also those among factors, in this model. Tucker has shown that these two further assumptions are not a necessary part of the factor model. However, when one proceeds to higher-stratum factoring the relations of higher- to lower-order factors have to be linear to fit the model.

The first restriction is not peculiar to factor analysis but occurs in all CORAN approaches, and in most such approaches, e.g., in multiple regression, the second assumption is also involved. By contrast, ANOVA designs escape these restrictions—nonlinear relations and interaction effects, from product terms, do not upset its conclusions. However, as pointed out elsewhere (Cattell, 1966), so long as auxiliary analyses show that in a given area the main linear and additive assumptions are essentially safe, factor analysis is enormously more economical of time and experimental effort in discovering and testing structural concepts. Commonly, it condenses in one multivariate experiment what could only be reached by a whole series of ANOVA experiments.

Three developments are needed for the continued effective use of factor analysis in areas where there is doubt about the fit of the simple factor analysis model: (1) an alertness to testing at each step of one's advance whether the assumptions—particularly the linear and additive— truly hold; (2) some creative modifications of the basic factor model and its extraction processes to develop a more flexible model that will better fit the scientific model when the latter clearly demands less restrictive assumptions; (3) the exploration of particular relations revealed in factor analysis by auxiliary analyses. These would be principally (a) ANOVA methods or (b) augmenting with special, limited-range factorings (p. 383) capable, as shown below, of demonstrating the existence of nonlinear, nonadditive (interactive) relations. The linearity assumption (along with homoscedasticity) is of course built into the correlation coefficient itself and so, if it is wrong, trouble begins right at the R matrix stage, before any factoring. However, there is plenty of evidence that significant nonlinearity is far less common than some critics suggest. A test (Cattell, 1955) of several thousands of correlations among very diverse objective personality and ability tests, on samples of 250 and 500 adults, yielded only 5% of correlations significantly nonlinear at the $p < .05$ level! For such testing see Kristof (1973).

Nevertheless, nonlinear relations, though less common than supposed, *do* exist, instances being the low score on ego strength found simultaneously at high and low extremes of motor perceptual rigidity (Cattell, 1935); and, though more doubtfully, of high and low anxiety with underachievement (Sarason, 1972). If the problem is one of nonlinearity between a pair of variables it is easily handled, when the relation is monotonic, by rescaling. If it is not monotonic a transformation of the form $x_t = x_o^3$ (o = observed, t = transformed) may bring x_t and other variables nearer to linearity in their relation. Unfortunately the rescaling of one, or a pair of variables, may then produce nonlinearity with many other variables. Mutual nonlinearity of variables thus has no sure solution. But when the nonlinearity is actually in the relation of a factor to a variable, the variable may be transformed with little effect on the score on the factor, which is largely derived from other variables.

Empirical results applying the normal extraction procedure and model to plasmodes with variables deliberately made nonlinear in their relations demonstrate the factor model to be quite hardy. Experiments with plasmodes made up to be curvilinear in the specification equation (below) have yielded the number and kind of factors put in, though not with high accuracy regarding loadings (examples have been circulated by Bargmann, by Saunders, and by the present writer).

The kind of relations tried were as in Eq. (13.1):

$$a_{ij} = b_{j1}F_{1i} + b_{j2}F_2^2 + b_{j3}F_3^{1/2}, \text{ etc.} \tag{13.1}$$

The quadratic relation of F_2 to a_j and the other variables did not prevent its emergence as a distinct factor.

A psychologist can also think of instances where the most probable scientific model to fit the data would require other than additive relations among the factor influences. The most likely instances are those where the level on the variable would be a product of two influences. For example, progress (as number of pages understood) through a mathematical textbook might be a product of time devoted and intelligence. In the physical sciences there are many instances, e.g., the amount of a chemical product produced being a function of temperature and concentration of reagents in the vessel; work done being a product of force and distance acted, and so on. Such a general illustrative specification equation is

$$a_{ij}^y = b_{j1}F_{1i} + b_{j2}F_{2i}^3 + b_{j(12)}F_{1i}F_{2i} + b_{j(1)}F_{1i}^2 + b_{j(2.1)}F_{1i}F_{2i}^2 \qquad (13.2)$$

with b as coefficients and the factors entering into products, as squares, and cubes would not be impossible. However, Bargmann's and Saunders' factoring of plasmodes in which factors are made to produce scores in such more complex ways showed that factoring of the correlations of such scores commonly does not lose the general factor structure. Product terms turn out to be approximated by sum terms, e.g., bF_1F_2 comes as $bF_1 + bF_2$. On the other hand the analysis does tend to break down when higher powers of factors are inserted. Even an F_x^2 term placed additional to F_x in the specification equation is likely to show up as a second, new factor, though its substantial correlation with F_x might direct the astute analyst to plot one against the other and detect the real relation.

Since anything as complex as or more complex than Eq. (13.2) is unlikely, we can proceed with reasonable confidence to assume that factor analysis will continue to yield a tolerable *first approximation* to the structure of determiners in most experimental data. From there one must proceed by segmental or bivariate methods to determine relations more precisely. Thus one would go on to estimate the factor scores and examine the relations among these scores. One would compare the relation of increments in F_1 to increments in a_j and, proceeding similarly to, say, the fourth factor in Eq. (13.2) would obtain a plot showing the relation to be with F_1^2. After all, such successive trial and error methods, hypothesizing in succession this relation and that are standard procedure in the general game of science, and we should not expect the factor method to yield immediately the final complex structure. Though the mapping of land from an airplane does not tell everything, the geographer is grateful that he does not have to depend on pedestrian methods from the beginning. Similarly, factor analysis gives an overview showing particular influences to exist. It is the task of other, more disarticulated methods, such as bivariate experiment, to shape more exactly the nature of each influence and the equation of its relation to variables. Incidentally, examples in current research of "unfinished business" through failure to follow up factor analysis with other designs are all too numerous. An important case lies in the theoretical center of psychodynamics, in the relations of dynamic structure, motivation, and vehicle factors (Cattell, Horn, Radcliffe, and Sweney, 1963; Cattell and Child, 1975), where much extraneous evidence suggests that the scientific model should accept a product rather than the additive picture initially given by factor analysis.

Some of the "tidying up" toward the true scientific model can be done within factor analytic procedures themselves, while others require ancillary steps with other designs. A first step within the model itself is, as indicated below, changing the scaling of variables suspected to have a different relation. This handles only the "y" on the left of Eq. (13.2); but, if, say, rescaling variable a to $a^{1/2}$ yields larger communality and better simple structure we are on the right track. And, as mentioned before (see Fig. 11.5) curvilinear relations due to exponents on the right might be revealed and evaluated by factorings carried out on samples of narrow range at successively higher levels on a given factor. Figure 13.1 puts this in simple form.

A frontal attack on the problem of nonlinearity, with a proposed generalized nonlinear factor analysis, has been created by McDonald

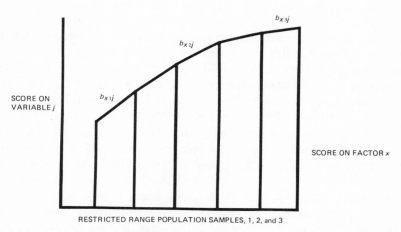

Fig. 13.1. Determining equation for a nonlinear curve from successive narrow-range factorings.

(1962). Other interesting but more oblique approaches are what has been called the "proximity analysis" by Shepard and Kruskal (1964), Kruskal (1964a), Shepard and Carroll (1966), Kruskal and Shepard (1974), as well as polynomial factor analysis (Carroll 1972) which solves for functions of the original exponential terms which have a linear relation. (It comes in the form of a component model, however.) These are complex in application and have so far handled only simple cases. They are theoretically important (see discussion in Ahmavaara and Markhanen, 1958) and hopefully will overcome the difficulties in the way of a general nonlinear factor analysis. The reader who wishes to pursue them must be referred to Gorsuch, Harman, Mulaik, and the original sources above. Meanwhile, the above discussion supports Ahmavaara and Markhanen's general conclusion that "the linearity of the model . . . (does not) restrict the accuracy and applicability" of the method, though we would modify it by substituting "general structure revealing power" for "accuracy."

13.3. Adjustments to Mixtures of Populations

The problem here is best followed initially through instances in ability and personality research, where experience is greatest. An environmental, institutional influence will produce through a common learning schedule over several variables, a personality pattern which we recognize as a *sentiment* structure, e.g., a sentiment to home. This reinforcement schedule must be differentially built into the social life of various individuals. However, the influence, e.g., that of a church, will in fact not be identical in different groups and social classes. Its uniqueness and identity as a factor will be only the degree of identity we have called a common species, with variation within the species. Similarly with genetic maturational influence. For example, though factor analysis finds a mating erg (Cattell, Horn, Radcliffe, and Sweney, 1964) the pattern of the behavioral expression tendencies would be expected, according to all we know from mammalian species, to be somewhat different in male and female, and also different in subcultures.

The strategy of factor analysis when resources are limited has generally led to factoring a large composite population sample before factoring samples of more homogeneous subgroups within it. This can lead, to dangerously ill-defined concepts though admittedly there is something to be said for asking what is present in gross terms before one enters with more refined questions. Certainly it is true historically that both the primary abilities and the primary personality factors (in the 16 P.F.) were obtained first by factoring males and females, and persons of diverse social status together. However, in almost every case where separate factorings have subsequently been done on subgroups, more uniform in learning background or genetic makeup, the resulting factors have been (a) more clear in simple structure and (b) larger in communality. As regards the latter, it is easy to see that if we mix subjects in whom the actual pattern of learning or maturation is different, the correlations will be attenuated, as by error. Indeed, it is extreme instances of such poor experiment which have caused some personality theorists, e.g., Mischel, to deny the existence of broad common factors.

Ideally, therefore, both the definition of a broad common factor and the scoring of it should be based on the better fit obtainable from factoring the most culturally and genetically homogeneous subgroup, and, when measuring a given individual, using the values for the most relevant subgroup. This implies in practical prediction work from the factor scores that the specification equations for the practical, concrete criteria should also be worked out afresh in the subgroup concerned. If we follow the psychological model to its ultimate implications—which would be that the habit pattern designated by a "common" source trait has in the last resort some uniqueness of form for every individual—then the only exact fit achieved in the factor model is when it turns to P technique. Thus the subgroup of one person leads to the closest fit—though its values cannot be used directly in a group prediction—while the small subgroup is the next best, and permits prediction by the group specification equation. There are various "applied" consequences of this fact that the discovered factor pattern is only a loose fit to whatever pattern would be produced if the influence at work acted in a truly uniform fashion (but in different degrees) on all members of the population, instead of differently in different subgroups. Further aspects of this question are taken up again in Section 15.6 on choice of the referee group in a factor analysis.

13.4. Alpha and Image Methods of Analysis

So far our discussion has concerned the aptness to the most likely scientific model of the most general factor model. In the next four sections we shall consider variations of the analysis model. They fall essentially within the general factor model as used above, but extend from rather radical modifications to others which in some cases come close to being only variations in extraction procedure.

The first two variants we shall consider—alpha factor analysis and image analysis—were developed primarily in terms of claims for some psychometric advantages, rather than (as in real-base factoring, for example) of any better fit to a scientific model.

The statistical aim of alpha analysis is to raise the homogeneity coefficients of the factor score estimates, and it derives its name from its relation to the alpha coefficient for estimating homogeneity of items within a test scale. Since, as we have seen, one estimate of the validity of a factor score is the square root of the correlation of the two-part estimates, i.e., two estimates each from one-half of the variables, this may also be regarded as maximizing the validity, assuming the criterion of validity is the factor as defined by the variables. In factor measures, incidentally, the equation relating the equivalence coefficient (typically form A with form B of battery for factor x) to the validity coefficient is $r_{xx} = r_{xf} r_{xf}$, where x is the factor score on half of the variables, whence $r_{xf} = r_{xx}^{1/2}$ though r_{xf} will, as validity, then have to be corrected back by Spearman–Brown assumptions (see Cattell & Radcliffe, 1962), to full battery length, i.e., to $r_{2x \cdot f}$. One should note that this assumes the two parts are equally valid and that they share no other common factor than the one in question.

The procedure as proposed by Kaiser and Caffrey (1965) is to calculate a new correlation matrix among the variables, which we will write $\bar{R}_v$, derived as follows:

$$\bar{R}_v = D_h^{-1} R_v D_h^{-1} + I_v \qquad (13.3)$$

Here $\bar{R}_v$ is the correlation matrix with the diagonal left blank, i.e., it has only 0's therein. D_h^{-1} is a diagonal matrix of the inverses of the square roots of the communalities—the h being obtained by any of the usual estimate methods.

The result is that $\bar{R}_v$ is a matrix of "inflated" correlations relative to R_v, the values becoming *bigger* for those variables which have lowest initial communality. Statistically, the alpha factoring procedure has the disadvantage also of assuming that the correlations among the variables are those true for the population rather than the sample. The $\bar{R}_v$ is then factored in the usual way, the unities inserted in the diagonal by I_v in Eq. (13.3) being iterated to communalities in the process. The resulting component matrix is then "deflated" again by multiplying each row by the communalities whose reciprocals were originally used in Eq. (13.3) to "inflate" the R_v, i.e.,

$$\bar{V}_0 = D_h V_0 \qquad (13.4)$$

It is argued that this $\bar{V}_0$ may be regarded as a more reliable basis of factor estimation than the V_0 inasmuch as greater weight has been given in this last calculation to variables of lesser uniqueness or error (though originally these "poorer" variables were inflated before being deflated). One effect, as Gorsuch notes, is that the first factor has the highest homogeneity–reliability and the later ones less, but this difference is lost in subsequent rotation to meaningful factors. As indicated earlier, the rationale which connects alpha factoring with the alpha coefficient of homogeneity and with the K–G index for number of factors is not as logical as it initially seems. Finkbeiner has also objected that it treats the factor as a function of the variables rather than the converse, which clashes with the model for which both Tucker and the present writer have argued. Bentler (1968) presents a variant more closely adapted to the common factor model and to the emphasis on the priority of the factor just mentioned.

Alpha factoring is not a modification of the common factor model (as image analysis is) but simply a way of implementing it through a different weighting in the computational procedure. It aims at a weighted least-squares fit of factors to variables, where the weighting differs from the usual principal axis procedure in weighting the residuals as the square root of the communalities of the given variables. This boosting of the common factor contribution has been questioned by leading psychometrists as making little sense, and the heightening of the alpha consistency of the factor scores regarded as an illusory "boot-strap" effect (see also criticisms by McDonald, 1970). But it is an interesting variation, though, as will be seen in comparisons of results later, the difference it produces in loadings from the principal axes is for practical purposes quite trivial.

Image analysis, due to Guttman (1953), is a more fundamental departure statistically, though its results likewise depart only trivially from an ordinary principal axis solution. It begins with the demonstration by Roff (1935) and Guttman (1953) that the multiple correlation of a variable in a correlation matrix with all the other variables is *a lower bound for the possible communality* of the variable found in the subsequent factor analysis. One should note—especially when using this in other connections as an estimate of the communality—that it is a *conservative* lower bound, reliably replicable, but not corresponding to the best, i.e., most likely, estimate of the communality from all sources of evidence.

Using the value for the communality of each variable, image analysis constructs, with the multiple R, what may well be described as an *image* of each variable as seen in the other variables. This image variance represents as much of the original variable as can be constructed by the multiple R, from what it shares with other variables in the R_v matrix. It then in effect constructs a new matrix (in covariance rather than correlation terms) which expresses the relations among these images. Thus, it is *not* a factoring of the ordinary R matrix (or covariance R) as in the above variants.

Since the image score for a variable will not reach the actual score (by an amount roughly equal to the latter's unique component) we can speak of an *anti-image* as the difference. The image score, which we will represent by $\hat{\underline{z}}_j$ to distinguish from that estimated from factors by the usual specification equation, is estimated from variables by multiple regression weights b_v (not to be confused with the behavioral index loadings, b_f in a factor equation). When this is applied to standard scores on the other variables, thus:

$$\hat{\underline{z}}_{ji} = \sum^{p=(n-1)} b_{jp}\, z_{pi} \tag{13.5}$$

where p is any variable other than the j to be estimated, it gives an estimate of j.

The anti-image for j, $\hat{\underline{U}}_j$, will therefore be

$$\hat{\underline{U}}_j = z_j - \hat{\underline{z}} \tag{13.6}$$

(It wears a hat because, since $\hat{\underline{z}}$ is an estimate, it also is an estimate.)

Image analysis now proceeds to calculate not simply a correlation matrix but, as stated, a covariance matrix from these image scores. The properties of covariance matrices will be more fully discussed under real-base factor analysis below, but it may be pointed out that such a matrix consists of the cross products of the raw-score deviations, not the deviations in standardized scores, divided by N the number of cases. The covariance c_{ab} is thus

$$c_{ab} = \frac{\Sigma xy}{N} \tag{13.7a}$$

where x and y are raw-score deviations from the raw-score means. Note that, by contrast,

$$r_{ab} = \frac{\Sigma xy}{(\Sigma x^2\, \Sigma y^2)^{1/2}} = \frac{\Sigma xy}{N\sigma_x \sigma_y} \tag{13.7b}$$

(where x and y are again raw-score deviations from their means). We have seen it is simple to shift from a correlation to a covariance matrix by pre- and postmultiplying the former by a D_σ matrix of σ_x, σ_y, ... σ_z entries thus:

$$C_v = D_\sigma R_v D_\sigma \tag{13.8a}$$

but here we do the reverse with the special C_v above:

$$R_v = D_\sigma^{-1} C_v D_\sigma^{-1} \tag{13.8b}$$

The image covariance matrix G_v below could be got by calculating all the image scores by Eq. (13.6) above and computing covariance relations between them. But this labor of image score estimation can be avoided by an algebraic equivalent:

$$C_v = R_v + DR_v^{-1}D - 2D \tag{13.9}$$

where the D are diagonal matrices containing the reciprocals of the diagonal elements in R_v^{-1}.

Below, in discussing covariance and real-base factoring, we shall see how there are conceptual and statistical equivalents at each stage in covariance factoring for those in the corresponding correlation factoring. Thus the V_{co} from the covariance factoring can be brought to the V_0 from the equivalent correlation by pre- and postmultiplying by a diagonal matrix of the inverses of the standard deviations, i.e., reversing what is done in Eq. (13.8a) above. Most psychologists using image analysis seem to rotate from the V_{co} as it is, but, as discussed elsewhere there are difficulties, e.g., over hyperplane widths or rotating to a usual PROCRUSTES target in using covariance loadings, so reversion to ordinary loadings is appropriate in rotating and at the end of the process. A practically useful, detailed expansion of the above essential description of process is provided by Rummel (1970), and other good descriptions occur in Gorsuch (1974), Harman (1976), Mulaik (1972), and others.

Since there is still dispute as to whether image analysis brings more gains than losses relative to a straightforward principal axis extraction it behooves us to consider in a broader view what is happening. In one sense image analysis could strictly be said to lie outside the common factor model and involve more than a difference of process. It implies a different basis for the linear model of predicting variables from factors; for whereas factors in ordinary factoring are the common part of the variables in image analysis they are the common part of the images. And we must not forget that the *specific factors* in all the variables have entered into the estimation of the images, so that the new factoring *is no longer in the old common factor space but in an "ersatz" space built up with admixture of the specifics*. This deserves italicizing because it is often said that image analysis "operates only in common factor space."

If the number of variables is large the effect of any single specific is small, and Kaiser (1963) and Harris (1967) have defended image analysis on the grounds that in the limiting case of an infinite sample of variables the image space and the common factor space become identical. Harris adds a second and different definition of an image, as the "perfect image" predictable from some "ideal population," and calls the ordinary mundane practice of image analysis "partial image analysis." Metaphorically, one may say that any actual image factoring uses a mirror too restricted for seeing the real object, and distorts additionally by the mirror's curvature the input of specific variances.

Empirical evidence is not yet organized, some users saying that they get cleaner simple structure in image analysis and others that *SS* becomes more obscure. The present writer is among those who question it on theoretical grounds. As we concluded in discussing the scientific model there is every reason to believe that some of the unique variance in a variable will not dissolve in time into as yet unknown common factors but will remain as obstinate evidence of a real specific trait. Here one recognizes truths in the reflexological learning theorist's position that some conditionings are quite specific. One can point to quite a host of particular skills (in R-technique specifics), such as arise from historically unique conditioning, where the specificity arises from a frequency of reinforcement that is found absolutely nowhere else. And in O and S techniques one can point still more readily to properties absolutely specific to occasions, etc. There will always be specifics contaminating the image common space, and no actual image analysis in real life is anything more than what Harris dubbed "partial image analysis." The image scores, that are considered more "ideal" representations of a variable, in fact vary with every change of company in the matrix, and the common space remains—more obviously so in small factor studies—a farrago of the real common space and the specificities of all variables.

Reverting to practical goals, one must recognize that though the common factor model, as in a principal axes solution, clearly theoretically separates out the common factor space, yet the actual *estimates of factor scores* from variables are as vulnerable to contamination by specifics as is the common space of image analysis, but at least the general structural analysis is clear. Finally, at the practical level, as Harris (1963) and others have pointed out, the image solution is very close to the solution by ordinary factoring using the Guttman–Roff lower-bound (multiple R) as the communality estimate, and, one may add, it frequently differs from standard factoring with iterated communalities only in the second or third decimal places in the V_0 matrix values.

13.5. Canonical Correlation and the Factor Concept

No account of factor analysis and its relatives would be complete without an account of canonical correlation and other treatments of regression analysis in relation to groups of variables. Incidentally, canonical correlation should not be confused with canonical factor analysis which is treated in the next section as a stage on the way to maximum likelihood factoring. Canonical correlation could be studied in Section 15 below when we glance retrospectively at the general relation of factor analysis to regression analysis, but it has sufficiently specific resemblances to factoring to be more aptly introduced here.

Canonical regression may be thought of as a two-sided multiple regression. In ordinary multiple regression we take a criterion variable and a bunch of predictor variables and calculate the maximum prediction—the multiple correlation—of the former from the latter. Psychologists, economists, and others have played with countless problems of this form in which a desired criterion estimate is to be maximally predicted from a set of variables. They have also played with what Horst, following Hotelling, called "the most predictable criterion" selected as a weighted composite of a whole set of criteria. We shall give no discussion to the latter here, since even as a factorially blind "applied" procedure it has no great appeal in our field. For it tells the psychometrist what arbitrary combination of criteria he can best predict, as a "makeshift," not the combination he would most like to predict. If there are several criteria and he knows what combination he would most like to predict, he can, indeed, proceed to estimate that as a single criterion and find the possible prediction thereon from the predictors by the ordinary multiple correlation.

But in canonical regression the aim is to find a weighted linear derivative of a set of variables $u = a_1, a_2, a_3, \ldots, k$ that will give a higher correlation with a weighted linear composite of another set $v = b_1, b_2, b_3, \ldots,$ p than will any other pair of weightings for the two sets of variables. The meaning of predictor and predicted vanishes here, except through ulterior purposes, since the relation is mutual and symmetrical. This is readily seen by turning to Fig. 13.4(c). Some have ventured to call what appears in Fig. 13.4(c) a "double regression analysis," since the weighted composite u_1 is followed by another weighted composite u_2 which predicts as much more as is possible of what u_1 [equivalent to the left-hand term in Eq. (13.10)] leaves unpredicted in the v's. Also we have in Fig. 13.4(c) direct predictive relations between the derivations themselves, u_1, u_2, etc., being mutually orthogonal (as v_1, v_2, etc., are) but maximally correlated u_1 to v_1, and so on.

The solution to the problem was given by Hotelling (1933), and since full reasoning and calculations have been set out excellently by Tatsuoka (1969b, 1971b), Rao (1965), Van de Geer (1971), Harris (1961), Horst (1962, 1965), Harman (1976), and others, we shall pause only to give an outline. The desired relation is that ($a, b, \ldots, k$, being variable scores and u's weights)

$$u_a a + u_b b + \cdots + u_k k = v_m m + v_n n + \cdots + v_p p \qquad (13.10)$$

shall give maximum mutual prediction.

The correlation between the two will be

$$r_{uv} = \frac{u C_{uv} v}{(u^T C_{uu} u \cdot v^T C_{vv} v)^{1/2}} \qquad (13.11)$$

where C_{uv} is a variance–covariance matrix of the $a, b, \ldots k$ values, i.e., cross products of score deviations, divided by N, and correspondingly for the other C. Introducing Lagrange multipliers and taking partial derivates

Table 13.1. Illustration of Canonical Variates[a]

U set; personality factors as variables	U Variates				V set; classroom behaviors as variables	V Variates			
	U_1	U_2	U_3	U_4		V_1	V_2	V_3	V_4
C, ego strength	.5	−.5	.3	−.2	Fighting	.7	−.2	.2	0
E, dominance	.4	.1	−.1	.1	Competitiveness	.4	−.1	−.1	.1
G, super ego	−.4	−.2	.2	.1	Anxiety	−.3	.6	.3	−.2
Q_4, ergic tension	−.6	.6	−.2	−.2	Tardiness	0	.4	−.1	.1

[a] The figures in this illustration are chosen purely for discussion purposes and do not arise from actual data or actual calculation.

results ultimately in two matrices of the form shown, for four variables, in Table 13.1.

The columns here are called *canonical variates*, and, as in principal components, there could ultimately be as many of them as there are common variables, i.e., in this case the number in the smaller set, if the variables are unequal in number. Variates u and v are the weights which give maximum possible prediction from one set to the other. The two sets can conveniently be set out in a single column canonical variate as in each of three covariates which follow:

$$
\begin{array}{ccc}
\text{I} & \text{II} & \text{III} \\
R_{c1} & R_{c2} & R_{c3} \\
\end{array}
$$

u weights $\begin{cases} x & x & x \\ x & x & x \\ x & x & x \\ x & x & x \\ x & x & x \end{cases}$

v weights $\begin{cases} x & x & x \\ x & x & x \\ x & x & x \\ x & x & x \\ x & x & x \\ x & x & x \\ x & x & x \end{cases}$

where the x represent various weights. The R_c at the top represent the value of the multiple correlation between the weighted u and the weighted v for the given canonical variate. Bartlett gives significance tests for such canonical correlations (1941).

Just as in principal components, when the first is taken out there remains a residual, which can lead to a second canonical variate, II above, the R_c of which will be smaller than R_{c1}. A very clear illustration of the working procedure is given in Lawlis and Chatfield (1974). It will be noted that the canonical variates therefore emerge as mutually orthogonal, which means among other things that the u values in I will give no significant prediction on either the u or the v values of II, and so on. However, the sum total of prediction from the u variables to the v variables will be increased as each successive new canonical variate is employed. Usually a canonical correlation analysis shows trivial contributions after the first two or three variates, and, though the R_c's may be just significant beyond that number, few researchers use them. Rozeboom's (1965) observations are valuable here.

The methodological doubts a thoughtful psychometrist must raise about canonical correlation are both statistical and conceptual, and they have some resemblance to those about discriminant functions. (Incidentally, Tatsuoka (1970, 1974) brings out the similarity of the statistical procedures of canonical variates and discriminant functions.) Statistically it is found that values from separate random split samples of a larger sample are apt to vary quite a good deal. Frequently, the first canonical is practically identical, the second a bit different, and the third or fourth barely recognizable as the same. (See Barukowski and Stevens, 1975.) Some users have claimed that even the order of variables taken affects values within one variate, but, if so it is a rounding effect for strictly this should not occur. The more important problem is conceptual: can any meaning other than a useful weighting scheme for a particular experimental combination of variables in a particular situation be ascribed to the variates? The answer of the present writer to this is "no"; but as usual in *post hoc* reasoning some psychologists claim to see permanent conceptual value, beyond computational value, in canonicals. It is reasonably clear, in, say, the first variate in Table 13.1, that individuals high in ego strength and dominance and low in super ego and ergic tension will tend to show in the classroom more fighting and competitiveness and less anxiety. Proceeding to the second, to the u_2 and v_2 weights, we see a relation of high-ergic tension and low-ego strength to anxiety and tardiness. However, the reasons for these connections in terms of personality concepts are seen with more clarity in the separate specification equations for, say, fighting, tardiness, etc. Tatsuoka coined the insightful term "double-barrelled principal components analysis" for canonical correlation, which both describes the statistical essentials and reminds us that as they first emerge the variates have no more freedom from the dictates of a particular experiment and set of variables than have principal components.

With these considerations, the question arises, as in multiple discriminant functions, whether it is appropriate to apply rotations to the canonical variates. The answer is that, mathematically, rotation to either orthogonal or oblique transformations is meaningful; but there is yet no general principle such as simple structure to guide to a unique position. (The total mutual prediction of course remains invariant, as in factor rotation.) What would probably be useful, when one has a certain set of u or v that are more practicable to measure or more relevant as a social criterion, is rotation to bring these to highest weights. An increase of meaning would also be gained if one used factor measures rather than variable measures (as in fact in Table 13.1) since one would at least not be operating with groupings of variables as such that are "accidental." Norman Cliff has recently developed useful suggestions on this "factor" use of canonicals.

When using factors in this context, one approaches a question frequently raised, for example, in the educational field as to how much of a certain criterion domain can be predicted from a given set of predictor factors within a certain psychological domain. For example, Cattell and Butcher (1968) predicted school achievement in a total of subject areas (English, arithmetic, etc.) from (a) ability tests alone, (b) personality tests alone, and (c) dynamic motivation tests alone. They then examined prediction from the three two-modality combinations possible from these three, and finally from the combination of all three. The question "What is the relative effectiveness of these three modalities?" is not easily answered since, owing to correlation among them, their joint variance contributions are not a simple sum of their single variance contributions. When two contributors thus overlap, one wonders whether to say that all the overlap belongs to one or whether to divide it equally (in our ignorance) between the two (since part-correlations would be one-sided). One sort of answer to this question of how much is common to two sets of variables is given by interbattery factor analysis (p. 359), where we look for the factors common to two sets of variables (p. 360), though this may deliver the psychological factors unrecognizably distorted. Another is through the present canonical approach, for applying canonical correlation to such a set of mutually correlated primary factors will give an answer in terms of uncorrelated components with R_c (mutual variance prediction) for each, which is one kind of answer to this problem. A third possibility is commonality analysis (Kerlinger and Pedhazur, 1973) which is described under general regression relations in Section 15 of this Chapter and is illustrated as just mentioned by the Cattell and Butcher (1968) methods.

In leaving canonical correlation one should note that if the domain of research is one of meaningful factors on the one hand and criterion variables on the other, the ordinary V_{fp} and V_{fe} of the factor analyst come close to doing the same job. The addition (squaring) of rows on the former will tell us what the set of factors contribute to the chosen set of "criterion" variables as a whole, and the addition of columns of the latter will reciprocally tell us the amount of estimation of a particular set of factors from variables. However, this will not give the sets maximally mutually predictive, as does the "double-barrelled" canonical approach.

13.6. Canonical and Maximum Likelihood Factor Extraction, Ordinary and Proofing Forms

The rationale for the basic expression "canonical" factor analysis, as for the above "canonical" correlation is that in both one is seeking weights on sets of variables. In the case of canonical factoring, it is k factors so chosen as to give maximum prediction of a set of n variables: the experimental variables (Rao, 1965). To be precise, this is both maximum prediction from the factors of the *scores* on the variables and of the *correlation coefficients* in the R matrix obtained from them. Maximum likelihood factoring will here be considered along with canonical factoring (Rao, 1965) because it grows out of (Lawley, 1940, 1943) the latter and for most uses today supersedes it. It is instructionally appropriate, therefore, to take a brief glance at canonical factoring before proceeding to the derived method which the psychologist is today more likely to use.

There is a sense in which canonical factoring is representative of several submodels just as components with communalities are of several variants. Indeed, it is helpful in a broad grasp of models in the field to think of a dichotomy into the Lawley–Rao model (Lawley, 1940; Rao, 1952, 1965) and the Thurstone model (1937)—the latter having roots in Hotelling (1933) and Truman Kelley (1928). Image analysis and alpha factoring may be thought of as variants on these. In canonical one starts not with the ordinary R matrix as it stands but with such a matrix "blown up" by increasing each relationship according to the communality of the two variables involved. This is done by multiplying by the inverse of the uniqueness, u, not by multiplying by the communality, h, directly. Thus we seek a V_0 such that

$$V_0 V_0^T = U^{-1}(R_v - U^2)U^{-1} \qquad (13.12)$$

where U is a diagonal matrix of square roots of unique variances, initially obtained by whatever approximating method one prefers for obtaining h^2 values (e.g., Guttman's lower bound; Burt's method; etc.). The correlation matrix R_v in the middle of Eq. (13.12) is, of course, the unreduced R matrix, but the total expression in parentheses is the reduced matrix, i.e., with communalities not unities in the diagonal. This formula is the basis both for canonical and maximum likelihood factoring, the latter having statistical estimation procedures added.

In comparing earlier the simple centroid method, which sums the columns of correlations in an unweighted fashion, with the principal axis or weighted method, which returns after each simple summation to weight the correlations according to the amount of common factor variance the first "pass" has shown each variable to have, we have pointed out one superiority of the latter. It is that the principal axis more quickly extracts the common variance (p. 29), i.e., it is better at restoring the correlation matrix from anything short of n factors. The canonical method goes one further in weighting the variables from the beginning with the totality of communality that the variable is estimated as likely to have. This is the function of the U^{-1} terms in Eq. (13.12). (It will be noted that this is the opposite emphasis from alpha factor analysis and, since its rationale is rather compelling, it raises again some doubts on the ultimate value of the latter.) Canonical factoring still does not go as far as maximum likelihood because it rests on a choice of communalities more arbitrary and fallible and uniterated than in maximum likelihood. It may be regarded as a "conditional" maximum likelihood solution, conditional on the choice of h^2's (as U^{-1}) being optimal—which it cannot be except by accident. Nevertheless, like maximum likelihood, canonical factoring, like a number of other factoring procedures, typically avails itself of iterative convergence. Its end result is a linear set of components from the observed variables giving the highest possible correlation with a set of linear components correspondingly derived from the other set of "variables," which are the common factors. But its iteration gives only a "maximum likelihood" conditional on the first choice of U^2.

Canonical correlation and factor analysis may be seen as having their greatest importance historically, as grandparent and parent of the maximum likelihood method. Today, canonical factoring is little used where maximum likelihood programs are available. Maximum likelihood may be regarded as canonical correlation in which the iterative procedure is pursued to further and more refined goals. It starts with the same formula as in Eq. (13.12) and with a guessed communality. It proceeds until a test of the correlation matrix residual shows it to be insignificant (Lawley's chi-square test, 1940). That is to say, it solves uniquely for U^2 for a fixed initial choice of number of factors. When the best possible fit is obtained (adjusting both F and U^2) for k factors, it tests for significance of fit to $(k + 1)$ and proceeds with new U^2 estimates, and so on iteratively with new k and new U cycles until an acceptable fit is obtained.

Thus the vital difference from the "standard process" we have described for ordinary principal axis work is that it *plays both with variations in the number of factors and in h^2 (actually in U) until a combination is reached which gives the best possible fit* (the "most likely" as defined below). As the reader realizes the standard method fixes the number of factors, e.g., by scree or K–G or other means, and then adjusts h^2 exactly to fit that number. Digman has aptly called maximum likelihood a

"wheels within wheels" action; on account of this alternation of adjustment of U^2 and of the number of factors. As to the starting point, the initial estimated U^2 first indicates a probable number of factors, iteration
against that number gives a new U^2, and then, as indicated above, the new
U^2 probably leads to a slightly different number of factors when the residuals are again tested. There is thus an optimization of the logarithmic
likelihood function simultaneously on U^2 and the number of factors.

Improvements to make maximum likelihood a less costly calculation
task have been introduced by Jöreskog (1967a, 1967b), Jöreskog and
Lawley (1943), Clarke (1970), and others, leading to its wider use. The
essential process in the search for maximizing the criterion of fit is a differential calculus employed on the curves of U values in the multidimensional variable-factor space. Newer methods permit anticipation of where
the maxima will be, without the full labor of curve-continuous calculation. The upshot is that the maximum likelihood method finishes with
that combination of number of factors and of communalities, as $(1 - U^2)$
values, which gives the best possible reproduction of the reduced correlation matrix, i.e., of $U^{-1}(R - U^2)U^{-1}$ as well as the best reproduction of
variable scores from factor scores. A recent method operates wholly on U
adjustments. Hartley (1958) has improved the method by making it applicable to incomplete data.

Since we mentioned similarity to the minres method, one should
note that minres gives the best possible fit to R_v, in the ordinary least
squares sense (diagonal not estimated) not in that of maximum likelihood. The minres ("*min*imized *res*idual") extraction method, conceived
by Thurstone (1955) (while awaiting a plane in Stockholm airport!) and
advanced by Harman, has been given relatively little emphasis here because
it is not an instance of the major differences of extraction principle we
have needed to clarify—though it is a useful practical procedure. Its novelty, within the ordinary factor model and iterative procedures lies in bypassing the various handlings of communality estimation. It takes the R_v
matrix with nothing in the diagonal and aims to get such a V_0 as in the
product $V_0 V^T = R$ will give the best possible fit to the off-diagonal values,
i.e., the given correlations of the variables. Although in Table 13.2 it is by
its nature placed in the second row and first column, several empirical
studies show that it comes closer than some other extraction methods to
the numerical solution reached by maximum likelihood. On the other
hand, Nosal (1977) points out that the method is a comparative failure if
the starting point of the iterations is not well chosen.

In maximum likelihood itself the statement that the method gives
the "best possible" reproduction of $U^{-1}(R - U^2)U^{-1}$ requires some
added precision. The expression "maximum likelihood" arises from the
fact that reproduced R_v is technically to be considered as the population

value of which the given R_v is the most likely sampling value. To operate anywhere with the maximum likelihood principle, one needs to know the form of the probability distribution, i.e., of samples from the population. One can then calculate the likelihood of some real value of the population parameter producing the empirically observed value in the sample. It is in this sense that we have just said that the R_v calculated from the given factor solution is required to be "maximally likely." In deciding when the fit to the R_v matrix is best, the distribution is compared with what is called a *Wishart distribution of covariances* which assumes normal distribution and multivariate surface normality among the variables themselves.

Fuller and Hemmerle (1966) have shown that with samples of 200 or more it seems unnecessary to remove variables departing from normal distribution. As stated at each extra factor admitted to the system the fit of the reproduced R_v matrix to the original R_v matrix improves up to the point where the number of factors and the associated communalities give the best fit, and this index of fit is therefore best plotted and watched as it climbs to a plateau. Thereafter, the addition of any further factors gives usually neither an improvement nor a falling off, and this "minimum" number for reaching a best fit is taken as the required number.

It has been pointed out at the beginning that the whole extraction procedure is carried out in a matrix "inflated" by the multiplications by U^{-1}. So the last step, once convergence is reached, is to "deflate" it again by multiplying each row of the V_0 by the final U value reached. Thereupon one has the final V_0 ready for whatever rotational procedures are called for in the given experiment.

The maximum likelihood method has gone through a series of developments, especially in programming time-saving algorithms culminating particularly in the work of Jöreskog and coworkers, which finally brought the process into a practicable (but still rather more costly) alternative to the standard principal axis method as described here. The developments can be followed through Lawley (1940, original), Lawley and Maxwell (1963), Rao (1952, 1965, developed canonical form), Harman (1976), Jöreskog (1967a,b), Jöreskog and Lawley (1968), and Jennrich and Robinson (1969). Excellent practical accounts are available in Mulaik (1972), Lawley and Maxwell (1963), and Gorsuch (1974) which spell out procedures in detail as also does Harman (1976).

The general superiority of the maximum likelihood to other procedures rests on its being a statistical rather than a purely psychometric method as most other communality and factor number estimations are, i.e., it does not assume that the given study is no other than the population value, but makes an inference from the given sample to real population parameters. However, this extra sensitivity entails some disadvantages. First, the estimation is based on large not small sample theory and is perhaps more correctly used, therefore, with 80 or more (better 200 or more) cases. Second, as Gorsuch (1974, p. 114) says "Maximum likelihood estimates may occasionally lead to biased solutions." However, since we have to compare with other available methods it must be said that all do! Third, with large samples, such as one may wish to use on general grounds, e.g., for sharpness of hyperplanes, it is likely to count relatively trivial factors as significant. Some subjectivity therefore remains in the experimenter's decision that he is not interested in small factors that show up evanescently only in, say, samples of 500 or more. Fourth, it is costly of computer time. Against these disadvantages must be balanced the production of factor scores that correlate somewhat higher with the true factor.

To bring practice into perspective, the present writer should therefore report that from his experience with samples of N around 200–300 he has found such good agreement between the standard principal axis (with scree) factor procedure and maximum likelihood that following the more elaborate procedure may well ordinarily by dispensed with. But where issues are in keen debate and greatest exactness is called for, it is certainly the preferred method—or, at the least, an indispensable checking method. At its best maximum likelihood is—barring, possibly the least-squares methods yet to be described—certainly the method of choice for the most refined factoring.

In conclusion, it seems desirable to focus a little more closely on the decision of when to stop and accept the numerical indication of the maximum likelihood function. This function of goodness of fit we have seen rises to a plateau, at k factors (the real number for that sample size), and then stays at the plateau. While this is a probability statement to satisfy one that the plateau is reached, one may occasionally, with a borderline probability, wonder whether just one more factor could be taken "for safety." Tucker and Lewis (1973), aware of some uncertainty in fixing this point and concerned about reliability, might understate the contribution of the method a little when they say that it provides "a useful statistic for rejecting overly-simple factor" solutions. Indeed, this it usefully does, for we can always be certain that anything obviously short of the plateau is wrong; but there is also a risk of going too far. One could debate "under" versus "over" estimation of k in maximum likelihood, since over-estimation may reduce replicability of factors and trueness of score estimates; but, in rotation, here as in any method, the failure to include a substantial factor that should be there disturbs the patterns more than adding an extra factor which is in any case trivial in size. Incidentally, several well-known personality structure theories and factor namings, e.g., of Eysenck, Howarth and Browne, and probably many others (even Digman's 8–11 personality factors in a domain which commonly goes to 20; Cattell,

1973) are almost certainly decidedly short of the true number by maximum likelihood standards, as well as by the scree test (Cattell, 1973), which, as stated below, tends to agree well with maximum likelihood with samples in the 200–400 range. On present empirical evidence, with known plasmodes, the only cautions for use of maximum likelihood are in Hakstian, Rogers, and Cattell's (1978) finding of underestimations of number of factors when communalities are low and overestimations (at least as regards factors operating in all subjects) when N's are very large.

In contrasting exploratory and hypothesis-testing factor analysis there has briefly been mentioned above the use of *confirmatory maximum likelihood* (as a more substantial alternative to a procrustean shift to the hypothesis, followed by congruence coefficients of matching). In English a better translation of Jöreskog's Swedish term would be *proofing factor analysis*, for it is by no means always confirmatory, nor should it be. In this book we shall indicate this by using "proofing" though with the author's term alongside. Among the comparatively recent developments in convenience of algorithms we shall therefore briefly describe the work of Jöreskog (1966), Gruvaeus (1970), and Jöreskog, Gruvaeus, and Van Thillo (1970) in developing confirmatory procedures.

The essence of proofing or confirmatory factor analysis is that it sets up a V_{fp} and an R_f (or alternatively a V_{fp} and an H^2, since the correlation of factors for a given V_{fp} determines the H^2 vector) and then asks if the residual from the given R_v, after extracting those factors, is significant. If it is then the null hypothesis that the factors as defined in the hypothesis matrix can fully account for the R_v has to be abandoned.

However, the method also has the interesting possibility that one may only partially define the target and ask if those fewer values can be matched, the blank values being free to reach whatever quantities would permit a fit on the more limited set of prescribed values, if that is possible. Such a design is actually a compromise between an hypothesis testing and an exploratory factor analysis.

The reader realizing that with rotation an infinite series of V_{fp}'s are equivalents will recognize that unless restrictions are added in terms of an R_f, and other ways, no very precise test of a hypothesis has been made. Proofing (confirmatory) maximum likelihood permits a variety of precisions of hypothesis statement to be tested according to the four combinations of what are described as (a) restricted vs. unrestricted and (b) unique vs. nonunique conditions. It is even possible not to put values for two loadings or two factor correlations but only to stipulate that they shall be equal. For a restricted solution the number of values fixed in the target must exceed the square of the number of factors. To be unique, i.e., to constitute a unique rotational position not to be mistaken for any other, it is necessary to distribute the defined values so that no factors are left free to rotate. As Bentler (1976) summarizes "a number [of values] smaller than the maximum number possible for unique identification are free parameters, and the remainder up to the maximum are some combination of fixed and constrained parameters." For most ordinary hypothesis testing clearly the most satisfactory procedure is to have a complete target both as to the V_{fp} and R_f. However, although a theory, or a previous result, is clear as to its salients, it may not be so clear as to which variables are just in the hyperplane and which are significantly out of it; so that some of these might be left blank (if need be down to the limit of the square of the number of factors). A recent development with considerable promise is Jöreskog's (1971) application to simultaneous hypothesis testing with the same variables over several population samples.

Clearly, proofing maximum likelihood has a great future in factor analytic research, particularly as the structure in various areas of ability and personality trait and state concepts gets filled in with reasonably sound hypothetical concepts from pure exploratory research. One may well fear, after seeing certain liberties taken with PROCRUSTES, that it may suffer similar abuse, e.g., by testing the targetting of one experimental result to another before the first has been well enough substantiated by two independent experiments independently rotated to simple structure. However, at present it is unlikely to be used in large studies because its demands on computer storage are very great. Gorsuch (1974), who presents an illustration of the procedures needed above, estimates that even the larger present-day computers are likely to handle typical studies easily only up to about 40 variables. For further detail see also Mulaik (1972).

13.7. Other Developments and Comparison of Extraction Models

Two new developments of recent times, mainly by those interested in maximally exact extraction methods as in maximum likelihood, are the weighted and unweighted least-squares fit methods [Jöreskog (1977), Jöreskog and Goldberger (1972)]. They operate within the ordinary factor model, as Table 13.2 shows, and aim at a fit to the R_v matrix based on the familiar least-square criterion rather than maximum likelihood. Thus, in general intention, there is no difference from older iterative approaches but only some sharpening of the criterion. Since little empirical research has yet proceeded from them, it is too early to comment on their effectiveness relative to effort.

A straight comparison of the efficiency, calculation cost, precision, and mutual similarity of results from different extraction methods is not possible, because the extraction methods are simultaneously involved with different models, such as components, factors, the image model, and real base factoring. We have asked the reader to recognize the difference between methods of extraction, such as centroid, principal axes, minres, alpha factoring, and maximum likelihood, and these scientific models *per se*. As pointed out, there is some "interaction" of the two, but in Table 13.2 we have ventured to set out a two-way diagram permitting a two-way *"analysis model* by *extraction method"* description.

The nature of the models can be briefly, but not fully, indicated by listing the properties of the R matrices with which they begin. Incidentally the symbol R on this left column always means the matrix with zeros down the diagonal whereas in the body of the table [in order to conform with and continue the use in Eqs. (13.3) through (13.12) above and elsewhere], R means an unreduced matrix, with ones down the diagonal. It should be noted that for a detailed survey one would need to repeat the table with a doubling of the rows for (a) solutions left at orthogonal unrotated factors and (b) solutions rotated to oblique simple structure. But this last step can be left as "understood."

It will be seen that only about half the possible $4 \times 6 = 24$ combinations in Table 13.2 have yet been tried or are viable. Actually, most of the empty cells represent impossibly inconsistent designs or those without obvious use or advantage. For example, in cell row 1, column 3 would a model carrying alpha to n factors (as many factors as variables, in a component model) make much sense, i.e., without having to use communalities? It is possible, since n factors can be taken out without going to unities in the diagonal. But what does it mean? A similar question can be asked at row 1, column 4 and the answer is that even with n factors it is really a factor model if $h^2 < 1$. These blanks we shall leave to speculative exercises; and we shall also leave out the row for a covariance analysis with the equivalents of unity in the diagonals, i.e., a components covariance extraction. Until the next section the fourth row and sixth column cell— real-base factor analysis—will be left unexplained, as there is need for special concentration on what is more than just a covariance factoring. For real-base factoring has features transcending simple representation as a matrix model and an extraction method, but it must nevertheless be inserted here as a combination of model and extraction method.

In the interests of a complete view of possible modern procedures, the various methods of extraction have been described in essentials above and are now summarized in Table 13.2. But the psychological researcher as such, in listening to the siren songs of enthusiasts for this and that particular analysis model (which have been loud in this decade), should retain his perspective by recognizing that the numerical outcomes (in the V_0) among the various extraction methods in the factor model row will differ in numerical end-result very little. The large divergences sometimes

Table 13.2. Perspective on Combinations of Models and of Extraction Methods (Weightings and Iterations)[a]

Models (in terms of reproduction relation to correlation matrix)	Extraction method treatments					
	Unweighted sums of r's as in centroid	Weighted sums of r's, in an iterative process, as in principal axis extraction	Weighted r's by inverse of the communalities	Weighted r's by inverse of the uniquenesses	Weighted r's on inverse of uniquenesses and explored for maxima across changing factor number	Weighted r's by ratio of raw-score sigmas to those of a "core" reference experiment
Component model $R = V_0 V_0^T$	Burt's "classificatory" use of centroid with u variables and u factors	Hotelling, Kelley, Burt and math statisticians' use of principal components				
Factor (common, broad) model $(R - D_{u_2}) = V_{fp} R_f V_{fp}^T$	Thurstone's centroid; minres; Jöreskog's unweighted least squares	Thurstonian most common factor analytic iterative extraction, Jöreskog and Goldberg's generalized least squares	Caffrey and Kaiser's alpha factoring $D_h^{-1} R D_h^{-1} + D_h 2 = R$	Canonical factor analysis $D_u^{-1}(R - D_u 2) D_u^{-1} = R$ in original form without iteration	Maximum (likelihood; same formula as canonical, with iteration over u values, or both u's and factor number	
Covariance (real-base) factor model $(R_c - D_{cu_2}) = V_{cfp} R_f V_{cfp}^T$	Ordinary covariance factor analysis (by centroid)	Ordinary covariance factor analysis (by components)				Real-base factor analysis, with ratios to core matrix preserved
Image factor model $(R - D_{s_2}) = V_{fp} R_f V_{fp}^T$	Guttman's image analysis, variety one	Guttman's image analysis, variety two		Harris's Rao-Guttman combination		

[a] R is the straight correlation matrix of variables with unities in the diagonal. What is often written u^2 for uniquenesses is here D_{u_2} to remind one it is a diagonal. D_{s_2} is the matrix of specific variances left after a multiple R estimate from other variables c, as in R_c means that raw-score deviations are involved in covariance factoring.

reported in the substantive literature arise sometimes from unwittingly using the wrong model, e.g., calling a components result a factor result through failing to insert communalities, or from gross neglect of a necessary step, usually of an objective decision on number of factors, and, above all, from poor rotation.

Although some large divergences can be quoted, they are usually due to nothing more than erroneous use of the methods concerned. They sometimes lead to false substantive conclusions that a real developmental or cultural difference exists. But it seems that whenever accurate work is done, the differences in numerical outcome in the V_0 are surprisingly small. (Rotated factor patterns are another matter.) Digman and Woods compared, for example, several R matrix analyses using four or five different extraction methods and found close resemblance indeed between principal axes, alpha, and image, methods in a *factor* framework, i.e., axes with h^2. Velicer (1977) concludes essentially the same. The former found, however, some moderate divergences from the others in canonical analysis when the realistic condition of a large number of variables in the R_v is met, and some excessive factors in maximum likelihood with very large N. Under rotation, slight differences in the V_0, they report, sometimes become more evident in matches on the oblique factors, but this could be true of any simple structure rotations from identical V_0's. This is not to say that one should not use more elaborate extraction methods when some particular emphasis is indicated, and when programs and machine time is available, but a factor analysis is quite unlikely to go wrong in anything but the second or third decimal place through using the standard method (in cell 2-2 in Table 13.2) with a scree or other psychometric test for number of factors. It is far more likely to go astray through poor choice of samples or variables, absence of any test for number of factors, generally negligent techniques, hasty rotation, and absence of a test for significance of simple structure.

As far as scientific use is concerned the more important differences remain those in the rows of Table 13.2, concerning whether (a) the scientific model shall, as in components, account for all the variance of n variables merely from those n variables; or (b) as in factors, admit the existence of influences outside what is common to those few variables (for uniqueness represents not only true specifics but also still unknown broad factors); or, finally, (c), as in real base, admit and set out to measure real differences of size among the broad factors. Related to this last is the "scale free" properties of some methods, discussed below under real-base factoring.

The manner of growth of computer programs and resources has been so spectacular in the forty years since Burt, Holzinger, Thurstone, Kelley, Saunders, the present writer, and many others used to carry out factoring on a desk computer and an IBM sorter, that it would be reckless to guess what extraction methods will be most prevalent over and after the next decade. But on the question of models, it would seem that much convergence of opinion has occurred, and that the relevance and virtues of various models are likely to remain as now understood. Thus the component model is included in Table 13.2 for basic completeness; but the situations where it is likely to fit as a scientific model are very rare. One may hope that the chief development in use of a new analysis model in the next decade will be in the real-base model. Except for using covariance factoring, this remains a regular factor model and can use whatever extraction methods are available and preferred. However, since collating the relations of results among different studies, systematically varied across sample and modulating situations, are essential to its development, its growth will depend on programmatic research. It will call for a more socially coordinated "mapping" use of factor analysis, especially for settling on factor sizes, than has existed up to this decade.

Reverting to procedures, we would anticipate that for deluxe jobs of top theoretical importance the maximum likelihood and unweighted least-squares programs are likely to emerge as favorites.[1] But researchers in well-endowed institutions must not forget "how the other half lives" and that for the foreseeable future many psychologists in, for example, small colleges, will continue, with good workmanship, but computers of limited capacity, to get excellent results simply with the principal axes program and iteration of communalities beginning with a careful estimate (by scree, K–G, etc.) of the number of nontrivial factors.

With the deluxe methods one must not overlook, however, the problem mentioned above that maximum likelihood can be almost unduly sensitive to the size of sample, frequently yielding, from the same variables, decidedly more factors with a larger than a smaller sample of people. That there are indeed likely to be some quite small but still statistically significant factors with a larger and more varied sample is, of course, true. But, especially at the exploratory stage, research coordination of different experiments is simpler if factors of truly trivial size are

---

[1] In favoring the above trend we may permit ourselves at this moment to wonder why, in view of the theoretical attractiveness of the maximum likelihood method, it has not been more widely used at the present time. The answer is partly in the realm of expediency; that the computations were originally long and costly, and partly from too great a susceptibility to sample size as discussed elsewhere. Even today, with the contributions of Jöreskog (1967a), Lawley (1940), Jennrich and Robinson (1969), Lawley and Maxwell (1971), and Clarke (1970), the reduced computer algorithms remain appreciably more expensive than the usual principal axis-to-a-decided-number-of-factors extraction above. Especially where really large R matrices, say 150×150 and above, are concerned, one may very reasonably decide instead on ordinary principal axes.

not included in the matchings, sortings, and collations. There is such a thing as being confused by too sensitive instruments—as if a car speedometer had three dials, miles, yards, and feet—instead of one. And the rougher, more "hardy" estimate of the number of factors by the scree or other "psychometric" type of test may better suffice for many purposes and conditions. Under purposes and conditions one includes the stage of research, as just indicated, and budgetary considerations of whether limited funds can better be directed to analysis or broader data gathering. More experience with comparisons of maximum likelihood and iterated principal axes is needed before advantages and disadvantages can be fully evaluated. In any case it is encouraging to find, as mentioned, that the scree verdict proves to agree with the maximum likelihood result on number of factors quite well over the range of sample sizes most commonly used (roughly 150–450 cases) [Hakstian, Rogers, and Cattell (1978)].

13.8. The Aim of Real-Base Factor Analysis

As factor analysis interacts increasingly with other kinds of experimental work, and particularly with various ANOVA uses of factor scores, it encounters a curious problem. Whereas with ordinary variables we can readily enter into an ANOVA comparison of one group with another, such comparison of means becomes impossible with two factored groups. It is impossible because by ordinary factor analysis the factors in any and all groups have a mean of zero and a sigma of one. This forbids comparisons not only of group means but of scores of an individual in one group with an individual in another. There are *ad hoc* escapes from this, such as we have discussed in joint factoring of the two group samples and in the equipotent and isopodic methods of scoring, but these still leave certain possibilities excluded. For example, we may want to study change scores in the same group or to recognize difference of size (of sigma) of the same or different factors. A scientific model requires that factors be able to assume different levels and sizes.

Real-base true-zero factoring is an attempt to gain these more universal and negotiable properties for factors and scores. It amounts to asking for equal interval and true-zero properties in factor score scales. These properties cannot be acquired by factors unless they also exist in the variables from which the factor scores are estimated. The extent to which we may approach these properties in variables is to be evaluated in Chapter 14. But even when these properties may have been gained in variables they are lost, by ordinary factoring, when we reach factors which must have zero mean and sigma of one, regardless of population and conditions.

The first step toward real-base factoring is to shift from *ordinary factoring* to *covariance factoring*, for this will permit differences in factor size. In what follows we shall add a small c to any familiar symbol for a factor or correlation matrix to indicate the corresponding covariance terms. Thus, as mentioned earlier, the covariance matrix among the variables R_{vc} is related to R_v, thus:

$$R_{vc} = D_v R_v D_v \tag{13.13}$$

which means that each row and column is multiplied throughout its length by the sigma of the variable concerned where D_v is a diagonal matrix composed of the raw-score sigmas of the variables.

As regards the factor pattern matrix which restores this $R_{v \cdot c}$ we can write either

$$R_{vc} = V_{fpc} R_f V_{fpc}^T \tag{13.14a}$$

or

$$R_{vc} = V_{fp} R_{fc} V_{fp}^T \tag{13.14b}$$

(counting R_v and R_{vc} as usual as reduced matrices, i.e., with communalities in the diagonal, not ones).

This means that the change in loading of a variable which comes with covariance factoring can either be written right into the covariance factor pattern matrix, keeping R_f the same which is preferable in some later dealings, or reciprocally, it can be absorbed into the expression of the relation among factors as covariances instead of correlations, leaving the V_{fp} unchanged. (It has been asked if a third alternative of absorbing the change into both an R_f and a V_{fp} change is viable. This absorbing through $D_f^{1/2}$ in both might be employed but is conceptually obscure.) Let us next study, however, the properties of covariance factoring more closely as such.

13.9. Covariance Factoring and the Law of Constancy of Factor Effect

It follows from Eq. (13.13) and the familiar fact that $R_v = V_{fp} R_f V_{fp}^T$ that Eq. (13.14a) can be rewritten as

$$R_{vc} = D_v V_{fp} R_f V_{fp}^T D_v \tag{13.15}$$

From this we realize that the covariance factor pattern matrix $V_{fp \cdot c}$ is simply the ordinary V_{fp} with each row multiplied by the raw-score sigma of the variable in question, thus:

$$V_{fpc} = D_v V_{fp} \tag{13.16}$$

This would hold only if the correlations of factors remain at the same value, and would be simply true as

$$V_{0c} = D_v V_0 \tag{13.17a}$$

However, it does not follow that a new V_{0c} having this property can be derived from a new R (reduced correlation matrix) in which the same factors exist as in V_0 but change their variance. As discussed more fully below, the following necessary inequality has to be observed

$$D_v V_0 \neq V_0 D_f \tag{13.17b}$$

where D_f is a factor variance changing matrix. By ordinary factoring (principal axis with communalities), Eq. (13.17b) can be made an equality only by including an orthonormal matrix L (related to D_f and V_0 such that $L = V_0^{-1} D_v^{-1} V_0 D_f$):

$$D_v V_0 L = V_0 D_f \tag{13.18}$$

as expressed slightly differently in Eq. (13.22). Only in the case of *scale free* extraction models which utilize U^2 or H^{-1}, as in canonical, alpha, and maximum likelihood is L in Eq. (13.18) unnecessary (an identity matrix), and Eq. (13.17b) true.

The principle of real-base factoring which we are approaching begins, however, at the opposite end, so to speak, of covariance factoring. It supposes that it is the factors that change their sizes, consistent with our determiner model, and that the resulting changes in variances of variables are due to this. This requires and brings out a vital assumption which we shall call the *postulate of constancy of factor effect*. This states that, with other factors not interfering, *the ratio of increment in a variable score to increment in a factor score remains constant, variables and factors being in raw scores*. (The particular ratio we shall later discuss and define as the potency of a factor relative to a particular variable.) The second assumption is that *when raw-score variances on any variables change from one population or situation to another they do so because the factors have altered in variance*. That is to say, changes in the variance of causal factors come first, so that factor variances determine variable variances, not the converse (at least not in any simple causal sense).

The changes in factor variance can happen either because of significant population selection effects, or through manipulative effects, as registered in the modulation values of "occasions." The change in the variances contributed by each factor to variables from any occasion or population 1 to another, written as 2, must therefore be predictable by multiplying each factor by some ratio. As we now have to consider, however, the full variance of each variable, the specific factor matrix U (we will omit error at this point) must also be involved. And the change in variance magnitude which applies to a broad factor is a happening to which unique factors must also be liable. The full V_{fp}, which we can conveniently reduce to an $n \times k$ order correctly in several situations above must here be itself again, as an $n \times (k + n)$ matrix, and the diagonal matrix which now multiplies it must also be $(k + n) \times (k + n)$. This must be understood as in the notation in Eq. (13.19):

$$V_{fpc2} = V_{fpc1} D_{f(1-2)} \qquad (13.19)$$

where D_f is the diagonal matrix of ratios of changes in factor sigmas.

Here one must revert to Eq. (13.17b) and the pitfall in conceptualization into which one may easily fall in the real-base model. If the transformation is by changes in covariance factor sizes, as in Eq. (13.19), then the change of variable sigmas from one experiment to another as Eq. (13.18) shows cannot be simply related to change of factor sigmas. It just might happen, as a special case under restrictions to be discussed that an equality exists, but normally it will not and so we write reverting for simplification to the orthogonal case rather than the special rotated case, i.e., dropping R_f and using V_0 instead of V_{fp}:

$$D_{v(1-2)} V_{0c1} [D_{v(1-2)} V_{0c1}]^T \neq V_{0c1} D_{f(1-2)} [V_{0c1} D_{f(1-2)}]^T \qquad (13.20)$$

That is to say, no D_v and D_f values can be found that will make this true. The reason is that the D_f changes will not cause a uniform change across the row for each variable. What will happen actually can only be understood by following our real-base model, in which we are supposing that the observed change between the two experiments is due to changes in the factor sizes, as shown in the expression on the right of Eq. (13.20). The observed (calculated) new covariance matrix $C_{v \cdot 2}$ consequently has the property

$$R_{cv2} = C_{v \cdot 2} = V_{0c1} D_{f(1-2)} (V_{0c1} D_{f(1-2)})^T \qquad (13.21)$$

This concept must be brought into relations with that of "scale free" factor analysis models, which is a term having an unfortunate ambiguity. Harris noted (1964) as others have, that an accidental and unsought property of some of the factor analysis models produced in the last fifteen years is that in Rummel's (1970, p. 309) succinct words "they result in factors that (within a scaling constant) are invariant of the units and scales of the raw data." A vital term is "within a scaling constant," which means, geometrically that within the configuration of tests the orthogonal factors come out in the same positions by these methods when used over different samples, but with extensions or contractions of the scale of the factors. (Note the resemblance to the relation in the orthogonal confactor resolution position.) The analysis models which have this property are canonical factoring with its maximum likelihood derivative; Jöreskog's version of image analysis; alpha factoring; Jöreskog and Goldberger's generalized least squares; unweighted least squares (to which minres closely approaches); and weighted least squares.

Letting the subscript (1–2) mean a ratio change in variable and factor sigma, we have recognized in ordinary factoring in Eqs. (13.17b) and (13.20) that

$$D_{v(1-2)}V_{0c1} \neq V_{0c1}D_{f(1-2)}$$

can be brought to an equality simply by introducing an orthonormal matrix L as stated in (13.18) so that

$$D_{v(1-2)}V_{0c1}L = V_{0c1}D_{f(1-2)} \tag{13.23}$$

The scale free methods of analysis are those which make L an identity matrix, through adjusting the loadings appropriately, but in ordinary factoring an L has to be interposed. The property is useful for factor recognition and matching; but it is not merely this that the real-base models sets out to achieve. The real-base model with more important goals expects an inequality at this stage [as in Eq. (13.23)] as a natural consequence of the law of constancy of factor effect. It holds that the factors, broad and unique, are the determiners and that they do retain a constant loading pattern from experiment to experiment, expressed as a relation between units on the non-unit-length covariance factors and the stable raw-score units on the variables. The accidental scale free (strictly not scale free since scaling is a question of attaching arbitrary translations to raw scores, but *variable-raw-score-sigma-change-free*) property of some analysis models is thus not to be confused with the fundamental scientific model property of real-base factoring. It should be noted too that the scale free relation with the above methods holds only for the orthogonal factors, whereas real-base factoring is concerned with maintaining a proportionality of factor pattern, with the usual freedom of obliquity, from study to study.

For some more general problems in resolving covariance matrices into variance–covariance structures the reader should see Bargmann and Bock (1966) and Srivastava (1966) on testing hypotheses in covariance structures.

13.10. The Conception of Standard Raw Scores at a Core Matrix Position

Factor size must by its nature be defined relative to some standard, which in real-base factoring resides in the concept of the *core matrix*. Before defining this, since the reader's mind is probably uneasy for practical reasons with covariance factoring, on account of the enormous differences that can exist in the raw-score matrices of variables, let us show that this practical problem can be reduced without losing the real variance difference basis. We need to retain the real differences in a variable raw score, as far as each variable is concerned as part of our keeping to a real base rooted in natural measures. But this can be achieved by taking some central, standard, preserved sample, and condition of experiment—an equivalent of the earth's circumference and the platinum bar for the meter scale—and calling each raw-score sigma *unity* at that point. Then in any other situation of greater or lesser factor variance the variable sigmas can depart from that unit sigma as required. Thus in all but the single unique situation the variances would deviate from unity and behave essentially as raw scores, getting bigger and smaller proportionally through changes from the central raw scores of each. They would essentially be raw scores brought to convenient comparable sizes, and referred to a scale existing in and defined by the one standard experimental occasion. We shall call these "comparable raw scores" *AZUS* units (pronounced á-zus), for *a*rtificial *z*ero, *u*niverse-*s*tandardized units. We say "universe standardized" for in the end we shall define the "specific occasion" which we have called the *core study* as the ideal central value averaged across a defined *universe of populations and occasions*. This is more reliable and widely acceptable among investigators than some arbitrarily designated single experiment. Even though this provides a value for the units of deviation from zero, it still leaves the zero lacking the properties of a true zero in a ratio scale. Later we shall discuss the possibility of a true zero.

The factor size problem is better initially approached in terms of size changes in one and the same factor from occasion to occasion than as a difference between two factors on the same occasion. Traditionally, e.g., when talking of larger and smaller latent roots, something roughly called size has been evaluated by squaring the loadings down a factor column, adding and dividing by the number of variables, i.e., by $\Sigma\, b^2/n$—for the square root thereof for factor sigma. However, if we consider that such a factor changes in size from occasion 1 to occasion 2 it will at once be evident from the discussion about Eqs. (13.18) and (13.19) that not all variables in the column of that single factor will change V_{fp} loadings by the same amount in an ordinary factor analysis. A variable has unit variance and if factor X has, say, a .8 loading on it, and becomes "larger" as a factor, .8 may not only fail to increase but may actually decrease *if two or three factors other than X should also happen actually to increase in variance*. Indeed, the sum of the whole column of ordinary loadings *could* decrease with a real increase in size of the factor.

In short, the ordinary factor model, with unit variance variables and unit length factors is incapable of representing change of real size (variance) for the influences and dependent variables involved. But in real-base, covariance factors, if, say, all other factors increased and this one did not, the latter would still keep its variance contributions to variables constant according to the postulate of constancy of factor effect. But the variance of all variables can go up if that factor and other factors all increase their variance.

Elsewhere, in a more extensive monograph on *real-base true-zero factor analysis* (Cattell, 1972a) to which the reader may be referred from time to time here, four ways of calculating the size of a factor (relative to the core study) are given and computationally illustrated, of which the two most illustrative are described here.

1. It is always necessary to have available the $V_{fp \cdot s}$ (s now for standard core position) at the standard core experiment, which is a referent for all others. We are out to compare the size of factors in a new experiment, at occasion–situation t. In the present calculation we will suppose that what we have on occasion t is the new variance–covariance matrix $R_{vc \cdot t}$. What we require is the $D_{f(1-2)}$ matrix of Eq. (13.19)—now written generally $D_{f(s-t)}$ in Eq. (13.25). It is shown (Cattell, 1972a, p. 26) that

$$R_{fct} = (V_{fps}^T V_{fps})^{-1} V_{fps}^T R_{vct} V_{fps} (V_{fps}^T V_{fps})^{-1} \qquad (13.24)$$

R_{fct} is the correlations among the factors in the t experiment, expressed in covariance terms. [Eq. (13.24) uses the generalized inverse to invert an oblong matrix V_{fps}.] Taking the square root of the diagonals of R_{fct} gives us the required $D_{f(s-t)}$ diagonal matrix, for this is the comparison of R_{fct} and R_{fs}.

2. More commonly we will have factored R_{vct}, for general purposes, and be in possession of V_{fpt}. (This R_{vct} will of course have sigmas for its variables proportional to unit variance at $R_{vc \cdot s}$.) The stretching (tensor) matrix $D_{f(s-t)}$ (which is also written for short $D_{f \cdot t}$) can then be calculated from the two covariance factor patterns by

$$D_{f(s-t)} = (V_{fpcs}^T V_{fpcs})^{-1} V_{fpcs}^T V_{fpct} \qquad (13.25)$$

Of the two other ways, one is the obvious empirical procedure of comparing the sigmas of actually estimated factor scores on the second and the standard occasions.

13.11. Relative Size, Potency, and Final Size of Factors

The more extended use of factors in various experiments, by real-base calculations, calls for recognition not only of changes in size of one and the same factor in different situations but also of differences in size existing among different factors on the same occasion of experiment. If we adopt a scaling whereby factors are all of unit size at the standard, "core" position, then it follows that in any but the core position the factors in a given study will in general be of different sizes, for each will have changed by a different amount from the core value. The shift to covariance thus achieves a basis for both kinds of size difference. However, the reader may object that although this permits us to talk of different factors having different sizes the whole thing rests on an arbitrary assignment of equality of size at the position of the core experiment. Is it just possible that some meaning can be given to difference of size even at the standard position where all variables have a sigma of one? As we noted above, factorists have long spoken of "larger" and "smaller" factors, e.g., when they say the later factors—latent roots of factors to be precise—extracted in a principal axis or centroid extraction are smaller than the earlier ones. In this case the meaning of size is the mean variance contribution of the factor to all variables, calculated through squaring the loadings or correlations [or alternatively taking the products of correlations and loadings to get the variance contributions as in the associated contribution matrix (Chapter 10, p. 235)], down the column, as mentioned above. The decision between the first and the two latter alternatives need not greatly concern us at this point. For the same reasons as given in Chapter 10 we favor calculation of potency in the b of the V_{fp}, thus:

$$\hat{P}_f = \frac{\sum\limits_{j=1}^{j=n} b_j^2}{n} \tag{13.26a}$$

But if one accepts the associated factor contribution matrix V_{afc}, then the expression would be

$$\hat{P}_f = \frac{\sum\limits_{j=1}^{j=n} rb_j}{n} \tag{13.26b}$$

It will be noted that if we confined the notion to the original V_0 this $\hat{P}$ would represent the latent root value.

It will be at once appreciated that this cannot be considered a precisely calculated property of the factor (in either expression), since it is fixed only for that sample of variables (and, less relevantly, for the same group of people) so that it is an estimate, as shown by the "hats" on P. To give maximum scientific value to P_f as more than a single estimated value it would be necessary to define a population of variables and people and take the mean score for two or three stratified samples therefrom. Accepting the estimate, what then would such a value mean? A low value would indeed mean that the factor has relatively small potency and importance with regard to the totality of psychological variables (or the totality in a defined domain, such as personality, or cognitive learning or social psychology). A high-potency factor could reach this high-potency score in part through affecting many of the world's variables, i.e., having many unit entries in the *factor mandate matrix* (p. 235) and partly through producing a large increase in score on the average variable for every unit increase in the factor as an influence. Though no empirical data on defined and acceptable stratified samples exist, there seems little doubt that factors will differ appreciably in this potency. For example, in the domain of plant growth annual rainfall is surely going to have greater potency than, say, hours of visibility of the moon, or percent of calcium in the soil. Potency p, in conclusion, is size at the core position, variables being in standard scores at that position.

Combining the meaning of size rendered possible by real-base, co-variance factoring with the concept of potency as thus proposed for measurement at the core position, we see that the magnitude of effect of a factor in any given situation (and matrix) as measured by Eq. (13.26), would be partly the effect of its inherent potency as just discussed, duly defined at the core position, and partly of the change of its size from the core position. All kinds of transactions and discussions would be clearer if we keep these two effects conceptually distinct, so it is proposed to define them as *potency P* and *relative size E* (expansion–contraction ratio) relative to the core position. Then the final size F due to the effect of both of these is

$$F_f = E_f \cdot P_f \qquad (13.27)$$

where all three are diagonal matrices. That is to say, if one sums the squared loading down a factor column in $V_{fp \cdot t}$ he will obtain a final size for some factor x which is partly due to its self-relative size change—a value in the diagonal of E_f,—and partly due to its original potency at the standard core position, as shown in P_f.

13.12. The Role of Modulation in Determining Factor Size

So far we have defined a core matrix or standard position to which all changes in size are referred. From what sources does such change in size emanate? In a broad sense there can be only two: (1) different selections from a population, or (2) the impact on the factors of other influences, either natural, such as environmental learning or maturation, or manipulative, as with factors that are states. These could operate either directly on primaries or on higher-order factors. Examples of manipulative effects would be exposing a group to a strong anxiety stimulus increasing both the mean and sigma of anxiety, or to an experimental learning experience which might increase the mean and sigma of some skill factor. In the case of state factors, showing a manipulative change associated with a single occasion k, the concept of modulators has been developed (Cattell, 1963c, 1972b, 1973), and since this turns out to be of broader theoretical impact than for psychological states alone we shall briefly define it.

Modulation theory states that an individual has a traitlike propensity to experience certain emotional or dynamic states, possessed in higher or lower degree by that individual than by others. This liability or proneness is represented by a personality factor L and is modulated for everyone by a value s_k, the magnitude of which is peculiar to the occasion (situation, ambient stimulus) k. The state S being at its given level on occasion k for an individual i, then enters into determining his behavior by the usual behavioral index b. Thus to treat the specification equation as if it had but one term (the others being analogous) we have

$$a_{jki} = b_{jx} S_{xki} \tag{13.28}$$

which is the equivalent of

$$a_{jki} = b_{jx} s_{xk} L_{xi} \tag{13.29}$$

when we recognize that a present state S_{xk} is the product of a liability L_x and a situational modulator s_{kx}.

The fuller account of modulation theory may be read elsewhere (Cattell, 1963c, 1972b) since we are here concerned only with its relevance to factor size. However, in passing one should note its relevance to three- and four-way factor analysis (p. 336). Therein two alternatives are possible in representing the analysis of the increase of variance in a variable due to taking in variance from another coordinate set. The most natural formulation of the analysis of the joint variance–covariance over people and situations (if we restrict for illustration to two set coordinates) follows the additive design of analysis of variance. It will take deviation as

$$a_{ij(k)} = \sum b_{jp} T_i \tag{13.30a}$$

over people at the mean situation, and

$$a_{jk(i)} = \sum b_{js} T_k \tag{13.30b}$$

over situations at the mean person. Where T_i is the factor score of an individual, T_k of a situation, b_{jp} a loading on a person factor, and b_{js} on a situation factor. Then a_{ijk}, as a deviation from the grand mean, is

$$a_{ijk} = \sum b_{jp} T_i + \sum b_{js} T_k \tag{13.30c}$$

if kept in covariance so that the T_i and the T_k have their appropriate sizes.

The alternative model, which we follow in using modulators, is to treat the situation effect as a ratio change in the b_{jp} loading, which can be done only if the factors over people and over situations are matchable in meaning. Psychological work on states and other modulable personality factors suggests that though such factors across other sets need not be matchable they will be so in many instances, at least over the person and situation sets. In that case a value s can be taken such that

$$s_k = \frac{b_{jp} + b_{js}}{b_{jp}} \tag{13.31}$$

again with allowance for any difference in size of the person and situation factors.

If this should prove feasible over all dimensions, i.e., beyond those for persons and occasions, it would point to the possibility, which we have not explored, that a general theoretical parallelism could be set up across the whole data box, such that the additive model as in Eq. (13.30c) would have a product equivalent essentially as in Eq. (13.32):

$$a_{hijk} = \sum b_{jh \cdot x} b_{jk \cdot x} b_{jp \cdot x} L_{xi} \tag{13.32}$$

The observer set is for simplicity omitted here and the response variables (j) are always retained as the relatives so that the same factor could emerge across people (b_{jp} loadings), across stimuli (b_{jh} loadings), and across situations (b_{jk} loadings), but the two last would be entered as ratios modifying b_{jp}. The reader should compare and contrast this formulation of three-way analysis with that of Tucker's *three-mode* analysis on page 339.

It can readily be seen that, as illustrated in two sets—people and situations—the same result—producing an s_k index—could be achieved by taking the mean b_{jp} value for each of a series of situation values, as illustrated in Table 13.3.

This is called a multisituation–multidemo design (*demo* being a useful term for a "demographic sample of people," made according to some orderly scheme for selecting groups of different variance). If a covariance factoring is now done for all of these $o \times y$ resulting V_{fp} matrices they will provide us with a basis for systematic evaluation of differences of factor size. The factor size change due to demo variation can be found by comparing the mean of the y situational matrices in each of the o rows. Alternative expressions of such means can be proposed but we take for a row—where d is a particular demo row and t is a particular situation column (Table 13.3). They are actually written pd and kt to remind us that ol is a particular p and t is a particular k:

$$V_{fp \cdot c \cdot d} = V_{fp \cdot c \cdot s} D_{f \cdot p \cdot d} \sum_{kt}^{y} \frac{D_{f \cdot k \cdot t}}{y} \tag{13.33}$$

Correspondingly, the mean matrix for a situation (kt) is evaluated by averaging the $V_{fp \cdot c}$ matrices down each column, thus

$$V_{fp \cdot c \cdot t} = V_{fp \cdot c \cdot s} D_{f \cdot k \cdot t} \sum_{pd}^{o} \frac{D_{f \cdot p \cdot d}}{o} \qquad (13.34)$$

The variance among them is then found to determine the $D_{f \cdot k \cdot t}$.

Up till now we have taken the calculation of the core matrix, which is to be the referent for all particular group and situation comparisons, on

Table 13.3. Design of the Multisituation–Multidemo (MSMD) Experiment for Determining Core V_{fp} and Demo (Selection) and Situation (Modulation) Indices[a]

Demos (N subjects in each)	Situations				Mean demo V_{fp}
	k_1	k_2	k_c	k_y	
p_1	a_{j1}	a_{j1}	$\cdots$	a_{j1}	$b_{j1}p_1$
	a_{j2}	a_{j2}	$\cdots$	a_{j2}	$b_{j2}p_1$
	.	.		.	.
	.	.		.	.
	.	.		.	.
	a_{jn}	a_{jn}	$\cdots$	a_{jn}	$b_{jn}p_1$
p_2	a_{j1}	a_{j1}	$\cdots$	a_{j1}	
	a_{j2}	a_{j2}	$\cdots$	a_{j2}	
	.	.		.	
	.	.		.	
	.	.		.	
	a_{jn}	a_{jn}	$\cdots$	a_{jn}	
p_c					
p_o	a_{j1}	a_{j1}	$\cdots$	a_{j1}	
	a_{j2}	a_{j2}	$\cdots$	a_{j2}	
	.	.		.	
	.	.		.	
	.	.		.	
	a_{jn}	a_{jn}	$\cdots$	a_{jn}	
Mean situation V_{fp}	$b_{j1}k_1$	$b_{j1}k_z$	$\cdots$	$b_{j1}k_y$	

[a]Elsewhere (Cattell, 1971b) the writer began by designating this an MRMS design. In broader perspective this proved a misnomer and the present designation—MSMD, for multisituation-multidemographic group—is distinctly preferable. The variables in this oy experiment are the same—a_{j1} through a_{jn}. The corresponding b's in the V_{fp}'s are shown at right and bottom for j_1 only. There is a hypothetical experiment representing a core situation at k_c and a core population at p_c. The core $V_{fp \cdot c}$ across both situations and demos will be considered to have variable sigmas of one and b values illustrated by that for j_1 as

$$b_{j1 \cdot c} = \sum^{p=0} \sum^{k=y} b_{p \cdot k}/oy$$

trust. But now, let us transcend the view of it as perhaps some arbitrary, relatively central matrix, and give it a definite home as the average of both rows and columns in some representative sampling of demos and situations as in Table 13.3. This can be written

$$V_{fp \cdot c \cdot \underline{s}} = V_{fp \cdot c \cdot s} \sum_{pd}^{o} \sum_{kt}^{y} \frac{D_{f \cdot p \cdot d}}{o} \frac{D_{f \cdot k \cdot t}}{y} \qquad (13.35)$$

If we now designate q_{xt} as the quotient or ratio expressing *total* change of the sigma of a factor from the core position to an experiment differing both in demo and modulation situation; and if d_{px} is the ratio from demo (population) selection only; and $q_{x \cdot kt}$ ($=s_{kx}$) that from choice of ambient situation (occasion) only, then it can be seen that

$$q_{xt} = d_{px} \cdot s_{kx} \qquad (13.36)$$

The b_c value—the literal factor loading behavioral index—obtained, however, from a covariance analysis in covariance terms, for any given demo–situation group, will then be related to that at the core position by

$$b_{c \cdot j \cdot x \cdot t} = d_{px} \cdot s_{kx} \cdot b_{c \cdot j \cdot x \cdot s} \qquad (13.37)$$

where j and x are variable and factor, respectively, as usual. The t and the s subscripts to the b indicate, respectively, the particular and the standard (core) values.

The above is an exploratory move in the realm of real-base factoring. It cannot be brought to final precision because it depends on more than statistical handling, namely, an empirical finding, e.g., as to the persisting identity of a factor across ranges of people and ranges of situations. The current absence of empirical indications may well be due to an "instinctive" recoil of conservative experimenters from what is felt to be the vagueness of the score on a covariance factor, where variables of different metrics are thrown into a composite. We have suggested above that the worst extravagances of this procedure can be avoided by setting the variables to standard score at the core position, whereupon they will not range over much more than 0.5 to 2.0 at most other positions. This solution to the metric problem means that errors in certain variables (those of large raw-score sigma) cannot have extravagant effects upon the precision of a covariance factor score.

As to the ranges of size change under modulation and change of demo, one might guess in the absence at the present time of anything but sketchy experimental data, that most real-base factor sizes will vary from occasion to occasion or factor to factor by ratios up to two or three, though potencies may stretch over a much wider range.

The secret of convenient handling of the covariance factoring connected with real-base factor work is to keep in mind the transformations which enable one to move smoothly in and out between ordinary, standardized factor analysis and covariance factoring. For example, when rotation is to be carried out it will be found to be a very clumsy procedure if plots are made for ROTOPLOT with one factor three times as large as another, and it is best to transform from $V_{0 \cdot c}$ to V_0 [Eq. (13.21)] before proceeding to ROTOPLOT. The same is true for use of most computer programs designed for ordinary factor analysis: they can be used on the values transformed to the ordinary factor matrix and then brought back again to final real-base form through multiplying by the sigmas of variables or of factors according to which is appropriate. It should be remembered that the sigma in real base, unlike that in simple covariance, is both for variables and factors, a value calculated relative to the core or AZUS (universe standardized) values.

The extension of real-base factoring to include a true zero, i.e., a ratio scale in factor scores, is taken up in Chapter 14.

13.13. Broader Models: Nonmetric Analysis and Bentler's Multistructure Model

Most actual factor analyses are made with variables with some claim to an approach to equal interval scaling properties, and although we recognize that perhaps even a majority of psychological variables can strictly claim only ordinal properties the results are generally given the benefit of the doubt, in subsequent estimations and predictions. Our present concern is to scrutinize this situation more closely. Turning to the four well-known levels of scales—nominal, ordinal, interval, and ratio—we recognize a form of factor analysis called *nonmetric* factoring which applies to the first two, while what we have called "ordinary factor analysis" might, for distinction, be called metric factor analysis, which assumes interval scaling and in some cases ratio scales.

Nonmetric factor analysis includes nominal scale factoring as well as nonparametric factor analysis. Nonparametric statistics of any kind do not depend upon or involve assumptions about the distribution parameters of the parent population, such as the mean and the various moments (sigma, skewedness), e.g., they claim no normal distribution, and in the case of factor analysis it means using some coefficient other than r and its derivatives which do have such assumptions. It will be pursued only briefly here and the reader is referred to J. D. Carroll (1961, 1972), Guttman (1955a,b, 1966), Shepard and Kruskal (1964), Kruskal (1964a), Kruskal and Shepard (1974), and Siegel (1956).

Little use or attention has been given, until recently by Rummel (1970), to the nominal scale form of nonmetric analysis, though it was persuasively advocated in connection with logical analysis of qualitative data by Burt, as long ago as 1950 and 1951, and developed by Coombs and Kao (1955) and Lingoes (1968). Useful discussions have ensued by Guttman (1955a,b, 1966) and others. When a variable takes categorically different forms, such as male or female, red, blue, and green, Christian, Mohammedan, Buddhist, Jainist, etc., which cannot be put on any continuum beyond cavil, it is presently possible to use in the linear factor model only the *two* category case. If this can be done the association of the two alternative qualities, which should best constitute the totality of possible qualities of that class, can be calculated with *other* dual divisions or with *continuous* variables by several suitable coefficients among those we have discussed, from ordinary r, and the tetrachoric to the point biserial (see Krause, 1966).

As to the interpretation of the loadings from such an analysis Rummel (1970) points out the similarity to the use of "dummy variables" in regression analysis (Suits, 1957; Johnston, 1963). In "dummy variables" where we have, say, six categorical alternatives, A, B, C, D, E, five of them are used as dichotomous variables—A against not-A, and so on—though of course this introduces some spurious (negative) correlation of A, B, etc., as in ipsative scoring. He writes, "The regression coefficient of X_1 regressed on X_2, when each is measured as 0 or 1, can be interpreted as . . . the probability that X_1 will be present when X_2 is present. . . . Analogously in factor analysis a loading for 0–1 scaled variables can be interpreted as the probability of its presence (if the loading is positive) or absence (if negative) given that the factor exists" (1970, p. 225). ("Exists" may here be taken to mean a 1 value if a purely categorical factor, or some positive value if continuous.) By his pioneer work in the use of factor analysis in social data where a qualitative dichotomy often exists, Rummel has provided us with extensive experience on the meaning and use of loadings on such variables and in factor score estimates. Naturally one tries to avoid extreme numerical frequency disparities, of the contrasted categories, as one does in the tetrachoric, if distributions of correlations are not to be odd.

The most systematic theoretical treatments of nonmetric analysis, however, have been in regard to the lesser but more common problem of data which yields only ordinal scales, and where the relations of variables can be depended upon to be monotonic though not linear. In this penumbra the issues of nonmetric factoring here get involved with those of nonparametric statistics on the one hand (Lingoes and Guttman, 1967) and nonlinear factor analysis on the other. The problems of representing nonlinear relations while keeping to the ordinary linear and equal-interval-assuming methods have already been discussed on page 381. It was a far-seeing conception by Thurstone (1947) which led him early to consider bypassing these difficulties by a nonmetric factor analysis.

There are two senses in which nonmetric factor analysis may be implemented: (1) employing rank coefficients of correlation, such as Spearman's rho, which ignore the equal intervals need, and proceeding by ordinary factoring; and (2) developing an approach from multidimensional scaling principles, as in the method of Kruskal and Shepard (1974). The former needs no description; the latter is complex, but will be described in essentials.

Instead of starting with an R matrix, the Shepard and Kruskal approach begins with the Eckhart–Young decomposition of a score matrix directly into factor scores and factor pattern values, thus:

$$S_v = S_f V_{fp}^T \qquad (13.38)$$

It then produces a monotonic transformation of the column scores in S_v to a new matrix $S_{v \cdot 1}$ and iterates the $S_f V_{fp}$ solution to give the best possible fit to it. The degree of failure of fit is called the "stress" index, and a fresh monotonic transformation is tried, testing for reduction of this stress index. The end result is a transformation of scales, within the restriction that they be monotonic, which permits the best prediction from the standard factor equation in Eq. (13.38). It should be pointed out finally, that as Bentler states (1976), the model should be considered "nonmetric principal components" rather than factor analysis since although the solution can have fewer factors than variables it does not utilize uniqueness contributions.

For factoring depending on ordinal properties in the variable scales, Bentler (1971) has used the term "monotonicity analysis." Such nonmetric factoring can use any of the rank correlation methods examined by Kendall (1961). Although an ordinal scale is necessarily monotonic, it is not necessarily linear in relation to its own hypothetical equal interval units or to other scales, so that ordered, monotonicity analysis and nonlinear analysis become related; although the latter can become precisely aware of its degree of nonlinearity if based on equal unit scales. Woodward and Overall (1976) have recently criticized Kruskal and Shepard's (1974) form of nonlinear factor analysis, arguing that, if one makes a rank order transformation of the original nonlinear (but monotonic) variable scales and applies principal components to the rank correlation matrix, a solution with better representation of the structure is obtained.

The interaction of scaling and nonlinearity problems has been discussed at two or three points in this book. It appears early in the notion that clearer structure is obtained by rescaling variables to normality of scores (and to presumed equal interval units according to one operational definition of the latter) and to transformations which maximize the loadings on the confirmed factors. The Shepard–Kruskal device is a more systematic attack on essentially the same goal. The relation of scaling and linearity of relation appears also in the concept of reaching equal intervals by the relational simplex property (p. 460) to which the factor results should also be related. Basically, of course, one has to recognize three possibilities of linearity and nonlinearity: (1) in relations among variables, (2) in relations of variables to factors (loadings), and (3) in the relations of factors to factors, as in higher-strata relations. One must further distinguish between apparent nonlinearities due to wrong scaling, correctable by suitable monotonic rescaling as here, and true nonlinearities of relations which persist when scaling is corrected. It will be a long time before psychology has researched enough to be sure which causes are operative in important instances of nonlinearity now known. And where nonlinearity is curvilinear, as in a parabolic relation, no method is likely to eliminate it and we need further development of the type of McDonald's nonlinear factor analysis.

Again, as in several theoretically interesting developments, e.g., in the more far-out models of factor analysis, we need to recognize that the effect of the more refined methods and "corrections" is in practice very small. Kruskal and Shepard themselves recognized that when the usual random error entered the data the method's restoration of the underlying true object configuration (the place of people in the score matrix) was no better than in ordinary factor analysis. This has recently been clearly demonstrated by Woodward and Overall (1976) who showed that the handling of a problem by the first, more elementary approach to nonmetric factoring, namely, by rank coefficients (discussed by Kruskal and Shepard, as they mention, as a "trivial" solution) yields, with the typical error-associated data, a better solution than the more elaborate method. For the sake of theoretical adventure the more elaborate method needs to be pursued but, as Laughlin pointed out when he programmed this method, it is quite expensive of machine time and the average applied psychologist would do better to meet cases of presumed poor scaling properties by using rank correlation coefficients.

One of the theoretically exciting features of factor analysis in the past decade has been the increasing perception of unifications and general principles among models that originally grew up in isolation. For example, it is only in this decade that it has been explicitly recognized that canonical factoring, alpha, Harris's form of image analysis (see Harris, 1963, on Rao–Guttman relationships), and the present writer's real-base factoring (in a more specialized sense) share the property of being "scale free," i.e., of yielding the same loading patterns no matter what raw-score sigmas or arbitrary variable scalings exist. Some still broader perceptions have been brought about in the recent work of Jöreskog (1973), Anderson and Rubin (1956), Jennrich and Robinson (1969), Kruskal and Shepard

(1974), McDonald and Burr (1967), Wiley, Smith, and Bramble (1973), and particularly Bentler (1968, 1970, 1976).

In his 1970 paper Bentler develops a general regression model covering minimum residual (minres), alpha, and maximum likelihood factoring. He adds (1970, p. 111) that "while existing alpha and minres solutions are theoretically unacceptable because they do not provide a Gramian residual matrix, the corresponding regression solutions guarantee the residual matrix of covariances of estimated uniqueness to be Gramian."

A still broader encompassing of statements of structure in variance–covariance analysis models has been made in approaches by Bock and Bargmann (1966), Wiley, Smith, and Bramble (1973), and Jöreskog (1973). The latter's includes the others as special cases and is in turn handled as a more restricted case in Bentler's (1976) model, for which we use his term *multistructure factor analysis*, and which is compressed in the following:

$$S_v = u + V_1 S_1 + V_1 V_2 S_2 + V_1 V_2 V_3 S_3 + \cdots + V_1 V_2 V_3 \cdots V_m S_m$$

$$(13.39)$$

Here S_v is an $n \times N$ matrix of scores on n variables and N cases. (Note it is transposed relative to our standard use here.) Bentler's symbols have been changed to those usual in this book and V_1, V_2, etc., and are to be "factor by variable" parameter matrices, respectively, $n \times k$, $k \times p$, $p \times q$, etc., as we have used for different strata in our general SUD (stratified uncorrelated determiner) model (p. 551). However, Bentler's is a statistical model less restricted than our scientific hypothesis SUD, since it permits V_1 to be diagonal while the score matrices, S_1, S_2, etc., need not be factor scores in the strict sense of our model. Further, Bentler adapts to the generality of all kinds of scores by making u the expected central values of the scores, expressible in parameters K, L, M of order $(n \times a)$, $(a \times b)$, and $(c \times d)$, respectively, while T, U, and W are known matrices of rank b, c, and d, respectively.

T, U, and W are known in that they are matrices of observed predictors used in fixed effects multiple regression (as are dummy variables in multiple regression). Thus in UMW [in Eq. (13.40)] we are in the familiar situation of having U as a matrix of known values, e.g., scores, and M as a matrix of unknown regression weights, to which W adds a set of restrictions on M from an agreed *a priori* view of the action. By contrast to T, U, and W, K, L, and M are parameters in the general model for which we are trying to get estimates, and K and L may again be thought of as regression weights. If we substitute values for u in Eq. (13.39) we can (the expected values for $S_1, S_2, \ldots, S_m$ being zero) extract values for the V^T, or if substituted in Eq. (13.40), values for K, L, and M. However, there is an advantage in doing both together since in that case we can use maximum likelihood methods, using a likelihood ratio for the fit over both simultaneously.

Bentler thus expresses the second relation of matrices discussed above as

$$E(x) = u = KLT + UMW \qquad (13.40)$$

The result is that the covariance matrix R_c is given by

$$R_c = V_1 R_{f1} V_1^T + V_1 V_2 R_{f2} V_2^T V_1^T + \cdots + V_1 V_2 + \cdots + V_k R_{fm} V_k^T$$
$$+ \cdots + V_2 V_1 \qquad (13.41)$$

The reader will recognize at once the similarity of this to the formula for restoring the correlation matrix from factors of different orders (p. 213 and Cattell, 1965c, p. 249, equation X1):

$$R_v = V_{fp1} V_{fp2} \cdots V_{fpm} R_{fm} V_{fpm}^T \cdots V_{fp2}^T V_{fp1}^T \qquad (13.42)$$

except that this is a standard factoring formula, whereas Eq. (13.41) is a covariance matrix so that the parts brought together in Eq. (13.42) can be handled as distinct additive contributions. The SUD model also keeps these additions separate. However, the SUD model was developed specifically as a factor strata model whereas Bentler's multistructure is free to represent a variety of causal contributions including the interesting case of temporal developments of new structure.

The multistructural model is shown by Bentler readily to break down into various more special models, such as Jöreskog's (1973) general covariance analysis [see our Eq. (13.40)]:

$$R_c = B(VR_f V^T + U^2)B + E^2 \qquad (13.43)$$

when T in Eq. (13.40) is null.

Similarly it reduces to the usual factor model ($R_v = V_{fp} R_f V_{fp}^T + U^2$) and with U^2 as zero to the components model. (It thus takes the general position we have here, that components are a special case of the factor mode; though so special that it becomes in most respects a new category.) The form which Bentler stands by as of most universal use in a true factor model as here defined is

$$R_v = Z(VR_f V^T + I)Z \qquad (13.44)$$

[Here Z is the diagonal scaling matrix, V an $n \times k$ matrix of relation of the k determiners to the n variables, R_f the correlation among the determiner (factors), and I is an identity matrix.] The special utility he points out (1976) for this is that "any method that estimates the parameters in [Eq. (13.44)] and the mean u can automatically generate unique scores and common scores that are completely consistent with the factor model. There is no need to invoke an additional estimation criterion nor to worry about the 'best' method of estimating factor scores."

The reader is referred to Bentler's paper (1976) for matrix calculations associated with the developments of the model. The interesting fact is that so simple a model as in Eqs. (13.39) and (13.44) can cover so wide an array of calculation forms as those of univariate and multivariate analysis of variance, multivariate regression, principal components, Markov models, path coefficient analysis, structural equations, reliability–homogeneity theories, and factor analysis. In effect it brings out the high generality of application of "factor analytic thinking" in psychometry and related forms of analysis in other subjects.

13.14. Path Analytic Factor Analysis and the Quantifying of Causes

In Chapter 9 a wide range of possible interaction models for factors and variables was studied. As scientifically acceptable probabilities we reduced them to two: the *stratified uncorrelated determiner (SUD)* and the *reticular models*. As we have just seen, Bentler's multistructure equations and the SUD model are mutually consistent and it may be that multistructure equations can also be adapted to the reticular model, at least in the path coefficient form we shall now pursue. In our domain of scientific models we have been forced to confess—reluctantly because of its calculation difficulties—that the only universally acceptable model is the reticular one, and that it contains the SUD within it as a special form. It seems to the present writer that though Bentler's multistructure model can fit SUD within it, it cannot embrace anything but restricted (special assumption) forms of the reticular model. For the reticular model not only includes diverse factors acting on variables, and factors acting upon factors, but also circular feedbacks in which A acts on B and C, B acts on C and D, and D acts on A, in various positive and negative contributions.

It would meet practical research procedures, and presently known factor methods, if we assumed at this point that ordinary factor analysis has indicated rough outlines of the main factors, at their various strata, in the given domain of variables. Because of nonlinear relations, and the more complex reticular interactions, however, these initial patterns and strata levels will, as we have seen, be somewhat blurred. The particular causal actions and even in some cases their directions will, however, often be problematical and require exploration (Rickard, 1972).

It is in this connection that developments in the last decade of path coefficients and path analysis [Blalock (1967, 1971), Duncan (1966), Morton and Yee (1974), Van de Geer (1971), Li (1975), Tukey (1954), Werts and Linn (1970)] —after some years of neglect of Wright's basic proposals (1934)—are important. It is hard to say whether at this juncture path analysis is more important to factor analysis than factor analysis is to path analysis. In any case, the approach to integration of the latter here is to treat path coefficients as a natural extension of that emphasis which we have placed on factor analysis as a scientific model, rather than as a statistical mill for reducing data.

Accepting the reticular model as the most general causal network, we find in path analysis a means of stating a causal hypothesis and testing it. In itself it is no different from proceeding to specific exemplification of the reticular model, except that it has grown up around variables of any kind, rather than simply factors and variables, i.e., it has dealt with directly observable variables. Inasmuch as factor analysis can supply through simple structure and confactor rotation some evidence of factor strata and of what entities act causally on others, it reduces the amount of trial and error that path analysis has to pursue in its largely blind, or comparatively speculative hypotheses about the actual causal chains. Secondly, it goes beyond the observable variables to those hidden causal entities we call factors, which in many cases will provide the basis for a truer model than can be stated in terms of variables alone. On the other hand, the available factoring processes cannot give us directly, by any automatic procedure, the factor structure in those many forms of the reticular model which have complex feedbacks. Neither can path coefficients provide any such total lead, but they can enable us to state what the correlational relations would be from various models, and thus permit us by successive trial-and-error hypothesis experiments to fit a reticular model to the facts.

The essential propositions and calculations in path analysis can best be stated in connection with a diagram, as shown in Fig. 13.2. To begin with we must note that a path coefficient is nothing more than a regression coefficient in which the variables have been put in standard scores (and in some settings is therefore simply and precisely a correlation coefficient) in a setting in which one variable is assumed to act causally on another (as shown by introducing arrows, as in Fig. 13.2). It therefore states how much one experimentally independent variable contributes to the variance of the variable on which it acts. The aim of path analysis is to define the correlations that should exist among variables when any particular network of causal actions is hypothesized to connect them.

In Fig. 13.2(a), we see first a simple chain relation in which v_1 contributes to v_2, but is not the sole source of v_2 variance, and so on through two more variables. If we represent a path coefficient by p, then the coefficient p_{14} expressing the contribution of p_1 to p_4 can be written

$$p_{14} = p_{12} \cdot p_{23} \cdot p_{34} \qquad (13.45)$$

The next important element after a chain is a chiasm, as in Fig. 13.2(b). Taking the part encircled on the left first, we have

$$p_{11.2}^2 + p_{12.2}^2 = 1 \qquad (13.46)$$

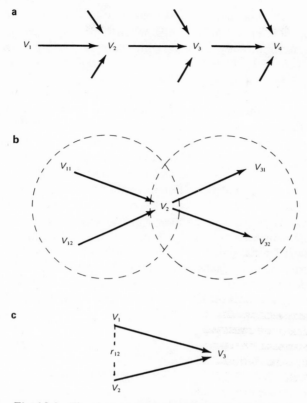

Fig. 13.2. The elementary propositions of path analysis.

since v_{11} and v_{12} entirely account for the variance of v_2, and by our initial definition of path coefficients all variables are at unit variance.

In the case of the second circle, since the whole variance of v_2 is again involved, we have

$$p_{2.31}^2 + p_{2.32}^2 = 1 \qquad (13.47)$$

In these cases the path coefficient is a standard regression coefficient which is also a correlation; but if the contributary variables v_1 and v_2 happen to be correlated, as shown in Fig. 13.2(c) *the equivalence of path coefficient between variables and correlation between variables no longer holds*, and path coefficients are then partial correlations.

The reader will notice that if we substitute factors for variables, the unity of factor analysis and path analysis is obvious, *provided the reticulum hypothesized in the path analysis is the same as that of the strata model in factor analysis*. Thus (*r* and *p* being the same) the correlation of

v_1 and v_4 to be expected from the given p is just what one would calculate by the Cattell–White formula when "vaulting" to the effect of a
third-order factor directly on a variable, namely, $r_{14} = r_{12} \cdot r_{23} \cdot r_{34}$. To
take just 3 variables, A, B and C, for a numerical illustration, if $r_{AB} =$
.71 and $r_{BC} = .60$, then where A is one of several causes of B, and B is one
of several causes of C, we know that A contributes 50% of the variance of
B, and B 36% of the variance of C. (Note incidentally that p's like r's
square to give the fraction of variance accounted for.)

Using Eq. (13.45) on this numerical example, we would conclude
$r_{AC} = .43$. From the 50% and 35% variance contributions above we would
conclude the fraction of variance of C accounted for is 50% of 36% = 18%,
which checks with the path coefficient (squared) result ($.43^2 = .18$).

The second example in Fig. 13.2 can similarly be perceived to fit a
factor model, where v_2 is a variable with no specific and v_{11} and v_{12} are
two factors loading it. It has a communality of 1.0 and the two factor
loadings squared must equal this. The second circle in the second example
is the case of only one factor in common to two variables, and we then
know that their correlation is the product of their correlations (loadings
in the orthogonal case). In path coefficient terms the correlation to be expected from path coefficients of this value, with arrows (causal connections) of this kind, is

$$r_{31.32} = p_{2.31} \cdot p_{2.32} \tag{13.48}$$

In the third case in Fig. 13.2 it is supposed that when the path
analyst chooses the hypothesis that v_1, v_2, and v_3 are related as shown he
also knows that v_1 and v_2 are not independent but related by a correlation
r_{12}. In this case the relation of path coefficients is

$$p_{12}^2 + p_{23}^2 + 2r_{12}p_{13}p_{23} = 1 \tag{13.49}$$

Again we see the identity of form (factor for variable, and specifics
nonexistent) with the strata-factor model where two correlated factors
contribute to one variable. The above are examples from basic elements
of networks but they show how the hypothesized connections can be
checked by results. For example, in the numerical example from Eq.
(13.45), if the empirically obtained r_{AC} correlation had shown itself to be
greater than 0.18, one would have to infer that A acts on C also directly
or by some additional path to that through B. The reader wishing to pursue path analysis further will find Li's primer (1975) a helpful next step.
The practical problem with path analysis is that as the number of variables
(which also means factors) increases, the number of possible reticular
hypotheses to be tried out in trial-and-error fitting to data increases out of
all proportion. Even the order of r variables in a chain has $n!$ possibilities,
e.g., 120 for five variables.

The extent of this problem but also of some gain from assistance from factor analysis can be illustrated by taking a concrete psychological example, which by its nature, inevitably involves seven variables. The assistance to path analysis from factor analysis, mentioned in passing above, is that factor analysis already has three methods potentially capable of tracking down some of the directions of causal action. The inference from and to causality is (1) explicit in confactor rotation, (2) implicit in simple structure, and (3) explicit in lead-and-lag correlation comparisons of factors on variables. Regarding (2), simple structure, our hypothesis is that the position in which a minority of a stratified sample of variables is loaded and the rest left untouched is that of a cause. However, we pointed out that the cause could be conceived as either (a) presently acting, or (b) the "archeological" residual pattern of an extinct determiner.

The real illustration we propose to look at is the hypothesis in Chapter 9 that the anxiety factor (indexed as QII in questionnaire and $U.I.24$ in objective tests) is not operating in the ordinary sense of a second order., i.e., being a determiner of some variance in first strata primaries, but even be conversely determined. We present this as a problem, not as a solution, for a full setting out of the inferences in a solution to Fig. 13.3 would take considerable space. Fig. 13.3 shows by interrupted lines the correlations that are known. These are not arrows, i.e., they have no direction. By continuous arrows it shows the causal connections for which path-coefficient values could be inserted for our hypothesis. It is possible that

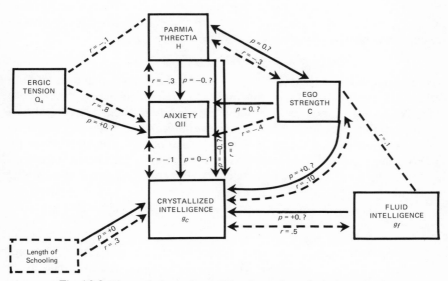

Fig. 13.3. A psychological example of a path-analytic hypothesis.

even with the defined correlations some different combinations of sizes and directions of path coefficients would fit those correlations, so the path-coefficient approach requires auxiliary experiment.

The first dilemma in path analysis occurs in deciding where to cut the network. In most cases it could go on indefinitely, as we are reminded by the fact that H, Q_4, C, g_f, etc., on the boundaries of Fig. 13.3 themselves have causal contributors, and could be joined to them. And where exactly is the dependent variable? We began with interest in anxiety; but we find anxiety is also correlated in the data we have (from schools) with school achievement (not shown) and with crystallized intelligence ($r = -.1$). It would be possible to treat crystallized intelligence as the dependent variable, with some shift of our interest, and in any case, since the hypothesis is that low intelligence in the school situation contributes to anxiety, we would naturally bring in the two chief contributors to crystallized intelligence, namely, length (with goodness) of school educations and endowment in fluid intelligence (by the investment theory, Cattell, 1961). The Cattell and Butcher (1968) high school survey indicates ego strength C also to be significantly correlated with crystallized intelligence, so we insert a positive path coefficient there.

At the moment only hypothetical signs are inserted in Fig. 13.3 for the p (path coefficients) and the next step would be to try hypothetical values. Although the precision of solution will usually be greater as more elements and evidence are brought in—much as a larger R matrix gives more precision of h^2 loadings, etc., in a factor analysis—it may be good strategy in path analysis to obtain first approximate solutions in parts of the network. Thus, for anxiety, taking the four direct contributors we can translate the figure into a formula:

$$QII = p_H T_H + p_C T_C + p_{Q4} T_{Q4} - p_{gc} T_{gc} \qquad (13.50)$$

where the T are the traits, identified psychologically by the usual subscripts (i for individual's score omitted). In the general use of path analysis, investigators have generally assumed that their hypothesis networks completely account for the given variable, so no U term for uniqueness (unaccounted variance) occurs in most path equations. However, in this case, since H factor correlates with Q_4 and C, and C with crystallized intelligence, the path coefficient statement of the variance of QII would be

$$\sigma_{QII}^2 = p_H^2 + p_C^2 + p_{Q_4}^2 + p_{gc}^2 - r_{HQ_4} p_H p_{Q_4} - r_{HC} p_H p_C + r_{Cgc} p_C p_{gc}$$

$$(13.51)$$

With this sketch of the aims, basic equations, boundary conditions, and relations to factor analysis, the student must be left to pursue technically more advanced developments, for which Li (1975), Morton and Yee (1974), and Van de Geer (1971) would be particularly valuable, the latter especially for "recursive" models. However, inasmuch as practical illustrations in diverse fields are useful, and path analysis has been much developed in relation to such special fields as genetics and sociology, he is recommended to read the still unsurpassed writings of the originator, Sewall Wright (1932, 1954, 1960) and to proceed to Blalock (1971), Blalock and Blalock (1968), Borgatta (1969), Borgatta and Bohrnstedt (1970), Lazarsfeld and Rosenberg (1955), and Duncan (1966). In mathematical statistics he should turn to the work of Jöreskog (1973), Goldberger and Duncan (1973), Simon (1957), Turner and Stevens (1959), and Tukey (1954); and in biology and genetics to the applications of Wright (1954) and Rao, Morton, and Yee (1974a,b). In some domains and contexts the student will find path analysis dealt with under the rubric of "structural equations," which betokens a linear model expressing both path analysis and factor analysis.

Advances may well occur more rapidly in one or another of the areas of biology, economics, electric and hydraulic networks, genetics, psychology, and sociology, and one should be prepared to glance across academic boundaries. In psychology, the new developments most demanding path-analysis treatment are those of the dynamic lattice (Cattell and Child, 1975) as it appears in motivation measurement and structured learning, with their use of multiple determiners (Cattell, 1977). There the most provocative problem is that of recursive models, in which feedback may even run full circuit. [What we have called spiral action in the development of introversion above (Cattell, 1973) is such a case.] Conceptually, it is important that factor and variable in the factor model become fully integrated with path analysis despite its being at present couched in variables only; and that joint use of factor and path designs be developed to lessen the amount of blind trial-and-error pursuit of initially too numerous hypotheses with the latter. Especially important is the coordination of this hybrid "path factor analysis" with R, dR, and P technique designs employing time sequences to penetrate causal connections. For example, in Fig. 13.2, the hypotheses could be more fruitfully developed with evidence that as between ergic tension (Q_4) and anxiety (QII), a rise in ergic tension, due to frustration, is followed, not preceded, by a rise in anxiety. In most fields such sequential evidence can be obtained and in some cases, e.g., in genetics, where the behavior of animal offspring is related to the genetic makeup of an environmentally separated parent, the sequential conclusion is beyond cavil. Future use of path coefficients is particularly indicated in systems theory.

13.15. Brief View of Factor Analysis in Relation to Regression and Other Models

Through the variety of models surveyed in this chapter, the reader is probably beginning to perceive generic relations emerging among statistical models previously learnt in isolation. The last decade has seen a remarkable growth of synoptic views in which one can instance (hopefully not invidiously where so many have contributed brilliantly) the writing of Burt (1966), linking ANOVA and factor analysis; of Anderson (1966) and Tatsuoka (1975), linking regression analysis and discriminant functions; of Cohen (1968), bringing CORAN and ANOVA into one framework of analysis and significance testing; and of Bentler (1976), viewing the regression and factor analytic approaches in more generalized formulas. It is appropriate, therefore, to finish this chapter with an extremely brief integrative comment on relations of factor analysis, especially to various regression analyses. For a wider overview see Kerlinger and Pedhazur (1973).

Enough has been said early in this book about the properties and relative advantages and disadvantages of ANOVA and CORAN, of which latter factor analysis can well be considered the crowning expression. And in the last paragraph we have referred to conceptualizations which bring CORAN and ANOVA into a common perspective. Sufficient attention has perhaps also been given to the frontiers which factor analysis shares with path analysis, discriminant functions, taxonomic analysis, and canonical correlations. It remains in this section to look more closely at the relation to various forms of regression analysis, especially in relation to practical problems where uncertainties might arise as to suitabilities of the alternatives.

No one can have followed the above manifold aspects of factor analysis without already having a good background of understanding of regression, and noting such things as that the V_{fe} gives regression of factors on variables and the V_{fp} of variables on factors. A brief rundown on domains of regression usage, before proceeding, may seem superfluous, but a systematic tidying is helpful at this point. Let us pause to note that the logically complete possibilities in regression analysis, speaking for precision in terms of linear regression, are:

1. Predicting (estimating) one variable from another.

2. Multiple correlation predicting one variable ("criterion") from several variables.

3. Predicting many from one, which is the same as 1, unless we add the notion of predicting some composite of many from one.

4. Predicting some weighted or unweighted composite of many from a composite of another set, which we have seen in canonical correlation above.

5. Various designs of partial correlations in which the prediction of X from Y is considered when a third variable Z is partialled out of both, or out of only one of the two (called, respectively, partial correlations and part or semipartial correlations).

6. The part or partial regression in 5, considered for groups of variables rather than single variables. This we shall look at as "commonality analysis" below.

An overview of the meanings of most of these, as they proceed from simple regression to factor analysis, is given in Fig. 13.4. Simple regression of b on a, i.e., estimation of b from a, would be represented by the upright arrow in (a); but (a) as a whole symbolizes *partial correlation*, by the abstraction of any c correlated variance in a and b from their rela-

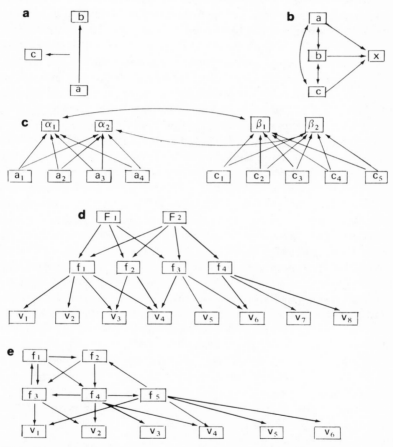

Fig. 13.4. Diagrammatic summary of relations of models in regression and factor analysis. (a) Partial correlation; (b) multiple correlation; (c) canonical correlation; (d) factor analysis (stratified); and (e) factor analysis (reticular) with path analysis and recursive relations. Except in (d) and (e), the arrows show dependence–independence in an arbitrary mathematic sense. In (d) they may also have a causal action sense (according to one's theoretical choice) and in (e) they are to be interpreted as entirely causal.

tionship. [It is surely unnecessary to pause here to give the formulas for partial and part (semipartial) correlation calculations; see Dubois (1957)]. Incidentally, it will be remembered that we encountered an unusual use of partialling in regard to dR technique, where it was debated whether to take out the initial or final score level from the difference score. However, the most obvious use of partialling in factor analysis is the partialling of the influence of other oblique factors from that of a given factor, upon the variables, in the V_{fp} matrix.

Multiple correlations, symbolized in Fig. 13.4(b) take account of the relations among a, b, and c in getting the best (least squares) weighted composite to estimate x. In factor analysis we get this in both directions of regression: estimating factors from variables, in a column of the V_{fe}, and estimating variables from factors, in a row of the V_{fp}. In either case the goodness of the estimate—the multiple R—is obtained by multiplying the vector of weights—b's in multiple correlation, w's in our notation—(a column from V_{fe} in the case of V_{fe} and a row in the case of V_{fp}) by the corresponding vector in the V_{fs}. Thus, in estimating variables from factors, the row for the given predicted variable from the V_{fp} lists factors' weights and the row from the V_{fs} their correlations with the "criterion" variable. The difference of ordinary from covariance factoring is here simply that between standardized weights (beta coefficients) and unstandardized weights (B coefficients). In connection with choice of subsets of predictors and the procedures of stepwise regression, the reader will find valuable slants, over and above textbooks, in chapters, respectively, by Hocking (1977) and Jennrich (1977).

A next step in regression designs shown in Fig. 13.4 is prediction of a weighted set of a u class of variables from a weighted set of a v class (say, criteria and predictors). The form of this, which simply aims at maximum prediction, we have already studied as canonical correlation.

In the present context we may well pause to look at some further possibilities in canonical correlation (Section 13.5). The step of partialling out one set of variables from the canonical relation between two other sets, envisaged in passing above, was achieved by Cooley and Lohnes (1971, see their Chapter 7) and would be useful in many situations. Other advances are Meredith's (1964) allowances for error in the data and Thorndike's and Weiss's (1973) examination of stability of canonical variates in cross validation.

Recently Thorndike (1977) has reacted to problems of canonical regression raised by Cronbach (1971) and by Cooley, in the broader perspective we are now considering. Cronbach points out that playing at once with several predictors and several criteria does not make for straightforward decision procedures, since the practitioner has his own notions of what criteria are more and less important to him. The difficulties of interpretation which we raised back in Section 13.5 also render the answers uninviting to the theorist, and, further, as Cooley mentions, there is necessarily a tendency to capitalize on covariations specific to the sample. Thorndike's proposal is to make stepwise calculation among the predictors in relation to the criterion or criteria of importance to the applied psychologist concerned. This partial and controlled use of either canonical correlation or multiple regression should recommend itself to the practitioner, but its aims and methods continue nevertheless to be restricted by the characteristic differences (arbitrariness, conceptual bareness) of regression and canonical regression methods from factor analysis. However, in applied factor analysis there exists the stepwise regression analogue in which one finds when a factor in a specification equation for a criterion (or chosen group of criteria) declines in contribution to the point where testing time is saved by dropping it.

A central and frequent problem in regression analysis with groups of variables, and one which needs to be related to factor analysis, is that which is sometimes called *commonality analysis* (Coleman, 1966; Kerlinger and Pedhazur, 1973; Veldman, 1975). Here one has certain meaningfully different sets of predictor variables, e.g., measures on personality ability and measures of educational experience, and a criterion variable (or group of variables). Unlike canonical correlation one is here concerned, not with maximal prediction *per se,* but with *partitioning the contribution into that due to set A and that due to set B relative to the joint contribution of A + B* sets. If sets *A* and *B* are uncorrelated, there is no problem, since the joint contribution is the sum of the two sources but almost invariably they will be correlated. A clear and detailed statement of equations and treatment is given by Veldman (1975) and other references above. Veldman (1975) says that through commonality analysis that part of the variance of a criterion that is predictable "can be partitioned with respect to the independent and joint contributions of the predictors." This states the essence, and is embodied in his computer program, but we should be careful to see just what uncertainties this leaves.

For in the last resort what we are able to conclude depends on the logic of the situation, not statistical calculation alone. The conclusions, in short, depend upon our scientific model. It can be seen most clearly in the case of single predictors *A* and *B*, though it applies also to sets and any number of sets. In a blind situation the only way to treat the covariance of *A* and *B* is to divide it equally between them, finishing with one-half added to the variance contribution of each to produce a statement of the total contribution due to each. The alternative is arbitrarily to give stepwise priority to *A* set or *B* set, which assigns contribution unfairly. If one deals with factor concepts, however, he will look for common or second-order factors first, in the two groups of variables, and finish with a specification equation predicting the criterion from whatever primary and secondary factors are back of the two sets of variables.

It is at this point that the reader will best begin to appreciate the logical, scientific, and historical-developmental differences which distinguish factor analysis from all other regression procedures. At first one might think that using regressions, full or partial, between sets of variables is virtually factor analysis. In statistical terms certain steps in the two may be identical, but the conceptual framework, and the uses which history has brought, are very different.

As to the former apparent similarity, one must recognize that the regression groupings are nearly always arbitrary, in the sense of being local to a situation. The notion of a causal direction, as distinct from a statistical prediction direction, is generally absent. On the historical side one can see that the more developed regression methods contrast with factor analysis in having developed very much in applied psychology, economics, etc., and it is easy to see why. An immediate, specifically adapted solution for maximum prediction is reached, untrammeled and unassisted by theoretical second thoughts. This is clear in the pioneer work of Hotelling (1935) on "The most predictable criterion." (A theoretical psychologist would never sell out his favorite theoretical variable or concept for a more predictable hash!) And in the maturation of regression analysis developments in the fine work of Horst (1961, 1965)—developments which spring from the interests of a life-long professional applied psychologist—one sees the same difference of emphasis. The basic laws of multiple and partial correlation are vital to a correct use of factor analytic concepts, but the latter involve decisions rooted in broader evidence and operating as continuing scientific notions. This latter calls for centering factor regression procedures on primary and higher-strata factors, correlated in whatever ways nature has prescribed.

Accordingly, reverting to the progression in Fig. 13.4 we come next to the standard stratified-factor model, which takes the regression form of serial regression equations (Tatsuoka, 1971b), and the SUD model with the Schmid–Leiman regression equations.

Finally, at (e) we have the most general reticular model, with recursive and feedback $(f_1 - f_3)$ relations, which, as we have seen, cannot be put into the "mill" of any straightforward regression or factor analysis program. Although each connection is one of a regression relationship the correct representation and solution of variance contributions can only be reached by path analysis, guided by trial-and-error hypothesis formation and testing. With the exception of the last, Bentler's multistructure model suffices to subsume all.

In summary, we may repeat the three basic differences of the scientific model of factor analysis from regression analysis: (1) that in common practice with the regression model one takes concrete variables on both sides (for the criterion is normally concrete), whereas in factor analysis one searches (by factor extraction methods) for a set of abstract variables on one side, (2) that the particular abstract variables—the factors—are fixed uniquely by some properties scientifically extra to the statistical system and not just any linear combination of variables such as could be set up in pure regression practice, and (3) that whereas the independent and dependent variables in regression are commonly (unless manipulation has taken place) only independent and dependent in a statisticomathematical sense (the direction being the choice of the experimenter) in the influence factor model at least the factors are considered as the causally independent variables. However, not all factor analysts follow this third clause, and the user is, in fact, statistically free to make either factors or variables the independents.

There remains perhaps a statistical question, namely, whether the weights giving maximum prediction in multiple regression are more or less stable from sample to sample than those for factor estimations. The stability of regression weights on variable sets has long been investigated (Mosier, 1943) and the general experience (McKeon, 1965) seems to be that cross validation is frequently poor. Lawlis and Chatfield (1973) find stability particularly weak in canonical correlation weights. A comparison of stability of factor weights against other regression weights would not at the moment rest on sufficient evidence, but both empirical impression, and the expectations from the nature of simple structure factors suggest higher stability in the former.

13.16. Summary

1. Some modifications of the factor model are attempts to make it fit better the indications of what the scientific model of latent influences in nature really needs to be. Others are based on mathematical and statistical possibilities of alternative assumptions, producing in some cases better consistencies in outcome and comparative irrelevances to a scientific model in others.

2. The built-in assumptions of linearity and additivity in the factor model fail to meet the full needs possible in a scientific model. However, this failure to fit occurs much less often than is supposed. It has been shown repeatedly and empirically that, even when nonlinear and nonadditive relations do exist in the data, the solution tends to give the essential factors, but relates them, for example, as apparent sums rather than the products they are in that case. A beginning has been made with non-

linear factor analysis; but ordinary factor analysis, applied with tactical skill, will normally permit more complex specification equations than linear and additive ones to be reached by trial and error, e.g., by factoring a succession of small-range samples. The fitting of product and exponential expressions of factors to curves for variables follows the usual course of exploration in science, once the factors as such are sufficiently located by factor analysis.

3. One misfit of the factor model to the operation of learning, genetic, and other influences arises from such factors operating with somewhat different patterns in variously selected subgroups. Differences of pattern for a real influence undoubtedly exist between social classes, ages, and the sexes. Consequently, it behooves the psychologist, if he wishes maximum correctness of pattern, highest possible communality, and greatest precision in prediction, to discover the forms of the patterns in such more homogeneous groups. This requires acknowledgment of the model of among-type and within-type dimensions, although the types may never be wholly separable. Comparisons of factor scores across groups offering different patterns require use of *equipotent* or *isopodic* scoring principles.

4. A distinction is drawn between half-a-dozen *computational modes of extracting factors* on the one hand and some four (eight admitting rotation) *ultimate models* on the other, even though some modes of extraction involve or preclude some models. The specific natures of some ten different combinations of extraction methods and models—centroid, principal, axis, alpha, image, canonical, generalized least squares, maximum likelihood, unweighted least squares, real-base factoring, and confirmatory maximum likelihood—are discussed, showing how they differ in mode of weighting variables, tests of fit, the role of iteration, and the form of the scientific model.

5. The method advocated here for highest accuracy is the maximum likelihood method of Jöreskog, Rao, and Lawley (though weighted and unweighted least squares have their virtues) and the model of widest verity is real base, in SUD or reticular form. Nevertheless, as an extraction method, maximum likelihood may sometimes be too expensive of time, too restricted in number of variables usable on certain machines and programs, and too sensitive to changes in size of N. The ordinary principal axis based on a scree determination of number of factors probably suffices for general use. A valuable offshoot of maximum likelihood is *proofing maximum likelihood* (often called "confirmatory") for hypothesis-testing use of factor analysis.

6. The general researcher should take fashions for certain methods of extraction (though not of models) relatively lightly, for the resulting numerical differences are generally trivial, relative to those from general

experimental design, e.g., effects of choice of variables, and factor number, and thoroughness of rotation, etc. The centroid is obsolete except for those without computers; alpha analysis pays attention to sampling of variables and homogeneity of factor scores but uses questionable weighting and is not a commonly recommendable method; image analysis steps outside the true factor model and brings specifics into the common factor space; canonical factoring is superseded by maximum likelihood; proofing (confirmatory) maximum likelihood is the best hypothesis-checking approach; but use of scree, principal axes extraction, and oblique simple structure meets most requirements of ordinary usage.

7. The aim of the real-base, true-zero factor model is to permit factors to differ in size (variance) and mean level from experiment to experiment, and factor to factor. It does so in order, among other things, to integrate results with ANOVA treatments, etc., and to relate to real units and real unities in the physico-social world. It involves covariance factoring along with the concept of a standard "core" factor pattern matrix at which "standard raw-score" units are fixed for all variables. This core matrix is the mean of matrices across a multisituation–multidemo set of experiments, among which changes in factor size are determined either by demographic or situational changes. The latter changes are governed by modulation theory, using the concept of the modulation index or coefficient.

8. Real-base factoring rests on the postulate of constancy of factor effect on variables, i.e., of a fixed ratio of factor and variable changes in raw scores. The calculations to this point defer the issue of a true zero but recognize differences of mean from the central AZUS (artificial zero, universe standardized) score. An analysis is presented of three concepts and measures of factor size: potency, P, relative size, E, and final size, F. Four means of calculating factor size are discussed and the various formulas for transformation to ordinary factor analytic values are presented.

9. The conception of *factor size change* by change in range of person scores and by modulation from situations can be handled either by n-way factor analysis in which each set is factored in succession using the mean score on the variables across the other sets, which gives an additive relation of factors and loadings, or by the modulation approach of the multidemo–multisituation, which handles the increase of variance by ratios instead of additions. The latter seems better to fit psychological concepts, e.g., of modulation of states by ambient situations; but the former cannot be translated into ratios, as in the latter, unless the recognizable factors among n responses from factoring across persons, situations, and other sets turn out to be the same.

10. Broader models of factor analysis arise in a simpler more con-

crete sense from different qualities (metric, ordinal, nonmetric, and non-parametric) in the variables and the association coefficients employed. We distinguish these from the more fundamental varieties due to the different statistical and scientific models. Nonmetric and nonparametric uses are illustrative of the former but have only restricted use. More embracing conceptions of statistical analysis models, devised to subsume various analyses as special cases have been proposed, notably by Jöreskog, Guttman, and Kaiser, and especially in Bentler's multistructure model. The latter is indeed an equation embracing factor analytic models, regression, univariate and multivariate analysis of variance, etc., and has close resemblance in its broadest expression to the SUD scientific model. However, the scientific model of greatest generality is the reticular model of which SUD is a restricted case, and the multistructural model cannot at present embrace the recursive forms of the latter.

11. Any overview of factor models requires that relations and relative utilities in regard to diverse regression practices be understood. A diagrammatic summary of relations to simple, multiple, partial, and part (semipartial) regression and to canonical correlations and commonality analysis is given. Factor analysis is two-way regression with abstract variables on one side and concrete on the other. It differs from most regression practices, however, in not resting on arbitrary groupings. Canonical correlation fits some applied psychology needs, but is no substitute for factor analysis, for in both it and commonality analysis the solution is local, and uninformed by wider scientific knowledge, unless combined with factoring.

12. Path coefficient analysis, including path factor analysis, is of considerable importance for further penetration of complex fields in a way to check causal understandings. It requires that correlations be known among factors and variables and that directions of influence be hypothesized by arrows in a reticular model. The expression of path coefficient solutions is ultimately as p weights in contributing to variances of factors and variables in various chain and lattice models, recursive and nonrecursive. The inference from observed r's to hypothesized p's requires considerable labor because of the large numbers of alternative possible p hypotheses arising with even a few dependent and independent variables in a reticular relationship, as in systems theory. Consequently, one restricts to a more probable series of hypotheses through preliminary guidance by tactical ancillary experimental aids such as SS and confactor resolution, and sequential and manipulative dR- and P-technique results.

CHAPTER 14

Distribution, Scaling, and Significance Problems

SYNOPSIS OF SECTIONS

14.1. The Five Domains of Distribution and Sampling in Factor Analysis

As the statistical interests of psychologists advanced, with expanding mathematical models, beyond those of psychophysics, reaction times, paired comparisons, stochastic processes in learning, and the like, they found themselves in domains which lacked the statisticomathematical landmarks to which they had been accustomed. ANOVA provided a tight, dependable system for examining the significance of experimental results in simple models; probability theory offered mathematically firm answers in other directions, and so on. It was largely in the factor analytic developments out of CORAN that psychologists found themselves having to

441

make shrewd, even "artistic" judgments, and the more "precise" investigators found these rough pioneer customs not easily tolerable.

The uncertainties showed themselves in such areas as decision on
unique simple structure, on the significances of rotated factor loadings,
the treatment of time series correlations in P technique, the decision on
number of factors, the goodness of fit of an obtained to an hypothesized
factor pattern, the matching of factors across researches, the separation
of species types in factor space, the reliability of a factor score, and
several other places. While one can sympathize with the requirement of
meticulous objectivity, and exact dependence on statistical assumptions,
this precision can be bought at too high a price. Research can ill afford
retreat from a scientific model that fits scientific concepts to the evidence,
broadly, if loosely, into the ivory tower of an internally "tight" but
patently incorrect model. A certain school of mathematical psychology is
in the position of those who urged transport planners in the 1920s to stay
with well-understood automobiles and steam engines rather than trust the
doubtful ways of airplanes. Fortunately, one by one the subjectivities
in decisions on factor analytic resolutions and significances have been
largely removed in the last twenty years, though few critics and not all
practitioners seem aware of this advance. One must still say "largely,"
because proofs for the mathematical bases of some decisions are bound
instead to remain for an interim period, as inductive or Monte Carlo
method "laws," rather than to stand explicitly on deductive and mathematical foundations.

Now up to this point our aim has been to pursue a comprehensive
view of what factor analysis is and what it means for science. Consequently, we have skirted some questions of statistical significance which
belong as a last refinement in the use of the method. Let us now turn,
however, to tighten the rivets in the structure of the ship, sealing some
still open joints. Those possibilities of leakage of error have not passed
unnoticed, inasmuch as we have given much attention to uncertainties in
the number of factors decisions, the uniqueness of rotational resolution,
the magnitudes of hyperplane loadings, etc., but the time has now come
for more formal examination. This is a tidying function, first in regard to
evaluating significance of parameters and secondly in regard to scaling and
miscellaneous matters related to these.

Significance tests actually concern themselves with two uncertainties,
those due to sampling error and measurement error, though the effect of
the latter is generally necessarily looked at within the framework of the
former. A stated probability of rejection of a null hypothesis depends on
knowledge of the distributions of the measures involved and that distribution in any given instance can be examined in terms of the sample size and
the reliability of the measures themselves. The study of distributions in

psychology has concerned itself largely with distribution of response measurements, over people, but recognition of the data box alerts us to the need in factor analytic work to broaden our concepts to consider distribution and sampling problems over five sets, not just this one. Indeed the experimenter with broader perspectives must be on his guard against unconsciously assuming that rules developed about people as referees do not get applied without further inspection to other sets of referees— such as occasions in P technique or stimuli in other facets. All this applies to sampling *referees*—the statistician's idea of population—but in multivariate research it becomes necessary also to consider distribution and sampling simultaneously on the *relatives*—typically "tests" in much psychological research. This becomes particularly important in *representative design* of experiments, as we have seen. Thus, if a personality structure researcher is to have confidence that his study is (a) comprehensively finding what can properly be called the main factors in a domain, e.g., personality factors, (b) possessed of sufficient hyperplane stuff to give a dependable basis for a hyperplane in relation to each of all the factors, and (c) not blowing up some narrow specific into the status of a broad common factor (as may occur when a specialist in some area proceeds from a population of his favorite intensely studied variables, without sampling guidance), he must heed the population sampling of relatives as much as of referees. Incidentally, the illustration has already been given of the empirically discovered limit in the operation of (b) above, that rotating with as many as six factors on as few as twelve variables is likely to leave two or three practically equally good simple structures for at least two or three of the factors.

Turning to population and sampling principles in the referees one finds, at present, very little discussion concerning sampling of occasions (ambient stimuli), focal stimuli, responses, and observers. In regard to stimuli and responses, Cattell and Warburton (1967) have developed an explicit basis for stratified sampling of stimuli and responses applicable especially to the domain of tests. This basis, to which we can only refer briefly, hinges on the principle that every response and every stimulus is an id that can be measured in a multitude of ways. Time sampling throughout the day, while adequate to define a total population for stratified samples of stimuli and responses as ids (pattern entities), does not give us the dimensions by which insightfully to define and stratify the personality sphere of measurable behavior. A reduced set of dimensions of stimuli and responses as such needs to be reached. The Cattell and Warburton examination of ways in which stimuli and responses can vary and be measured attempts to provide a logical basis for this definition of a population of behavior measures. Meyer, Kaiser *et al.* (1977) have pursued the question from a more statistical viewpoint.

Probably the next most important sampling need arises in the set of occasions (ambient situations). Here the work of Sells (1963), certain sociologists, Barker (1963, 1965), Cattell (1935, 1963a), and others offer a reasonably encyclopedic basis for a population, and a stratified sample thereof, in this area. Sampling of observers is an untouched subject and obviously a survey of role relations would need to precede any plan for a stratified sample on this set of ids.

This last comment brings us to an aspect of distribution little considered in classical distribution and sampling theory, The latter systems generally set up a somewhat unreal, academic notion that there exists a homogeneous population that is essentially normally distributed, but from which random samples may be taken. It is true that unless such assumptions are made our neat answers about distributions of sample means and variances, and much else in ANOVA, for example, become invalid. Nevertheless, it behooves us to ask how likely the assumptions are, and to develop some rational, if not equally simple, solution in regard to greater complexities if they exist.

Principally, we ought not to forget that the biologists' taxonomy of phylum, class, order, family, genus, and species almost certainly exists structurally in sociopsychological as in biological variables. If the statistician's "population" concept is not to deceive us we must recognize that any actual population—the population in a city, a country, or a classroom—certainly does not fit the assumed definition. It may—and probably does—show a fairly good normal distribution on any one variable (though Burt's observations on exact determination of intelligence mental age distribution show that even in so common a variable it is not exactly true). But, as Fig. 14.1 reminds us, the demonstrated existence of a normal distribution on each of two uncorrelated dimensions does not guarantee the existence of a "bivariate normal distribution." There may be several different species so coincidentally thrown together that they yield practically normal distributions on various single dimensions despite the population being segregated in species types. This becomes more likely with more dimensions. Alternatively, the single variable distributions may not be as normal as our sampling formulas require.

As regards factor analysis we have already pointed out that a factoring should ideally be done in samples from an essentially homogeneous group, and that factors get blurred in pattern and reduced in loading when this is not done, and hence, where different types are thrown together. But how does one locate such homogeneous groups if they are not necessarily revealed by distributions on single variables? This question brings us to the need for a method of discovering "types" in psychology. Typological concepts and methods must be a momentary digression from our main concern with distributions and significances. Fortunately, it is a subject now with a technically substantial literature and since the best of it deals with typing procedures subsequent to factor analysis, it logically belongs here. We shall therefore take a brief view of the essentials of typing methods and concepts, and become equipped to return to significance problems with a more realistic view of populations.

14.2. Taxonomic Principles in Recognizing Homostats and Segregates

It has been pointed out that to use the concept of "type" for the extremes on a single dimension, e.g., extraversion–introversion or intelligence, is a superfluous concept—and a vague one inasmuch as an arbitrary cut has to be made in the distribution to separate such "types." On the other hand a type as a species, segregatable from others, is a well-known natural phenomenon. It implies difference from other species occurring simultaneously on several dimensions, and the operational possibility of recognizing the species as a denser group of points in multidimensional space separated by relatively empty space. Such groups are shown in two dimensions in Fig. 14.1.

A type is thus basically a mode—a value at which individuals stand with unusually high frequencies—but a multidimensional mode in multidimensional space. It is not our business to pry far into causes for this, but in natural species one can see that only certain combinations of trait

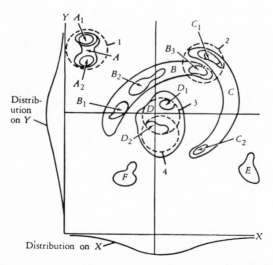

Fig. 14.1. Types recognized as modal frequencies in the form of homostats and segregates. (From R. B. Cattell, *Handbook of Multivariate Experimental Psychology*, Rand-McNally, Chicago, 1966, p. 312.)

levels are biologically viable—winged elephants belong to legend—and in human societies only certain combinations of skills fit certain occupations and other role positions. The space in which persons are located can be one of variables or of factors, but using the former is unattractive because it places variables that are generally much correlated as orthogonal coordinates. Such a space permits excessive weight to be given when many equally weighted variables actually represent virtually only a single factor, and it has a dimensionality unrelated to the true number of dimensions in the usual functional factor analytic sense. Accordingly, we shall assume that individuals are first measured in factor scores—usually on factors not deviating much from orthogonality and easily transformed into orthogonal equivalents. Each individual will thus have been given an economical profile (which is technically a vector) of standard scores on meaningful factors. This vector, at its tip, places the individual person as a point in k dimensional space.

The grouping of people in such a space can be found either by (a) a computer counting of frequencies in each multidimensional cube of space, systematically sequentially pursued through all possible cubes, or (b) by a Q matrix (not a Q technique matrix) bounded by people and having in each cell entry an index of the closeness of those two people in the factor space (Cattell, Coulter, and Tsujioka, 1966). Now the index of nearness of one person to another in such a space can be determined either by Mahalonobis' generalized distance function (suitably inverted of course), or by the pattern similarity coefficient r_p, applied to the two-factor score profiles. (Note that use of a correlation coefficient to determine the similarity of the two profiles is quite unsuitable for this purpose: it rewards for similarity of shape but ignores the absolute distance apart of the two profiles.) Since the generalized distance function permits no reliable comparison across systems with different scaling and numbers of factors, the pattern similarity coefficient is preferred for similarity entries in the Q matrix. Its formula is

$$r_p = \left(2k - \sum_{}^{k} d^2\right)\bigg/\left(2k + \sum_{}^{k} d^2\right) \qquad (14.1)$$

where k is the number of factors in the profile and d is the difference in standard score between the two people on each of the k factors. r_p ranges like the familiar correlation coefficient r from +1.0 for two people completely alike, to 0.0 for chance resemblance only, and so to −1.0 for two people as opposite as possible. The distribution is not precisely the same as for r, but it is near enough for the psychologist in practice to depend, in evaluating an r_p, upon his familiar judgments from correlation coefficients, and this also is a practical advantage over using d, the distance function. [On the question of distance functions discussed elsewhere here (p. 451) see also Holman (1972).]

Now the forms of groupings that can be recognized for a Q matrix are of two kinds: (1) What has been called a *homostat*, derived from "homo" as "like" and "stat" as "place." A homostat is thus a group of people standing in much the same place in scores in factor space, and (2) a *segregate*, to be discussed below. The homostat ("stat" for short) is defined as a small sphere in the space and it is worth recording only when it contains an unexpectedly large number of people for its size, i.e., is of different density from the average space. Operationally, the locating of a homostat means finding in the Q matrix a number of people all of whom resemble one another above some agreed upon high positive value of r_p. Such a group would be like those shown in the spheres (circles here) marked 1, 2, 3, and 4 in Fig. 14.1. The people in $A, A_1, A_2, B_1, B_3, C_1, C_2,$ $D_1,$ and D_2 would meet the same standard r_p criterion, standing as they do in spheres (1, 2, 3, and 4) of diameter fixed by the agreed r_p limiting value. By contrast, a segregate is a collection of people such that members are in a continuous chain of resemblance. But spread beyond a stat circle, as in the "islands" B and C in Fig. 14.1. Note that in this case *those at one end may not greatly resemble those at the other*. This failure to meet the homostat criterion in B and C in Fig. 14.1 stands in contrast to the tight packing of $A, E,$ and F. Although not all resemblances are high in a segregate (henceforth abbreviated to "ait," as in a river island), it is a segregated set of people inasmuch as the spaces around them are relatively empty of people. The objective delineation of an ait is more difficult than for a stat but a formula is possible. Thus, both stats and aits are operationally definable concepts which provide initial bases for the idea of types.

The means of finding such types have been worked out as the *taxonome principles* (Cattell, Coulter, and Tsujioka, 1966; Sokal and Sneath, 1963) and as a computer algorithm and program, TAXONOME[1] (Coulter and Cattell, 1966). A variant developed by Bolz (1972) is also very promising. In both a Q matrix of r_p's is first worked out from the individual's profiles on factor scores, or orthogonally transformed scores. A cutoff value of r_p for stats is fixed, and the program lists the stats of the sizes that are found. At this point the Cattell and Coulter program calls for a cutoff level, in terms of a minimum number of people considered to constitute a worthwhile stat. Since number is density this simply says that only stats above a certain density are significant, and what is significant must be decided in relation to the "texture" revealed in the field, i.e., the fraction of the sample of people that will be taken care of by stats of that

---

[1] A distinction should be made between the taxonome principle (Cattell, Coulter, and Tsujioka, 1966), which includes the homostat and segregate concepts, and TAXONOME program, since other computer programs could presumably be written operating with the same taxonome principles.

minimum size and over. In the second main step (proceeding to aits), one takes the stats as entities bounding a Q matrix and finds the amount of linkage among them, by fixing a minimum size of overlap (in persons) between them to constitute a link. The isolation of aits proceeds by matrix multiplication using Boolian algebra. The aits of various sizes are then listed by the program output.

One must recognize that there is going to be arbitrariness in any search for clusters, but one can at least make this explicit and adjusted to the general "texture" presented by the data. The TAXONOME program requires arbitrary decisions to plug in parameter values into the program at three points. (1) The r_p to fix the diameter of the stat sphere; (2) the size (number of people) of stat considered nonnegligible; and (3) the amount of overlap between two stats to constitute a link for an ait. It has not proved difficult to judge these on the basis of the mean r_p and mean stat sizes, i.e., by the texture of the material, and the program has proved more effective than most. Tried on plasmodes of known type structure, it has successfully classified ships (Cattell, Coulter, and Tsujioka, 1966), breeds of dogs (Cattell, Bolz, and Korth, 1973), musical instruments (May, 1972), though it has fallen down somewhat, in classifying psychiatric patients (May, 1971). Among nations it yielded (Cattell, 1957c) essentially the same cultural groups as independently reached by Toynbee (1947) in his "civilizations." Slightly different type analysis procedures have been worked out by McQuitty (1963) and by Sokal and Sneath (1963), and a recent summary (but called "Cluster Analysis") of some related procedures is surveyed in Anderberg (1973). Sokal and Sneath, and others have sought also a dendritic or hierarchical taxonomy, such as biology definitely needs. In TAXONOME this hierarchical placement in species and genus, etc., is reached by taking the mean profile of persons in each of the first-discovered types and then treating these as representing the individual profiles in a fresh type of search, a procedure which, like higher-order factoring, can proceed until only one type embraces all.

14.3. Within Type and Between Type Factor Dimensions

The above excursion into typology is essential if we are to refer intelligibly and precisely to types in what follows. It has been pointed out that though high homogeneity merely in the sense of small sigmas on individual variables is not tactically desirable in factor analysis (since larger variance means clearer factors), a typal homogeneity *is* desirable in most research in the sense that having all ids of a common species is desirable. Physical organ factors—for horns, wings, etc.—would be obscure if one factored cows, birds, and cabbages together. But if the only objective way to locate types (where one's senses, or past "clinical" attempts at typing, are not dependable) is by the above taxonomic principles, then one is compelled initially to obtain the dimensions for using r_p by factoring what is probably an appreciably heterogeneous population. The entrant to an entirely new field may well ask "if good factor analysis requires species homogeneity I must work with the relatively pure species types, and to find these must I work with dimensions first obtained from a poor factor analysis?"

The question brings out two important answers. First, one must distinguish the goals and conditions of factoring across types from factoring within types. Second, one must recognize that, as in most other areas of exploratory research, an iterative and therefore programmatic approach is indispensable. What needs to be done is reasonably obvious on common sense grounds. To separate horses, cows, sheep, crows, and dogs, the taxonomic investigator has first to choose a set of variables that are measurable on all and that are relevant with a common operational definition, such as length of body, nature of skin cover, length of legs, number of teeth. [A good discussion of relatively objective approaches to "relevant and irrelevant variables" is available in McQuitty (1963).]

These cross-species variables will suffice to generate cross-species factors, and on cross-species factor profiles the use of r_p and the TAXONOME procedure will yield the component species types. At that point some obscurities and difficulties in clear separation of the species will probably suggest some new or modified cross-species variables for a second, iterative factor experiment. With species separated, one is ready to seek the new domain of variables for the separate factorings within each species. Initially, however, it is inadvisable to throw away all the cross-species variables in the within-species factorings, for they give some orientation to the new intraspecies factors to the old interspecies factor and having been used on the individuals of the whole population they are certainly amenable to being continued as measures within species. An example of within and between factoring approximating the above design can be seen in Linn, Centra, and Tucker (1975). The reader should also consider Aiken (1972) on types and dimensions.

A third possibility is to take, in the cross-species variables, the scores presented by the means of each species, assuming there are enough species for correlation. This should give essentially the same factors as across the population, but with some changes of emphasis because the within-group variance–covariance when we include several ids in each species is now omitted. The analogy to ANOVA and ANCOVA within and between group variances on single variables will be readily seen.

If we look at the purely logical concepts underlying the long-employed biological taxonomic systems we see that they are neatly paralleled by statistical concepts. Just as the biologist places the individual in his species and genus, so we represent the individual by deviation scores on the dimensions within his species, and his species by a profile of deviation on cross-species dimensions from the mean values of the genus. These deviations can be summed to a final "population" score, as in Eq. (14.2), but only on those variables and factors that happen to be common to within and between species. So that if F_c are the factors for measuring the species mean within the genus, and F_w those same factors now measured over individuals within the species, we have

$$a_{ipj} = \Sigma b_{jw} F_{wi} + \Sigma b_{jc} F_{cp} + \text{uniqueness} \qquad (14.2)$$

where i is the individual and p his species. The total information we have, however, is much richer than is usable in Eq. (14.2), for the within-species (and sometimes the between-species) dimensions will contain much not in the cross-species dimensions.

However, from a factor analytic standpoint the important consideration about first locating "species" with some measurable degree of homogeneity of pattern and then factoring within is that more useful and powerfully predictive factors (i.e., factors of high communality contribution and stability of pattern) are likely to emerge. For, as pointed out earlier, if the same determiner has operated in different subgroups the fit of the factor model patterns to the real pattern of the determiner, when a conglomeration of such subgroups is factored, will be poorer. For one thing, the presence of an uneven subspecies distribution, in regard to density, in the population space, will tend to produce curvilinearity of relation absent in the separate more homogeneous groups and perhaps departures from homoscedasticity and other conditions needed for good use of the correlation coefficient. In short, examination of the multidimensional distribution is desirable for understanding obtained factor structure and for reaching clearest results. At times there will exist subspecies of which we would be unaware without such examination, but factoring would often be improved simply by making separate attacks within the groups we do know: the sexes, age groups, occupations, classes, races, ethnic groups, and national cultures. (Humphrey and Horn's examinations of intelligence factor patterns in Negroes and whites offer a recent illustration in this area.)

Parenthetically, it is the writer's experience that students often confuse the locating of natural typal modes (species types) by TAXONOME with the use of multiple discriminant functions to get greater separation of types. Space forbids us to define the latter here, other than to say that it assigns different weights to each variable (say, each between-group factor) such that it will reduce within-group variance of people on the total combined score relative to between-group variance (for a brief but excellent treatment, see Tatsuoka, 1970). On a composite single score so derived people are more clearly grouped than on any one variable or on an unweighted composite. However, it is important to recognize that the multiple discriminant function can only work when the groups and their membership have already been given to the investigator. The multiple discriminant is not a means of finding types. It has been suggested that taxonome and multiple discriminant functions be combined, the former locating the naturally existing groups in the unweighted undistorted space, and the latter then pulling the groups apart more completely by a revised scoring. This is a promising technique; but one must remember in using it that such weights are not "permanent" values. The multiple discriminant functions (the dimensions of which are one less in number than the number of groups) change each time a new group is added to the set among which separation is to be maximized, though this change becomes less as more groups are involved in the calculation.

14.4. "Multidimensional Scaling" or "Dimensional Integration" Concepts

One would have to wear blinkers to write on factor analysis without reference to an organic connection with its neighbor, multidimensional scaling. Yet we propose to sketch the latter only very briefly, for it is scaling in the broader sense which has more fundamental importance for factor analysis.

We have just finished a discussion of distributions and distances apart in factor space. However, this applied to individuals (in general terms, referees), whereas in multidimensional scaling it applies to tests (in general terms, relatives). That tests can also be placed in factor space we know full well from discussions on simple structure. In the space of TAXONOME research the coordinates of the point are the scores of the given referee but in the present approach the projections are the loadings or correlations of the tests.

In either case if we had some way of directly measuring the distance apart of pairs of points (people or tests)—some way independent of first getting their projections in the factor space—it should be geometrically possible, from such data given for several such pairings, to decide the dimensionality of the system necessary to contain them (see elementary instance, Cattell, 1957b). Seemingly, it has not interested psychologists to do this in any precise sense with person distances, but a highly elaborate branch of psychometry—multidimensional scaling—has grown up around doing so for tests and stimuli.

Unfortunately the greater part of this development has been built on subjective judgments of the likeness of two stimuli or the "distance" apart of two kinds of performance. But when it is done with the best behavioral operations for estimating such distances, the number of dimensions found necessary for encompassing the data from all the paired comparisons comes out encouragingly close to that from factor analysis of the same data (Tucker, 1972; Dunn-Rankin, 1970; Carroll, 1972; Overall, 1976).

It seems reasonable to conclude, therefore, that multidimensional scaling should be logically viewed as a branch of factor analysis, in which the distances of tests (in R technique) rather than their correlations (or people's, in Q technique) are made the basis for erecting the dimensional system. Rummel (1970), in reacting to this reciprocity of projections on dimensions and interid distances, has suggested that multidimensional scaling would better be called *dimensional analysis*, for it has nothing to do with scaling in the ordinary sense. The present writer would agree that "scaling" is a less than apt term and would prefer distance integration, since the method moves from given id distances, i.e., that between specific variables, to an integration of the space necessary to contain them. Indeed, if we may paraphrase Rummel's comment (1970, p. 507), which we quote elsewhere, it would read: "What integral calculus is to differential calculus, distance integration is to factor analysis."

Parenthetically, lest anyone feel an inadequacy through our omitting the exact expression for distances between tests, let us here do so. It is the formal equivalent of Mahalanobis' expression for distances between people, as discussed but not set down in the last section.

In the general oblique case:

$$d_{jm} = \left[\sum^{x=k} \sum^{x=k} (b_{jx} - b_{mx}) c_{xx} (b_{jx} - b_{mx}) \right]^{1/2} \tag{14.3a}$$

where c_{xx} is an element in a covariance matrix. This simplifies in the orthogonal case to

$$d_{jm} = \left[\sum^{x=k} (b_{jx} - b_{mx})^2 \right]^{1/2} \tag{14.3b}$$

where the b's are loadings, as usual for tests j and m, on orthogonal factors. This could have practical research value especially in O technique where occasions are factored over responses and would help in objectively deciding the similarity of the psychological meaning of any two occasions. The general distance function d is just another way of evaluating pattern similarity, which can be used for any type of id in any kind of orthogonal or oblique space, e.g., to state the similarity of scales (tests, stimuli), of people, of responses, and of observers. It is, in fact, part of the formula for the pattern similarity coefficient r_p, but, as we have seen, it is less apt for some purposes.

Pretty though multidimensional scaling is in its mathematical elaborations, one must never overlook its at present unremedied weaknesses.

Commonly, the first is that the determination of distances is a human, subjective guess on "degree of similarity" or some other unguaranteed operation. In cases where one is interested in this very subjectivity, wishing to discover the perceptual dimensions of the subject's world, this is no objection. Even there, however, direct factor analysis could do as well by operating on subjective ratings or rankings, as in Dickman's (1960) study of children's dimensions of perception of ballistic behavior or Osgood's semantic differential. It seems unlikely incidentally, that the small dimensionality of Osgood's semantic space would be sustained by multidimensional scaling.

However, for the behaviorist in psychology, and for objective work in other sciences, it is quite possible that distances could be based on objective operations and the results be of wider use than for private perceptual worlds. Check and countercheck by factor analysis and multidimensional scaling could then function effectively. The present writer has suggested evaluating similarity by degree of function substitution possible, as determined by experiments extra to any factor analysis, e.g., how many times do people take (1) a bus, (2) a plane, (3) a train, when they cannot use a car? But Rummel (1967, 1970, 1972), in sociology and political science, has suggested several real bases for distances, e.g., those between nations as defined by frequencies of interaction, etc.

14.5. Equal Interval Scales and True-Zero Factor Scores

As in every other kind of psychological calculation and analysis, factor analysis has had to take on trust the assumption that raw-score units are sufficiently close to truly equal units to cause no serious disturbance in using the model on the assumption of true equal interval properties. Faced with this uncertainty the investigator has two alternatives: (1) to discover a means of reaching equal interval (and perhaps even true zero, absolute) scales or (2) to be willing to throw away information and use rank order or nonparametric measures. The second alternative (which also arises without throwing away information, when the original data has no claim to continuous properties) is more aptly discussed under coefficients of relationship in Section 14.8.

The attack by factor analysts on the first problem—the lack of guarantee of equal interval properties—has so far been either piecemeal and provisional, or more of an evasion than an attack. There have been few criticisms of results from studies ignoring the examination of equal interval or true-zero properties. Meanwhile, sporadically in certain areas, provisional advances have been accepted, as when Nesselroade (1966) correctly pointed out that true-zero properties exist in the otherwise unjustly derogated difference scores, or when others point out that rank correlations give very adequate results so long as the relation of raw to true scores remains monotonic. However, in some passing comments above, we have noted that lack of equal interval properties in absolute scores may seriously upset the use of difference scores, causing even the monotonic property to vanish. We have also paid tribute to the importance of scaling when we pointed out that normalizing skewed raw-score distributions before correlating has been found, empirically, to improve the simple structure and general clarity of the factor solution.

Since equal interval scaling properties are unquestionably necessary to preserve the linear and additive properties required for a good fit by the model, and are rarely guaranteed in the raw data, let us now consider the scaling variables. For if variables are scaled the scaling of the factor scores will in general take care of itself. Two basic principles have been invoked (Cattell, 1962a, 1973, p. 383) to ensure equal interval properties: (1) pan-normalization and (2) the relational simplex.

As to simple normalizing of raw scores, the assumption that a normal distribution should exist in true scores is consistent with the postulate of factors as influences, and the general multivariate model whereby every actual variable is distributed as the outcome of a large number of independent influences. For a summation of many small nonnormal influences will produce an approximate normality. On the other hand, since a factor is a single influence there is no guarantee of normal distribution of its true scores, though, as the summation of many variables, its estimated score will tend to be more so than its true score.

What we have called the pan-normalization principle starts with the assumption that in a typical total population the true variable scores will, in general, be normally distributed. However, in actuality the raw scores, through floor and ceiling effects in the scales and sporadic distortions over certain ranges, will often not be normally distributed except over limited ranges of the scale. The aim of the principle we call *pan-normalization* is so to expand and contract the translation of raw scores into true scores that the manipulation will emerge with a translated score that will be normal no matter what sections of the raw-score scale are taken. To ensure this it is necessary to get a series of populations or population samples which have different means but overlapping distributions, as shown in Fig. 14.2. These successive samples and populations could be successive age groups on the same spatial ability test, successive social classes on a verbal ability test, or successive developmental ranges on a personality test. In successive age groups it would be assumed that at each age the true sigma is the same and that those all standing at one age are in true scores normally distributed. The crucial check in pan-normalization is whether the raw-to-true score translation value for one

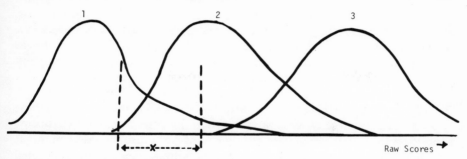

Fig. 14.2. The pan-normalization principle in achieving transformation to an equal interval scale.

group is also holding for the adjacent overlapping group. There might arise some discrepancy between the normalization requirements of two neighboring groups, as in the raw-score interval x in Fig. 14.2. Here a more normal distribution would be obtained in population 1 by a transformation of the raw score in the direction of a reduction from raw to ideal units, whereas for 2, a better normal distribution would be obtained by an expanding transformation over the same range. The problem is to find a translation value, from raw to true units, changing somewhat from a group in one part of the range to that in another, which will produce the best compromise, i.e., the nearest simultaneous approach to normality in the distributions of any two or more near and overlapping groups.

By *pan*-normalization, as the name implies, the transformation from raw score to ideal, equal interval score must be one which thus *maximizes the closeness of the resulting locally transformed distributions to normal curves over all population samples considered together*. Although an expression showing how far such maximization has been attained can be found fairly readily, it would seem at present that the transformations at various intervals on the scale necessary to achieve it have to be found by trial and error, and iteratively. We will shortly return to the actual methods required.

The second main principle—that of the *relational simplex*—rests on the postulate that the majority (but certainly not all) of the relations among variables that exist in nature are linear when the variables are truly scored. One may debate how probable this postulate is, but at any rate one can set it up as a condition and define what is a "true" score as that which conforms to it. At least such scores would have the characteristic of maximum simplicity (and convenience) in the whole domain of regression of variables, and we could take our scientific stand on Newton's *natura est simplex*. Now, where there exists a linear correlation between the true scores on two variables any deviation of the raw-score units from the true corresponding equal interval scale units will act like random error, and reduce the magnitude of the correlation calculated. Conversely, we would argue, any transformation of raw-score units which produces an improvement (beyond chance) of the correlations—among many variables—betokens a closer approach to the use of true equal interval units over all variables.

The exception to this generalization would be that any stretching or contracting, making a departure from the true units, which occurs simultaneously in the two variables in a range of mutual regression might go undetected by any fall in the correlation. However, the relational simplex principle explicitly states that the condition must hold over many variables, where the probability of such mutually coordinated distortion would be very low. Thus in practice this risk can be avoided by dealing with correlations of variable x not just with y but with a whole set of variables, whereupon, as stated, it would become extremely unlikely that the raw scores and their translations would all represent a stretching or contracting over just the same range in the mutual regression. Moreover, correlations would occur where a group is in a high range on one variable and low on another.

If the transformation from raw-score units toward the ideal equal interval units needs to take place in all variables together, for the quickest convergence, how are we to apply the necessary trial and error variations to discover the pattern of optimum translations for the various variables? Presumably the algorithm and program for this will take each variable in succession in relation to the rest, and test transformations (systematically of successive parts of its range) in terms of success in maximizing its average correlation with all. It will then shift to the next variable and repeat. The procedure is obviously one of trial and error, without possibility of an analytical solution, and it is likely to be fairly costly of machine time. Soon after the relational simplex was first proposed (Cattell, 1962a), a program was worked out by Lingoes (1966) to produce systematic changes in raw-to-ideal score transformation over the range of any variable, and to test iteratively the improvement of mean correlations in order to select the best transformation. However, this program, demonstrating the practicality of the relational simplex for obtaining equal interval scales, was ahead of its time and few have availed themselves of it to give us further experience in this area. An application to the 16 scales of the 16 P.F. showed that n-stens (normalized sten units originally deriving essentially from use of the pan-normalization principle) coincided very closely with those indicated by the relational simplex.

In this relational simplex theoretical framework, one must distinguish between equal interval and normal distribution properties which in pan-normalization are by definition united, but not necessarily here. We have given reason to postulate that multiply determined variables will tend to have a normal distribution. The pan-normalization and the relational simplex principles might therefore reasonably be expected, in general, to lead to the same results in raw-to-true transformation expressions, as far as variables are concerned. On the same principle that many contributors lead to a normal distribution, the contributions of many variables should produce equal interval and normal distribution properties in the factor score estimates. But this tells us nothing about the true score distribution of the determiner we recognize as demonstrating itself in the factor, and concerning this the relational simplex approach is our only reliable guide.

From a practical point of view, however, it can already be repeated that experience indicates that attempts to set equal interval properties in variables improves the properties of a factor analysis. Contingently, this suggests that better scale properties exist when normality is accepted, and that the assumption of normal distribution in factor scores is, in general, supported. (However, in one well-investigated case—intelligence—Burt and Howard (1956) showed that a significant deviation from normality did exist.) Nevertheless Thurstone's discovery that cleaner factor structure is generally revealed by rescaled, normally distributed measures is supported by results of the present writer and others. As a strictly practical suggestion, in concluding this discussion of scaling, the psychologist is advised not only to seek equal interval properties but also to "groom" his raw scores for accidental, erratic values, anomalous in being way out beyond the distribution, before handing them to the computer. A "quality control" procedure when there are, say, 80 variables and 400 subjects, will almost invariably reveal some very skewed distributions and instances of single scores which, either through clerical error or an anomaly of measurement, lie far out of the distribution, with dire effects on the correlations.

One situation (over and above that of difference scores, already mentioned) where equal interval properties become highly important is that of comparing scores across different populations. In describing the equipotent correlation for weights in such a comparison (p. 318) we indicated that a more complete equivalence could be brought about by using the isopodic ("equal footed") addition. The isopodic transformation adds equality of the "footing" in variables to the already achieved equivalence of weighting in the V_{fe}, by the equipotent transformation. If the reader will return to Chapter 11, p. 311, he will note that the concern of the equipotent scoring system is to recognize that the same determiner in two culturally and parametrically different population samples will need to be estimated by different weights (V_{fe}'s) because of the different patterns of expression. The isopodic method goes beyond this, as just indicated, for on throwing the two distributions of the variables together (step 2 in the equipotent method) it applies the pan-normalization principle throughout and reaches an equal interval scaling on the assumption of that basis (or if need be on relational simplex principles). It then proceeds through 3, 4, and 5 as in the equipotent method, but basing each step on the "revised" isopodic scaling of the variables.

14.6. The Theory of a True Zero

If factor scores are to be used freely in all kinds of experimental situations and forms of calculation it is desirable that they have three properties: (1) the real-base regard for absolute size, already proposed, (2) some approach to equal interval properties in the estimated scores, as just discussed, and (3) possession of an absolute zero. The last may sound utopian, for psychologists are lucky when they can progress from a rank to an equal interval scale and they know that progress to a ratio scale is practically unknown. Yet if factor analysts are to get in business with those aspects of scientific work which require product and quotient calculation —and most aspects of most models do—this last is also a requirement.

The theoretical bases proposed for a true-zero concept are essentially two: (1) the relational simplex implication for true zeros and (2) the inherent change properties principle. For a more adequate treatment of these the reader must be referred to the monograph on real-base true-zero factor analysis (Cattell, 1972a) since space permits us here to handle these logically and methodologically complex concepts only in brief essence.

Let us first handle the inherent change properties approach which is simpler. This rests on observations of modulation effects, of growth changes, and of other properties, such as oscillations, of measurements which are likely to be a simple function of the true absolute magnitude of the thing measured. Let us illustrate it by a simple physical example, and, in psychology, by the modulation of states. If a meter ruler were clamped to a table at a point hidden behind a curtain we could infer the distance to that point of support by hanging a weight at various points on the part available to us using the simple calculations associated with Young's modulus. That is to say, the bending would be proportional to the distance from the true clamping point, and we could infer that in terms of the units on the part of the scale visible to us. In the case of states we have the concept of a liability to state x by individual i, written L_{xi}, and of the potency of a situation k in modulating the state, as expressed by the modulating index s_{kx} which leads to the equation $S_{kxi} = s_{kx} L_{xi}$. From this it follows that by observing the mean and sigma on the S_x measures at two different situations, k_1 and k_2, on an occasion scale of known units of stimulation level, we can infer the position on the k scale where S_{kx} has zero sigma. At this point we have located both a zero for provocative-

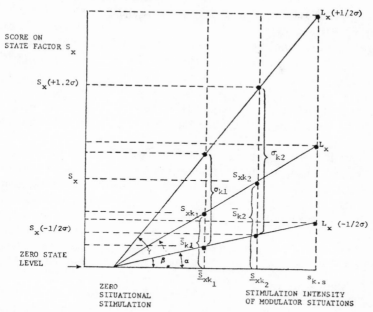

Fig. 14.3. Calculation of a true zero from the modulation model, illustrating that $s_{k_1}/\bar{S}_{k_1} = s_{k_2}/\bar{S}_{k_2}$, Eq. (14.4), i.e., the law of constancy of coefficient of variation for states, across different ambient stimulus intensities (simple case). (From R. B. Cattell, *Real Base True Zero Factor Analysis*, Multivariate Behavioral Research Monographs No. 72-1, Society of Multivariate Experimental Psychology, Fort Worth, 1972, p. 82.)

ness of situations and score level on a state. This is discussed elsewhere (Cattell, 1972a) in relation to Fig. 14.3, and also to other properties, such as learning, oscillation, and growth rates, all of which could have an organic inherent relation to a true absolute level.

A similar principle was used by Thurstone (1928) in his rather neglected paper on an absolute zero in intelligence measurement. What is general to these more special instances of growth, modulation, oscillation, etc., in what is here defined as the *inherent change principle* is that *some secondarily measurable feature of the trait, state, or process can be assumed to be a function of its absolute magnitude.* By plotting changes in the former, one may extrapolate to the value at which the effect would be zero, and the absolute magnitude of the cause or necessary determiner is therefore zero. There are, of course, some problems in that equal interval properties are necessary in the associated scale—in the stimulus strength of the occasion environment in the above case. As the footnote to Table 14.3 explains, this problem is lessened by taking more than two occasions, k_1 and k_2, at which the sigma of the dependent "manifest" variable is measured.

The relational simplex principle as it is extended and modified to determining true zeros we shall call, for reasons that will become apparent, the principle of *common lowest point of relation termination* or CRT for short. The above two principles—inherent change properties and CRT—are not methodologically independent. For in the former, it is an assumption that the secondary properties of absolute score present in other variable scores will all converge on zero as the absolute score sinks to zero. The CRT methodology is a check on this probably reasonable and certainly rational hypothesis.

The CRT principle begins with the recognition that the operations of measurement on most variables, i.e., the raw scores, can either cease to be carried out before the true, absolute zero is reached or can be continued beyond it. The former is easily appreciated by such examples as a domestic mercury thermometer ceasing to register before absolute zero ($-273°C$) is reached, or reports of pitch by a human subject ceasing below a frequency of about 30 Hz. The latter is not quite so readily recognized but can be seen when pulse rate is taken as a measure of anxiety, where measures can continue downward after the anxiety as such has ceased, or in a multiple choice intelligence test where 100, say, is the top raw score but a score of 20 is expected by chance, i.e., with complete lack of intelligent insight. These two possible relations of raw score and true score zeros are shown in (a) and (b) in Fig. 14.4.

The second part of the CRT principle is that, when a significant correlation—and therefore an underlying linear or other relation—exists between two variables, it will fall to insignificance when the raw-score operational measures are carried below the true zero in one or both of the variables. Except coincidentally, the termination of the relationship will occur through just one dropping in its raw score below its true zero, as shown in Fig. 14.4 in the relation of Y to variables X_1 and X_2. The plots in Fig. 14.4 may be considered regression lines of Y on X (in situations where other variables diffuse the $X - Y$ relation producing scatter) or (with complete experimental control) the literal plot of X against Y values with no scatter. In the case of X_1 and X_2, the X's reach their true zero before they reach their operational raw-score zero, and the significant linear relation with Y fails, as shown by the horizontal line, though correlations with Y have been sought all down to their, and Y's, operational zeros. With X_3 the opposite has happened: the raw score zero is reached before the true zero and XY correlations consequently persist down to the raw zero.[2] If the operations were carried further on X_3, a continuation as at (1) would ensue.

_____

[2] For the sake of familiarity of concept, regression coefficients have so far been referred to in Figure 14.4(b) but actually one must shift to an "equivalent" or "reciprocal regression" line (Cattell, 1972a) to avoid the dual representation of the relationship.

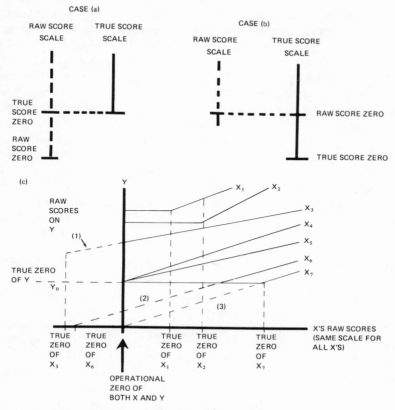

Fig. 14.4. The relational simplex approach to the true-zero problem.

In X_6 the failure of correlation, as shown by the horizontal line, is due to the other member Y having reached its true zero before its raw zero. If Y had any true substance in its operations below the true zero, the plot would extend as in (2). X_7 is the uncommon case where X reaches its true zero at the same time as Y, so that (3) cannot exist because both X and Y have moved into a "meaningless" raw-score range, where no further fall in X will produce any change in Y. X_4 and X_5, despite an apparent unusual coincidence, represent, if our theory is correct, a moderately numerous class of variables in which raw operational and true zeros coincide. They therefore retain a relation to Y all the way down to their own raw zeros but not down to Y's raw zero. In fact they both cease relationships to Y at the same level of raw scores on Y, which designates Y's true zero.

Some psychological illustrative examples of such relations will be deferred to the discussion of the related problem of the *staggered onset factor model* below, but a physical example may suffice, namely the discovery of the true zero of temperature. If we take iron at several thousand degrees C, we shall find a significant correlation of its expansion with the intensity of light emitted, but, like X_1 and X_2 in Fig. 14.4, light will cease to be emitted, as temperature falls, and reach a true zero long before the contraction effect does. Conceivably, measures of malleability would correlate over some range but cease before absolute zero is reached through difficulty, as in X_3, in getting operational measures below a certain point. Ultimately, we find a set of X measures, like X_4 and X_5, such as contraction, certain chemical properties and certain electric properties all reach zero relations with the criterion at their own zeros, and no other variables can be found that will maintain a relation below that level. Some others may themselves go below that level, as X_6 does in Fig. 14.4, but relation with the main concept indicator Y ceases below that operational point in Y. This is considered the absolute zero of temperature.

The general CRT (common lowest point of relation termination) principle thus leads to a method of determining true zeros on variables (or factors) based on the possible relations in Fig. 14.4. One expects in any random choice of variables that correlate with the one needing a zero determination, Y_1, that many, like X_1 and X_2 above, will cease to correlate at different levels of Y, but that ultimately the experimenter will find a relatively numerous set that will agree in showing a significant relation down to the same value Y_0. They will coincide in ceasing at this point no matter whether they operationally cease simultaneously themselves (as in X_4 and X_5) or not (as in X_6). Y_0 then remains the true zero for Y unless later research discovers a variable which succeeds in correlating with Y below the raw-score level Y_0.

The procedures are bound to be essentially iterative, keeping records of the results with various pairs of variables, and, within any one pair, correlating systematically over different combinations of segmental ranges in each. Trial and error alone can fix the lowest point of persisting significant relations. A labor-saving beginning has been suggested by Hakstian in the form of regressing each variable on the rest $(n - 1)$ of a set of n appropriate, likely, variables. In that lower range of raw score of the variable where the multiple R falls toward zero (insignificance to be exact), finer exploration would then follow more exactly to detect the true zero.

In a work on factor analysis, it is not appropriate to study a wide array of scaling diversities, as in a psychometric text. Factor analysis can get along, with the approximations described above, with rank-ordered raw data and can adapt its forms to qualitative data, ipsatized data, etc. But for full realization of its potential, it needs the two properties we have considered the important ones in the last two sections, namely (1) equal interval properties, and (2) the true zero that permits ratio scales. Methods exist for approaching these, but, as indicated in the discussions above on isopodic scoring, real-base factoring, dR technique, and various factor estimation methods, the good interval and zero properties gained in the variable data may yet be lost in the factor scores unless appropriate methods, e.g., the isopodic, are studied.

It must be assumed that the reader is already familiar with the more elementary conventional transformations of raw scores, which in any case we have mentioned at appropriate points in passing. They include simple standardization; normalization (fitting to a normal curve) with or without standardization (see s-stens and n-stens in Cattell, Eber, and Tatsuoka, 1971); centering or semistandardizing (same means only); bounding (getting the range of all within some limits, such as +1 to −1, without full standardization); null scaling (changing scores to deviations from some theoretically expected value); and, of course, the same scalings across different referees, as in ipsative, abative, and performative standardizations. Among transformations one commonly recognizes linear, as in most standardization (e.g., s-stens), and nonlinear, as in changing to a logarithmic derivative scale. The latter transformation has been popular among factorists with ratio scores and others which in their immediate form give a badly skewed distribution. Discussion of the typical scaling procedures in and of themselves is adequately dealt with in numerous psychometric texts, and a very insightful treatment in regard to factor analysis appears in Rummel (1970).

14.7. Permissive, Staggered Onset, and Continuous Action Factor Models

The departures from the standard factor model here to be considered encompass four possibilities: (1) the ordinary moderator variable situation, in which variable a cannot be affected by factor X until variable b reaches a certain level; (2) that where the level on variable a decides when factor X can come into action upon it; (3) that where, reciprocally, the level on factor X decides when it can come into action (acquire a significant regression) on a; and (4) that where the level on factor Y decides when factor X can act. Case (1) can be subsumed under (3) and (4) if we suppose variables entirely accounted for by the common factors. As far as factor action is concerned the present writer has called (4) [and by implication (1) and (2)] the *permissive model*, because it is the level on another factor or factors that permits X to come into action. (Coombs and Kao, 1960, who seem the only other authors to have dealt with this model, call it the *disjunctive model*, contrasting it with conjunctive for the ordinary "additive over all ranges model.") We shall call (2) the *staggered onset model*. This yields three in all, the third being the ordinary *continuous action model*, and the two first the permissive and the staggered onset. Staggered onset is moderator action in which action is all-or-nothing.

Examples of permissive action are not hard to find. In the ability field, for example, on a test of "problem arithmetic" it is evident the N factor cannot come into action until V factor has reached a level where the subject can read what the problem is about. And in the personality field a certain level reached on the anxiety factor may cut off, say, the action of a need for sociability (for the permissive model, of course, supposes upper cutoff levels as well as starting levels). As an example of the delayed action which is part of the general staggered action model [(3) above] one can instance the inability of rising temperature to hasten plant growth till it reaches a certain level, namely, above freezing, or of intelligence to correlate with rate of learning complex mathematical ideas until it reaches a certain upper IQ range.

When we turn to special factor analysis problems that do not arise from difference of raw-score and true zero points, as above, but from differences in point of onset of real influences, some resemblances of required method will appear which makes the above a useful introduction—provided the fundamental differences are also kept in mind.

In a purely logical perspective on the possible effects of influences as influences, we would have to recognize two types of measure—factors and variables—offering four types of relation—factor on factor, factor on variable, variable on factor, and variable on variable. These must be further multiplied according to whether the attainment of a certain level (1) on the acting influence, (2) on the dependent variable, or (3) a third-party variable, decides when a significant relation shall be switched on or switched off. The term *moderator variable* (Digman, 1966) has been used for some time for the third case—though in the sense of the level of X modifying the magnitude of the relation of Y and Z rather than switching it on or off completely—but we will retain it.

As a preliminary introduction we have just mentioned instances three paragraphs above, but we will clarify the nature of these by physical and psychological examples, confining ourselves, however, to those actions of factors on variables most encountered in real data, namely, (a) action beginning on a variable only at a given level of the factor, (b) action of a factor beginning only at a level of a variable, and (c) action of a factor beginning only at a given level of a "third-party" factor or variable. The rise of pressure in a locomotive boiler related to rate of movement instances the first, for the steam pressure rises without corresponding movement to a certain point and then becomes positively correlated to it. Psychologically, a case that might illustrate the reverse direction is the relation of fear to the amount of effort to escape, where a high level may be initially paralyzing.

An example of (b), in which the level on the variable decides, would be the influence of hours of conversation upon the gain in a foreign language, where poor command of the language would initially make available hours ineffective. Moderator action (c) is shown in chemical catalysis, and in, say, the relation of number of books read to books on shelves, the third term, hours of leisure available, actually deciding how high the correlation of reading to available reading shall be.

Knowing that data illustrating these models undoubtedly exist, the factor analyst has to create experimental designs and methods of analysis that will permit him to disentangle the mechanisms and to quantify them. It may help to consider first a psychologically important case even though not as typical of the above models as might be. This is one in which a series of state liability (proneness) factors are activated each at a different level on some continuum of potency of ambient situations. As Fig. 14.5(a) shows, a variable a_j, loaded by these three factors, will increase at a slow rate over the first ambient situation intensity range, and faster as first one and then another factor come into operation. In Fig. 14.5(b) we suppose that the situational intensity has to increase to s_{k1} before even the first state liability factor begins to generate a state, so that only the steady effect of a trait T operates up to the situational level s_{k1}.

Thus, Fig. 14.5 illustrates the general conception of "staggered onset" in the familiar psychological setting of modulation effects. (The three curves illustrate the modulator effect simultaneously upon the mean and the sigma of the variable.) It shows that the obvious general methodological approach is to factor a_j and associated variables at each of several (three in this case) levels on either the dependent or independent variable. Then, at level s_{k3}, one will find two factors, L_1 and L_2, instead of just one, as at level s_{k1}, and so to three factors in (a) and (b) at the highest level of s_k. Instead of situational intensity, the horizontal axis could represent age (time for maturation) as a "moderator variable" (if a "moderator" may bring in extra factors as well as stronger action of an existing factor).

A k-dimensional plot extension of Fig. 14.5 would be necessary to indicate what generally happens in the case of a staggered onset, when the point of onset depends on the level of a factor not a variable. A test vector loaded on two factors would then shift into a third dimension, acquiring a loading on a third, but only when the third reached, in absolute score, a critical value, which could be different for different variables. Since the axes just spoken of represent loading, an extra set of axes would be necessary to represent these absolute score onset points. And in the specification equation, the b coefficients would ultimately have to be couched in so strongly curvilinear a form that increases in the factor up to a certain point would give virtually no increase in the variable. Incidentally, these staggered onset effects are as likely to be

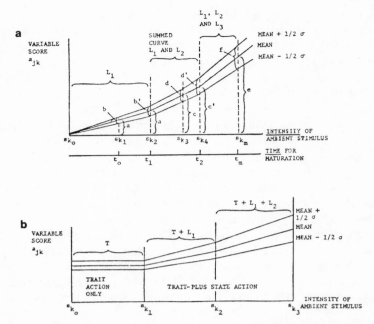

Fig. 14.5. Model of staggered action of factors. (a) Case of multiple factor action. Since $a/b = a'/b'$ and $a'/b' = c/d$, then $a/b = c/d$. Similarly, $c'/d' = e/f$, whence $a/b = e/f$. (b) Case of states and traits. Multiple and staggered action of factors in relation to a variable (couched in modulation situation). Primarily the top portion, (a), is meant to be comparable to Fig. 14.3, in which the vertical axis is all state measurement, and it shows how both the mean and variance take a new steeper upward path as a second and third factor come into action, at $s_k(2)$ and $s_k(4)$, respectively. However, the basic constancy of sigma to mean persists despite these additions as the geometric analysis in this figure shows. (From R. B. Cattell, *Real Base True Zero Factor Analysis*, Multivariate Behavioral Research Monographs No. 72-1, Society of Multivariate Experimental Psychology, Fort Worth, 1972, p. 111.)

found in P technique as in R technique, depending, in the former, on certain ranges being reached in the set of occasions for the individual and, in the latter, on the sample of people extending into certain ranges.

No systematic presentation in the literature of factor analysis has yet addressed itself to the concept of staggered onset and the methodology of detecting and evaluating it, so the reader must proceed on his own from the present compact sketch of models and possible methods. The methods are bound to begin with that design of factoring cross-sectionally at each of several ranges which we advocated for finding curvilinear relationships. Since our model admits that the change in regression may occur either through a variable or through a factor reaching a particular level (in either case directly or as a moderator) and since factor scores are derived from variable scores, it is evident that the research will zero in on these relations only by successive programmatic studies varying the company of variables from which the factor scores are derived and finding samples that will stand at relatively divergent scores on different variables.

14.8. The Effects of Different Coefficients of Score Relationships Including "Cross Products" in the R_v Matrix

Factor analysis has been pursued with quite a variety of coefficients of association among the relatives. A brief glance at their comparative aptness for various data on the referees is required. The two most basic departures from the use of the correlation coefficient in its various forms are the use of the congruence coefficient r_c and of the pattern similarity coefficient r_p (see Bolz, 1972)—or the very similar intraclass coefficient r_n.

The last two—r_p and r_n—are essentially similar in giving higher resemblance values if the scores in the two columns are not only in the same rank order, but closer in mean. Thus whereas r, by standardizing columns, takes account only of their agreement in shape (see p. 507), r_p and r_n (as we may symbolize the intraclass coefficient) also take into account agreement on level of the score vectors. Although very little seems to have been done with factoring these coefficients, the expectation would be that they would tend to restore the first factor thrown away in R and Q techniques (p. 507; see also Davies, 1939, and Holley, 1970). That is to say, the general size factor, thrown away when factoring people by Q technique, would, with r_p or r_n, reenter the available covariance, since people similar in "size" would generate a higher coefficient. (And reciprocally, of course, regarding the appearance of the "species pattern," i.e., agreement with the human type, factor in R technique.) In comparisons with usual parameter values in factor analyses some statistical values need to be reappraised since r_p, for example, does not have exactly the same distribution as r (see Horn in Cattell, Coulter, and Tsujioka, 1966).

The effect of using the congruence r_c coefficient instead of r is to produce still larger modifications. Where X and Y are the raw scores this is

$$r_c = \Sigma XY/(\Sigma X^2 Y^2)^{1/2} \tag{14.5}$$

The procedure has sometimes been called *direct score factoring* and also the *factoring of cross products*, because it works directly from the scores rather than from the deviations. Some reading on this will be found in Saunders (1950), Gollob (1968b), and Corballis (1971) in recent years, and Tucker (1968) has brought out explicitly some important modifications that occur. However, perhaps the most brief and sure statement of essentials was given by Ross (1964); while Nesselroade (1966) and the present writer stated the principles of exchange in cross products versus r and explored them in empirical data. The initial allure of what we may call descriptively *joint mean and deviation factoring* (since "direct score factoring" has ambiguity) using r_c is that retaining both of these pieces of information promises to bring into the same factor system—and probably to explain by the very same factors—both the variation about the means and the absolute means themselves. This would be valuable in developmental studies where one might hypothesize that the growth factors presently accounting for individual differences are also responsible for the absolute growth of all from zero.

Unfortunately, one is at present balked of such a solution by a fact which the unwary should note. As Nesselroade points out, the obtained r_c matrix can be systematically analyzed as a matrix of products of means (of raw scores, of course) to which is added a (raw-score) covariance matrix. However, in this there are actually three sources when analyzed in factor terms: (1) a product of the broad factors affecting the means, (2) a covariance matrix representing the common broad factors in deviations, and (3) a covariance matrix of specific factors with the above common factors and with themselves. It is this last that causes the trouble since we cannot equate specific factor products to zero and proceed with factoring in the ordinary sense without the unlikely assumption (inconsistent with ordinary factoring) that specific factor scores are not simply positive values but, instead, are equally distributed around zero. Thus at present the method calls mainly for a warning sign against simple interpretations, though it will be seen that it has necessary connections with real-base true-zero factoring (Cattell, 1972a) which might someday lead to an intelligible use of the approach. Meanwhile, if we want to check whether factors involved in means are the same as those in individual deviations it can be done with separate experiments. Nunnally (1967) has presented an interesting argument that the factoring of cross products should reach similar factors to those from factoring distances (p. 451).

Almost all the treatment of factor analysis above (except in Chapter 13 where we explicitly consider the nonparametric model as such) is based on dealing with continuous variables, as indeed one does in nine-tenths of factor analysis. Returning now to the usual factoring of deviations our concern is to consider a broader range of indices of association beyond the most commonly used product moment r that can be entered into cells of the "correlation" matrix R_v. This involves consideration both of the properties of continuous and discontinuous variables. For although the nonparametric model has been discussed, some coefficients will involve us again in what factor analysis can do with nonparametric and discontinuous, "qualitative" variables.

For the guidance of most readers it is probably useful to divide the indices of association that may enter into what we may now more broadly call the "relation" matrix R_v into two classes: (1) those which are fairly close relatives or derivatives of the product moment correlation and (2) more "far-out" measures of association, sometimes positively extravagant, concerning which relatively little experience has accumulated. A list of these with brief reminders of their nature (available in any general psychometric text) is given in Table 14.1.

It is appropriate to study these indices directly after scaling because their differences have mainly to do with assumptions about scaling. However, as Rummel well points out in a more extensive analysis of the coefficient problem, the aptness of a coefficient needs to be examined not only in relation to the scale and its units, as just recognized, but also in relation to (1) which facet is involved in the data box; (2) what splits (percentages above and below a cut in the distribution) occur in the data; (3) what the general distribution is; and (4) what one considers the nature of the sampling and measurement error to be. An excellently clear and

Table 14.1. Indices of Association of Two Variables Employable in Factor Analysis

Family of product moment	Other coefficients
1. Pearson product moment r	(a) Less remote from r
2. Covariance (raw deviation) products	1. Biserial
3. Phi coefficient Φ and point biserial (r reduced to two categories, on one or both variables)	2. Intraclass
	3. Tetrachoric (assumes cuts on a normal distribution)
4. Phi-over-phi max (Φ divided by the largest Φ possible at the given cut)	(b) More remote from r
	4. Kendall's T (and its extensions, T–B, etc.)
5. G-coefficient	5. r_p pattern similarity coefficient (requires prior factor space)
6. Spearman's rho and some other rank coefficients	6. d, distance function (uncertain validity of operational definition of distance)
	7. Cross products (congruence coefficient r_c) and other implicit coefficients in direct factoring of data matrices
	8. Coefficients of association (nonparametric, qualitative)
	(i) Information theory coefficients, joint probability, contiguity, etc.
	(ii) Contingency coefficient, for categorical variables

profound analysis of adjusting choice of coefficient to the data is given by Carroll (1961).

Consideration of the varieties of coefficients and their properties in relation to a given purpose often proves bewildering, but at the cost of some judgments without detailed support we offer the following condensed perspective on the outline in Table 14.1.

The product moment correlation assumes homoscedasticity but not normal distribution. However, dependable inference about its significance and its derivates, e.g., partial correlations, is only possible with normal distribution and repeated reminders have been given above of the desirability of normalizing in the interests of good factor properties from a factor analysis. Nevertheless, rescaling variables to normality admittedly presents problems later, in regard to factor scores and the specification equation. For example, one may be interested in estimating from the specification the number of barrels of oil likely to be imported into a given country, but if that variable is badly skewed over countries and has been normalized, what one predicts is the new redistributed variable. It will have to be scaled back to barrels to give a prediction in barrels—a back transformation which often gets overlooked in stating specification equations. Two considerations which have sometimes arisen regarding use of r concern the correction for grouping and correction for attenuation. Shepard's correction for coarse grouping, when finer grouping is known in principle to be possible, is desirable when one gets down to a dozen steps or less in any variable. It will not alter the general form of the factor patterns themselves, but it will tend to raise salient loadings and give better factor score estimates.

Somewhat similar gains may be made by correction of correlations by attenuation from known degrees of unreliability. (Note this should strictly be stated as "undependability" to avoid confusion with "low homogeneity.") That correction is

$$r_{12 \cdot c} = r_{12 \cdot o} / (r_{11} \cdot r_{22})^{1/2} \qquad (14.6)$$

where o is observed and c corrected. Spearman showed long ago that the essential factor structure (number and nature of factors) is unaffected by this translation (though loadings are expanded) and writers since have wondered if anything is to be gained by it. The $r_{12 \cdot c}$ value is, of course, an estimate, with the possible errors of the dependability coefficients r_{11} and r_{22} thrown in, relative to the observed correlation, $r_{12 \cdot o}$ value. Nevertheless, any matching of factors which involves absolute levels, e.g., by r_p, is likely to be better done with this correction. However, its greatest usefulness is probably in exploratory factor analysis attempting to cover a wide span of variables (which must therefore be short and of low dependability). The researcher is out to test an implicit hypothesis as to whether a certain variable is a good salient and might, if perfectly measured, be a valuable addition to a battery, or, in other hypotheses, he may be out to ask if a variable could possibly be fully accounted for by a certain factor (see also p. 359).

The use of the covariance "index," i.e., simply the product of raw deviations, divided by N, is nowadays fairly widespread. The value of this approach has been sufficiently dealt with as the avenue to real-base factoring above. It creates no difference in the final rotated result and it is only in the rotation from the principal axis values that one may experience some difficulty due to the larger variance variables more decidedly determining the nature of the initial components and perhaps presenting more difficulty in rotating away from them. The relation of covariance derived from r-derived factors is of course a simple one, namely

$$V_{fs \cdot c} = D_s V_{fs \cdot r} \qquad (14.7a)$$

or

$$V_{fp \cdot c} = D_s V_{fp \cdot r} \qquad (14.7b)$$

where D_s is a diagonal matrix of raw-score sigmas of the variables, and c and r subscripts denote covariance and correlation matrices, respectively.

The phi coefficient necessarily came into play early in factor analysis, in situations where variables could only be initially crudely categorized in regard to two or so levels or where yes–no responses are involved. It is, of course, a form of the product moment r but since it goes far below the limit indicated above for Shepard's correction, namely to two categories, it brings in some inaccuracy and it has, further, the property of not being able to reach a correlation of -1.0 or $+1.0$ when the "slices" are not 50/50. One can see this quickly from Table 14.2 when it will be evident the value can reach 1.0 only if the marginal totals are all equal, i.e., the cut is such that 50% are considered high and low.

Table 14.2. Problem of Handling Dichotomous Data by the Phi Coefficient

Variable B	Variable A		Marginal Frequency
	Score A−	Score A+	
Score B−	w	x	$(w + x)$
Score B+	y	z	$(y + z)$
	$(w + y)$	$(x + z)$	N = Total

$$\Phi = \frac{wz - xy}{\sqrt{(w + x)(y + z)(w + y)(x + z)}} \qquad (14.8)$$

To overcome this difficulty it was proposed to divide the obtained phi by the maximum possible phi for the given "slices," and this in fact eliminates some of the bias to be discussed (Carroll, 1945; Cureton, 1959; Ferguson, 1941; Wherry and Gaylord, 1944; Cattell, 1952b; Horst, 1965). However, Carroll's more recent evaluation (1961) shows it to have some "vicious" features, principally that the regression surface presented is bizarre. The main distortion from handling dichotomous data by the ordinary phi is the appearance of "difficulty factors" (or, better, "level" or "extremity" factors if we have a broader view than the ability field alone). These factors appear, over and above the proper number of substantive factors, as a result of the higher correlation existing between two variables with uneven splits which, by other correlation methods (respecting the full contribution) would correlate more (or actually) equally. Consequently "instrument" factors appear loading variables which cluster together in regard to "level" alone, regardless of their content. Gorsuch (1974, p. 260) gives illustrative values of the maximum phi at different cutting positions, showing how one can, by "phi-over-phi-max," obtain results closer to the r coefficient analysis of the same data not dichotomized. An attractive way out of the uneven slice problem has been suggested by Holley and Guilford (1964) in the G coefficient. This simply takes the score matrix and makes a double entry for each subject, but reversing his scores (relative to the mean) in the second. Thus if he is in the top slice on A and the bottom slice in B he is also entered conversely. Over the whole score matrix the number of people in the top and bottom slices of A will be equalized to 50–50. The procedure is exactly analogous to the double entry of pairs of twins in getting an intraclass correlation. A good account of its use is given in Gorsuch (1974, p. 263).

The tetrachoric correlation is an alternative for variables that come in dichotomous form or continuous data suitable for dichotomizing. Both it and the ordinary biserial yield the same correlation as would result by r from continuous data, if those continuous data were normally distributed before being sliced. Both avoid the reduced values and the intrusion of "level" (difficulty) factors found with phi and the point biserial. The problem remains, however, that when cuts are extreme and slices only correspond to small percentages the tetrachoric values become randomly unreliable. When this happens these coefficients can yield non-Gramian matrices, which may be described, in this context, as due to inconsistent correlations as if each r came from a different sample, yielding a number of factors (with ones used in the diagonal) not equal to the number of variables. Practically every possible position regarding the relative goodness of the various dichotomous coefficients has been held by one analyst or another. At one time the present writer was enamored of the "phi-over-phi-max" coefficient, but Comrey and Levonian's (1958) finding that it inflates communalities and Carroll's evidence above would indicate that the tetrachoric is superior in dichotomous material.

Rank correlations seem to the present writer to be, like the dichotomous indices, something to be avoided if the data permit anything else. The reason is that in any form (Spearman or Kendall) they lose information relative to r, and they give undue weight to the middle range of subjects since, in say 100 subjects, 50 and 51 are considered as far apart as 1 and 2 or 99 and 100, which denies a normal distribution.

Spearman's (1927) formula is

$$r_r = 1 - \frac{6\Sigma d^2}{N(N^2 - 1)} \tag{14.9}$$

where d is the rank difference of any person across the two series, while Kendall's (1957) is

$$r_r = \frac{2S}{N(N - 1)} \tag{14.10}$$

where S is the number of pairs that are in their natural order on A when ranked in order on B, minus the rest of the pairs. Tied ranks are generally handled by using the middle rank. Where data *can* only be obtained in rank form these coefficients can be used and cause no problem in factor analysis.

The first two of the less used and justified coefficients in Table 14.1, namely the pattern similarity coefficient and generalized distance d are appropriately considered together. The pattern similarity coefficient has been applied in two ways: directly to the score matrix, computing similarities between two variables or to the rows of test vector loadings after the two variables have been fixed in space by a preliminary factor analysis. We shall consider the latter here and the former in connection with the intraclass correlation below.

As discussed (p. 452), the distance apart of two tests or two persons in factor space can readily be calculated, but since here the coefficients should express nearness, one is tempted to use r_p. [The problem of bringing d into some form useful in this connection is discussed more broadly by Young and Householder (1938).] As we have seen under multidimensional scaling, or *dimensional integration* (to paraphrase Rummel), the original solution to finding the dimensionality from distances (Young and Householder, 1938) has been developed in depth by Torgerson (1958), Coombs (1951, 1964), Ross and Cliff (1964), and several others. A setting out of the computational steps necessary to get an expression that is high when variables are close, i.e., a reverse function of d, to enter into an "R_v" that uses inverse distances, and to which principal components analysis is applied, is given by Rummel (1970), as well as by multidimensional scaling tests. Since we have given some reasons why r_p is the most useful "nearness" function one may wonder why it cannot actually be used here. The answer is that an orthogonal dimensional system must first be set up in which the pattern similarity of tests, like that of people in TAXONOME, can be calculated. Such a calculation of closeness of two tests can be made by getting r_p between their row values in the V_o matrix, but except for some matters of detail, this would be proceeding in a circle because by dimensionality integration principles it would yield essentially the same factor structure as that from which it started.

A meaningful use of r_p, sometimes proposed, is to apply it directly to the score matrix, and then its properties are such that it can be considered along with the intraclass coefficient, since both do not stop at shape similarity alone, as pointed out earlier, but include "level" or "distance" effects. As pointed out above (p. 362), this retention of information about level comes close to yielding a factor structure which does not omit either the first factor omitted by R technique or that omitted by Q technique. In this connection there will be a difference of outcome according to whether r_p is calculated on scores first ipsatized, i.e., standardized over rows (persons) of the S matrix, or not; for if so standardized the difference associated with the first, "general size" factor will have vanished. Bolz (1972) has an interesting exploration of the use of r_p in a Q technique (transpose R) factoring. However, one must remember that the r_p formula calls for differences on orthogonal measures. This makes ordinary Q-technique use only approximate since variables are correlated, but raises no obstacle to the present R-technique use.

Discussion of the remaining coefficients in Table 14.1, the coefficients for nonparametric or qualitative data, and the coefficients implicit in direct factoring of score matrices could lead us into extensive digressions and must be only briefly commented upon.

If qualitative data is dichotomous, e.g., pink versus green, member of a club versus nonmember, psychotic versus nonpsychotic, there is no problem. Chi-square or some derivative of chi-square such as the contingency coefficient, can be used:

$$C = \left(\frac{\chi^2}{N + \chi^2} \right)^{1/2} \tag{14.11}$$

where χ^2 of course is

$$\chi^2 = \sum \frac{(O - E)^2}{E} \tag{14.12}$$

When one variable is continuous and one qualitative the point biserial is appropriate. But if qualitative data have more than two alternatives, e.g., as in several alleles to a particular gene, we are in trouble, for any serial coefficient we use would require ordering of the categories, and by

definition a qualitative collection cannot be ordered. Intensive consideration to nonparametric factor analysis has been given by Lingoes (1966)
and Guttman (1959, 1967), as mentioned in the preceding chapter, but
their concern has been rather with the nonparametric case in the true
theoretical sense of assuming no mean and sigma in a population from
which the sample is taken. But the possible "metricizing" of unordered
data is a part thereof.

Although some writers, e.g., Horst (1965), have called the restriction
of factor analytic work to the product moment coefficient "unfortunate"—
and indeed any routine restriction of imagination may be unfortunate—
an inspection of other coefficients, as done above, reveals several pitfalls
and distortions in a fair number of them. The present writer's main argument (1972a) would be that the factor analysis of the future will increasingly be real-base true-zero factoring. That is to say, it will be (a) based on
covariance terms in the R_v matrix and (b) referred to a basic scaling at the
core of a multidemo–multisituation system (Cattell, 1972a). Such an advance will require more programmatic research, either by individuals
working for many years in one field or by coordinated groups since the
concepts around "size of factor" require careful comparisons of use of the
same variables over different demographic groups ("demos") and situations. One may perhaps venture the judgment that the only major departure from such use of r and its covariance equivalent will be when categorical—principally binary dichotomized—data has to be used, and to this
we have suggested the best choices in principles stated above. A useful
discussion of other practical aspects of the above coefficients is available
in Fruchter (1954, p. 200), and there are further valuable observations in
Gourley (1951), Henrysson and Thurber (1965), and Koopman (1968).

14.9. The Statistical Significance of Correlation Matrices and Factors

The question of determining statistical significance of values found in
factor analysis must have occurred to the reader at numerous incidental
points. Yet until the questions of scaling of distributions and of indices of
association had been considered in this chapter we were scarcely ready to
attack the problems involved. As usual significance involves reference to
the "swamping" effect of experimental error of measurement as well as
the distortion from sampling error, though the latter, based on expected
distributions, is the primary issue. The areas of main concern to the
factorist are as follows.

1. The significance of the total pattern of correlations (or covariances) in an R matrix as such, also extended to the significance of the difference of two such matrices.

2. The significance of factors extracted from the R_v matrix and the significance of residual matrices. One gets concerned with this both for rotated and unrotated factors.

3. The significance of *correlations* and *loadings* of a specific variable with a particular factor. That is, how big does the value have to be before we can confidently say there is any loading at all? This also splits into orthogonal and oblique cases, and primary and higher-order factors.

4. The significance of *differences* of loadings or correlations, as when we compare two research results in which it is hypothesized that the loadings will have changed.

5. The significance of correlations among factors.

The significances of differences in factor scores can be handled by ordinary ANOVA methods, and, except for the problem of changing loadings and estimation weights, requires no comment here.

For the first problem—the significance of the original correlation matrix—there is both a theoretical answer and a practical answer in the form of R_v matrices of various sizes produced by Monte Carlo methods by Horn (1965a,b), Humphreys *et al.* (1969, 1970), Cattell and Gorsuch (1963), Cattell and Vogelmann (1976), and others through correlating random normal deviates. There is no real difference between the indications of the two approaches, though Humphreys has stressed the occasional quite large correlations from random deviates, the existence of which one is apt to overlook in taking the standard error value of, say, a zero correlation.

As to the significance of individual correlations and their differences nothing need be said that is not given in an elementary text. As to the standard theoretical expression for testing significance of a correlation matrix *per se* we have the suggestions of Hoel (1937) and the chi-square test of Bartlett (1950), involving examination of significance of the value

$$\chi^2 = -\left(N - 1 - \frac{2n + 5}{6}\right) \log_e |R_v| \qquad (14.13)$$

where, as usual, N is the number of people, n the number of variables, and $|R_v|$ is the determinant of the correlation matrix. The degrees of freedom for evaluating chi-square are $n(n - 1)/2$. A clearly worked-out example is provided by Gorsuch (1974). Gorsuch also sets out Bartlett's (1950) test for the significance of a residual correlation matrix in a principal components extraction. However, the components matrix to that point is not quite the matrix with which one would be beginning a factor analysis, because of 1.0 instead of h^2 in the diagonal. The reader will note that though the testing of the significance of a residual correlation has been important in the past and is still important for checking the number of factors decided by some independent approach, it is not one of the most commonly relevant significance problems. As far as "completeness of extraction" is concerned, the main methods were presented in Chapter 5 mentioning Bartlett's (1950), Burt's (1952), and Sokal's (1959) tests, while the maximum likelihood test has been described in the last chapter. As a less demanding computer test than maximum likelihood, the present writer has found that Sokal's (1959) test aligns well with other evidence. This considers the correlations in a residual as partial correlations, the correlations with the earlier extracted factors being partialled out. The distribution of correlations in the residual is then compared with that for partial correlations of zero.

In the present context the principal "significance of an R_v matrix" issue arises in domains where an exploratory factor analysis is being done with a deliberately very diversified choice of variables, e.g., chosen on the personality sphere principle. Critics, e.g., Humphreys, have been prone to condemn factoring of highly diverse, randomly sampled variables on the ground that they approach random normal deviate correlations. Of course, one could choose variables so remote from one another that, although real enough as variables, they would be expected to have negligibly small, though not truly "random" correlations with one another. The criticism is in some respects naive, first, for with sufficiently large samples such "weak looking" R matrices are still significant. Indeed, even with the N commonly present in a decent factor experiment, Burt's and Bartlett's tests commonly reveal significance in matrices rejected by some psychologists as insignificant. Secondly, even with an hypothesis checking design covering, say, each of 30 factors with 4 variables (120 variables in all), the overlap of factors may be so little that only 180 (from 4 markers per factor) of the 7,140 correlations, i.e., about $2\frac{1}{2}\%$ would have significance! Thirdly, even the correct application of the significance of R_v test is open to suspicion in that unusual combinations of factors can produce very small correlations. The pattern of factors, for example, in Table 14.3, is not very different from some empirically obtained, and is one in which 3 out of 6 variables reach a communality of 0.64. Yet it can be derived from an R matrix with only 2 out of 15 r's nonzero!

Admittedly it would leave the choice of estimated communalities in such an R matrix at once speculative and crucial. But with relatively slight correlations substituted for the "freak" zeros of this example, one would still have an R matrix well short of significance by the above traditional treatments (even with an N of 300 or 400) yet indubitably having substantial factors in it.

To complete the survey R_v matrix significance we need to refer finally to the case of the significance of the *difference* between two correlation matrices. The proposals for this purpose by Boruch and Dutton (1970) and Maxwell (1959) deal with covariance matrices. However, if one has re-

Table 14.3. Comment on the Significance of Correlation Matrices

(a) Factor matrix

Variables	Factors			
	1	2	3	4
1	.4	.4	-.4	-.4
2	0	0	-.4	.4
3	-.4	-.4	-.4	-.4
4	.4	-.4	0	0
5	.4	.4	.4	.4
6	-.4	.4	0	0

(b) Resulting correlation matrix (with communalities as known)

	1	2	3	4	5	6
1	.64					
2	0	.32				
3	0	0	.64			
4	0	0	0	.32		
5	0	0	-.64	0	.64	
6	0	0	0	-.32	0	.32

tained the raw-score sigmas of the variables—and one usually has—it is the briefest of calculations to shift from the R_v to the R_{cv} covariance matrices, thus

$$R_{cv} = D_v R_v D_v \qquad (14.14)$$

where D_v is the diagonal matrix of sigmas of variables.

The testing of R-matrix significances as residuals in the factor extraction to help decide the number of factors question is not often done nowadays except within the maximum likelihood method, the vital role of which in deciding on the number of factors has been discussed in Chapter 5. It essentially asks what the other (Burt, Bartlett, Sokal) tests do: "Do the residual r's *in toto* differ significantly from zero?" but does so in the special maximum likelihood context.

In these questions of significance of a residual or difference R matrix, and of the associated factors, psychologists sometimes ask whether the formulas apply to the rotated as well as the unrotated, for which they are originally derived. The answer is that the significance decision holds equally for any linear transformation of V_0, whether orthogonal or oblique.

14.10. The Significance of Correlations and Loadings of Particular
Variables on Particular Factors

Factorists have concerned themselves with statistical significance in several ways but mainly in regard to (a) the number of factors, which amounts to asking if a final factor is significant, and, by implication, if a whole residual correlation matrix is significant, and (b) the significance of the loading of a particular variable upon a particular factor. Other issues, as mentioned above, have been the significance of correlation between two factors and the significance of score differences on factors. The factorist's concern, however, has not been as deep as mathematical statisticians have wished and some statisticians have even retreated from factor analysis because, in default of accurately known distributions, it becomes more of an art. Another response to the situation, among some factorists, has been to adopt what we defined as a *psychometric* rather than a *statistical* position which means that results are considered simply as descriptive of the sample without inference to any population.

This essentially unfortunate situation has begun to clear in the last decade with development of some real theoretical foundations for inference, but not before the more serious workers had in desperation begun to solve it empirically by Monte Carlo methods. The question of significance of factors, which could be handled theoretically at a fairly early stage in orthogonal systems, e.g., by Burt's and Bartlett's formulations (p. 476) and has recently been broadly solved by maximum likelihood, is not the concern in this section. We are concerned now with the second issue, that of the significance of particular loadings and by implication, of the significance of the difference of two loadings, e.g., of the same factor on the same variable in two experiments. This is obviously of considerable importance scientifically, and one can point to many recent theoretical debates in psychology, economics, and the sociology of national cultures which could have been illuminated by a firm answer on this question. However, it is incompletely solved if one considers the vital case of oblique simple structure factor loadings and, instead of a definite answer, here we shall direct the reader to the main contributions in progress.

The main stream of theoretical inference has been associated with the maximum likelihood method in the work of Rao (1955), Lawley (1940), Lawley and Maxwell (1963), Lawley and Swanson (1954), and Jöreskog (1962, 1969). Jennrich and his co-workers have developed this powerfully in recent articles (Archer and Jennrich, 1973; Jennrich, 1973, 1974; Jennrich and Thayer, 1974). A second lesser source has sprung from ordinary regression principles in Gorsuch (1974), Harris (unpublished), Lawley and Swanson (1956), and others (see Anderson and Rubin, 1956).

As to the latter, Gorsuch advocates using the ordinary textbook formulas for the significance of a correlation or, alternatively (in the V_{fp}), of a weight (loading). Some have argued that this is adequate only for the r or loadings of a first component.

In the framework of principal components, an expression for standard error of a loading has also been proposed by Burt and Banks (1947) as follows, but applies only to orthogonal components in order of extraction:

$$\sigma_p = \sigma_r \left(\frac{n}{n+1-f}\right)^{1/2} \tag{14.15}$$

where σ_r is the standard error of a correlation of the given size, n is the number of variables, and f is the order of the factor in the extraction series. Child (1970, p. 99) gives a useful table for the above. An early proposal, suitable for rotated factors according to its author's arguments and not given sufficient trial and discussion by psychologists, is that of Saunders (1948):

$$\sigma_{b \cdot jk} = \frac{(1 - r_{rj}^2)b_{jk}}{2r_{rj}\sqrt{N}} \tag{14.16a}$$

where r_{rj} is the reliability of the measured j, and b_{jk} the loading on the factor k.

When Rummel (1970, p. 23) exclaims with some justified reproach that in factor analysis "tests of significance are practically unknown in application" he refers, however, to the real situation of "finished" rotated factors, to which Eq. (14.15) is not applicable. A formula directed to the real need was circulated, but not published in a journal, by Harris, incorporating the obliquity of the factors (by a derivative c_{ff} from a diagonal element of the factor correlation matrix) in a test of the null hypothesis that the loading is zero:

$$t_a = a_{jf} \left/ \left[\frac{(1 - h_j^2)c_{ff}}{N - n - 1}\right]^{1/2}\right. \tag{14.16b}$$

where a_{jf} is the primary factor pattern coefficient (loading in V_{fp}) of variable j on factor f; h_j^2 is the communality; N is the number of people and n the number of factors; and c_{ff} is the diagonal element (column and row of factor f) in R_f^{-1}—the inverse of the correlations among the factors. t_a is distributed as the statistic t, so one needs only to decide the cut thereon, e.g., that for $p < .05$, to determine if the loading is significant at that level. (In the V_{rs} one can use the equivalent correlation of reference vectors.) It will be seen that this is formally analogous to, and has the same logic as, the test of significance of a partial regression coefficient, familiar in textbook treatment of multiple correlation where

$$t = \frac{B_j}{[i(1 - R^2)/df]^{1/2}} \tag{14.16c}$$

i being the value, in the inverted correlation matrix, of the given variable with itself, and the expression $(1 - R^2)$ expressing, as in $(1 - h_j^2)$, the variance unaccounted for. The degrees of freedom df are similarly $(N - n - 1)$, n being the number of predictors. The logic of treating the factors as correlated predictors seems sound enough, and the only loophole for doubt arises from the uncertainties of accuracy of rotation discussed below. Part of the importance of the above four formulas is, of course, in defining the hyperplane width (which we may take as $\pm 2\sigma$ of the zero r) in such tests of simple structure significance as that of Bargmann (Appendix, p. 555).

Meanwhile some advance heartening to empirically minded researchers was made through the Monte Carlo attacks by Horn (1973), Humphreys, Ilgen, and McGrath (1969), Linn (1968), Finkbeiner (1973), and several others, including the present writer, in less available publications. One approach, illustrated by Humphreys' work, is to take random normal deviates, correlate them, and spin the axes to a large number of positions. Humphreys concluded from this that quite large loadings could sometimes be generated and that the standard error from such reasonings as the above was an underestimate. However, one could argue that random deviates are an unsound source and that the variation in various real correlations, from different samples, better represents the actual problem. "Monte Carlo" can in general represent two things: an artificial play with random numbers or the taking of some empirically given distribution and deriving, e.g., by random notation, some notion of the distribution of the derived values. Finkbeiner and the present writer (Finkbeiner, 1973; Cattell and Finkbeiner, 1978) took a variety of real data of varied N, n, and k, and rotated with extreme care to a maximized simple structure. At that position the width of the hyperplane was ascertained as the width of the "bulge" on the otherwise normal distribution of loadings, using the Kolmogoroff–Smirnoff statistic to determine the point of departure. Widths (2 sigma, plus and minus) were found to run from about .06 at 200 cases to .04 to 400. These results point to factor loadings being significant at decidedly lower values (probably the sigma of a zero loading being about .03 to .02 here) than yielded by the artificial basis of the Humphreys Monte Carlo studies and, incidentally, also to values closer in accord with the Jennrich and Jöreskog theoretical approach below.

While on the topic of essentially Monte Carlo methods one should also consider the possibility of taking a large number of rotation alternative positions on a given study and building up a distribution of these, instead of going to tabled or theoretically calculated significance values for studies in general. The method is developed by Mosteller and Tukey (1968) (see also Tukey, 1954), and by Pennell (1972) on a "jackknife" principle to build up distributions for confidence intervals. A similar approach, taking successive different samples to produce R_v from the given available sample, and then factoring and getting loading distributions, has been worked out by Finkbeiner (private communication). Such approaches, "custom built" for the given experiment are, of course, costly because many factorings have to be done but it may be justified in critically important studies.

To consider the theoretical basis the reader should refresh his concepts of maximum likelihood and confirmatory maximum likelihood above, especially the hypothesis-testing "philosophy" of factor analysis as broadly stated by Fruchter (1966) and statistically by Jöreskog, Gruvaeus, and Van Thillo (1970). The confirmatory or checking method first begins with the hypothetically given target matrix of factors and takes out these factors from the correlation matrix to determine whether each successive residual matrix still has significant variance. A very clear account of practical use of the confirmatory (proofing) maximum likelihood method is given by Gorsuch (1974, p. 117).

Real progress on a theoretically adequate treatment of standard errors of loadings in the maximum likelihood has been made by Archer and Jennrich (1973), who extended to rotated factors, but still only to orthogonally rotated ones; and by Jennrich (1973) and Jennrich and Thayer (1974). Jennrich then proceeded to oblique factors (Jennrich, 1974) but now with the restriction that they must be rotated by an analytic (not topological) automatic program with a known function in its criterion. In a more general context, factor correlations with variables have been well discussed as to significance by Kraemer (1975) and by Olkim and Siotani (1964).

A problem theoretically distinct from that of evaluating the standard error of a zero loading, but which in practice becomes thoroughly entangled with it, is that of the effect on hyperplane width of random error in rotation. If we suppose that at a perfect (oblique) rotation position approximately 95% of the true (zero) hyperplane variables will fall within $\pm 2\sigma$, say, $\pm.08$, what increase in the range of the hyperplane loadings should we realistically expect from uncertainty in rotation? In other words if $\pm.11$ is a significant loading at $p < .05$ probability, what loading is needed to reach that significance when the "standard error" of rotation is also allowed for? Since in rotation we are now dealing with human error, either directly in ROTOPLOT procedures or indirectly in the computer devices humans put into rotation programs, there is no answer to this except to discover the distribution of human errors in rotation.

Some investigators using Monte Carlo methods, like Humphreys, but including the human error, have concluded that "capitalizing on chance" can yield quite large (say 0.2 to 0.4) loadings on variables that should really be at zero, and in any case in the hyperplane band. Others, notably Cliff and Pennell (1967) and Cliff and Hamburger (1967) show empirically that with proper precautions, and regard for communalities, etc., the danger of falsely assuming a loading to be significant is much less than supposed. Supporting this, in the rotational setting, is the above-mentioned finding of Finkbeiner that a properly maximized simple structure will show a sharp bulge in the otherwise normal distribution of loadings, and that this bulge, made by the truly zero loadings, is seldom in fact broader than $\pm.05$ to $\pm.10$, with the typical size of N, n, and k values, and a topological rotation program, e.g., MAXPLANE or ROTOPLOT.

Some progress toward a theoretical basis may be made by hitching the sampling effect to the properties of a particular analytical or topological program, as Jennrich (1973, 1976) has ingeniously done. But decisions on the cut separating significant and nonsignificant loadings, when both sampling and rotation error are involved—which is the common real situation—will probably long continue to rest on experience and Monte Carlo studies with constructed plasmodes. In practice one can only advocate, as has been done all along, the importance of rotoplotting beyond any automatic program to a demonstrably unimprovable (by "history of hyperplane" plot) simple structure. If a Bargmann test shows a $p < .05$ or $p < .01$ significance of the simple structure on a factor, one can with some confidence apply to any particular variable's loading on such a rotated factor Jennrich's test (assuming generalizability from any particular analytical program), of the significance tests in Eqs. (14.15) and (14.16). Alternatively, one can use the determinations above of the width of hyperplane as a basis for deciding what is out of the hyperplane, i.e., significant and deserving of being called a salient.

Beyond the question of the significance of a single variable's loading lies the still more complex question of the significance of the difference of a whole factor loading pattern from a straight set of zeros or from some hypothesized pattern. Gorsuch argues reasonably "when variables defining all the factors are stated before the data are examined [factored] . . . no capitalization upon chance occurs" and reemphasizes his conclusion which we gave above that for checking the significances in the experimental result "the formulas [using the usual zero order r and loading distributions] are appropriate" (1974, p. 184). The key word above is "all" for one can rotate to match a single factor at the expense of the other V_{fp} matches. Regarding the use of zero-order r distributions for testing significance of factor correlations, we have expressed some doubt, and Gorsuch concedes that when these are compared with factor correlations found by

Monte Carlo methods the latter are found to be "about double" the theoretical standard calculated for a zero r. The Finkbeiner result would reduce "double" to "half as large again."

We shall return to this vexed question of a single loading for some supplementary comments in a moment, but it will be noted that if the whole matrix has to be considered, as Gorsuch points out, in testing the fit of a particular hypothetical loading to a particular experimental one, then we have unavoidably been pitchforked into the matching of factors question, introduced in Chapter 6 and studied in Sections 10.7, 10.8, and 10.9. We are now moving from the study of significance of a loading to that of the difference of two loadings—repeated over a pattern. In the matching methods so far discussed one is not concerned with the significance of each precisely calculated loading difference. For example, in the use of the salient variable similarity s index, one needs only to mark the salient (significantly loaded) variables in each; and in the congruence coefficient r_c method we are again not concerned with differences in absolute loading. For the congruence coefficient recognizes that the same factor may stand at different factor variance magnitudes in the two studies, so that loadings in one, of, say, $-.6, -.4, 0, .2$, and $.8$ could be the true equivalent in the other of $-.3, -.2, 0, .1$, and $.4$. In an interesting article on this matching topic Pinneau and Newhouse (1964) express some dissatisfaction with this property of r_c and suggest alternative indices, but as the present writer sees it, the above assumption is usually appropriate and the issue of whether the same factor should be significantly larger in a second study is a separate one, i.e., constancy of size may or may not be needed, after identity of pattern has been established. In any case the matching by r_c should ideally be done on covariance factors, as stated in Chapter 10, and the size comparison invariably made on that basis (see real-base factoring).

The title of "confirmatory" or proofing maximum likelihood factoring may suggest that it also can be involved in this testing of the significance of the difference between two factor patterns and therefore of their degree of matching. But there are some subtle yet important differences of purpose here. The significance of loadings, of differences of loadings, and of the resemblance or matching of two patterns are commonly between two experiments that have reached an ultimate resolution independently in each, i.e., that are comparing rotated positions. Maximum likelihood, in its proofing use, asks how far a set of correlations on a sample of the given size will yield (a) the same number of factors, (b) the same communalities, and (c) the same latent structure in the sense of a V_0 capable of being rotated to agree with some hypothetical rotated solution. One could, however, rotate that V_0 to many other positions, and what is being tested is not rotation, but the potentiality of matching the rotated hypothetical pattern.

The maximum likelihood method, in either of its two uses, is concerned with the significance of loadings in the sense of (a), (b), and (c) above, but in many situations in which a researcher is concerned with loading significances he is asking more specific questions, needing the more specific significance tests discussed above.

Tied up with the question of significance of loadings is the question of using the common expression "salient loading," "prominent loading" or "salient variable" with some precision. Sometimes the term is used for very highly loaded variables, but one may venture to suggest that its origin better suggests simply a variable "leaping out" of the hyperplane to join a company of significantly loaded variables, i.e., we would make it, as in the s index usage, synonymous with significantly loaded (on the given factor). The rough practical tradition has been to put that hyperplane width at ±.10, but as indicated, it depends on N, n, and k and, in rotated studies on the significance and perfection of the simple structure.

Among *salients*, it may be useful to speak of a subclass of *prominents*, but the dividing line cannot be drawn with any objectivity. Moreover, one surmises that concern with "prominents" is based on the illusion that a factor is not entitled to serious consideration unless half-a-dozen variables can be found with loadings upon it of 0.7 or over. (A kind of objectivity is given to 0.7 because 50% of the variable's variance is then accounted for by the given factor—if both V_{fe} and V_{fs} have 0.7.) But the fact is that a factor rotated and defined by a sharp and emphatic hyperplane is deserving of recognition as a firm construct even with salients of no more than, say, 0.3 to 0.4 loadings if these are significant. It may yet be objected that with such loadings it cannot be validly scored and interpreted. But, as shown in Chapter 11, ten salients each at 0.3 can give as valid an estimate as two at 0.7. And, as to interpretation, though it is a greater mental feat to abstract a concept common to ten variables than two, the interpretation can usually be more safely made from correlations of the scored factor with external criterion, which should be as characteristic for the score estimated from ten as from two variables.

Finally in considering the assignment of significance, in the fullest sense, to loadings one must agree with Rummel that a calculated significance value for a single study has less real importance than an average of loadings across replicated studies. That is why programmatic researches, with and without changes of population, and hinging on careful "cross-validation," are of unusual importance in factor analytic work.

14.11. Error of Measurement

Error of measurement will affect the interpretation of practically every value we have considered above in the context of sampling error, e.g.,

the correlations of variables and factors, the factor loading, the communality, the size of a factor, and the correlations among factors.

As to the correlation coefficient, if we take Carroll's (1961) systematic examination of the four kinds of error that can contribute to it, we
have already handled "errors of scaling" pointing out that although they
affect several coefficients, the rank formula and the tetrachoric are largely
though not entirely immune. Errors in correlations due to Carroll's "selection effects" are sampling errors, and their specific effect is best considered
in the next chapter. Scedastic and other errors remain.

Carroll points out that all errors, other than scedastic errors, in correlations can change the rank of the ensuing correlation matrix. It was
stated earlier that correction of correlation for attenuation was found by
the work of Spearman and his co-workers not to alter the main factor
structure obtained (except, of course, to raise all loadings proportionately), but this is not entirely supported in more refined work and it does
not, in any case, deny that when measurement error reduces or (more
rarely) augments correlations it can add to the rank of a matrix. (Error can
do this to a matrix, regardless of whether factoring is done before or after
correction of r for attenuation.) Porter (1976) argues that correction for
attenuation can itself change the rank (the number of factors) except in
the case of alpha factoring which Glass (1966) has shown to be immune to
this effect.

It is assumed the student is familiar with the correction for attenuation due to error of measurement (and of scaling), but we will repeat the
formula here:

$$r_{ab \cdot c} = r_{ab \cdot o}/(r_{aa} \cdot r_{bb})^{1/2} \qquad (14.17)$$

where o is the observed and c the corrected correlation. In matrix terms
this means that the corrected R matrix, R_t (t for estimated *true*) is calculated:

$$R_t = D^{-1} R_o D^{-1} \qquad (14.18)$$

where D is the diagonal matrix of roots of reliabilities. It follows that the
corrected unrotated matrix V_{ot} from such a corrected R_t will be

$$V_{ot} = D^{-1} V_{0 \cdot o} \qquad (14.19)$$

[Glass (1966) points out, however, that with factoring methods that are
not "scale-free" this $V_{0 \cdot t}$ is not exactly what one would get from factoring
R_t.] This attenuation correction will, naturally affect correspondingly the
rotated matrices from the $V_{0 \cdot t}$. It also means that for any individual variable its communality also becomes greater in the corrected matrix in that
$h_{at}^2 = h_{ao}^2/r_{aa}$.

In any sample, but not in the population, the errors of measurement will become to some extent correlated and since we have seen that random deviates can produce appreciable common factors, they will as stated, increase the rank of any matrix of rank less than n (with ordinary communality estimates) and will spuriously increase the communality. This effect will be greater the smaller the N, and, as we have noted, a common error factor with small N may come to assume a magnitude that places it in the ranks, as regards size, of substantive factors. This is the real danger of small sample factoring. For in theory the sampling effects are on the real factors, which affect their mutual correlations and their sizes, but the error factors with small samples are large enough to intrude and distort the maintenance of the true factor structure. Nevertheless, at any rate with covariance factoring, the true influence factors will keep their number and their loadings down to quite small samples. Consequently, if the error factors can be rotated out, being distinguished by lacking the typical hyperplane (possessing only the modal value of a normal distribution at zero) bulge, the result would (except for subsequent higher-order factoring) be acceptable.

If instead of accepting the effect of measurement error in this way, and attempting to rotate out common error factors, we attempt to correct for it by the attenuation correction, we do not really get rid of it and, especially with small N, we are likely to produce non-Gramian matrices and r greater than unity. All in all, therefore, correction for attenuation is not to be advocated with small N matrices and with any R matrix it is better to apply it to the V_0, as in Eq. (14.19), to see what it does there rather than do it with R and lose this immediate evaluative view of its effect. Probably the only useful purpose in applying it is that of getting a better idea of the real size of the salients, thereby shaping better any hypothesis for factor interpretation. One also can thereby judge if a particular test would have promise of good validity if extended in length from the given experimentally short form, or if its reliability were improved in any other way. This last tactic in research of using relatively short tests to explore over a wide field and then correcting their loadings for attenuation is not to be despised. For many a valuable test would have been passed over if judged by loadings seen only on brief forms, i.e., if one had not corrected (by the attenuation correction) for its relative brevity.

The argument put forward above that with certain conditions and skills smaller samples than are normally ideal may throw light on at least the number and nature of factors has to be considered, however, with one or two other questions, in the light of the existence of interactions between measurement and sampling error. Our point was that neither the pattern of the factor, in covariance real-base matrices, nor the variables marking the hyperplane (which are, anyway, part of the V_{fp}) should alter in sample selection.[3] Consistent with our influence model, this is saying that a factor as an influence has the mandate, as recorded in the factor mandate matrix, to go on affecting certain variables and leaving others severely alone. It will continue to leave these alone, no matter what the sample size, and only its variance and correlation, not its basic pattern, will alter in small samples.

However, if one then asks why is it not as good in general practice to do a factor analysis on a small sample as on a large one (provided one avoids the extreme indeterminacy of using no more people than variables), the answer is (a) as we have seen that the sizes of factors and correlations among factors will "wobble" to the determinable extent expected of correlations on a small sample and make inferences to the population on these matters uncertain to that extent; and (b) the hyperplane is also likely to get more blurred for if, say, a loading of .10 is the standard error of a zero r for the given n, a correction for attenuation "across the board" will also throw this up to, say, .15 and make simple structure more difficult to find.

The principle to be reiterated here in regard to error is that sampling error is, by the model, to be regarded as sampling on factors, whereas measurement error is initially on variables. If sampling selection in a population is carried out on one variable, e.g., to get a sample of reduced variance thereon, it operates on the factors concerned in that variable and does not affect the essential structure. As Thurstone (1947) pointed out, it affects the factor correlations but not their patterns. If it is carried out in irregular fashion on several variables, it is no longer possible to translate the effect simply into factor modifications (Thomson and Ledermann, 1939). Introduction of low reliabilities over an arbitrary set of variables will act, however, only on the factor structure of those particular variables. It should be noted that intrusion of measurement error, enlarging the uniqueness, also leads to poor evaluation of the sizes and roles of true specific factors.

---

[3]The same is true of the effect of error on the hyperplane position. It should be reiterated, however, that the above statements about constancy of factor pattern in small samples were made conditional on covariance factoring and real-base representation. From what we called for brevity the "Brazil nut effect"—the effect of the variance taken by one factor upon that left for another in a unit sigma variable—a change by selection on one factor will affect the other factors in any ordinary factor pattern matrix in a way that will not simply bring changes (through factor selection) down any column as by a simple multiplication. Since it does not matter by what a zero loading is multiplied the simple structure should remain on the same variables *even in ordinary factoring, but the loading patterns will not retain exactly the same form, when a factor variance change occurs*, i.e., they will not be simply proportioned down the column to the old values.

14.12. Summary

1. When distribution, sampling, and significance are considered in factor analysis they need to be considered (a) with respect to both relatives and referees, since in multivariate methods and representative designs the representativeness of the relatives is also vital to a correct view of the factor structure of a domain, and (b) with respect to the populations in five domains (sets) of ids: persons, ambient situations, responses, stimuli, and observers. Some inroads have been made on the meaning of population and population sampling in the last four sets (Cattell and Warburton, 1967; Guttman, 1956; Kaiser, Cerny, and Green, 1977) but the fullest statistical understanding and formulation has long developed mainly around people as referees.

2. The mathematical statisticians' "population" is a convenient fiction, usually realizable and used only as an untested abstraction. If the criteria for such a population were rigorously applied the results would force us to much more restricted generalizations of our results than we commonly practice. Actual populations (in the ordinary social sense) tend to be complex collections of species types, often ill defined, but nevertheless making homogeneity a fiction. The nature of most real populations remains, in any scientific sense, to be explored, but some modal densities can be found suggesting the prevalence of species type, apparently yielding within themselves approximate normal distributions on the factor variates, and, with mutual overlap, often giving normal distributions on single variables in the total population.

3. The much-needed discovery and mapping of types aims eventually at (i) the location of types, (ii) the definition of the type-common dimensions by which the central types are themselves recognized and placed ("between type factors"), and (iii) the definition of the dimensions by which individuals are placed within types ("within type factors"). The discovery of types is pursued by the taxonome principle embodied in the TAXONOME program. This principle distinguishes two operational steps in locating types: (a) finding homostats (or stats)—individuals all mutually related (in factor score profile pattern) above a given r_p—and (b) finding segregates (or aits) each consisting of a set of stats all overlapping by a given number of individuals. Since the TAXONOME program works well with populations known on other grounds to have subgroups ("types" and "breeds," as in ships and dogs, respectively), but so far has revealed few discrete psychological types in humans, one might conclude the latter to be less distinctly broken into types than is widely supposed. However, research with precise TAXONOME and similar programs is still scant, especially on psychiatric classifications.

4. What has not too happily been designated "multidimensional scaling" has so systematic a relation to factor analysis that it deserves adequate but brief relating. In the former we begin with the distances between members of all pairs in a set of variables and infer the dimensionality necessary to accommodate those distances. In factor analysis, of course, we begin with dimensions and can end with distances. The difference has aptly been called by Rummel analogous to the difference between integral and differential calculus and if not too late it would be helpful to students to call M.D. scaling *dimensional integration*, (of distances) as a more apt description of the essentials of the method. Unfortunately, most of the psychological operations by which distances are actually evaluated have been subjective judgments, though it should be possible in some areas to provide more behavioral measures. Meanwhile, factor analysis remains the less questionable and more exact method of reaching these dimensions.

5. Real-base true-zero factor analysis requires for its completion, beyond the reaching of a real base for factor size initially discussed (Chapter 13, Section 13.8), the introduction of equal interval properties in scales and the discovery of a true zero. Actually all forms of factor analysis (except nonparametric) imply dependence on equal interval scales. Equal interval transformations from raw-score scales can be approached by two procedures: (a) pan-normalization and (b) the relational simplex. A true zero is implied and used in ordinary differential R technique, but in all other factor techniques the absolute zero on a factor can only take its properties from absolute zeros on variables, just as equal interval factor properties depend on equal interval variable properties.

6. Two methods of search for an absolute zero on a variable are described. The first follows as a logical extension of the relational simplex principle applied to equal intervals. Both the condition where the initially assumed zero is above and that where it has been assumed below the true scale zero can be recognized by collating a sufficient sample of regressions of the given variable on other variables, using the principle of lowest determiner in staggered onsets. The second rests on "inherent secondary properties."

Related to this, as an addition to the factor model as used hitherto, some advance is now necessary by two elaborations of the model. These are the concepts of (a) staggered onset and (b) permissive action models. These deny that the zero on a factor should be equated by regression to the "average zero" on many added variables. Both consider any variable to be the product of factor influences that may come into action at different points in its range. To find these points it is necessary to factor at various levels. And to find a factor zero it is necessary to discover instances of "limiting variables," i.e., variables in which a steady regression relation of variable to factor continues below the factor score previously found on any other variable. The second model extends the modulation model, first used with states, to include traits, becoming an "inherent change" principle in all factor structures and permitting determination of a factor zero as the raw-score point at which modulation influences cease to produce effects.

7. Before any question of significance of correlation matrices and factor matrices is approached, a brief overview has necessarily been taken of the properties of coefficients of association between variables that are usable in an R_v matrix. At least 6 coefficients of the product-moment family and 8 others have been used on various data. With continuous variables cut at "pass–fail" points, or dichotomous variables based on an underlying assumed continuous normal distribution the tetrachoric has the best properties while phi-over-phi-max suffers from some weaknesses. With continuous properties on one variable the point biserial and others are to be used. When all is said, the product moment and covariance are the most satisfactory, but raw data should be normalized and correction for coarseness of grouping is desirable when the real scoring is actually more refined. Nothing effective can be done with unordered qualitative variables with more than two phases. Nonparametric factoring is discussed elsewhere.

8. Using cross products of raw scores (sometimes called direct score factoring) seems at first to offer the most comprehensive and least artificially distorted factoring, since it promises to cover in one set of factors the covariance among individual differences on the variables and the relations of the absolute levels (means) of the variables. However, products of common and specific factors systematically enter the result, and without scarcely tenable assumptions about specifics it cannot be used.

9. Statistical significance tests are required for whole correlation matrices, factor loadings, differences of loadings, and correlations among factors. Those for R_v matrices have long been known in connection with deciding when residuals in extraction are negligible. The maximum likelihood method adds a new basis for this general approach. Debate exists because empirical correlation matrices of random normal deviates yield loadings on factors such as are commonly evaluated as significant. On the opposite side, it is pointed out that real matrices with very small correlations can yet contain substantial factor structures. The two streams contributing to evaluation of the significance of a loading are the ordinary regression arguments and the confirmatory ("proofing") maximum likelihood method. Through Jennrich and others adequate solutions now exist for orthogonal and oblique loadings. The fact that these are different for each factor and variable complicates certain assessments, e.g., of hyperplane width. Significance of the usual (rotated) loadings has to take into account also the accuracy with which the rotation is completed and the reliability of the variable as determined by measurement error.

10. Expressions are known for effects of univariate and multivariate

sampling selection on factor patterns, the effect of the former being typically to alter correlations among factors or factor variance, but not essential factor pattern.

11. A distinction can be drawn between V_{fp} values signifying (a) hyperplane variables, i.e., those which could be truly 0; (b) salient variables defined as of *statistically significant level* ($p < .05$); and (c) prominent variables, as a subset of (b), above some arbitrary value like 0.7. The last is useful in some contexts, the others in many. In setting the width of hyperplane by the theoretical formula for loading significance, i.e., for the range of a truly zero r, one encounters the difficulty that it differs for every factor, and to work with an average seems the usual practical compromise. Finkbeiner has shown that in a thoroughly rotated study the hyperplane width can be discerned by the "bulge" in the otherwise normal distribution of loadings and that this agrees reasonably well with the theoretical formula value. However, both sources suggest that one reason for the poorness of analytic as contrasted with topological programs is that the former do not work to a sufficiently narrow bandwidth. For in reasonably well-designed experiments (say $N = 300$; $n = 60$; $k = 20$; reliability around 0.75), the width is almost certainly narrower than the traditional ±.10 and more in the range of ±.04–±.08. Monte Carlo tests for significance of simple structure have been developed on this basis.

12. Error of measurement has to be considered in its interaction with sampling error. Together they determine the magnitude of common error factors. In the absence of error of measurement the effect of sampling is merely a sampling on factor variances which leaves simple structure and factor loading as precise in a small as in a large sample (descriptively in that sample). With error it is necessary to rotate to set aside common error factors, which are detectable by absence of the significant congruences with cross validating studies shown by the substantive factors. Only with low reliabilities and smallish samples (say below 200) do common error factors present such problems of intrusion into substantive factor rotation. Granted that skilled rotation has been able to set aside common error factors, small samples should yield true factor patterns as reliably as in large samples. But the factor sampling effects in small samples mean that factor variances and correlations among the factors will be undependable and unfit for conclusions regarding the higher-order factor structure in the population.

CHAPTER 15

Conducting a Factor Analytic Research: Strategy and Tactics

15.1. The Choice of Experimental Design

It is proposed here to run through the essential steps of a factor analytic research experiment in the context of the general strategy of investigation. This will permit integrating the application of the principles and concepts treated in preceding chapters, with the touch of a summary.

It seems desirable here to give rather more emphasis to the exploratory use of factor analysis, since there are still substantial areas of psychology, and still more substantial in social and biological sciences, where the uncovering of central concepts and structures remains a major task for this method. At the same time we shall interject due regard for those somewhat different procedures more involved in hypothesis testing. The needs of the latter will be considered at each appropriate point of departure.

Texts on factoring usually indicate the first step as that of choosing the variables. This is correct if still more prior aspects of design of experiment have been decided. But if we are to begin truly at the beginning we must ask what the basic design is that calls for the choice of variables. It is difficult to cater to all readers at this point, since investigatory interests are infinitely Protean. Yet all are covered by the dimensions of experimental design we set out in Chapter 1, and the possible relations to be studied in data are all encompassed by the basic data relation matrix (p. 322). As to the former, it will be seen that among other things the researcher has to decide whether or not to include manipulation or to arrange definite time sequences among his variables such as will permit causal inference. Either inclusion will allow dependent and independent variables to be designated in more than a mathematical sense, namely, as causes and effects. The manner of incorporating manipulated and sequentially placed variables has been discussed. Incorporating them into the correlation matrix can take place either by (a) allowing the independent variables involved to assume the naturally existing correlations of those independent variables one to another or (b) by factorial design, forcing these independent variables to be orthogonal. For example, if temperature, humidity, and daily hours of sunshine are entered in an experiment with, say, some thirty mood variables, it may be impracticable or undesirable to try to make the "causes" uncorrelated. But if number of trials, magnitude of reinforcing reward, and quickness of reward after response are entered in a factor experiment with a dozen measures of different aspects of response, it becomes practicable and desirable to make the former mutually orthogonal since in any case there is no natural structure among them.

One sees researches where the existence of a square matrix of correlated variables has hypnotized the experimenter into factoring it as a single matrix when segmenting would make better strategic sense. Every argument of good design calls for at least an initial separate factoring of independent and dependent variables, followed by correlating of factors from one with factors from the other. And, as will be seen in a moment, most design arguments would call for factoring variables from different domains (domains being defined as below) at first separately, even when they have no distinct status as dependent and independent variables.

After deciding on the manipulative, sequential proportional representation, etc., of measured variables and the degree of control of unmeasured variable aspects of the experiment, the next—or perhaps the simultaneous—issue concerns what facets or grids are to be used in the data box. Naturally, this is almost immediately decided by the initial research interest. If one is interested in personality traits he will use R technique;

if states, P or dR technique; if in the "Zeitgeist" dimensions existing among many historical occasions, S technique; if in the dimensions which observers add to the things they see, then he will use a factoring of responses over observers; and so on.

Research strategy may call for a combining of various techniques. For example, if trait view theory supposes (Cattell, 1970) that misperceptions are a function both of the observer and the real characters of the subject, a combination of two kinds of facets in a grid may be called for. Or if one's theory is that traits change and develop in the same shapes in which they appear as individual difference patterns, then an R-technique experiment is to be combined with a P- and dR-technique experiment.

Not only will a significant research—as distinct from an experiment—call for some planned sequence of combinations of some of the different correlational techniques—R, P, O, S, T, dR, dS, etc.—and for repetitions of the same techniques with a succession of samples or circumstances, but it will also call—more fundamentally—for combinations and alternations of multivariate designs and bivariate designs outside factor analysis. Discovered factors are not for a museum, but for functional use as vital concepts in further experiments. And in the intensive study of a particular factor its use simply as a scored "variable," measured by a battery for that factor in a bivariate experiment with some second, dependent or independent variable, will often be demanded. A nonmanipulative instance is the plotting of personality and ability factor scores in relation to age changes. A manipulative example would be the change of ego strength factor scores under tranquilizers (Rickels *et al.*, 1966), or, using the factor as independent variables, the change of cognitive performances under various induced levels of anxiety.

Although it is true that all of the infinite variety of possible experiments which a psychologist can think of doing can be reduced to CORAN or ANOVA treatments, and then to the limited number of facets, faces, and grids of the data box, nothing short of a vast series of illustrations could show how that would be done. But the above may suffice to remind the psychologist that a research, which means a programmatic series of experiments (even when the program develops like a hunt and is not completely set down at the beginning) needs to keep in mind the possible techniques of the data box and the possible dimensions of experiment—bi- and multivariate, manipulative, and nonmanipulative, etc. (see Chapter 1).

15.2. The Choice of Variables: Markers and Matching

It is at the choice of variables that the cleavage between explorative use and hypothesis testing first and most explicitly expresses itself. Incidentally, it is sometimes rather pedantically said (in an attempt to deny the importance of the exploratory vs. hypothesis-testing distinction) that *no* experiment is conducted without a hypothesis. Naturally, if one is willing to call it a hypothesis when a researcher says, "There must be some structure in personality" or "Some kind of change must always occur through rewarding a response," then this is true. In short, even the choice of exploratory variables always implies some very general hypothesis. However, one recognizes how tenuously general such "hypothesis testing" is when one comes to a representative design, where the experimenter deliberately seeks a stratified sample of a domain merely in the belief that some structure will be found in it. Such research is better explicitly recognized as exploratory.

Nevertheless, the selection of variables follows different principles in the two cases. In representative design one defines some population of variables comprehensive of a domain. They may be personality responses, as in the *personality sphere concept*, or ambient stimuli in structuring the motivational occasions in a culture, or verbal counts in therapeutic sessions, and so on. In hypothesis testing, by contrast, one designates "salient" or "prominent" marker variables relevant to the hypothesis, recognizing that the hypothesis may often be only that a certain factor pattern exists. It is surprising but true that factor analytic researches continue to appear that are essentially impotent and non-contributory through observing neither the one nor the other of these vital aims in choice of variables. These lame researches "explore the factors in a domain" without any explicit definition of a population of variables, or any demonstration of sampling operations applied to them. More frequently they set out to check hypotheses regarding concept X or Y without even carrying the salient markers which factor analyses in the past literature have shown to mark the central characteristics of those patterns.

In regard to this last, let us be realistic enough to recognize that factor analytic research is still at a stage in many fields where firm replication of a given factor *simply as an empirical pattern* at a sheer descriptive level apart from any explanatory hypothesis is the most needed achievement. Ability primaries and secondaries are now generally infallibly replicable; concepts of personality and motivation primaries and secondaries are less matured. They have been replicated half a dozen times, it is true, but not to everyone's satisfaction. And at the extreme of experimental chaos, we find many social attitude and value factors and some physiological state factors which seem to change with every investigator. To factor a set of arbitrarily chosen variables in any domain, no matter what the hypothesis behind them, without introducing markers consistently found in that same domain by two or more good previous researches, is an act of scientific irresponsibility. The investigator has encapsulated himself in a private world, precluding any possibility of an architectonic growth of scientific knowledge.

15.3. The Choice of Variables: Numbers, Matrix Divisions, and Combinations

As to the number of markers needed to make a factor match possible, the mathematically necessary minimum for a common factor is, of course, two, since one variable would not be able even to make a common factor appear. But psychologists with experience of the ambiguity which sometimes plagues matching and interpretation will advocate three, four, or even five markers, as Gorsuch does. The difficulty with this more proper, ample treatment is that in a well-planned research, aiming at sufficient hyperplane stuff for good rotation and enough variables to avoid confusion of factors that happen to be somewhat alike, the research will often need to cover as many as 20 factors or more. Five markers for each brings the total to a hundred and this leaves little room (if one must stop at, say, 120 variables) for the new, hypothesis-testing variables one normally introduces at each new experimental round in a programmatic research, and which are necessary for convergence on the most clear and stable patterns. Actually, three or even two markers per factor will usually suffice, if need be, provided (a) the loadings are above .7, so that no other single factor elsewhere is likely to load them that highly, (b) the markers primarily chosen for one factor characteristically have some small but significant loading on another that will help in its recognition. Thus, say 40 prime markers for 20 factors may actually be averaging three markers per factor. It is also helpful in identification to have some variety of sign in the factor markers for a given factor, e.g., + − + − instead of + + + +.

A special problem related to markers is that which arises when one needs to resort to "dovetailing" two or more factor analyses together. To survey a domain with proper coverage it may be desirable to span, say, 300 variables. But, usually subject time (and occasionally, even nowadays, computer size) forbid encompassing them in a single experiment. One solution to this problem is to divide the 300 variables in equal parts, a, b, and c, each of a hundred, and do three factorings $(a + b)$, $(b + c)$, and $(c + a)$. By Tucker's method (1951, 1960), each of the three possible pairs of factor matrices $[(a + b)$ with $(b + c)$, and so on] can be brought to maximum congruence (*on the 100 variables in common* to each two adjacent studies) and then rotated to the best simple structure. Alternatively, as preferred by Saunders and the present writer (Cattell and Saunders, 1950), each can be separately rotated to simple structure and the factors identified across studies by matching patterns on the 100 overlapping variables. The putting of the three together, to cover 300 variables, is in this case a relatively approximate procedure but can be made more exact by setting out the average target pattern (across separate experiments) for each matched factor and rotating each study thereto by PROCRUSTES, i.e., to a maximum congruence of pattern.

As to the absolute number of variables for an ideal experiment, first it should be said that the number should be larger than is often the custom. It is rare indeed for phenomena in any domain to be accounted for by only three or four factors, and cutting down the number of variables to, say, seven or eight, is not going to simplify proportionately the number of factors acting in the experiment. Consequently, factors are going to be ill defined. In definitely controlled "late in the game" experiments, where "irrelevant" and hyperplane variables are known and salients and hyperplane markers can be confidently inserted and recognized such economy may be practicable. But in most "natural setting" experiments there are likely to be many factors and therefore many required variables. Thirty to forty is a reasonable minimum in any such experiment.

In seeking some rational basis for the number of variables to be chosen in relation to the number of expected factors one counsel of perfection has been that on an average a test vector should be available for every quadrant of the expected factor space, i.e., every combination of positive and negative loadings (poles) on the factors. The importance of deliberately introducing variables for hyperplane stuff has been stressed in Chapter 7. In connection with "quadrant representation" we now see the desirability of the "stuff" being such that simple variables with a zero loading on a given factor shall have an appreciable loading in turn on all factors against which it is to be plotted, and this is best met if there is at least one variable in every quadrant. (In short, hyperplane stuff is not useful if composed of variables that correlate with nothing else, i.e., no other factors.) But with k factors this "quadrant" design means 2^k variables; for instance, 256 variables with 8 factors. This ideal counsel calls for a perfection of experiment that is impracticable as one moves toward 20 factors, but fortunately a much more modest total will, for various reasons, work well. Experience has suggested a rule of thumb of about three times as many variables as factors, which is likely to give the required minimum of two markers per factor even for the less well-represented factors, but five times is sometimes needed. Meyer (1973) has presented arguments for estimating desirable ratio of variables to factors on a basis of factor determinacy.

What chiefly deters many experimenters today from a sufficiency of variables for good factoring is not fear of the computations, which the electronic computer has removed, but the amount of testing time or data gathering required. In effect this means the experimenter is often forced to make a painful decision between having a sufficient number of variables by the above canons, and having variables long enough to be individually reliable. Probably more than half of Ph.D. theses in multivariate experiment fail of clear conclusion through an insufficient n of variables and an insufficient N of people. The two deficiencies are bound together, because the more testing time demanded, the fewer subjects will volunteer. The solution of that problem must be left to the perseverance of the experimenter and positive enlightened policies in professional psychology for financing subject pools. Meanwhile the dismal refrain at the end of many theses persists. "The results suggest that if some future investigator obtains more subject time and subjects, statistically significant results will be obtained." In the thirty years of programmatic factor research on objective personality test devices by associates of the present writer the subject time ranged from three to nine hours. Since it was vitally necessary to cover 15–20 factors with 40–160 tests (some markers, some newly devised hypothesis testers), the time per test device was likely to be 4–5 min, which seldom brought reliabilities of above 0.75 and sometimes left them quite low. Although such researches are reported with their actual loadings, the insightful pursuit of better kinds of tests and of clearer conceptions of the factors was successfully pursued in our laboratory by inspection of the *attenuation-corrected* loadings. On at least a dozen occasions the promise revealed by such examination of a test has vindicated itself in a well-replicated substantial loading on longer tests of higher validity in a shorter battery of selected markers so discovered. Furthermore, some tests ultimately of great value, would never have had a chance to appear at all if the common psychometric demand for immediate high reliability—often vociferated by critics who missed the programmatic plan—had forced the designer to a long test or none.

What one chooses for relevant variables in the experiment and what one chooses to factor in a single matrix should often be different things. Too frequently one sees a study covering ability tests, personality measures, home background variables, and various criteria all factored together. The result is commonly chaotic, in the first place because not enough variables exist to pull out clear primaries in each domain, and in the second, because different *strata* (orders) are almost certainly being thrown into the same matrix. Such a "melting pot of variables" procedures has most frequently been followed merely with some such object as "reducing the number of variables." Though this "reduction" is what factor analysis can undoubtedly accomplish in statistical terms, that is not what the scientist uses it for. A mixed set of factors of different orders, in which some correlations in the V_{fs} are between factors and some between variables and factor does not help very much the psychologist who is seeking functional understanding in terms of scientific concepts.

It is admittedly practicable, once structure and markers begin to get known in the given domains, to factor different instrument approaches in the same modality, e.g., personality, by Q data (questionnaire), T data (objective "performance" tests), and L data (life observation ratings). But throwing the three instrumentalities together before factor perspective has been obtained in each almost inevitably leads to chaos, as the present writer and Saunders (1950) found in the early days of personality structure research. Knowledge of the usual instrument factors to be recognized and set aside also helps (compare Cattell, Pierson, and Finkbeiner, 1976, with Cattell and Saunders, 1950). Some problems though less severe, occur in factoring different modalities—abilities, temperament traits, dynamic traits, and states—together. However, sooner or later this always needs to be done, for otherwise we may not recognize that a facilitator of certain ability performances, which we have called "ability X" is actually the same as a more highly loaded factor we have been calling, in temperament tests, "temperament trait Y." With increasing replication within modalities and instrumentality domains separately, the factoring of variables stretching across these can be both successful and theoretically important. On the other hand the relative blind mixing of different sets in the data box, e.g., occasions and observers, stimuli, and response characters, rarely makes sense, even in an otherwise well-designed three-way or three-mode analysis, unless some knowledge of structure in the separate sets has led to clear thinking on what is to be got out of the design. As with the special case of manipulated variables and responses discussed above, this means that separate matrices are best analyzed initially.

Whenever one is in doubt about factoring diverse variables together, because some variables are in a different domain or belong to different data box sets, a useful and frequently used adjunct is a computing device, the Dwyer extension (Dwyer, 1937; Mosier, 1939b). This is also often used simply because the number of variables (as in items in a questionnaire), though all are in the same domain, is too large for economical factoring of the lot in one matrix. One takes a subset of suitably limited, practicable size in a square matrix, $n \times n$, containing the variables, e.g., markers, likely to contain the main dimension in all. The correlations of these n variables are then calculated, adding a long oblong extra matrix, with the large number m of "extension variables." Let us call this $(m \times n)$ matrix E_{mn}. The calculation of the correlations of the extension matrix variables with the factors found in the $(n \times n)$ "core" matrix is then obtained by the following matrix operations:

$$\hat{V}_{fs} = E_{mn}(V_0 V_0^T)^{-1} V_0 (L_r^T)^{-1} D \qquad (15.1a)$$

or, in simpler form, but less available matrices

$$\hat{V}_{fs} = E_{mn} R_v^{-1} V_{fs} \qquad (15.1b)$$

where $\hat{V}_{fs}$ is the $(m \times k)$ estimated factor structure matrix for the extension, $E_{m \cdot n}$ is the $(m \times n)$ correlation matrix of the extension with the factored n variables, V_0 is the usual $(n \times k)$ unrotated matrix, L_r is the transformation matrix to the reference vector simple structure position, and D is the usual diagonal of tensors (see pp. 161, 184) going from factors to reference vectors.

For comparison, an explicit correlating of the estimated factor scores themselves with the new variables would yield a $\hat{\underline{V}}_{fs}$, as follows:

$$\hat{\underline{V}}_{fs} = \frac{1}{N} Z_m^T Z_f$$

or if we want unit-length factors in a $\overline{V}_{fs}$,

$$\overline{V}_{fs} = E_{mn} V_0 (V_0^T V_0)^{-1} (L_r^T)^{-1} DD_{SF}^{-1} \qquad (15.2)$$

where Z_m and Z_f are the standard score matrices for the m variables and k factors, respectively, and D_{SF} is the matrix which, as seen earlier, contains the standard deviations of the factor score estimates, necessary to bring $\hat{\underline{V}}_{fs}$ to unit length factors. Thus the two approaches are essentially equivalent except for the factor sigma correction needing to be applied to the score estimates approach, i.e.,

$$\hat{\underline{V}}_{fs} D_{SF}^{-1} = \overline{V}_{fs} \qquad (15.3)$$

The transition from V_{fs} to V_{fp} can be made in the usual way by knowing R_f.

The important thing to keep in mind here is that the m extension variables can be given correlations only with the factors which exist among the n variables. If there are new factors in the m set that are not in the n set we shall know nothing about them. Consequently, one should take steps to ensure that the chosen n cover the known domain well. Second, one should note that the new m values are correlations, not loadings, though they can be converted in the usual way ($V_{fp} = V_{fs} R_f^{-1}$) as just stated. Third, it should be noted that the correlations are usually somewhat lower than they would be if the m variables had been factored in directly with the core. (For in the Dwyer case their specifics would also get into factor score estimates which are implicit in the method.) The alternative to the Dwyer extension, estimating the factor scores and correlating them with the extension variables as in Eq. (15.2), yields, as

Nesselroade and others have pointed out, exactly the same values, but it is rather more troublesome.

15.4. The Choice of Variables: Instrument Factors and Perturbation Theory

A special problem of selection of variables, which is as important in hypothesis testing as in hypothesis creating (exploratory) use of factor analysis, has to do with *instrument factors*. In one and the same domain one may expect, if different kinds of instruments are used (e.g., questionnaire, observer rating, and objective tests in the personality domain), to find instrument factors—perhaps as many as two or three running across any one instrumentality—expressing different "biases" in the measurement *per se*. Such factors can arise from the scoring form of tests, e.g., normative and ipsative scoring in different subgroups of scales; taking response variables, such as dominance response, all in one life stimulus area; common situations of test administration over a set of several variables; and so on. The choice of style, situation, scoring, etc., of measures should obviously be so distributed as to attempt to aid separation of instrument and substantive factors by much the same factor-differentiating patterns of markers as in seeking ordinary substantive factor separations. Campbell and Fiske (1959) attempted simply to reduce instrument factor bias by the multitrait–multimethod (better "instrumentality") design—essentially a "factorial" design in which every kind of trait is combined with every type of instrumentality— except that "every" is impossible. This is a practical gain in simple psychometry, but it does not lead to clean location and separating out of the instrument factors. Its ready comprehensibility has tended, not through any bias of the authors, to stand in the way of experiment with the more realistic but more complex concepts of *perturbation theory*. This systematic theory analyzes effects of mode of observation and measurement upon error and therefore upon (a) the recognition of separate substantive structure and instrument factors and (b) the estimation of substantive factor scores (Cattell and Digman, 1964; Cattel, 1961b). This is not the place to enter into the whole treatment of perturbation theory, with its subtheories of *trait view* (Cattell, 1968, 1973) and *test vulnerability*, and we shall leave Fig. 15.1 to give an overview.

Perturbation theory has reference to more aspects of strategic design and choice of variables in multivariate research than those required by instrument factors alone. For example, II(ii)(c) in Fig. 15.1 suggests that personality measures on the observers are variables needed for allowing for observer trait view effects. But, in general, the theory pre-

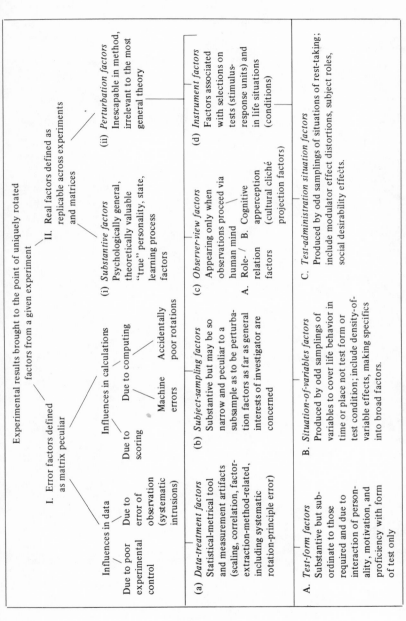

Fig. 15.1. Perturbation theory: Schematized in terms of effects of observation error or intrusions on factor structures. Note that a systematic relation is possible between certain categories here and those of the five dimensions in the data box in Chapter 14. Thus, II (ii) (b) deals with sampling on the subject coordinate II (ii) (c) with the observer coordinate; II (ii) (d) A and B with stimulus-response coordinates; II (ii) (d) B and C with selections on the occasions coordinate.

Table 15.1. Alternative Meaningful Resolutions When Instrument Factors Are Involved[a]

(a) Seven factor solution separating instrument factors

Variable		D_1	D_2	D_3	D_4	D_5	m	s	Specifics
D_1	$A_{1.m}$	X					X		Not entered. A diagonal
	$A_{1.s}$	X						X	matrix of one factor
	$A_{2.m}$	X					X		and one loading to each
	$A_{2.s}$	X						X	variable.
D_2	$A_{1.m}$		X				X		
	$A_{1.s}$		X					X	
	$A_{2.m}$		X				X		
	$A_{2.s}$		X					X	
D_3	$A_{1.m}$			X			X		
	$A_{1.s}$			X				X	
	$A_{2.m}$			X			X		
	$A_{2.s}$			X				X	
D_4	$A_{1.m}$				X		X		
	$A_{1.s}$				X			X	
	$A_{2.m}$				X		X		
	$A_{2.s}$				X			X	
D_5	$A_{1.m}$					X	X		
	$A_{1.s}$					X		X	
	$A_{2.m}$					X	X		
	$A_{2.s}$					X		X	

(b) Ten factor solution with "refraction factors"

Variable		$D_{1.m}$	$D_{1.s}$	$D_{2.m}$	$D_{2.s}$	$D_{3.m}$	$D_{3.s}$	$D_{4.l}$	$D_{4.s}$	$D_{5.m}$	$D_{5.s}$	Specifics
D_1	$A_{1.m}$	X										Plus specific factors.
	$A_{1.s}$		X									
	$A_{2.m}$	X										
	$A_{2.s}$		X									
D_2	$A_{1.m}$			X								
	$A_{1.s}$				X							
	$A_{2.m}$			X								
	$A_{2.s}$				X							
D_3	$A_{1.m}$					X						
	$A_{1.s}$						X					
	$A_{2.m}$					X						
	$A_{2.s}$						X					
D_4	$A_{1.m}$							X				
	$A_{1.s}$								X			
	$A_{2.m}$							X				
	$A_{2.s}$								X			
D_5	$A_{1.m}$									X		
	$A_{1.s}$										X	
	$A_{2.m}$									X		
	$A_{2.s}$										X	

sents the experimenter with a guide to starting with an enlightened selection of mutually "spaced" variables calculated to ensure separation and recognition of various instrument factor, observer, and substantive factor effects (Cattell, 1973). Parenthetically, the line between substantive and instrument factors is in the end one of research position and focus. Just as dirt is "matter out of place" so an instrument factor is often a substantive factor intruding where the focus is on a different substantive area, e.g., a factor of "ability to read" appearing in personality questionnaires.

A puzzling alternative sometimes presents itself in rotational resolution where certain instrument factors are known to be present and have appreciable variance. It can be illustrated by a problem in motivation measurement. Here the two attitude markers known to represent each dynamic factor well are each measured by two different devices or instruments, in this case memory and speed of decision—m and s in Table 15.1. Thus, there are four markers for each of the five dynamic structure factors themselves. There are, *a priori*, four rotational solutions that could appear here, the first two of which are shown in Table 15.1. They are: (1) the appearance of the unitary dynamic factors, responsive to the two (doubled) attitude markers present for each. In this case the special instrument variance of m, general goodness of memory, and s, general quickness of reaction, will separate as broad instrument factors as shown on the right of Table 15.1(a). (2) The true dynamic interest strength and the instrumental expression can run together, in what have been called "refraction factors" (Cattell, 1961) because they show the same pattern twice as through a glass prism, in this case giving an interest-through-memory and an interest-through-speed factor in place of each D. In this case there are no instrument factors (Table 15.1b). (3) Since two markers occur for each attitude it would be possible for each to appear as a factor, but in that case the m and s instruments might appear again. This would require 12 factors from 20 variables and is unlikely; but the other require, respectively, 7 and 10 factors, and in actual experiment are found to be tolerably good alternative simple structure solutions, the former being clearer with 7 and the latter with 10 abstracted. A last possibility is (4) the gross

---

[a]The X in the above are salient loadings; the rest are in the hyperplanes. The variables in (a) and (b) are the same. The D as in D_4 indicate a particular dynamic factor, e.g., the fear erg; the m and the s indicate two different devices (memory and speed of reaction) through which the given attitude variable is being measured. The attitudes A_1 and A_2 used to measure each dynamic factor are markers known to have prominence of loading on the given D. Thus, in the columns, the factors in A are indicated as dynamic factors each running over all markers for the factor regardless of their mode of measurement (devices m and s). In (b) the existence of two pairs of device measures for each attitude has permitted each dynamic factor to emerge as a factor in device m, and a factor seen through device s.

solution for five factors, which may sometimes appear. However, depending in part upon whether one chooses to start with a seven or a ten-factor solution, empirical research shows that either a resolution as at (a) in Table 15.1 or as at (b) will be reached.

Here we meet a real case of the possibility of alternative simple structures discussed earlier, and we also find in these cases an ambiguity of the scree test for number of factors. Modern methods of assessing number of factors, it is true, rarely leave such doubt, and one solution is usually a little more compelling than another. In the 12-factor case, one will need to seek the real dynamic factors in which one is primarily interested at the second-stratum level, and in the five-factor solution some of the extra variance will go into garbage factors (Cattell and Digman, 1964) and the hyperplanes will be a bit coarse. However, in the two main alternatives which simple structure may yield, we have in (a) the substantive and instrument factors separated. In (b) the "refraction factors" absorb the instrument variance in a special pattern. Meaningful psychological interpretive models can be thought of to fit either solution, e.g., an additive action of separate substantive interest and device endowment in (a) so that a linear additive model delivers them separately, and a psychological product action in (b) such that interest strength and expression are multiplicative, separation is difficult, and essentially there are ten factors. Experience shows that if variables expressing some relatively pure measure of the instrument device are added to the 20 variables, the instruments are more likely to be brought out on their own in the (a) pattern.

In this case, the existence of perturbation factors, such as instrument factors, calls for a concern about choice of variables additional to the principles of choice operating in the usual substantive design. As Fig. 15.1 reminds us, perturbation factors can appear from several sources, such as the form of measurement, e.g., forced choice, multiple choice, as well as from common ambient or stimulus situations, and from the methods of scoring. The general and basic issues on scoring have been dealt with in Chapter 14, on scaling, but what was there said on ipsative scoring can be appropriately developed a little further here in connection with dynamic trait measurement, where the concept of ipsative scoring was born. Ipsative scoring, like normative or performative scoring is in statistics a purely computational arrangement, but it can also be rooted in a psychological theory. When the vehicle for measuring interest is, say, memory, a subject with a powerful memory will tend to be higher on *all* of a full circle of interests measured in this device than would a person with a weak memory. By bringing all memory-interest scores for all persons to the same mean (semi-ipsatization) or the same mean and sigma (full ipsatization) differences due to the vehicle are eliminated and we have what may be called a "solipsistic" set of scores. That is to say, the circle of each individual's world is divided up into interest sectors showing their proportion for him. If we score in this way we consciously or unconsciously adopt the postulate that all people are equal in total interest. If this is wrong, then in trying to get rid by ipsatization of differences in the vehicle by which interest is measured—memory, or GSR response— we have also thrown away the real difference in total interest. Actually, as Radcliffe shows with algebraic precision (Cattell, Horn, Radcliffe, and Sweney, 1963), in losing by ipsatization a first unrotated factor one loses a certain combination of the total interest and the total vehicle strength factor, but some of each is left behind in later factors and theoretically rotation could thus reproduce them, with diminished variance, provided there is due hyperplane stuff to rotate them by.

Perturbation theory has probably not even yet clarified the full "instrumental" effect of ipsatized scoring, despite the shrewd contributions by Clemans (1966), Ross (1963, 1964), and some others we touched on in comparing results of R and Q techniques (Section 12.3). There we pointed out that the automatic standardization of columns in R technique loses a "general species shape" factor and of rows in Q technique a general "size" factor. Ipsatization, and R-technique procedure, is calculated, as we have just seen, to lose also the general size factor, i.e., whatever is immediately general to all measures.

The natural impulse of the discerning psychologist is to avoid ipsatized scores, but the stirring up of the issue in motivation (Cattell, 1935; Cattell, Horn, Radcliffe, and Sweney, 1963) brought to light that psychologists had unwittingly been doing ipsatized factoring for some time with resultant bias of results and interpretations. Any interest test in which a choice is made among alternatives or, indeed, any multiple choice test in which alternatives are treated as separate variables is ipsatized. It can be shown that with n variables in a matrix such that choosing one precludes choosing another, or where the total individual score over n must be held constant, the correlations when random normal deviates are used will center closely on $-1/(n - 1)$. That is to say, apart from any "extraneous" positive structure in the data, the matrix will be wholly negative. The effect of this when structure exists apart from the ipsatization has been investigated by Brennan and the present writer (1977) and is complex; but in essence it loses a general unrotated factor. The effect when a set of variables are factored in which a subset only is ipsatized is still more complex. A rough estimate [from the $-1/(n - 1)$ mean r] of what covariance is lost, and could be added to bring unavoidably ipsatized data "back into line," has been tried, but it belongs to an "artistic" correction. The issue deserves further study for much animal research, e.g., choices in a free situation, and human interest measures based on fractions of 24 hours or $1000 spent on this or that, rests on unavoidably ipsatized scores.

15.5. The Choice of Number of Referees in Relation to Relatives

The first consideration, after deciding the data box coordinate sets over which relatives and referees are to be chosen, and their mode of scoring, is settling on an appropriate size of sample. There are a few purely descriptive researches in which, as Stephenson pointed out, the size of sample is irrelevant and can be reduced to a quite trivial number. However, in the vast majority of researches we are not interested just in describing some highly local sample, but in scientific generalizations requiring inference about what happens in some general population.

The number of ids in a given data box set of referees—henceforth the number of people, since this is the commonest kind of id for psychologists—has to be considered in regard to both the number of variables and the number of factors to be expected. As to the latter, if we follow the argument in Section 15.3 we would ideally demand 2^k persons but in anything over 9 or 10 factors that becomes exorbitant. The choice of size of referee sample can be based on no single rule but on a combination of indications. Experience and factor definition considerations indicate a desirable lower limit of 3 to 6 times as many variables as factors (2 is, of course, the absolute limit, possible only when 2 good markers can be chosen for each with absolute confidence). From the number of indicated factors and variables one may, as far as the latter is concerned, also point to 3 to 6 times as many referees as variables (relatives) as the experientially indicated desirable lower limit. Five times is commonly considered to constitute an acceptably good study, and 3 times often has to be accepted. Claiming that "necessity knows no law" a substantial number of factor studies appear every year, however, with a ratio of two or even less. This would not be so bad if "reviewers of an area" did not quote such findings indiscriminately without checking against studies on a far sounder basis.

As we have seen, the one situation that must be avoided is a close approach to a square matrix in which relatives and referees are equal in number; for artificial dependencies are then forced among the correlation coefficients and, in spatial terms, specific and common factor space become confused. In some clinical data and when working in social psychology with groups as referees, there may be no escape from the number of referees actually falling below the number of relatives, and in that case it is best to "go the whole hog" and increase the relatives still further, switching to Q technique. This solves the factoring problem, but population generalizations about the factors are still limited according to the sample size. An alternative is, as discussed on p. 497, to split the relatives into two groups yielding the desirable R-technique oblong matrices, and to put together the factors from each through the common subjects, risking missing some factor present only in inter-action of the two sets of variables.

A special case, provoking almost philosophical considerations, is that where the referee group is what would be unacceptably small as a sample, yet can be argued to be itself the population. For example, the present writer has factored 80 variables over some 110 countries at a time when 110 countries were all that existed in the world. The counter argument is that some "ideal" population of countries could nevertheless be *conceived* of which these are a sample. It is true that one might con-ceive such a population as that of these countries repeated over many yearly occasions. But the statement of factors at a given point in history remains completely handled by the 110 cases despite the relative smallness of the "sample."

This discussion of size of referee sample relative to variables assumes that the reader also recognizes the necessary respect for absolute size. That has been handled under standard error of a loading. When down to 40 cases a loading has to be about .3 for even a $p < .05$ significance (two-tailed), and, what is worse, the hyperplane width will be so blurred that good rotation will be almost impossible.

Working in the range of 40–80 variables the present writer has com-pared resolutions from 100, 250, 500, and 750 subjects. It seems that quite good stability is reached by about 250 subjects and that thereafter there is no noteworthy change. Doubtless this conclusion depends partly on good sampling, reasonable test reliabilities, and, especially thorough rotation. Certainly 500 subjects would be a good number to aim at; but there is no discernible gain in simple structure or stability (as shown by congruence coefficients) in going beyond. However, in the context of most problems, even 250 or 200 is acceptable for most purposes, though below that, and certainly below 100, the product can deteriorate rapidly. On the other hand, one sometimes finds great weight irrationally attached to factor analyses done on 5000 subjects or more relative to, say, 800. Especially when accompanied by evidence of poor determination of number of factors and poor rotation the patterns from the former could be altogether less worthy of inference to the total population.

The extension of population sampling to large numbers, it is true, may do more than bring the trifling improvement in stability usual in going, say, from 500 to 1000. It is able to do so because classes and species of individuals vital to descriptions of the population may be brought in the larger composite which are simply not represented in the smaller sample and which may introduce dimensions actually absent in the smaller one-locale sample. In this connection we have pointed out that although the maximum likelihood method agrees (on admittedly still limited experience) quite well with the scree test as to number of factors when samples are in the 150–450 range, it is apt to indicate more on larger samples. But, after rotation, these added factors tend to be quite small and suggestive of imports from quite small odd subclasses. How one should regard these lesser factors is discussed in the next section.

The problem of sampling people, or occasions, or stimuli is commonly discussed in relation to a *facet score matrix*, i.e., with a uniform type of id, but one should recognize that it also exists in relation to "composite" ids in a *face* and a *grid*. In a grid we have a sample of people over a sample of occasions (or other set of ids) and the number of entries is the product of the number of people by the number of occasions. In the case of a face score matrix additional reliability is given to each id's entry by its being an average over a third set (coordinate) in the data box. For example, one might factor ability tests over people, each person's score being an average over, say, 20 occasions of testing. The formulas for population inferences in these cases have not been worked out, but if the factorist keeps logically clear what he is aiming to discover, ordinary rules can generally be applied. For example, averaging ability tests of 80 people over 20 occasions is still an inference from only 80 people, but it deals with a different entity—a trait with minimal state variance—from that in the more common practice.

15.6. Purposes in Manipulating the Selection of Referees

An investigator frequently encounters the choice between using a population or sample of large variance on the relevant variables compared to one of a more homogeneous nature with small variances. It was the argument of Thurstone that the former is to be preferred because the larger variance and larger loadings will bring out the patterns and the natures of the factors more clearly. The magnitude of the raw-score variance as such will of course disappear in ordinary correlation, i.e., not covariance, factoring, but it is nevertheless true that correlations will be larger in a larger range than a smaller range sample from the same population, if the latter is the result of selection.

To clarify desirabilities here it is necessary to distinguish such selection as just mentioned for narrow or wide range in a type-homogeneous population, from that involving mixing of species (modal) types as occurring in the later population. The latter has been sufficiently handled under within type and among type factor dimensions (p. 448). The clearest results are from doing the one or the other (the other being a deliberate factoring across well-sampled types each represented by a mean type entry). If many species types with several ids within each are mixed, then loadings will appear from the mixed within and between types covariance that are often confusing and extraneous to the domain of influences in which we are interested. The intruding interspecies factors are often of the same kind, incidentally, as second-order factors in psychological data, in the sense that they can easily be intruding "extraneous" sociological or physiological influences.

For example, an ability factoring of a group composed of adult mental defectives, top athletes, and freshman students is likely to bring into correlation with the intelligence factor, social status, and education (students), lower age (adult defectives), and physical fitness (athletes). The principles to keep in mind here are thus those of separating between type and within type dimensions (as discussed on p. 448). On the other hand, if we are constructing a set of test scales to be used in diagnosis of neurotic and psychotic conditions, we might argue that these are not species types but extremes in a normal distribution. Then the factoring of a rightly proportioned mixture of normals and neurotics representative of that borderline group, say, a fifth of the population, likely to come for diagnosis at a clinic, is appropriate. For that should give better definition of the relevant factors in the group likely to come for examination than either a wholly clinical or wholly nonclinical group.

Regardless of whether or not there are distinct types or continuities the investigator should ask what is likely to be of the essence of the factor he is investigating and what is merely situationally local and incidental covariance. For example, the nature of the primary abilities is one thing in individual difference terms and another in terms of differences of abilities due to age. If no selection is practiced for age, i.e., if the investigator merely takes in referees over a wide age range, then verbal and numerical ability, for example, may show correlations with stature, weight, etc., which are no intrinsic part of primary abilities.

The question of what one should hold constant in an experimental sample, by selection (or later by partialling out), and what really ought to be allowed to vary goes deep, in terms of method and scientific philosophy. There is no general answer except to be alert to the nature of the goals and the selection effects. In getting at the essence of many human traits it would be appropriate, for example, to select all to one age, and one sex, but not necessarily to partial out social status or years of education, since by doing so one restricts the range without necessarily gaining through dropping truly extraneous expressions of the trait.

It should be noted that in the above we have been careful to say "correlations" rather than "loadings." For, in many situations, if good simple structure is achieved, partly through a good choice of variables to define the factor *per se*, what would be local or spurious associations arising through too varied a population will, in fact, not appear as loaded variables, but only as *variables or factors correlated with the given factor*. For example, if age correlates with verbal ability, numerical ability, and spatial ability, rotation may yet leave the age variable vector in the hyperplane of all of them, and, if other diverse age variables are present, it will then appear as an age factor correlated with the ability factors. Thus large age variance in the sample will ultimately express itself not in distortion of the factors as such but in an effect upon the correlations among factors due to its being partialled out as a factor.

The sample selection questions should also be related to what has been said on the effects of univariate and multivariate selection upon factor structure. As regards the former, we accepted the solution of Thurstone (1947) and Thomson and Ledermann (1939) that it will not change the factor patterns but only the correlations among the factors. For example, if, as above, we suppose several primary abilities to be all affected by age, then a wider age range will produce greater correlations among them. This, of course, affects the second-stratum factors but only, as far as we yet know, in terms of their variance size. Again this links with the question already raised, about higher-strata factors (p. 200), as to whether some of them may not in fact correspond to and be best interpreted as influences like age, social status, etc., which bring about correlations among primary personality factors without themselves being of a psychological nature.

A special case, where selection to a restricted range is definitely indicated to be desirable, is that encountered earlier in connection with possible curvilinear relations of factors to variables. Nothing needs to be added to what was said on p. 381. The argument calls for a deliberate selection of samples of narrow range at well-separated intervals on the variables already known normally to be loaded on the factor or factors under investigation. A linking of the separate tangents—for the *b* (loadings) are tangents—should provide the curve to which nonlinear functions can be fitted.

In general, we have recognized that a clearer factor structure, and one with higher loadings and better factor score estimation, will be obtained by factoring a group whose members are largely of one type. However, we have also described procedures in common use when a type-common pattern is aimed at. These include throwing the deviation scores in each of several groups into the same score pool, factoring the mean values for types, etc. They aim in different ways to obtain the dimensions of a generic or even a composite population, ignoring the species variations. This has its dangers, but in the history of research it

is probably true that in general investigators have been first interested
in the generic patterns. They first asked "What is general intelligence?"
not what is the specific pattern in middle-aged women, or undergraduate
males, and so on (except when limited resources accidentally dictated
the latter).

15.7. Correlating Variables and Extracting the Unrotated Factors

On the subject of choosing among alternative coefficients, in proceeding to the next step—the R_v matrix—enough has been said (Chapter
13, p. 469) to suggest that for regular use the product moment r, or its
equivalent, the covariance product, is to be preferred and that the data
should, if practicable, first be transformed to scale to a normal distribution. In dichotomous data the tetrachoric and in mixed data the biserial
or point biserial (depending on assumption possible about the dichotomous
distribution) are indicated.

It is appropriate to consider here—though it could perhaps equally
well have appeared under choice of variables—the question of correlating
"parcels" or small relatively homogeneous scales or batteries instead of
"items." By items one means the typical questionnaire item, or any
other response of the (0 or 1 or 2) type that is very brief. Any psychological response based on a couple of seconds exposure to a complex and
often somewhat ambiguous stimulus—as questionnaire items sometimes
are—is bound to be of relatively low reliability. This issue has been well
examined by Nunnally (1967) who favors avoiding the factoring of
items *per se*. [On the other hand, Eysenck (1971) and Howarth and
Browne (1971) have argued for superiority of item factoring, though
with no support from statistical principles or empirical results.] And
although short item-type variables may yield relatively clear simple
structure on one occasion of factoring (representing the factor structure
of their meaning at that moment to that sample), their loadings will
fluctuate from occasion to occasion and sample to sample. When the
present writer introduced and systematically tried out parcelled factoring
(1956) it seemed to be rather widely assumed that the reason was to get
factor loadings for, say, 300 items at a time when computers could handle
the factoring of only about 50 variables. But the reason goes technically
deeper, namely, to the superior factor analytic properties of solutions
from the stabler parcel variable, as a result of throwing four to eight
relatively homogeneous items together to average out distortions of
understanding specific to each.

It can be shown (Cattell, 1973, p. 295) that the dimensionality of
the factor space when parcels with relative internal homogeneity are used

is the same as that of the original pool of items. This holds theoretically up to the limiting case where the parcels get too few ($<2k$), to define the factors, and has been experimentally demonstrated (Cattell, 1973). It is also empirically demonstrated that the simple structure is superior, that salients are more highly loaded, and that any discrepancy in the scree indication of k is in favor of a slightly higher number in the parcels, due to accumulated variance in parcels from borderline factors too weak to show up in items. The reader will find a demonstration of the above in Fig. 5.7, on page 82.

However, these advantages exist only if two conditions are met that have frequently not been met in parcel studies. (1) The homogeneity required for putting items together in a parcel is different from that defined only by a good positive correlation of items, which is a fallacious basis, and requires instead radial parcel construction, described below. Even more fallacious and destructive of clear results is putting parcels together on a basis of items merely assumed to have similar psychological meaning. (2) Correct methods must later be used for "undoing" the parcels, i.e., finding the correlation of each item, as an item, with the factor.

As to the first step, the radial parcelling method requires a double factor analysis. One factors the items and then, on the unrotated matrix, correlates the rows of loadings. A high correlation indicates two variables which lie close to the same radial line from the origin. Cattell and Burdsal (1975) and Cattell (1974) have presented a purely mechanical computer program procedure for determining the grouping of variables in parcels of equal size according to their radial closeness and have pointed out (1) that r does not perform this function and (2) that to get simple structure among parcels they must be of equal size. A second factoring is then done with the pooled scores of items for parcels made to the agreed uniform size, e.g., 2, 4, or 8 items. Parenthetically, since the labor of this double factoring sometimes inclines psychologists to the easier path, as in Comrey and Duffy (1968), of putting items together by "psychological inspection" one must point out that the only safe pooling by this "method," i.e., the only one others are likely to agree with, consists in asking virtually the same question again and again in the parcel members. This has the defect of piling up the specific variance into a pseudo general factor (see Cattell, 1973, p. 361, on "bloated specifics"). On the other hand, if one should succeed in what is intended as a broader representation, as in Comrey's FHIDs (1973), one runs the risk of the parcel variables becoming primary factors themselves, so that the resulting factor analysis yields a mixture of first- and second-order factors (Barton and Cattell, 1975). Radial parcelling is objective in its grouping and avoids these problems.

As to the second problem, when one undertakes to "undo" the

parcel and correlate the individual items with factor scores—either literally or by the Dwyer extension—one encounters the problem that the obtained correlation is boosted by both the item and the factor score containing the specific factor contribution from the item. If the factor is estimated from fifty or more items this effect may be neglected, but with smaller numbers—even if one goes to a dozen or so—it is surprising to see how much the contamination of the estimate of the factor with the item's specific boosts the correlation of that item with the factor estimate. If one is comparing such a correlation with the correlation of an item with the pure factor, as in the unit length rotated factor in factoring of items, some correction is therefore desirable. Even without a deliberate correction however, one can take comfort from the fact that the upset of comparison by the boosting just mentioned is to some rough extent compensated by a reduction of the r through the factor estimate being an underestimate of the true factor variance and of less than unit validity. Guilford provides (1965b) a correction for correlation of an item with a factor when the item has been included in the factor estimate, but it is a somewhat tiresome procedure to repeat it for all items. A second approach is to determine the communality of the given item with the sum of items in the parcel—essentially its correlation with the principle component of the pool—and multiply this by the correlation of the parcel with the pure factor. A third is to estimate the factor without the contribution of the given parcel or item, i.e., from $n - 1$ items or $p - 1$ parcels and correlate the item with that, but this means $(n - 1)$ or $(p - 1)$ different factor scores and therefore as many different V_{fe} calculations. For most item analysis purposes, however, one is comparing the item validities only across items in the given study, and then the approximation of assuming all item-factor correlations about equally boosted is generally acceptable if large weights, on twenty or more items, are contributing to the factor score estimation.

A second practical problem widely encountered at the correlation matrix and factor extraction stage is that of incomplete data. For example, some subjects may have missed some tests. Such score matrices are far more widespread than is commonly confessed, though most of such studies probably run to only 5–10% missing, and the problem only becomes serious methodologically when one gets to 30–40%. Parenthetically, Finkbeiner reports that an examination of a dozen texts on factor analysis and item validation, shows most not dealing with this very practical problem at all. Probably the most thorough treatment is in Rummel (1970).

Solution (1) to this, if it may be called a solution, is the rigorous one: that of crossing out every row or column that has a zero left in it. Note there is some choice here between losing subjects and losing variables, which is decided by the experimenter's N, n, and intentions.

Solution (2) consists in correlating each pair of variables over the referees they happen to have in common, and for this most computing libraries have a "missing data" correlation program.

Solution (3) consists of inserting mean values (across all subjects getting a real score) for the blanks in the column. When variables happen to be in standard score and the variables present a positive manifold, there is something to be said for a value midway between the mean for the column, which is zero, and that for the particular subject's row.

Solution (4) consists in using a multiple regression on the other variables to esimate an individual's missing score on the given variable.

Solution (5), developed by Rummel (1970), carries (4) to the refinement of an iterative procedure. This inserts the estimated missing values for variable 1, based on its multiple R with the rest. It then takes variable 2 with missing values and does the same, including the new values for 1 among the predictors, and, so on until all variables have been covered. It can be iterated until some agreed parameter standard has been met.

Solution (6), suggested by Gleason and Staehlin (1975), takes principal components out of the first rough R matrix and, if the first component is large, simply estimates the missing values from that, though it can be carried to more components.

Generally in the choice among these six one is pitting time, trouble, and cost against accuracy desired. Method (1) would surely be most researchers' choice when the blanks are, say, down to 5%. Method (2) has been the most popular, as the prevalence of programs shows, but it has very real risks. The referees in common to variables x and y are generally a different sample from those common to y and z, and so on, and this different sample basis for the different correlations can easily give incompatibilities among correlations that result in a non-Gramian matrix. This occurs in the writer's experience, in perhaps one matrix in ten, when the blanks are not above 20% and are fairly random. So one might catch the uncommon case by a test of inverting the matrix. Not that an uninvertable matrix is the only consequence of such disturbances. A non-Gramian matrix, as seen earlier, has somewhat disturbing effects on ordinary principal axis extraction, but it completely stops a maximum likelihood process. Method (3) is harmless, but is likely to reduce loadings generally. Methods (4) and (5) would theoretically seem best, and Rummel reports very good results from (5); but it is very costly. Method (6) is a compromise in cost and Gleason and Staehlin report well of it. However, in an exploratory study and a more intensive study now continuing in connection with direct maximum likelihood factoring of score matrices, Finkbeiner gives the tentative conclusion that the superiority of the more complex methods is often trivial. Methods (4) and (5) seem clearly best. Nevertheless when correlations in R_v are substantial, and for a quick, safe method, whose attenuating effect is generally quite small, solution (3) recommends itself and, in Finkbeiner's work, shows better stability from sample to sample. But, in crucial studies the cost of others may have to be borne, and if one is going on to estimate factor score afterwards they have the advantage.

As to extraction of factors, this review need add little to what has been said in our systematic treatment of the main methods. However, one should recognize that we stand in a period when new methods—mainly those due to Jöreskog and a few others—though best on theoretical grounds, remain to be evaluated in all their practical aspects. And, in most work, it is not simply a question of which is the theoretically best, but which has good theoretical properties along with "handiness" or economy in relation to computer resources available and goals in a particular research. On this practical basis four or five methods deserve highest consideration. (This list, of course, leaves the old centroid method aside as, in general, an historically interesting teaching avenue to initial understanding.)

1. *Principal axes:* This leads, via an initial principal components solution, to a subsequent plotting of roots to determine the number of factors by scree, K–G, Sokal's, and other "significance of residual" tests and so on. Wood's programmed scree (1977) should be sufficient for most number decisions today. This is followed by iteration of communalities to the given factor number. Almost all laboratories have the necessary programs for these steps, whereas the following demand less available programs.

2. *The maximum likelihood method:* This was formerly a "luxury" procedure, by reason of its quite excessive demands on machine time, but Jöreskog's (1967a) and Jöreskog and Van Thillo's improvements (1971) have made it more practicable.

3. *Generalized least squares:* Like (2) this has the advantage of containing its significance tests for number of factors directly in the calculation. The reader must be referred to Jöreskog and Goldberger's recent article (1972) for details of procedure.

4. *The Newton–Raphson derivative of the above:* It sometimes happens (with what frequency in practice we do not yet know) that maximum likelihood and generalized least squares cannot reach a solution with certain, possibly "ill-conditioned" data, or data departing somewhat from the model. The present method by contrast works on *any* matrix, Gramian or otherwise. It proceeds by Newton–Raphson methods to minimize the trace of $(R_v - V_0 V_0^T)^2$ simultaneously on V, Λ, and U^2 (where $R_v - U^2 = V \Lambda V^T$) *precisely*, i.e., not in the coarse manner of estimating h^2 from raw sums of squares of the preceding V_0 matrix for a given number of factors. The Newton–Raphson procedure is a development of a method of finding maxima and minima used by Isaac Newton. The search of space

takes note of the second as well as the usual first derivative, thus recognizing the approach of maximum or minimum position by acceleration and deceleration of the curve as well as by the actual climb rate (first derivative) being zero. Applied in factor analysis here it is part of a successive approximation procedure (see Jennrich and Robinson, 1969).

5. *Proofing* (*"confirmatory"*) *maximum likelihood:* In the case of a field where enough is known to test a likely hypotheses one would naturally use proofing maximum likelihood. It has the attractive quality, as explained elsewhere in this book, of permitting various combinations of "fixed" and "open" elements in the hypothesis statement.

In addition to the above major alternatives there are minor alternatives in what is essentially the principal axis method, notably minres, which operates without communalities down the diagonal (Harman, 1976; Harman and Jones, 1966), and the methods of Comrey (1973) and of Overall (1974). Extensive empirical comparisons of the outcome of the various methods by Digman and Woods—comparisons extending even to the changes of model in alpha and image (Chapter 13) analysis, show, as stated, surprisingly little difference in the ultimate V_0 from the different extraction methods. Commonly, only values in the second and third decimal place are affected, though in some cases differences may extend to the first decimal place in the loading. Mathematical statisticians may find these differences of interest, and may show some liking for the convergence of the "scale-free" methods, alpha, canonical, etc., but except for the contribution of maximum likelihood to a decision on the number of factors, psychologists will find all of these methods giving a sufficiently accurate determination of the V_0. The vicissitudes of samples, of test reliabilities, and, above all, of rotation contribute an altogether greater variation to the ultimate psychological interpretation of the V_{fp} than do any of these extraction methods.

15.8. Rotating and Testing the Significance of a Rotational Resolution

At rotation we encounter the alternative of the hypothesis checking versus the discovery-oriented use of factor analysis. If the former, PROCRUSTES or confirmatory maximum likelihood methods are called for. In either case, we would stress that the matching of the resolution to the hypothetical target matrices *should be examined not only at the primary but at the higher strata patterns.* The latter can follow only when primaries have reached some temporary best position, and so an iterative procedure is involved in trying different combinations. The same, of course, is true of the matching of two independent "discovery" or "hypothesis-creating" solutions, which need to be tested for agreement after the best simple structure or confactor procedures have brought each to its own unique position.

In this connection we must raise a suspicion, a little too sophisticated to have been discussed in the first treatment (p. 142), of rotation to simple structure. It is that simple structure itself may "capitalize on chance" (if we seriously consider this somewhat overworked bogey). One sense in which this could happen is that the pursuit of lower loadings on low-loaded variables could add some variables to the hyperplane which, except for chance variation due to test unreliability, would not normally and typically be in the hyperplane. This is particularly likely to happen through the extra gadget which Digman has added to analytic programs, like the H–K, which gadget at the end takes all variables between ±.15 and deliberately tries to force them to zero loadings. Some of these may rightly so belong, but others should perhaps stay at .15 or rise to .2. The same could happen, however, in religiously pursued ROTOPLOT procedures and a possible case in the writer's experience is that Gorsuch succeeded, by extremely long and thorough rotation of a single large experiment, on second-stratum factors in the questionnaire area (Cattell and Gorsuch, 1967), in bringing into the hyperplane a few primaries which in eight different independently rotated studies by Nichols (see Cattell, 1973) never lost significant loading (even as a mean value).

The discussion of "geometer's hyperplanes" (p. 169) indicates that in any study with many variables a very large number of such purely geometrically necessary hyperplanes exist. It is not surprising, therefore, that when the psychological and the geometrical hyperplane are close, a search for maximum count may by a slight angular change include the latter in the former. This may explain the above observation that a high-powered oblique program, such as the Harris–Kaiser class of programs (some permitting high and others only low correlations among the factors) will in the latter case bring certain variables into the hyperplane at the cost of erratic correlations among factors.

In short, it must be added as one more vice to be watched in simple structure search (and one which adds to the need for an oblique confactor program) that programs may overshoot the true hyperplane maximum and gather in other vectors that do not rightly belong. This being so, the goodness of a resolution should be examined finally not by the primary hyperplane count alone but by the hyperplane count at the second stratum and even higher strata as well. Just how this overall maximization—which would penalize "capitalizing on chance" at the first order at the expense of good simple structure at the second—can be reached economically is not yet entirely clear but it can at least be pursued in iterative cycles as indicated above. One way to shed light on the first order is to compare simple structure loadings from several independently rotated studies and average them. If low values at the fringe of the hyperplane are sometimes plus and sometimes minus the mean will place them in the hyperplane, but if the supposed "error" is always in one direction we may suspect that we have been trying to force into the hyperplane what should really be a significant but low loading. But the test of which variations in the primary rotation give the better simple structure at the second order, as suggested above, is a good independent check on such cases.

Incidentally, this possible danger of "overshooting" the true hyperplane by ruthless mechanical programs which arises in the recent suggestions of Digman and of Kaiser needs careful regard at the present stage of success in mechanization. In the form proposed by Digman the program does not try to force *all* x variables in the $\pm.15$ range from the preliminary method (in his case the H–K) into a closest approach to a zero hyperplane, but goes by trial and error through the various $(x - 1)$, $(x - 2)$, etc., combinations to see which *can* be so forced [using the method originally proposed by Thurstone (1947)]. The amount of trial and error could be enormous with more than, say, 100 variables, and the present writer would be inclined rather to seek this final tidying of a hyperplane by the use of "hobbled MAXPLANE" described above, i.e., a MAXPLANE restricted to small angular trial shifts and given an objectively calculated narrow hyperplane width as its final criterion.

If one is to follow a two-stage convergence on a tight simple structure then there is much to be said for taking Hendrickson and White's (1964) inexpensive and effective PROMAX as the routine analytical program step and following with hobbled MAXPLANE as the topological program. The advantage of finishing with a topological search is that, whereas all analytic programs are blind to the width of hyperplane they are working with and cannot define within what bandwidth they are maximizing, a definite width can be plugged into the MAXPLANE program, or held to in visual ROTOPLOT shifts. Furthermore, in this last step the empirical work by Finkbeiner and of Vaughan, as well as the theoretical formulas for the standard error of a zero loading by Jennrich, now permit us to settle on a sound estimate of the true width of the hyperplane appropriate to a given sample and experimental conditions.

It seems to be frequently overlooked in published studies that a psychologist evaluating the result needs proof not only of the maximum simple structure having apparently been reached, but also of that simple structure count being statistically significant. It has already been pointed out that by Bargmann's test, which may be somewhat severe on account of its assumptions, perhaps fifty percent of published alleged unique resolutions would have to be dismissed and their impact on psychological theory declared void. Such studies are numerous among relatively amateur one-shot studies; but at the risk of invidiousness it must be pointed out that by Bargmann's test (see Appendix, p. 554) the questionnaire factors of Guilford, Comrey, and others using complex or advanced mathematico-statistical methods nevertheless can also, by reason of inadequate pursuit of simple structure, stand below significance (which can be judged when sufficient of the V_{rs} has been published to permit examination). Those inadequacies have been brought out in the survey of past researches by Sine and Kameoka (1978) who are responsible for the improved and extended Bargmann SS significance tables given in the Appendix. By the somewhat less exacting significance tests from the independent Cattell–Vaughan approach certain of these studies still fail, but others, e.g., the German and Japanese factorings of the 16 P.F., which averaged roughly 12 to 16 factors significant by Bargmann, show essentially all significantly reaching simple structure by the Vaughan empirical tables.

Occasionally the question may be raised whether in some domains simple structure should be expected. The general answer is that wherever distinct determiners are operative there should be simple structure; though in a badly designed experiment in which all variables are likely to be affected by each of one or two factors it will obviously register a poor significance. However, in today's more sophisticated designs this cause is uncommon, and most instances of nonsignificant simple structure turn out to be improvable (see Cattell, 1945, 1973) and therefore show that the cause is insufficient application of the investigator to discovering an existing simple structure. The statistical significance of loadings at simple structure has been discussed in Section 14.10 and in the Appendix.

The alternative to simple structure—confactor rotation (p. 99)—is as yet theoretically sound only for orthogonal rotation and it must be carried out on the covariance V_0 matrices. Surprisingly little experience yet exists as to its practicability when the departures from orthogonality are only moderate, but the illustrative results in Appendix A3 show a true confactor solution is possible if obliquities are identical. If the number of higher-order factors could be settled from the beginning, confactor rotation could be applied to determine a Schmid–Leiman solution, which is orthogonal (p. 226) and Brennan and the present writer are now investigating this. Perhaps trial and error with confactor solution over a small range of number of factors extracted (around the most likely value) would not be prohibitive. Carrying out the comparison in covariance values should cause no problem inasmuch as the sigmas of variables at one position could be given values of unity, and the raw-score variances at the second position expressed in terms of multiples of the sigmas at the first position. The computer program for confactor presents no special complexity (see formula on p. 100).

No discussion has yet appeared in the literature over the significance of a confactor rotation, but in principle it involves asking if the congruence coefficients or salient variable s indices between the paired factors in the two studies reach certain extremities of cut on the distribution of such congruences. One has, of course, deliberately sought the positions of maximum congruence. Congruence significant tables are given (p. 253; Appendix 7, p. 568) but one would have most confidence in the evaluation if distribution of congruence coefficients were determined, say, by Monte Carlo methods, for various pairs of truly random V_0 of the same N, n, k, and h^2 values as in the given experiment rotated randomly and also to maximum possible congruence.

It remains, under rotation, to discuss the concepts of *progressive rectification* and of *artificial simple structures*, which are related. The former, left to be described elsewhere (Cattell, 1973), involves a succession of experiments progressing to higher loaded variables and aimed at better determination of interfactor angles. Item and parcel factoring may well be alternated in good programmatic work. Certainly, as in the radial parcelling method, the necessary initial step is to determine the dimensionality in items, and a first rough determination of the simple structure by rotation in items may be preferred. But since the simple structure position cannot be easily determined accurately in items, it is then desirable to proceed, perhaps with additional items, to a more accurate determination in parcels. Item and parcel factoring thus alternate in a good programmatic convergence on structure in a given domain. The principle of progressive rectification is demonstrated at greater length elsewhere (Cattell, 1973). The method rests on the fact that as higher loaded items are gradually discovered for known factors, the determination of unbiased correlations among those factors will be improved simultaneously.

The issue of artificial simple structure concerns whether inadequate, erroneous early research, followed by multiplication of scales aligning about the original erroneously placed factor vectors, can compel later hunts for simple structure with those scales to remain tied to the original erroneous positions. The issue obviously has relevance to progressive rectification. One must recognize that a slowness to rectify will occur if, for example, only a couple of highly correlated equivalent scales are used to mark each factor. However, in time the factoring along with other scales than those first constructed, and with more than two scales for each factor, would tend to correct the situation. Some optimism is justified on the basis of the difference between clusters and hyperplanes. As pointed out earlier, it is easy to create an artificial cluster in a desired position simply by repeating a number of very similar items or scales, but it is difficult artificially to create a hyperplane, since the scales must then meet the condition of defining a whole lot of new dimensions while maintaining zero correlation with the artificial factor. Granted a sufficiency of scales, old and new, the number of possible hyperplanes becomes very great, so that the rotator is unlikely to be held down any longer to an original hyperplane defined by few variables, if that hyperplane was erroneous.

15.9. Matching and Interpreting Factors: Programmatic Design and the Need for a Universal Index

Some aspects of matching have already necessarily crept into the discussion of rotation above, because confactor is a form of matching, yet in both confactor and simple structure rotation the issue of matching with other experiments and other theoretical patterns is and should be independent of finding the unique rotation, and carried out subsequent to it. (The two "subexperiments" in confactor rotation must count as one independent piece of evidence, and it still remains to be checked whether the joint solution is one agreeing with a dozen other experiments or with one of several specific theoretical positions stated in target matrices.)

It should surely go without saying, in this day and age, that "intuitive matches" by mere inspection ("eye-balling") of patterns is worthless, in view of the readiness of most of us investigators to see what we want to see. Some five methods of objectively evaluating matches of factor patterns can be entertained. The three most dependable and widely used have already been set out and discussed in some detail (p. 263). It remains only to look briefly, in this final extension of discussion of matching, at two more and to show how all five can be usefully applied in practice. To the above three—the congruence coefficient, the salient variable index, and the configurative methods, we would now add, and urge for further experiment, the intraclass correlation and the pattern similarity coefficient r_p. The latter, particularly, has the virtue of not counting as identical a factor of major size and one of minor size in the same area which happen to have some similarity of loading shape. The sigma to be used in calculating Σd^2 in r_p is, of course, the sigma of loadings along a row of the V_{fp} for study A combined with the row of study B for the same variable. This $2k$ basis may give no more than 20–50 cases on which to base the estimation of the sigma, but it is enough for most practical purposes.

Regardless of the choice among the five methods of pattern comparison, the innovation we would urge here for an ideal procedure is that *all strata of the comprehensive resolution (first-, second-, etc., order patterns) be simultaneously involved* in assessing the fit of the outcomes of two experiments. For that matter not only the evaluation of matching but also the prior evaluation of simple structure count just discussed should be made simultaneously over all levels. The general goodness of simple structure and that of matching are not unrelated because according to our basic theory a good match is quite unlikely to be obtained until both studies have reached a high and significant level of simple structure count in the hyperplanes. However, we have also just considered the possibility that sometimes the simple structure count can actually slightly overshoot the true value to the detriment of the trueness of angles among the primaries. For this reason it has been suggested that the goodness of a resolution be examined by simple structure simultaneously at all levels.

A less remote danger than undue regard for an artificial "overfit" at the primary level, in evaluating matching, is the use of PROCRUSTES to force a second study to a close match to a prior one, when the prior one was not patiently brought to maximum simple structure in the first place and is perhaps an easier but not a true target. Pretty good matches (in terms of congruence and s indices) can be obtained in such circumstances, but it will be found that this is achieved at the cost of primary correlations which are not only high but different in the two experiments and with a simple structure as poor in the second as the first. This fact that goodness of match at one stratum can be artificially (one could well say illegitimately) gained, but at the cost of distorting correlations to unusual values among the factors, is one further reason for demanding that matching be simultaneously done at all strata. Such matching of matrices covering two or three strata can be tried either on the patterns formed by the loadings of all strata on the variables, as in the Schmid–Leiman transformation, or on the patterns in what we have called the comprehensive SUD solution, in which the loadings of each stratum of factors is taken only on those immediately below it. An evaluation of the total goodness of match between two experiments can then be obtained by the mean of the congruence coefficients across all strata, when the factors presumed by hypothesis to be matches are paired.

As stated earlier (p. 252), it is possible for congruence coefficients between factors that are definitely known to be different, i.e., different real influences, to reach statistical significance. The fact that this can occur at several places in a matching index matrix between two studies, presenting a greater frequency of near significances than would be expected for random factor patterns on statistical grounds, is puzzling. The only solution to the riddle that the present writer has to offer is the phenomenon of *cooperative factors* (p. 200) as a scientific rather than a statistical principle. According to this principle a set of variables uniformly susceptible to one influence have more than chance likelihood of being simultaneously susceptible to another because of similarity of properties. Such overlap must be assumed to be in the nature of causal action in various fields of science, and its result will be the appearance of more than one significant congruence coefficient for a given factor in study A with factors in B, as well as significant congruences among certain factors in A itself. (Often they are the same factors as achieve pseudomatching in B.)

Both of these are actually observed and in ordinary experiment, and the test of matching, therefore, is whether one can arrange the factors in the two studies in such an order that the highest congruences (r_c) or salient variable similarity indices (s) fall down the diagonal, i.e., the highest r_c in the column for factor X in study A, is also the highest r_c in the row for the assumed match for X in study B. It is not unusual, however, even when all extraneous evidence, and the presence of a diagonal ridge, point to a match, for this diagonal high value not to be much above the next lower values, or even not much above $p < .05$ significance level for the given pair of matched factors.

The reason for reduction of the level of the pattern similarity when two factors are definitely the same is, as pointed out earlier (p. 266), that we are normally working in the standard factor model, i.e., with the unit variance factors and variables where the true factor–variable relations are distorted by what we called the "Brazil nut effect." In a study on which important theoretical inference depends, one should take the trouble to shift to a real-base factor system, changing the V_{fp} in the second study to covariance values by premultiplying by a D_v matrix giving the variable sigmas relative to raw-score sigmas in the first study. That is to say, one divides the raw-score sigma of each variable in the second study by the corresponding value in the first, and multiplies the V_{fp} row values in the second by these quotients. The r_c calculation should then be applied to these new V_{fp} values. In matrix terms if D_A is the diagonal matrix of raw-score sigmas in study A and D_B in study B, and $V_{fp \cdot A}$ and $V_{fp \cdot B}$ are the ordinary factor solutions, then we compare $D_A V_{fpA}$ with $D_B V_{fpB}$ or, as just suggested, $V_{fp \cdot A}$ with $D_A^{-1} D_B V_{fp \cdot B}$.

It is a common experience in such matching to find, say, 80–90% of the expected matchings satisfactorily in the $p < .05$ and $< .01$ regions, and then to find that the last one to three factor pairs fall at an altogether lower—indeed negligible—congruence level. The explanation (offered above) for this is that the last factor or factors in each series (last in congruence, not necessarily in size) represent common error factors, i.e., factors due to correlation of random measurement error in the data. If everything were perfectly measured, such factors would not appear, even with a very small sample. In this connection, we would emphasize the implication of what was said in the last chapter about the difference of sampling and experimental error, namely, that sampling is best considered to occur essentially upon factors, not variables. Consequently, with no experimental error of measurement, the patterns of factors and the clarity of their hyperplanes should be the same in a small as a large sample. However, when experimental error enters, i.e., reliabilities fall from unity, then the size of sample matters, for the correlations of random normal deviates become greater with small samples and the error common factors larger.

Thus two things would be expected to happen from the joint action of decreasing sample size and decreasing test reliability: (1) the congruence coefficients for factors that really are the same will drop, though they still offer a basis for matching decisions through one's finding whether they can be arranged to give a diagonal ridge, and (2) common error factors will move up in variance size to appear close to the end of the fairly large factors. They will nevertheless reveal their nature by the complete absence of significant matches across researches. Indeed, regardless of what other evidence we may consider, it is finally through the sizes of congruence coefficients that we recognize, and separate out, the error factors from the substantive factors.

The margins of error being what they are in factor analysis, especially through the prevalence of poor rotations, a real regard for scientific standards requires that at least one experimental replication, i.e., a significant congruence for each factor with another study, be demonstrated before an investigator takes a particular factor pattern seriously. In fact, the present writer would venture to speak of a definitely replicated factor only when significant matches have been shown among at least three studies. In the book summarizing programmatic research in *Personality Factors in Objective Test Devices* (Hundleby, Pawlik, and Cattell, 1965), the source traits considered established are patterns resting on from five to sixteen replicating studies placed side-by-side. Progress to overall matching in that domain has thus required 25 years of research and as many distinct experiments and samples.

It is appropriate to include in the matching section of this survey of research strategy the practices needed in programmatic work. For the implication of the vices and virtues of the factor analytic method that have been discussed is that, more than most other research methods, it calls for sustained programmatic work with faithfully repeated markers, systematic advance through interpretive hypotheses into new inferred variables for hypothesis testing, strategic coordination of samples and methods, and systematic recordings of matching. In connection with the last a good deal of waste, and lost theoretical opportunity, has occurred in the last forty years through investigators concentrating on only the few best matched factors, relegating the rest not merely to a limbo on the research shelf, but to the wastebasket and the janitor. Now even with studies which deliberately carry exactly the same variables, and certainly in those with only say 90% identical, it is possible to find factors without a match which are *not* common error factors. As we have seen, both the scree and the maximum likelihood test will vary a little from sample to sample in the cutting point for number of factors. The rotation process may, further, sort the variance out a little differently. The result is that there will be one or two factors among the smaller at the end of the rotated series that will come and go from one study to another. One can view a set of factors as a fleet of ships still climbing the horizon, and these more remote "hull down" cases will sometimes be visible and sometimes not. Consequently, no matches for them may turn up in comparison of a mere two or three studies and they will be treated as common error factors, without any further attempt to recognize and interpret them.

Because of this, and still more because of the psychologist's tendency to attend only to the larger factors which interest him, any programmatic work by one laboratory—or, indeed, work among several laboratories that have the sense of scientific community required for carrying markers from one to the other—will contain over the years toward the end of its published matrices a considerable number of smaller factor patterns left uninterpreted and neglected. Yet like the less rich "tailings" of a mine, left for years as a dump, these actually contain much metal and there comes a time when it is profitable to work them. Such retrieval, however, requires a lot of careful recording and systematic indexing over the years. As examples of the success of retrieval when the index system for variables and for factors is carefully worked out one may cite the objective personality test work of Hundleby, Pawlik, and Cattell (1965), that of French (1953) in the same area even though the indexing of variables in the latter is not precise enough for him to have used congruence coefficients, and that from Spearman, through Thurstone, Horn, Hakstian, and others in the ability field. Barton, Delhees, DeVoogd, Gibbons, Watterson, and the present writer similarly picked up the "shavings from the work bench" in the questionnaire factor field. Those efforts confirmed what had become suspected, from impressions and records of the "hull down" factors, to be no fewer than "seven missing factors" from the personality sphere domain used for the 16 PF test, plus some extra pathological factors (Delhees and Cattell, 1971; Cattell and Bolton, 1959; Cattell and Watterson, in press, 1978; Kameoka, in press, 1977). Such smaller factors rarely possess more than one or two tolerably good markers in one research. But when the retrospective survey and tentative matchings in the twilight zones are through, there will usually be enough salients from several studies to raise the variance of the factor to good levels over sufficient markers to meet the requirements of the final checking study.

It has been pointed out here that such "cleaning up" is possible only when research is kept programmatic enough for markers to be identified systematically and used in a progressive fashion, either through the coordinating efforts of a single laboratory in a single domain—which has so far historically proved the only sure way in some fields—or through sound principles accepted through a comity of laboratories in a broad field of scientific endeavor. Lest some reader think it inappropriate in a technical statistical text to include discussion of the social organization of science, we must point out that the battle for research advance can be lost or won on that flank, and that good techniques and good communication interact. In factor analytic research in any one domain of science, it is important, in the first place, to have a good (hopefully universal and international) indexing system for both marker variables and replicated factors. Marker variables can advantageously be indexed at an early stage, but until a sufficient core of researchers has attained agreement on extraction methods, models, and rotation principles, a list of standard primary and secondary factors would do little good. However, within groups working with agreed standards in psychological research the availability of a list of more than 1000 objectively defined, precisely administered, and scored behavior variables (Cattell and Warburton, 1967) has proved very functional. Similarly, a well-defined list of questionnaire markers has accumulated through the work of Derber and Harman and the researchers on the 16 PF, HSPQ, CPQ, ESPQ and PSPQ series (Coan, Digman, Dielman, Dreger, Porter, Schaie, and others). A set of some 40 rating markers developed by the present writer (Cattell, 1946, 1957) has been cross checked and expanded by Goldberg, Norman, Lorr, Thorne, and others. All these efforts have provided the necessary basis of a standard marker variable index upon which an index of replicated factor patterns could be erected.

Granted marker coordination, in all three measurement media, the second step is the organization of an index of factors, universally applicable. French (1953) accomplished this in the ability field, while the present writer, with the help of a dozen colleagues, has worked out an indexing for replicated factors in personality and in motivation fields (Cattell, 1957c, 1977; Hundleby, Pawlik, and Cattell, 1965; Sweney, 1967). A similar index will soon be necessary to coordinate the findings of Rummel and others in regard to the dimensions of national culture patterns and small groups. In individual differences the present writer's scheme has presented the alternatives of building up lists in the three modalities separately— abilities, personality–temperament, and dynamic—or together. But regardless of final choice in that issue, it is certain that psychologists must begin with separate lists for different media of measurement, and different strata of factoring. If it is later discovered that factors in two modalities can be assumed to be the same, when extracted as instrument-free cross-media factors (p. 503)—as seems true, for example, in the case of anxiety, Q II, appearing as a second order in questionnaires, and as a first order, U.I. 24, in objective, T-data—then a supramedia list can be built up. But we are operationally on much firmer ground if we initially keep to separate series of replicated factors in each medium.

The more detailed principles and features of a factor indexing system in being are given in the Appendix. The need for such a system springs from the very nature of factor analytic research. It takes several years of hypothesis-checking research as a rule for a factor to grow reliably from its discovery and initial status as a replicable and measurable pattern to a reasonably well-established theoretical concept which interprets it dependably as a scientific influence. The often excessive theoretical enthusiasm of different investigators while the evidence is actually still at the earliest stage constantly leaves onlookers in doubt as to whether two writers are speaking of two different factors or two different interpretations of the same factor. In such a melee the important fact to hold on to is that a certain *pattern of loadings on variables* may or may not be confirmed. To have a definite index symbol for a factor when established as a pattern but still in the limbo of disputed interpretation enables disputants and others to be clearer about the issues and alternatives. In the history of this domain there have been many wild goose chases based on nothing but the connotations of a misapplied labeling word. Other and older sciences have benefited greatly from careful indexing—for example of the vitamins, before their chemical structure was known; of elements by atomic numbers and recorded isotopes, before their properties were fully known; of blood groups in physiology before their genetic nature could be clarified; and so on. For these reasons there is real value to a factor index, and it is gratifying to note that the system in the Appendix has, over fifteen years, been adopted in many articles and several books.

The indexing as such is essential, but human nature seems also to crave a name for conversational use. It is here that trouble begins, and here that factor analytic research is still struggling with a legacy of confusion. Much of it arises from an undisciplined use of purely popular terms for new factors and a failure to recognize that a truly new discovery requires a new name. Popular terms such as "sociability" for exvia come with trailing clouds of vague and unproven associations, disruptive of precise meaning and discussion. Frequently the person who attaches such terms is actually using the name merely of a variable, yet a factor is an abstraction from many variables. Except in a few cases like intelligence or anxiety or some clinical concept like super ego or projective tendency, the factor is, one must repeat, a new concept and needs a new name. That name can well be an attempt at maximum description rather than theoretical interpretation until much criterion experiment has been done. Medicine has long recognized the need to avoid popular terms, e.g., rheumatism or a "cold," and has recognized more specialized entities with diverse Latin names. Psychology, as it recognizes this problem, will need to proceed similarly with Latin derivations or acronyms to give precision to the new concepts it has reached by advanced analytical methods.

The process of interpretation of a factor once it is well replicated as an indubitable and indexable pattern proceeds (as discussed on p. 231

above), along a variety of directions. Comparison of high positive, high negative, and zero loadings gives a first clue to the nature of the required abstraction. There must follow, still in the factor analytic experimental domain, the trying out of *new* variables inferred from the tentative hypothesis to have positive, negative, or zero loadings, and entered in the next experiment as tests of hypotheses. But quite ulterior evidence and experiment to that which is purely factor analytic must come in at the same time. What other clinical or laboratory behaviors can be observed in people very high or low on the factor? With what concrete life criteria does the factor score correlate? How does the variance of the factor change with this or that type of group and with this or that modulating circumstance? What are the sociological and physiological associations? If it is an individual difference factor, what is the nature–nurture ratio for it? And if nurture proves important, what are the learning histories that can be shown to go with differences of score? Such investigations require the factorist to seek the cooperation of colleagues who are not factor analysts, working in diverse fields, but who would be happy to be given important factor scores to use in ANOVA and other designs.

15.10. The Use of Factor Scores and Taxonomic Type Classifications

If factors produced to meet the conditions of a scientific model are no longer mere mathematical abstractions but meaningful determiners, then the end of factor analysis is not to produce a taxonomic museum but to give the experimenter measures of the most meaningful concepts with which to work in his field by bivariate or other experimental methods. This means providing the right collection of variables ("subtests" if the psychometrist prefers that term) and the most valid score from them. The purely statistical bases of the main procedures when one comes to the practice of assigning scores have already been discussed (p. 274). The choice in practice will lie between some kind of computer synthesis, on the one hand, using graduated weights on many items, variables, and even, possibly, other factor scales along with the main scale, and on the other, using just a few good subtests simply unit-weighted, i.e., equally, as "salients" or, rather, "prominents." The bugbear of either which one needs to guard against is the accidental production of quite substantial correlations among the factor scores, different from those among the true factors. At the present stage of research the above choice often has to be decided in favor of computer synthesis because we simply have not found the few subtests or items highly enough loaded for the second "limited set of prominents" procedure and must therefore gather a little from as many variables as possible. Moreover, as pointed out in the more intensive discussion (p. 289) it may prove to be *to some extent an inherent property* of a broad personality factor or the dimension of a culture, etc., that it cannot show itself in some single—or even two or three "magic" highloaded manifestations, but has to be estimated from many signs and variables each only very moderately determined by that factor.

In this obstinate fact lies the basis for the argument for proceeding with low battery or scale homogeneity. And, as we have seen, the true principle is still in battle with an obstinate myth that a good test scale or battery has high homogeneity (see several instances in Buros, 1972). But whenever good validity has to be sought by computer synthesis, i.e., scoring different factors by different columns of weights applied to the same variable scores, then the second main problem (the first is low validity) in getting factor scores, that of spurious correlations among them, becomes a threat. Of course, even with small, unit-weighted and nonoverlapping, sets of variables per factor some correlations may arise among factors different from the true ones. This, incidentally, explodes the notion that it is simple to work with orthogonal scales, as is seen when Guilford recognizes that his orthogonally rotated questionnaire scale factors give scale scores that are quite appreciably correlated.

Our comparative ignorance, at the level of empirical findings and experience, with regard to the use of factor score estimates in further stages and kinds of experimental work has several roots. First, the factorist himself arrives as it were exhausted at the threshold of this last experimental demand in the use of factor analysis. Second, mathematical statisticians, excepting Schönemann's and Guttman's articles on indeterminacy, have been little concerned in what lies beyond extraction alternatives. Third, the typical applied psychologist and experimenter, apparently not realizing the enormous difference in validity between measuring a concept by some single arbitrary variables versus measuring it by a factor battery of definable validity, has shown relatively little interest in advancing to the more demanding calculation of factor scores. Yet, this last step would, in fact, take him through the door to enter on a higher level of theoretical precision in all kinds of experimental work. If substantial theoretical development is to take place, the important thing to do at this juncture with factor scores is to use them in all kinds of applied psychological predictive work and manipulative sequential bivariate experiments. Only thus can the inherently great possibilities of interpretive development about these functionally unitary and clear constructs be brought to fruition.

Finally, as we have seen in some detail, the future use of factor analysis has to recognize that it cannot continue to operate with the myth of a grand, homogeneous, "statistician's population." Almost all actual populations are structured into species types, and sometimes into dendritic diversions of types within types. Preliminary work with the TAXONOME and other programs using essentially the objective TAXONOME principles for locating types, has shown, it is true, that these types are practically never as clean cut and neatly separable as the ideas of psychologists would like them to be. For example, work with plasmodes consisting of such diverse material as mixed breeds of dogs (Cattell, Bolz, and Korth 1973), ships (Cattell, 1968, p. 140), and musical instruments (May, 1972) shows that TAXONOME is inherently capable of separating discrete species types when they exist, but when it is applied to the 28-factor profile of the Clinical Analysis Questionnaire on clinical cases it does not nearly yield the categories used by psychiatrists. A close approach to such a typing result has, however, been reached by Cattell, Schmidt, and Bjersted with objective test factors (1972, p. 72). The disagreements among psychiatrists show, nevertheless, that the "criterion" itself does not consist of a true nonoverlapping world of types, for many of them are in fact abstract modes in the mind of the psychiatrist or even the less subjective biologist (Hartmann, 1976). Typing should also consider nonmetric grouping methods (Peay, 1975).

Recognizing that in most clinical, educational, and industrial domains types merge and overlap in such a way that no objective location of the modal frequencies of patterns will permit a formula to segregate more than perhaps a half of all given individuals, we nevertheless must see that factor analysis has to adjust itself to the concept of among types and within type factors described (p. 448). It also has to keep in mind the reciprocity between description by types and by dimensions, such that a given individual can be fixed by scores on among type and within type dimensions or alternatively by his distances from the modal profiles constituted by various types.

Lack of resources if not lack of theoretical concepts has led us hitherto to define an intelligence or a surgency factor by its pattern in the general population, but we recognize that the pattern would more correctly correspond to the source trait structure and yield score estimates of higher validity if it were determined separately for men and women, adults, and children, and various minority groups. Relatively clear species types may be expected, as research with taxonome proceeds, in the skills of occupational groups, the attitudes of members of political parties, well-diagnosed psychiatric clinical groups, and perhaps in ethnic groups. Where the species types are clear, factor analysis should obviously refocus its procedures on obtaining distinct sets of within-type and among-type dimensions. And as indicated in the technical discussion earlier, this will require principles for sampling variables that are as explicit as those for sampling subjects and coordinated with them.

15.11. Summary

1. A well-designed factor analytic experiment begins with consideration of choice among the possible general dimensions of experimental design as such: of the role best assigned to manipulation or absence thereof; control of conditions or absence thereof; temporal sequences (for causal analysis) or absence thereof; of the BDRM relations that are most apposite; and so on. The anticipated factors should have their anticipated meanings focused within this framework.

2. Consideration of which coordinate sets in the data box are most relevant to the relations to be studied under the given hypothesis or theory concerns the alternatives of using facet, face, or grid score matrices and of employing R, P, O, S, etc., correlational techniques.

3. Having fixed the kind of data box matrix and correlational technique the experimenter looks at the relatives and considers choice, range selection, and sampling questions. In the typical R technique this means the choice of response variables, in relation to the domain to be covered in the factor analysis; and whether that choice is for an exploratory (hypothesis-creating) or a hypothesis-checking design. In the exploratory design "sampling" means mainly the principles for representative sampling from a conceived population of variables, and in the latter "choice" means the selection of good markers, two, three, or more to the factor, from past research and current theory as well as "hyperplane stuff."

4. The decision as to sheer number of variables and the breadth of the range that can most profitably be covered depends partly on the number of expected factors and partly on exigencies of subject time. Since several considerations favor factoring with many rather than few variables, subject time is the chief determiner of what balance must be accepted between number of tests and reliability (length) of tests. When variables are available from different domains there are risks in mixing them in a common factoring, due to possible confusions of different strata levels, etc. This suggests that separate domain factoring is generally best done before joint factoring. The best way of dividing such matrices for separate or successive factorings and ultimate integration is discussed.

5. Beyond the above broad considerations, the choice of variables needs to take account also of (a) experimentally exact setup of marker variables, for reliability of subsequent factor matching, (b) provision of hyperplane stuff suitable to the more difficult rotational separations anticipated, and (c) provision of variables to locate anticipated instrument factors and other masking effects expected from perturbation theory. Multitrait–multimethod designs have also called attention to this, but the approach through instrument factor concepts is more sophisticated and yields more definite results.

6. Attention turns next to the selection of referees, where the first requirement is a due ratio of number of referees to the number of relatives. In the most frequent kinds of research, referees are people, and here one should begin with something between a 3 and a 5 to 1 ratio of N to n (n, the number of relatives) though the absolute magnitude of N is also important. Limits of stability of factor analyses with different N are discussed. It is essential to avoid a square score matrix and with an oblong matrix one should resort to R or Q technique according to which transpose makes the referees longer.

7. The second kind of decision required on referees concerns questions of their homogeneity as a type, their range, etc. Usually the aim is to get adequate variance of people (referees) on the relevant variables, and the control, reduction, or removal of their variance on others. However, the removal of variance from extraneous sources that affects the substantive factor variance, e.g., of age range variance upon intelligence and its variables, or from mixing of diverse types, will bring out factor loadings more clearly. Choice of groups that are homogeneous in the sense of not mixing species types but which retain large variance on the factors under study will also aid clarity, but there are dangers of distortion of factor correlations if one arranges larger variance on some factors than others. Usually, the argument for gaining higher and clearer loadings on a factor by taking a greater range (relative to the standard population) are apt only when a single factor is the focus (for one cannot increase loadings simultaneously on all factors) and when one is unconcerned about the correlations with other factors. When the goal is to portray with adequacy all features of the factor structure, obviously a well-stratified sample of the master population is the desideratum. Two other special aims that call for departure from this, however, are the best determination of within and between type dimensions in a typically mixed population (see No. 13, below) and the determination of curvilinearities of relation by factoring over successive reduced ranges.

The experimenter may think in terms of factor selection but, of course, has to translate into selection on likely variables. Formulas are available for allowing for the effects of univariate and multivariate selection, and it has been shown that the former changes factor variances and intercorrelations but not patterns.

8. In the variable relation matrix itself the product-moment r, or the covariance products applied to normalized score distributions, are to be preferred to other covariation indices where the data permits their use (otherwise tetrachorics are usually the next choice). In the ensuing factor extraction, maximum likelihood and generalized least squares are the deluxe methods, for the majority of situations and resources the principal axis method, iterating communalities to the number of factors indicated by the scree, etc., will probably long remain the standard procedure.

9. Rotation is either hypothesis creating or hypothesis checking. For the latter we have PROCRUSTES rotation and confirmatory ("proofing") maximum likelihood methods. For the former we have simple structure and confactor rotation. As a routine it is suggested that an approximation by PROMAX or the Harris–Kaiser–Digman analytical programs be followed by tightening with ROTOPLOT or MAXPLANE topological methods, using the hyperplane width indicated as most likely by the sample size and other features of the data. The position reached should then be tested for significance, by congruence or s coefficients in PROCRUSTES and confactor methods, and by Bargmann (see Appendix) or Cattell–Vaughan tables in the case of simple structure rotation.

10. For matching of a given factor against another factor pattern either in a second experiment or against a hypothesis defined as a target matrix, as in No. 9, above, we have essentially five resources: the r_c coefficient (congruence), the s index (salient, variable similarity), the configurative match, the intraclass coefficient, and the pattern similarity coefficient r_p. The possibility of arranging the coefficients or indices in the matching matrix from all combinations of factors in the two experimental pattern matrices so as to produce maximum values down the diagonal (diagonal ridge pattern) is considered crucial. Common error factors are detected by coming at the end of the series of acceptable matches, and by being at a totally lower, essentially unmatchable level.

11. A brief review of the bases of factor interpretation is given here. In the inductive hypothetico-deductive (IHD), spiral toward theoretical interpretation it is highly desirable to have two stages of factor recognition. In the first, the status of a replicated pattern is assigned to it, based on procedures in No. 10, above. It is then assigned an index number among known factors and a title essentially at the descriptive level, to "hold" it unambiguously while the research on interpretation proceeds. Use of a "popular" label from common speech is misleading, both because (a) it leads to contamination with fashionable associations and to degeneration of theoretical clarity and (b) because such terms commonly describe variables, whereas factors are abstractions from variables. A danger of a different sort arises with labels prematurely attaching the factor to some highly developed theory; for a factor needs first to be explored and given inherent clarity as an empirical construct. The best solution is a noncommittal indexing and descriptive labeling of the pattern as such, followed in due course by an interpretive technical term. Because of the dependence of firm progress in factor analytic research upon coordination among studies in the various respects here indicated, progress converging on valid measures and clear theoretical interpretation demands programmatic work more insistently than with many other methodologies. Such work calls for extensive indexing of variables and of factors, the latter initially separated for different media and modalities.

12. The deeper interpretation of a factor—after the hypotheses coming from the pattern itself—can only arise from its wide use outside factor analysis *per se* in other kinds of experiment. This requires that batteries and scales be available for scoring factors with high and known validities. In scoring, the alternative extremes are to use many variables with moderate and unequally weighted contributions as few variables with large and equally weighted contributions by each. At present, in many fields, variables with high enough loadings for the latter purpose are scarcely attainable, and it seems that using more numerous variables with moderate weights (with the advantage of low homogeneity) remains the safer alternative. The main second target, after validity, in obtaining factor scores is that the correlations of factor score estimates should depart as little as possible from the correlations of the true factors. Especially with computer synthesis which compels factors to share some specific and common factor variance beyond their own, the correlations of score estimates can become appreciable and are, of course, different from those of the true factors. However, methods are available to transform scores, with some loss, to scores correlating as the true factors do. The use of factor scores outside factor analysis, which both contributes to interpretation and permits more effective design in other experimental areas, takes many directions, e.g., manipulation of the factor, use of the factor as a dependent variable, correlation with life criteria, study of extreme sources, etc. These are extensively illustrated in, for example, the 16 PF and O-A (objective-analytic) personality factor measures.

13. The dimensional treatment of psychometric undertakings by factor analysis and the typal treatment by recognizing distinct species have been shown all along to be interdependent. Factor measures are better than variables for the employment of TAXONOME principles in finding types, but one must distinguish between type and within type factor dimensions. Using the former a skilled use of the TAXONOME program will recognize species types where they exist and proceed from homostats to natural segregates. But segregating types appear not to be very common in psychology outside occupational roles, political groups, and perhaps psychiatric syndromes. Nevertheless, the goal of description and measurement in factor terms has to be conceived as an iteratively reached schema of types placed within broader, cross-type factor dimensions and factor dimensions within types. Although clear segregation of species types is uncommon, factor analysis needs to drop the fiction of a homogeneous total population and to recognize the greater factor clarity obtainable by dealing with between type and within type factors, thus approaching closer fit of factors to the action of the natural determiners. Both a taxonomy of dimensions and a taxonomy of types are ultimately necessary.

14. The scientific use of factor analysis requires the investigator to have a broad acquaintance with the properties of various methods, and knowledge of where to go for aid in greater technical detail. While the psychologist or other researcher should thus be aware of the procedures made available by mathematical statisticians, it is equally important that he be able to make an intelligent and controlled use of them in the broader perspective of scientifically likely models and of effective research strategies.

Appendixes

SYNOPSIS OF SECTIONS

A.1. Proposed Standard Notation: Rationale and Outline
A.2. An Indexing System for Psychological Factors
A.3. Note on Utility of Confactor Resolutions with Oblique Factors
A.4. Transformations among SUD, SSA, and IIA Strata Models: Reversion from the Schmid-Leiman Matrix (IIA)
A.5. A Practicable Minimum List of Computer Programs
A.6. Tables for Statistical Significance of Simple Structure
A.7. Tables for Significance of Congruence Coefficients in Factor Matching
A.8. Tables for Significance of Salient Variable Similarity Index s
A.9. Recovery of V_0 and R_v from a Given V_{fp} and L Matrix

A.1. Proposed Standard Notation: Rationale and Outline

Mathematics is a language, and although, as Lewis Carroll reminds us, the mathematician is free to let any concept be represented by any symbol, there is no sense in turning freedom into caprice. The student's already great difficulties should not be wantonly increased by changing the language on him with every new writer. Actually, a firm language has long appealed to the logical mind of the mathematician and wherever possible he has settled down to agreed symbols—to π and e, χ^2, i, dx, and a whole "dictionary" of other symbols. These symbols, moreover, have the advantages of an international language.

Tolerably successful attempts have been made by Anderson (1966), the present writer, and Gorsuch (1974) to converge on such a system for factor analytic work. Such a dictionary naturally has to be a compromise, but it can be a compromise among clear principles, as follows.

1. As far as possible it should have the mnemonic advantage of

matching the symbol to the name. Thus we choose V for a variable dimension matrix, and if it is a factor pattern it is V_{fp} or if a reference vector structure V_{rs}. So also M is a mean, C a covariance matrix, df is degrees of freedom, S is a score matrix, a is an act or response measure, b is a "behavioral index"—the psychologist's meaning of a loading, akin to a beta weight, while in the reverse direction w is a weight in estimating a factor, and u a uniqueness component. Similarly, f is a factor and D a diagonal matrix.

2. Where usage is already irretrievably fixed, the first principle must bow to tradition, so that r is a correlation, R a correlation matrix, Z a matrix of standard scores, h a communality, F a variance ratio, and χ^2 a distribution test. However, even then such rules can be followed as using capitals for matrices and lower case for scalars.

3. As the above examples show, factor analysis has to dovetail its symbolism into those of neighboring areas, such as regression, ANOVA, etc., and convenience must then be considered across the whole statistical field.

4. Since most articles have to come from typewriters with fairly ordinary keyboards non-Greek symbols are preferable except where essential, as in the convention of population values being in Greek and sample values in our common type. Thus, we use L instead of lambda for a transformation matrix.

5. Unless subscripts are used fairly freely, the retention of standard symbols quickly becomes impossible, since a fraction of the full needs of representation already consumes the alphabet. The man who writes on the back of an envelope may have to use *ad hoc* brevities, but in texts there is much to be gained from using V_{fp} rather than A for a variable factor pattern matrix, and V_0 rather than A for an unrotated factor matrix. Some ambiguous use of simple alphabetic symbols is occasionally unavoidable, as when F is used for a factor and a variance ratio, and N is used both for the size of a sample and a matrix of n. But in the present system these are reduced to cases where the context is so different that no confusion is likely.

6. It is desirable that symbols be readily distinguishable also to the ear, for conversational use. Thus we abandoned n and m for number of variables and number of factors in favor of n and k. Although there is some problem from time to time in using a capital for a matrix and lower case for a single scalar entry, e.g., when N is number of people—such as R and r, C and c—the matrix traditions of capitals (often in bold) is indispensable.

7. Despite much progress that has been made toward uniform uses (Cattell, 1966a), there is still room for new developments in the "language" of statistics to facilitate communication. Instances suggested here are the balancing of the term ANOVA with the term CORAN, defining

general correlation–regression analysis method (and including factor analysis). It is also proposed, though now in the domain of the psychologist only, to represent a response act as a_{hijko}, in which h is the stimulus, i the individual, and so on through the data box sets as described in the text above. The use of R and Q, P and O, T and S, etc., for particular facets or faces (pairs of relatives and referees) factored is now a pretty thoroughly standardized shorthand. I have suggested elsewhere an immediately intelligible symbol system for more refined development of this nomenclature (Cattell, 1966a, p. 94). However, this and other developments are omitted from the following, which is confined to some main illustrations, and in areas where there has hitherto been insufficient crystallization.

I. Sets in the data box

		Range	*Specific Instance*
P	Persons or organisms	$1 - N$	i
E	Environmental occasions or conditions ("ambient situations")	$1 - v$	k
R	Response or act	$1 - n$	j
S	Stimulus (focal)	$1 - q$	h
O	Observer or instrument of measurement	$1 - x$	o

Note: S and R are frequently considered combined, in a test or learning bond and in that case the number is n and the specific case, jk.

In psychology the most common *attribute* entry in a data box cell is a *response magnitude*, a, which in the full data box case is a_{hijko}.

II. Score matrices

(a) Variables:

	Symbol	*Order*	*Entry*
Raw score variable data matrix	S_v	$N \times n$	a
Standard-score variable data matrix	Z_v	$N \times n$	z
Estimated-score variable data matrix	$\hat{S}_v$ or $\hat{Z}_v$	$N \times n$	$\underline{a}$

(b) Factors:

	Symbol	*Order*	*Entry*
Raw-(covariance) score factor data matrix	S_f	$N \times k$	a_f or f
Standard-score factor data matrix	Z_f	$N \times k$	z_f
Estimated-score factor data matrix	$\hat{S}_f$ or $\hat{Z}_f$	$N \times k$	$\underline{z}_f$

Parameters of scores: M mean, σ population sigma, s sample sigma.

III. Correlation and other relation matrices (relations of *relatives* over *referees* in data box)

(a) Elements:

    r correlation coefficient (see Table 14.1 for varieties, but chief are: ϕ for phi coefficient and ρ for rank r)

    r_c congruence coefficient

    r_p profile similarity coefficient

    c covariance product

Note also use of r_v for validity, r_h for homogeneity, r_d for dependability (reliability), and r_t for transferability (r_v, r_d, and r_t are thus forms of r_c consistency).

    d a difference or distance score

(b) Matrices:

    (i)   Variables or factors

(Dimensions)	*Symbol*	*Order*	*Entry*
A varible correlation matrix	R_v	$n \times n$	r
A variable covariance matrix	C_v	$n \times n$	c (or r_c)
A reduced correlation matrix (with h^2 in diagonals)	$R_{v \cdot 0}$ and $C_{v \cdot 0}$	$n \times n$	c (or r_c)
A correlation matrix of reference vectors	R_r	$n \times n$	c (or r_c)
A correlation matrix of factors	R_f	$n \times n$	c (or r_c)

    (ii) Variable dimension matrices in frequent use:

        (a) Broad, common factors

Orthogonal, unrotated standard factor matrix	V_0	$n \times k$	v
Orthogonal, unrotated covariance factor matrix	$V_{c \cdot 0}$	$n \times k$	c_f
Rotated matrix from V_0 (to simple structure, confactor, hypothetical PROCRUSTES position, etc.). Oblique with freedom to go orthogonal. Left in reference vector structure form (i.e., r with RV's).	V_{rs}	$n \times k$	r
Reference vector pattern matrix	V_{rp}	$n \times k$	b
Factor structure matrix (oblique; orthogonal as special case)	V_{fs}	$n \times k$	r
Factor pattern matrix	V_{fp}	$n \times k$	b
Factor estimation matrix (conventional)	V_{fe}	$n \times k$	w

Note: This means b is the weight of a factor estimating a variable; w is the weight of a variable estimating a factor.

Factor estimation matrix treating factors as the nonfallible variables	$\underline{V}_{fe}$	$n \times k$	$\underline{w}$

A component matrix is not often used in this text, but it is suggested it be written P, which is $(n \times n)$—since C is in use for covariance matrix, also (n × n).

(b) Unique, specific, error factors:

Factor loadings (correlations) of unique factors (assumed orthogonal in population) diagonal matrix	U	$n \times n$	b_u or u
Corresponding specific factor loadings (assumed orthogonal in population)	S	$n \times n$	b_s or s
Corresponding error factor loadings (assumed orthogonal in population)	E	$n \times n$	b_e or e

It is suggested that when the broad and unique factor variance contributions are written in the same matrix—which is the only truly complete matrix in this area—the above V be used, but with an underscore, thus $\underline{V}_{fp}$, $\underline{V}_{fe}$, etc. These will be of $n \times (n + k)$ order or even $n \times (2n + k)$ if the uniqueness matrix is split into specific and unique error matrix components.

(c) Some more special VD matrices:

Any nonfactor or matrix of multiple regression weights for estimating several criteria (q) from several predictors (t)	V_{qt}	$q \times t$	w
A factor matrix—the *associated factor* contribution matrix—for calculating multiple r of factors on variables	V_{afp}	$n \times k$	v
The dissociated factor contribution matrix (Schmid–Leiman only at variable level)	V_{dfp}	$n \times k$	r
Variable contribution matrix for calculating multiple r with F, in estimation of factors	V_{vc}	$n \times k$	v
Factor mandate matrix	V_{fm}	$n \times k$	1 or 0

IV. Operating, Transforming Matrices (Naturally only a few most used instances)

Diagonal matrix of latent roots	D_1	$k \times k$	λ
Any diagonal matrix of tensors (usually ratios or sigmas)	D	$n \times n$ or $k \times k$	d
Transformation matrix (e.g., rotation:			
For reference vectors	L_r	$k \times k$	r
For factors	L_f	$k \times k$	r

V. Higher-Order Factors

Affixing of subscripts I, II, III, etc., to factor scores ($F_{\text{I}i}$, $F_{\text{II}i}$, etc.) to correlation matrices, $R_{f\text{I}}$, $R_{f\text{II}}$, etc., and factor pattern matrices, $V_{fp\text{I}}$, $V_{fp\text{II}}$, etc., indicates the strata levels of the factors concerned. Similarly for loadings: b_I, b_II, etc. In specification equations simultaneously using different orders, there will be "stub factors." In such cases the loadings and stubs are distinguished by underlining—$\underline{b}_\text{I}$ and $\underline{F}_\text{III}$—and the corresponding V_{fp} (Schmid–Leiman) matrix by italics and underlining. Thus $\underline{V}_{fp\text{I.II.III}}$ is a three part Schmid–Leiman matrix with uniqueness.

A.2. An Indexing System for Psychological Factors

An indexing for all areas of factor analytic application is obviously a matter for committees in particular academic domains. Fortunately, the importance of indexing in psychology has been well recognized, both as to variables and to factors, by several leaders in the field, notably by Buss, French, Fruchter, Goldberg, Hakstian, Horn, Nesselroade, Norman, Royce, Sells, and Wardell in North America; Beloff, Butcher, Child, Eysenck, Saville, and Warburton in Britain; Häcker, Pawlik, Schmidt, Schneewind, and Weiss in Germany; and Reuchlin in France. The more difficult task of a standard population and stratified sample of general variables and marker variables to which Burt, the present writer (1957a,c), and Kaiser also gave attention has recently begun in symposia discussions (Society of Multivariate Experimental Psychology, Annual Meeting, November, 1975). Meanwhile considerable cross reference, matching research, and utilization in applied psychology (see bibliography in Cattell, Eber, & Tsujioka, 1971) have been given to the indexing system briefly laid out in what follows.

Five main principles need to be observed.

1. That factors will come not only from the most used techniques, e.g., R technique, yielding individual difference traits, but also from other techniques across other facets, faces, and grids of the data box. At present,

though there is a little work on factors in observers, situations, and variations in modes of response, there is a real accumulation of findings in state and trait change factors. We therefore have proposed two compendia, one marked A, of traits, and one B, of states and trait change patterns.

2. One purpose of a symbol system is to avoid premature interpretation, while being precise as to identity. This is achieved by resting the identification on an indexed loading pattern (plus correlation with other factors, etc.). Since establishment of identity across the three media of observation—L-data (record and observation rating in everyday life), Q-data (questionnaires and consulting-room self-revelation), and T-data (objective behavior, sometimes in laboratory)—is a second stage of research, at present quite incomplete and questionned, the factors should be indexed separately in the three media, with L, Q, and T prefixes.

3. The question of whether factors should also be indexed separately by modality—ability, temperament (general personality), and dynamic (motivational)—must await debate. In the instance of T-data French's list of 15 abilities and the present writers list of 20 T-data personality traits the two series have been run together, on the argument that drawing the line is difficult. On the other hand, as with different media, there may be advantages in listing in separate modalities, for which case A, P, and D have been prefixed for abilities, personality temperament, and dynamic traits.

4. Decisions as to factor order in the index series were made within the present writer's laboratory on size of factor variance, over a stratified set of variables, on a defined population (adults and older students), on factors which had their patterns replicated in at least four independent researches. *This order is bound to be approximate*, and dependent on the samples both of people and variables, as is illustrated by the fact, for example, that I factor in personality, premsia (see below) climbs from ninth place in adults to first or second in variance in ten-year-olds.

In other sciences, e.g., the indexing of vitamins, the order of symbols has been inexorably fixed by the historical order of discovery. Nevertheless, there is something to be said for inherent properties, and though the elements in chemistry are named by their discoverers and given symbols accordingly, the atomic number order in terms of which one often thinks of properties, turns out to be quite different. In psychology, fortunately, size and order of historical discovery tend to be tolerably aligned, since larger factors have generally been discovered earlier.

5. As findings on factor *order* are checked and crystallize into statements on factor *strata*, the latter can be respected by the same device we have suggested for matrices, etc. Arabic numerals for primaries, Roman for secondaries, and Greek letters for the comparatively rare tertiaries.

An illustration of the above can be given in T-data (objective test medium) by French's (1953) listing of ability factors and the present writer's continuation of it into personality factors (1946, 1957, 1965). First comes U.I. for universal index, then T, for objective test medium, then A, for R-technique individual difference trait factor, then A, D, or P, for ability, personality, or dynamic and then an Arabic or Roman number.

Since the main proposed index, though adequately precise, is sometimes cumbersome there has grown up side by side therewith a "vernacular" symbol understood in the right context. Thus the questionnaire and rating factors in Table A.1(c) above are well understood and commonly described in some 200 publications by the alphabetic series, which is in order of size and partly of order of discovery. Similarly, the second-order ability factors U.I.T.A.A.I. and U.I.T.A.A.II are commonly handled in papers as g_f and g_c, deriving from Spearman's g.

Table A.1.

T (objective test data factors)

(a) First-order (primary factor series)

Universal index no.	Vernacular symbol	
U.I.(T)A.A1	A	Aiming (eye-hand coordination)
U.I.(T)A.A2	Cf	Flexibility of closure
U.I.(T)A.A3	D	Speed of closure
U.I.(T)A.A4	D	Deduction
U.I.(T)A.A5	I	Induction
U.I.(T)A.A6	If	Ideational fluency
U.I.(T)A.A7	Ma	Associative memory
U.I.(T)A.A8	Mk	Mechanical knowledge
U.I.(T)A.A9	Ms	Motor speed
U.I.(T)A.A10	N	Number facility
U.I.(T)A.A11	S	Spatial relations and orientation

.
.
.

U.I.(T)A.P16	Ca	Ego strength
U.I.(T)A.P17	Rt	General inhibition and control
U.I.(T)A.P18	Hy	Hypomania
U.I.(T)A.P19	In	Independence
U.I.(T)A.P20	Ev	Evasiveness
U.I.(T)A.P21	Eb	Exuberance
U.I.(T)A.P22	Nr	Cortertia
U.I.(T)A.P23	Nr	Regression
U.I.(T)A.P24	Ap	Anxiety
U.I.(T)A.P25	Ar	Realism-vs.-psychotic tendency
U.I.(T)A.P26	Ci	Cultured, introspective self-control
U.I.(T)A.P27	At	Apathy

Table A.1. (*Continued*)

Universal index no.	Vernacular symbol	
U.I.(T)A.P28	Ee	Emotional evasiveness
U.I.(T)A.P29	Se	Super ego
U.I.(T)A.P30	Is	Stolidity
U.I.(T)A.P31	Wr	Wary realism
U.I.(T)A.P32	Ex	Extraversion

(b) Second-order (secondaries) (Illustrated by abilities, from Horn, Hakstian, and others)

U.I.T.A.AI	gf	Fluid intelligence
U.I.T.A.AII	gc	Crystallized intelligence
U.I.T.A.AIII	s	General cognitive speed factor
U.I.T.A.P.I	i	Integration
U.I.Q.A.P.I	Ex	Exvia–Invia
U.I.Q.A.P.II	An	Anxiety as a trait
U.I.Q.B.P.II	Ax	Anxiety as a state

Q- and L- (questionnaire and life record media)
(a) First-order (primary factor series)

In life record data			In questionnaire data
U.I.(L)A.P1	A	Affectia	U.I.(Q)A.P1
U.I.(L)A.P2	B	Intelligence	U.I.(Q)A.P2
U.I.(L)A.P3	C	Ego strength	U.I.(Q)A.P3
U.I.(L)A.P4	D	Excitability	U.I.(Q)A.P4
U.I.(L)A.P5	E	Dominance	U.I.(Q)A.P5
U.I.(L)A.P6	F	Surgency	U.I.(Q)A.P6
U.I.(L)A.P7	G	Super ego strength	U.I.(Q)A.P7
U.I.(L)A.P8	H	Parmia	U.I.(Q)A.P8
U.I.(L)A.P9	I	Premsia	U.I.(Q)A.P9
U.I.(L)A.P10	J	Asthenia	U.I.(Q)A.P10
U.I.(L)A.P11	K	Acculturation	U.I.(Q)A.P11
U.I.(L)A.P12	L	Protension	U.I.(Q)A.P12
U.I.(L)A.P13	M	Autia	U.I.(Q)A.P13
U.I.(L)A.P14	N	Shrewdness	U.I.(Q)A.P14
U.I.(L)A.P15	O	Guilt proneness	U.I.(Q)A.P15
(Many of remaining	P	Sanguine casualness	U.I.(Q)A.P16
factors found in the	Q_1	Radicalism	U.I.(Q)A.P17
questionnaire medium	Q_2	Self-sufficiency	U.I.(Q)A.P18
have not yet been	Q_3	Self-sentiment strength	U.I.(Q)A.P19
matched in L-data	Q_4	Ergic tension	U.I.(Q)A.P20
U.I.(L)A.P21	Q_5	Group dedication	U.I.(Q)A.P21
	Q_6	Social panache	U.I.(Q)A.P22
	Q_7	Explicit self-expression	U.I.(Q)A.P23
	D_1	Hypochondriasis	U.I.(Q)A.P24
	D_2	Suicidal disgust	U.I.(Q)A.P25

Table A.1. (*Continued*)

In life record data		In questionnaire data
	D_3 Brooding discontent	U.I.(Q)A.P26
	D_4 Anxious depression	U.I.(Q)A.P27
	D_5 Low energy depression	U.I.(Q)A.P28
	D_6 Guilty depression	U.I.(Q)A.P29
	D_7 Bored depression	U.I.(Q)A.P30
	P_a Paranoid tendency	U.I.(Q)A.P31
	P_p Psychopathic deviation	U.I.(Q)A.P32
	S_c Schizophrenia	U.I.(Q)A.P33
	A_s Psychasthenia	U.I.(Q)A.P34
	P_s General psychosis	U.I.(Q)A.P35

(b) Second-order (secondaries)

Life record (L) data (ratings)		Questionnaire data
U.I.(L)A.P.I	Exvia–invia	U.I.(Q)A.P.I
U.I.(L)A.P.II	Anxiety	U.I.(Q)A.P.II
U.I.(L)A.P.III	Cortertia	U.I.(Q)A.P.III
U.I.(L)A.P.IV	Independence	U.I.(Q)A.P.IV
U.I.(L)A.P.V	Discreetness	U.I.(Q)A.P.V
No match as yet in	Prodigality	U.I.(Q)A.P.VI
L-data for Q-data	Intelligence	U.I.(Q)A.P.VII
secondaries	Control	U.I.(Q)A.P.VIII
	Humanism	U.I.(Q)A.P.IX
	Toughness	U.I.(Q)A.P.X
	Schizoid temperament	U.I.(Q)A.P.XI
	Hypomania	U.I.(Q)A.P.XII
	General depression	U.I.(Q)A.P.XIII
	General psychosis	U.I.(Q)A.P.XIV
	Inhibition	U.I.(Q)A.P.XV

(c) State factors, from P and dR techniques questionnaire medium (S = state)

	Exvia	U.I.(Q)B.S1
	Anxiety	U.I.(Q)B.S2
	Depression	U.I.(Q)B.S3
	Arousal	U.I.(Q)B.S4
	Fatigue	U.I.(Q)B.S5
	Stress	U.I.(Q)B.S6
	Regression	U.I.(Q)B.S7
	Guilt	U.I.(Q)B.S8

These lists are given here as practical illustrations of matching "invariance" of factors from at least six researches in each case. The later items in, for example, (b) above, are not given as indubitable, and await further research. However, reference to such a factor matching and indexing, though only in one area in which factor analysis is concerned—human traits and states in three media of experimental observations—may suffice to show its practicality, to bring out the technical problems it faces, and to illustrate solutions. Since many traits are paralleled by states, for example, the question arises whether the same number could be kept, with A or B preceding. The answer is that caution and research accidents of sequence call for separate lists, respecting the empirical findings in each area and leaving interpretations until later. Though practicing psychologists are committed to the "vernacular" of single letters for certain personality factors, it is undoubtedly necessary to preserve in the background the full systematic reserves of the universal index series, e.g., A, P, D, for ability, personality, and dynamic modalities, Arabic and Roman numerals for primaries and secondaries, and so on.

A.3. Note on Utility of Confactor Resolutions with Oblique Factors

As stated (p. 100), confactor resolution has succeeded excellently with orthogonal factors, Brennan having shown recently (1977) by extension to a wider range of study sizes and degrees of experimental error that it can be relied on to find the factors put into a plasmode in all varieties of case yet tried. No basic solution has yet been found, however, to the oblique case, and since most real data have obliqueness, and simple structure is a relative failure with studies where due "hyperplane stuff" has not been included, there remains a pressing need to program confactor resolution for general use, even if solutions are only approximate.

The studies of Brennan and Cattell show that (a) if the obliquity of the primaries is *slight*—say, ranging from r's of .05 to .25 and averaging about .15—a solution remains possible, though rough. (b) If the correlations among the primaries are the same in the two experiments, then a definitely good solution for the factor patterns is obtained by Cattell's orthogonal confactor, even though the factor correlations are appreciable. Table A.3.1 shows in (a) the two V_{fp}'s, and D_f, and the R_f, as used to create the correlation matrices A and B (not shown). At (b) we have the output from the Cattell confactor program (p. 101) applied to the two given variable correlation matrices with the usual orthogonal output. It will be seen (a) that the proportionalities in D_f as reached and D_f as put in are very similar, and (b) that the factor matrices for A and B match the originals very adequately. What is not given is the obliquities, i.e., R_f, which, however, is in any case frequently not considered worthy of report in reaching solutions for primary factors, but it can be derived.

Table A.3.1. Application of Orthogonal Confactor Program to Oblique Cases

(a) Plasmode as made up

A Experiment: $V_{fp \cdot A}$ | B Experiment: $V_{fp \cdot B}$

	1	2	3	4	5	6		1	2	3	4	5	6
1	00	-01	19	16	-03	-67	1	00	-01	17	18	-04	-86
2	18	-05	20	02	-05	-54	2	13	-04	18	03	-05	-07
3	65	-02	-06	-06	23	18	3	45	-02	-05	-06	25	23
4	18	03	03	-02	41	-03	4	13	03	03	-02	45	-04
5	18	16	-37	-57	05	-05	5	13	13	-34	-63	06	-07
6	00	13	26	-63	-03	05	6	00	10	23	-69	-03	07
7	46	00	-01	22	14	-10	7	32	00	-01	24	16	-13
8	-53	-24	02	33	-31	-01	8	-37	-19	02	37	-34	-01
9	-56	-27	-01	03	01	-15	9	-39	-22	-01	04	01	-20
10	-52	06	-18	25	-07	37	10	-36	04	-16	28	-08	48
11	04	18	31	34	-44	-15	11	03	14	28	38	-48	-20
12	-56	-22	02	-02	02	04	12	-39	-18	02	-03	02	06
13	-07	-03	47	01	-05	-51	13	-05	-03	43	01	-05	-67
14	07	55	-29	-04	07	-02	14	05	44	-26	-04	07	02
15	-57	-16	01	02	-02	02	15	-40	-13	01	03	-02	02
16	01	-16	-19	05	28	19	16	01	-13	-17	05	31	25

(b) Output from program

A Experiment: $\underline{V}_{fp \cdot A}$ | B Experiment: $\underline{V}_{fp \cdot B}$

	1	2	3	4	5	6		1	2	3	4	5	6
1	-26	15	-16	18	-07	61	1	-17	12	-16	20	08	85
2	-36	01	-16	06	-11	51	2	-24	01	-14	06	13	71
3	-56	-25	-00	-08	25	-05	3	-37	-20	-00	09	39	-07
4	-25	05	-13	-05	39	09	4	-16	08	-12	03	84	13
5	-12	-00	48	-50	-32	04	5	-05	00	42	35	36	05
6	04	07	-09	-50	-25	-06	6	02	07	08	-35	28	07
7	-50	-09	-07	17	24	18	7	-33	-07	-07	19	28	26
8	53	-07	-05	30	-17	10	8	35	-05	02	33	-19	-13
9	47	-05	-01	04	-07	02	9	31	-08	-01	05	-08	03
10	61	16	12	19	09	-40	10	40	12	10	21	10	56
11	-10	20	-24	33	-18	17	11	-07	16	-21	36	-20	23
12	55	-03	-04	01	-03	-13	12	36	-03	-03	-01	-08	-21
13	-15	k5	-45	07	-06	46	13	-10	12	-38	07	-07	65
14	-11	46	29	-06	02	08	14	-07	36	26	-06	02	11
15	55	03	-02	03	-05	-12	15	36	03	-02	03	-06	17
16	04	-17	07	01	27	-17	16	03	-11	06	01	31	-21

	1	2	3	4	5	6		1	2	3	4	5	6
1	.7						1	.66					
2		.8					2		.78				
3			.9				3			.88			
4				1.1			4				1.10		
5					1.1		5					1.13	
6						1.3	6						1.40

Table A.3.1. *(Continued)*

	D_f as given [Relating A & B factors in plasmode (a)]					$\underline{D}_f$ as reached (Relating program output A & B)

	1	2	3	4	5	6	
1							Plasmode construction
2	-29						
3	16	-01					$R_{v \cdot A} = V_{fp \cdot A} R_f V_{fp \cdot A}^T$
4	-21	-38	07				$R_{v \cdot B} = V_{fp \cdot A} D_f R_f D_f V_{fp \cdot A}^T$
5	15	-02	43	06			Confactor program applied to $R_{v \cdot A}$ and
6	26	09	54	17	53		$R_{v \cdot B}$, as experimentally given, yields (b).

$$R_f$$
[As set up for plasmode (a)]

Note: factors in (b) come out with random directions, 1, 2, 3, and 6 being reversed. This is irrelevant.

A.4. Transformations among SUD, SSA, and IIA Strata Models: Reversion from the Schmid–Leiman Matrix (IIA)

In Chapter 9, it has been shown that if strata are recognized in the stratoplex model there are two main models to be considered: the stratified uncorrelated determiner (SUD), in which factors at each stratum are uncorrelated with one another or other strata but act by contributing to the variance of factors of the next lower stratum, and the S–L model in which the calculations, at least, are as if the higher strata acted in all cases directly on the variables.

One distinguishes between models and convenient forms of resolution and calculation. Thus, in the typical resolution by simple structure at each stratum, the first result one gets is a matrix of oblique primaries on variables V_{fpI}, then of oblique secondaries upon primaries V_{fpII} (from R_{fI} the correlations of primaries), and so on until a single higher stratum tops the hierarchy. This basic oblique primary resolution, as named in Chapter 9, may perhaps better be called the *Serial Stratum Analysis* (SSA), but it cannot easily be considered a final model because it is redundant. It is redundant because part of the variance of the primaries is again given in the secondaries, of the secondaries in the tertiaries, and so on. But it is usually our first statement of a factor analytic resolution, and it contains the full necessary information on structure for other models or calculations. One such alternative statement is to give the projections of the higher orders directly on the variables by the Cattell–White formula (p. 209). This may be called the *Serial Overlap Analysis* (SOA) as in Table 9.4(a); but though useful in interpreting factor hypotheses it is just as redundant as SSA, as shown by the summed squared loadings exceeding h^2.

The application of the Schmid–Leiman (S–L) formula to produce what we have called (Table 9.4) the *independent influence analysis* (IIA) we have preferred to consider a method of analysis rather than a model, though some researchers may perhaps make reasonable claims for it as a model. It accepts the SUD model, of independent stub factors, but proceeds as if they act directly on variables instead of the factors in the next lower stratum. This again is at least useful in factor interpretation of the SUD factors and it is often also a convenience, e.g., in understanding questionnaire item composition, since the squared loadings on the orthogonal stubs projected on the variables add up to the item communalities across all strata.

The transformations from one system to another—from the initially given SSA, to SOA, to SUD, and to IIA—are relatively straightforward and are given in Chapter 9 and later. Only one calculation presents difficulties and that is the reversionary transformation from the Schmid–Leiman IIA matrix to the SUD and SSA. If Cattell and Brennan (1977) are right in indicating that use of confactor on an (orthogonal) Schmid–Leiman may sometimes prove an avenue to the underlying oblique confactor structure, then the reversionary transformation from the discovered IIA to the SSA and SUD becomes an important device, but the researcher may find himself with yet other occasions of being forced to start with a IIA matrix and, in any case, will wish to understand the reversion.

Different calculations all successfully reproducing the factors in the SSA as tried by congruence coefficients have been put forward by Cattell and De Young (in preparation), Schönemann (1964), Tucker (1970), and Meredith (see Brennan, 1977). All assume one knows which columns of the IIA represent, respectively, primary, secondary, etc., factors in the SSA. In addition to putting forward a new solution, Cattell and De Young compared the three methods then available for accuracy of reproduction of the original SSA plasmode, with 5%, 10% and 20% of error superimposed on the error-free IIA, Schmid–Leiman solution. The results showed Schönemann's method one best, his next second, and Cattell–De Young's third at all three levels of error. The method here set out is therefore that of Schönemann, later checked by Tucker (since all three program with equal facility).

As shown in Chapter 9 [Table 9.3 and Eq. 9.8(c)] the Schmid–Leiman IIA matrix is given may be viewed as a super matrix:

$$V_v = (V_{v1} \, V_{v2} \, \cdots \, V_{vj} \, \cdots \, V_{vc}) \tag{A.1}$$

V_v signifying a factor matrix (orthog) with projections directly on variables.

In this matrix

$$V_{v1} = V_I U_{II} \tag{A.2}$$

while $V_{v2} = V_{II} V_{II} U_{III}$, and so on to a single factor at the ith order when $V_{vi} = V_I V_{II} \cdots V_i$.

V_I, V_{II}, etc., are the first, second, etc., order factor matrices (V_{fp}'s) in the original SSA solution and the U's are corresponding matrices of unique factor loadings. We shall suppose k first orders, p second orders, and so on through s and t, so that the Schmid–Leiman matrix V_v is an $n \times x$ matrix where $x = (p + q + \cdots)$.

This method aims first to clarify the nature (in the SSA original) of the first-order factors by finding the V_{0I} (unrotated factor matrix for V_I) and the transformation matrix L_v of direction cosines which places the Schmid–Leiman reference vectors orthogonal to the original V_0 orthogonal frame. With these two matrices in hand, one can proceed to obtain V_I by finding R_{rv}, the correlation matrix among the reference vectors, and the normalizing transformation matrix L_f from $[L_f = D(L_r^T L_r)^{-1} D]$ which rotates the $(n \times k)$ matrix V_0 to the factor solution in the SSA defined by V_I.

To achieve the first aim, Schönemann proceeds as follows: compute a new matrix M, which is $(k + p +$, etc.) square, by taking the Schmid–Leiman II A matrix V_v:

$$\underset{(x \times k)}{M_1} = \underset{(x \times n)}{V_v^T} \underset{(N \times x)}{V_v} \tag{A.3}$$

Now determine the characteristic roots and vectors of M (as in the process of factoring). Let P_1 be an $x \times k$ matrix of the latent vectors, corresponding to the k longest roots. Then

$$V_0 = V_v P_1 \tag{A.4}$$

Next compute a matrix $(x \times k)$ square, and take its diagonal by

$$K_1 = \text{diagonal } (P_1 P_1^T)^{-1/2} \tag{A.5}$$

Then determine $W_1 (k + p + \cdots + s + t)$ by

$$W_1 = K_1 P_1 \tag{A.6}$$

Therefore, there is a column vector in W_1^T for each Schmid–Leiman factor. The column vecotrs in W_1^T corresponding to the k first-order factors may be assembled as column vectors in the matrix L_r. Then the first-order analysis may be completed by the following steps:

$$R_{rv} = L_r^T L_r \tag{A.7}$$
$$D = \text{diagonal } (R_{rv}^{-1})^{-1/2} \tag{A.8}$$
$$R_f = D R_{rv}^{-1} D \tag{A.9}$$
$$L_f = D L_r^{-1}$$
$$= D R_{rv}^{-1} L_r^T \tag{A.10}$$

and one obtains the desired matrix V_1 by

$$V_1 = V_0 L_{fp}^{-1} \tag{A.11}$$

Let V_{vx} be a matrix formed from V_v by deleting column selected for first-order factors. A matrix $\underline{V}_v$ is formed by

$$\underline{V}_v = (V_1' V_1)^{-1} V_1^T V_{vx} \tag{A.12}$$

$\underline{V}_v$ is a Schmid–Leiman factor representation of the first-order correlation matrix after the first-order unique variance has been subtracted from the diagonals of R_f. The first-order unique variances, which are diagonal entries in U_2^2, may be found by either of two routes:

$$U_2 = D\ K^{-1} \tag{A.13}$$

where K is composed of the elements of K_1 corresponding to the selected first-order factors; alternatively,

$$U_2^2 = R_f - \underline{V}_v \underline{V}_v^T \tag{A.14}$$

The development of $\underline{V}_v$ brings us to a recursive point at which a further cycle can begin. The analyses for the first-order factors on V_v may be repeated for the second-order factors on $\underline{V}_v$. This involves returning to Eq. (A.3) and replacing each matrix with its second-order analogue. For the third order, $\underline{V}_v$ is formed from $\underline{V}_v$ in an analogous fashion to the formation of $\underline{V}_v$ from V_v. This plan could be carried to the fourth and higher levels when appropriate.

A.5. A Practicable Minimum List of Computer Programs

1. Calculation of a correlation matrix.
2. Principal axis extraction program iterating to the number of factors decided by scree plot of latent roots.
3. PROMAX or HK oblique analytical rotation, supplemented by
4. MAXPLANE topological rotation program.
5. ROTOPLOT (Cattell & Foster, 1963).
6. Computation of congruence coefficients and/or s indices.
7. Matrix inversion program large enough to invert R_v for the calculation of V_{fe} and factor scores.
8. A proofing maximum likelihood program, if feasible.
9. A confactor rotation program.

A.6. Tables for Statistical Significance of Simple Structure[1]

Bargmann (1955) provides tables for testing the statistical significance of simple structure for 2 to 12 factors over a range of 70 variables and for six probability levels ranging from .50 to .001. One enters the

[1] Copyright by Larry F. Sine and Velma A. Kameoka, the University of Hawaii, 1977. Reprinted with permission of the authors.

tables with the number of variables appearing in the hyperplane of width ±.10 corresponding to a specific probability level, the number of variables, and the number of factors, to get an estimate for any one factor. Given the size of contemporary factor analytic solutions, the number of factors and variables comprising Bargmann's original tables is no longer sufficient. Furthermore, the original article in German has had limited circulation and, except by Cattell and co-workers, appears to have been neglected in practical use.

In view of the foregoing, a computer program (Kameoka Sine, 1978) was written using Bargmann's equations and probability expression (with minor modification) to generate exact probabilities for at least r variables appearing in a hyperplane of width ±.10 for n variables and k factors. Due to the amount of space required to report exact probabilities for numerous factors, the following tables are presented for six specific p points (.50, .10, .05, .01, .001, .0001) of the probability distributions, for 2, 4, 6, 8, 10, 12, 16, 20, 24, and 28 factors, and for a ±.10 hyperplane width. The indicated r values (the number of variables in the hyperplane) should be interpreted as having probabilities less than the designated p values. Bargmann himself has not extended the argument to the significance of a study as a whole.

In using the tables, one must examine the number of zero-loadings (within ±.10 actually) contained in each factor of the factor pattern. However, the literal variable loadings must be divided by the square root of their respective communalities in order to determine whether the loadings, upon thus being adjusted for communality size, still fall in the ±.10 hyperplane. The number of variables whose adjusted loadings fall within the hyperplane of a factor is used to enter the probability tables to derive the statistical significance of the obtained simple structure for the factor.

To illustrate use of these tables, suppose one wishes to determine the statistical significance of simple structure (i.e., whether the hyperplanes are significantly determined) of a factor matrix containing 64 variables and 16 factors. Examination of the table for 16 factors with variable loadings adjusted for communality size indicates that at least 31 adjusted zero-loadings would occur for any factor by chance, i.e., at a probability level of .50. There would be 35 at $p < .10$, 36 at $p < .05$, 38 at $p < .01$, etc. If in this problem a hyperplane contains 37 adjusted zero-loadings, it is very adequately determined and justifies proceeding to matching and interpretation. Interpolation between tables may be performed by the user for a matrix containing a number of factors or number of variables n not presently reported.

As pointed out elsewhere, the empirical studies of Cattell and Finkbeiner show that for sample sizes of about 100–500 subjects the width of the band containing variables that are truly at loadings of zero is about ±.006. However, Kameoka and Sine have not yet recalculated their tables to this width, so that for the present the user should calculate with the more traditional ±.10 (after adjusting loadings for the communalities of the variables concerned).

2 Factors

	Probability (p) levels					
n	.50	.10	.05	.01	.001	.0001
5	2	3	3	4	4	5
6	2	3	3	4	5	5
7	2	3	4	4	5	6
8	2	3	4	4	5	6
9	2	3	4	5	5	6
10	2	4	4	5	6	7
11	2	4	4	5	6	7
12	3	4	4	5	6	7
13	3	4	4	5	6	7
14	3	4	4	5	6	7
15	3	4	5	5	7	8
16	3	4	5	6	7	8
17	3	4	5	6	7	8
18	3	4	5	6	7	8
19	3	5	5	6	7	8
20	3	5	5	6	7	9
21	3	5	5	6	8	9
22	3	5	5	6	8	9
23	3	5	5	7	8	9
24	3	5	6	7	8	9
25	3	5	6	7	8	9
26	3	5	6	7	8	10
27	4	5	6	7	8	10
28	4	5	6	7	9	10
29	4	5	6	7	9	10
30	4	6	6	7	9	10

Note: Entires are number of loadings in the hyperplane, taken to be written (but not including) ±.10; n is the number of variables in the study.

4 Factors

	Probability (p) levels					
n	.50	.10	.05	.01	.001	.0001
8	4	6	6	7	8	8
9	5	6	6	7	8	9
10	5	6	6	7	8	9
11	5	6	7	8	9	9
12	5	6	7	8	9	10

4 Factors (*Continued*)

	Probability (p) levels					
n	.50	.10	.05	.01	.001	.0001
13	5	7	7	8	9	10
14	5	7	7	8	10	11
15	5	7	8	9	10	11
16	6	7	8	9	10	11
17	6	7	8	9	10	11
18	6	8	8	9	11	12
19	6	8	8	10	11	12
20	6	8	9	10	11	12
21	6	8	9	10	11	13
22	6	8	9	10	12	13
23	6	9	9	10	12	13
24	7	9	9	11	12	14
25	7	9	10	11	12	14
26	7	9	10	11	13	14
27	7	9	10	11	13	14
28	7	9	10	11	13	15
29	7	10	10	12	13	15
30	7	10	10	12	14	15
31	7	10	11	12	14	15
32	8	10	11	12	14	16
33	8	10	11	13	14	16
34	8	10	11	13	15	16
35	8	11	11	13	15	16
36	8	11	12	13	15	17
37	8	11	12	13	15	17
38	8	11	12	13	15	17
39	8	11	12	14	16	17
40	9	11	12	14	16	17
41	9	12	12	14	16	18
42	9	12	13	14	16	18
43	9	12	13	14	16	18
44	9	12	13	15	17	18
45	9	12	13	15	17	19

6 Factors

	Probability (p) levels					
n	.50	.10	.05	.01	.001	.0001
12	7	8	9	10	11	12
13	7	9	9	10	11	12
14	7	9	10	10	12	13
15	8	9	10	11	12	13

6 Factors (*Continued*)

n	Probability (*p*) levels					
	.50	.10	.05	.01	.001	.0001
16	8	9	10	11	12	13
17	8	10	10	11	13	14
18	8	10	11	12	13	14
19	8	10	11	12	13	14
20	8	10	11	12	14	15
21	9	11	11	13	14	15
22	9	11	12	13	14	16
23	9	11	12	13	15	16
24	9	11	12	13	15	16
25	9	12	12	14	15	17
26	9	12	13	14	16	17
27	10	12	13	14	16	17
28	10	12	13	14	16	18
29	10	12	13	15	16	18
30	10	13	13	15	17	18
31	10	13	14	15	17	19
32	10	13	14	15	17	19
33	11	13	14	16	18	19
34	11	14	14	16	18	19
35	11	14	15	16	18	20
36	11	14	15	16	18	20
37	11	14	15	17	19	20
38	11	14	15	17	19	21
39	12	15	15	17	19	21
40	12	15	16	17	20	21
41	12	15	16	18	20	22
42	12	15	16	18	20	22
43	12	15	16	18	20	22
44	12	16	17	18	21	22
45	13	16	17	19	21	23
46	13	16	17	19	21	23
47	13	16	17	19	21	23
48	13	16	17	19	22	24
49	13	17	18	20	22	24
50	13	17	18	20	22	24
51	14	17	18	20	22	24
52	14	17	18	20	23	25
53	14	17	19	21	23	25
54	14	18	19	21	23	25
55	14	18	19	21	23	25
56	15	18	19	21	24	26
57	15	18	19	21	24	26
58	15	19	20	22	24	26
59	15	19	20	22	24	26
60	15	19	20	22	25	27

8 Factors

	Probability (p) levels					
n	.50	.10	.05	.01	.001	.0001
16	10	11	12	13	14	15
18	10	12	13	14	15	16
20	11	13	13	14	16	17
22	11	13	14	15	16	18
24	11	14	14	16	17	18
26	12	14	15	16	18	19
28	12	15	15	17	19	20
29	12	15	16	17	19	20
30	13	15	16	17	19	21
31	13	15	16	18	20	21
32	13	16	16	18	20	21
33	13	16	17	18	20	22
34	13	16	17	19	21	22
35	14	16	17	19	21	22
36	14	17	18	19	21	23
37	14	17	18	20	22	23
38	14	17	18	20	22	24
39	14	17	18	20	22	24
40	15	18	19	20	22	24
41	15	18	19	21	23	25
42	15	18	19	21	23	25
43	15	18	19	21	23	25
44	15	19	20	22	24	26
45	16	19	20	22	24	26
46	16	19	20	22	24	26
47	16	19	20	22	25	26
48	16	20	21	23	25	27
49	16	20	21	23	25	27
50	17	20	21	23	25	27
51	17	20	21	23	26	28
52	17	21	22	24	26	28
53	17	21	22	24	26	28
54	17	21	22	24	27	29
55	18	21	22	25	27	29
56	18	22	23	25	27	29
57	18	22	23	25	28	30
58	18	22	23	25	28	30
60	19	23	24	26	28	31
62	19	23	24	26	29	31
64	19	23	25	27	30	32
66	20	24	25	27	30	32
68	20	24	26	28	31	33
70	21	25	26	29	31	34
72	21	25	27	29	32	34
75	22	26	27	30	33	35

10 Factors

	Probability (p) levels					
n	.50	.10	.05	.01	.001	.0001
20	12	14	15	16	17	18
22	13	15	16	17	18	19
24	13	16	16	18	19	20
26	14	16	17	18	20	21
28	14	17	18	19	21	22
30	15	17	18	20	21	23
32	15	18	19	20	22	24
33	15	18	19	21	22	24
34	16	18	19	21	23	24
35	16	19	20	21	23	25
36	16	19	20	22	24	25
37	16	19	20	22	24	26
38	17	20	21	22	24	26
39	17	20	21	23	25	26
40	17	20	21	23	25	27
41	17	20	21	23	25	27
42	18	21	22	24	26	27
44	18	21	22	24	26	28
46	18	22	23	25	27	29
48	19	22	23	25	28	30
50	19	23	24	26	28	30
52	20	23	25	27	29	31
54	20	24	25	27	30	32
56	21	25	26	28	30	32
58	21	25	26	28	31	33
60	22	26	27	29	32	34
62	22	26	27	30	32	34
64	23	27	28	30	33	35
66	23	27	28	31	34	36
68	23	28	29	31	34	36
70	24	28	30	32	35	37
72	24	29	30	33	35	38
74	25	29	31	33	36	38
76	25	30	31	34	37	39
78	26	30	32	34	37	40
80	26	31	32	35	38	40
82	27	31	33	36	39	41
84	27	32	33	36	39	42
86	28	33	34	37	40	42
88	28	33	34	37	40	43
90	29	34	35	38	41	44

12 Factors

	Probability (p) levels					
n	.50	.10	.05	.01	.001	.0001
24	15	17	18	19	21	22
26	16	18	19	20	21	23
28	16	19	19	21	22	24
30	17	19	20	22	23	25
32	17	20	21	22	24	25
34	18	21	21	23	25	26
36	18	21	22	24	26	27
38	19	22	23	24	26	28
39	19	22	23	25	27	28
40	19	22	23	25	27	29
41	20	23	24	25	28	29
42	20	23	24	26	28	30
43	20	23	24	26	28	30
44	20	24	25	27	29	30
45	21	24	25	27	29	31
46	21	24	25	27	29	31
47	21	25	26	28	30	32
48	21	25	26	28	30	32
49	22	25	26	28	31	32
50	22	25	27	29	31	33
51	22	26	27	29	31	33
52	22	26	27	29	32	34
53	23	26	27	30	32	34
54	23	27	28	30	32	34
55	23	27	28	30	33	35
56	23	27	28	31	33	35
57	24	28	29	31	33	35
58	24	28	29	31	34	36
59	24	28	29	32	34	36
60	24	28	30	32	34	37
61	25	29	30	32	35	37
62	25	29	30	33	35	37
63	25	29	31	33	36	38
64	25	30	31	33	36	38
65	26	30	31	34	36	38
66	26	30	31	34	37	39
67	26	30	32	34	37	39
68	26	31	32	34	37	40
69	27	31	32	35	38	40
70	27	31	33	35	38	40
71	27	32	33	35	38	41
72	27	32	33	36	39	41
73	28	32	34	36	39	41
74	28	33	34	36	39	42
75	28	33	34	37	40	42

12 Factors (*Continued*)

	Probability (*p*) levels					
n	.50	.10	.05	.01	.001	.0001
76	28	33	34	37	40	42
77	29	33	35	37	40	43
78	29	34	35	38	41	43
79	29	34	35	38	41	44
80	30	34	36	38	41	44
82	30	35	36	39	42	45
84	31	35	37	40	43	45
86	31	36	37	40	43	46
88	32	37	38	41	44	47
90	32	37	39	41	45	47
92	33	38	39	42	45	48
94	33	38	40	43	46	49
96	34	39	40	43	47	49
98	34	39	41	44	47	50
100	35	40	42	45	48	51

16 Factors

	Probability (*p*) levels					
n	.50	.10	.05	.01	.001	.0001
32	21	24	24	26	27	28
34	22	24	25	26	28	29
36	22	25	26	27	29	30
38	23	26	27	28	30	31
40	23	26	27	29	31	32
42	24	27	28	30	32	33
44	25	28	29	31	33	34
46	25	29	29	31	33	35
48	26	29	30	32	34	36
50	26	30	31	33	35	37
51	27	30	31	33	36	37
52	27	31	32	34	36	38
53	27	31	32	34	36	38
54	28	31	32	34	37	39
55	28	32	33	35	37	39
56	28	32	33	35	38	40
57	28	32	33	36	38	40
58	29	33	34	36	38	41
59	29	33	34	36	39	41
60	29	33	35	37	39	41
61	30	34	35	37	40	42
62	30	34	35	38	40	42
63	30	34	36	38	40	43
64	31	35	36	38	41	43

16 Factors (*Continued*)

			Probability (*p*) levels			
n	.50	.10	.05	.01	.001	.0001
65	31	35	36	39	41	43
66	31	35	37	39	42	44
67	31	36	37	39	42	44
68	32	36	37	40	42	45
69	32	36	38	40	43	45
70	32	37	38	40	43	46
71	33	37	38	41	44	46
72	33	37	39	41	44	46
73	33	38	39	42	44	47
74	33	38	39	42	45	47
75	34	38	40	42	45	48
76	34	39	40	43	46	48
77	34	39	40	43	46	48
78	35	39	41	43	46	49
79	35	40	41	44	47	49
80	35	40	41	44	47	50
81	36	40	42	45	48	50
82	36	41	42	45	48	50
83	36	41	43	45	48	51
84	36	41	43	46	49	51
85	37	42	43	46	49	52
86	37	42	44	46	49	52
87	37	42	44	47	50	52
88	38	43	44	47	50	53
89	38	43	45	47	51	53
90	38	43	45	48	51	54
92	39	44	46	48	52	54
94	39	45	46	49	53	55
96	40	45	47	50	53	56
98	41	46	48	51	54	57
100	41	47	48	51	55	58
102	42	47	49	52	56	58
104	42	48	50	53	56	59
106	43	49	50	53	57	60
108	44	49	51	54	58	61
110	44	50	52	55	58	61

20 Factors

			Probability (*p*) levels			
n	.50	.10	.05	.01	.001	.0001
40	27	30	31	32	34	35
42	28	31	31	33	35	36
44	28	31	32	34	36	37

20 Factors (*Continued*)

n	Probability (p) levels					
	.50	.10	.05	.01	.001	.0001
46	29	32	33	35	37	38
48	30	33	34	36	38	39
50	30	34	35	37	39	40
52	31	35	36	37	40	41
53	31	35	36	38	40	42
54	32	35	36	38	41	42
55	32	36	37	39	41	43
56	32	36	37	39	42	43
57	33	36	38	40	42	44
58	33	37	38	40	42	44
59	33	37	38	40	43	45
60	34	38	39	41	43	45
61	34	38	39	41	44	46
62	34	38	40	42	44	46
63	35	39	40	42	45	47
64	35	39	40	43	45	47
65	35	39	41	43	46	48
66	36	40	41	43	46	48
67	36	40	41	44	46	49
68	36	41	42	44	47	49
69	37	41	42	45	47	50
70	37	41	43	45	48	50
71	37	42	43	45	48	50
72	38	42	43	46	49	51
73	38	42	44	46	49	51
74	38	43	44	47	49	52
75	39	43	45	47	50	52
76	39	44	45	47	50	53
77	39	44	45	48	51	53
78	40	44	46	48	51	54
79	40	45	46	49	52	54
80	40	45	46	49	52	55
81	41	45	47	50	53	55
82	41	46	47	50	53	55
83	41	46	48	50	53	56
84	42	47	48	51	54	56
85	42	47	48	51	54	57
86	42	47	49	52	55	57
87	43	48	49	52	55	58
88	43	48	50	52	55	58
89	43	48	50	53	56	59
90	44	49	50	53	56	59
92	44	50	51	54	57	60
94	45	50	52	55	58	61
96	46	51	53	56	59	62

20 Factors (*Continued*)

	Probability (*p*) levels					
n	.50	.10	.05	.01	.001	.0001
98	46	52	53	56	60	62
100	47	53	54	57	60	63
102	48	53	55	58	61	64
104	48	54	56	59	62	65
106	49	55	56	59	63	66
108	50	55	57	60	64	67
110	50	56	58	61	65	68
112	51	57	59	62	65	68
114	52	58	59	63	66	69
116	52	58	60	63	67	70
118	53	59	61	64	68	71
120	54	60	62	65	69	72

24 Factors

	Probability (*p*) levels					
n	.50	.10	.05	.01	.001	.0001
48	33	36	37	39	41	42
50	34	37	38	40	42	43
52	35	38	39	41	43	44
54	35	39	40	42	44	46
56	36	40	41	43	45	47
58	37	40	42	44	46	48
60	37	41	42	44	47	49
62	38	42	43	45	48	50
63	39	43	44	46	48	50
64	39	43	44	46	49	51
65	39	43	45	47	49	51
66	40	44	45	47	50	52
67	40	44	45	48	50	52
68	40	45	46	48	51	53
69	41	45	46	49	51	53
70	41	45	47	49	52	54
71	42	46	47	49	52	54
72	42	46	48	50	53	55
73	42	47	48	50	53	55
74	43	47	48	51	54	56
75	43	47	49	51	54	56
76	43	48	49	52	54	57
77	44	48	50	52	55	57
78	44	49	50	53	55	58
79	44	49	50	53	56	58
80	45	50	51	53	56	59

24 Factors (*Continued*)

n	Probability (p) levels					
	.50	.10	.05	.01	.001	.0001
81	45	50	51	54	57	59
82	46	50	52	54	57	60
83	46	51	52	55	58	60
84	46	51	53	55	58	61
85	47	52	53	56	59	61
86	47	52	53	56	59	62
87	47	52	54	57	60	62
88	48	53	54	57	60	63
89	48	53	55	57	60	63
90	48	54	55	58	61	63
91	49	54	55	58	61	64
92	49	54	56	59	62	64
93	50	55	56	59	62	65
94	50	55	57	60	63	65
95	50	56	57	60	63	66
96	51	56	58	60	64	66
97	51	56	58	61	64	67
98	51	57	58	61	65	67
99	52	57	59	62	65	68
100	52	58	59	62	65	68
102	53	58	60	63	66	69
104	54	59	61	64	67	70
106	54	60	62	65	68	71
108	55	61	62	66	69	72
110	56	62	63	66	70	73
112	56	62	64	67	71	74
114	57	63	65	68	72	75
116	58	64	66	69	73	76
118	59	65	67	70	74	77
120	59	66	67	71	74	77
122	60	66	68	72	75	78
124	61	67	69	72	76	79
126	62	68	70	73	77	80
128	62	69	71	74	78	81
130	63	70	71	75	79	82

28 Factors

n	Probability (p) levels					
	.50	.10	.05	.01	.001	.0001
56	39	43	44	46	48	49
58	40	44	45	47	49	50
60	41	45	46	48	50	52

28 Factors (*Continued*)

n	Probability (*p*) levels					
	.50	.10	.05	.01	.001	.0001
62	42	46	47	49	51	53
64	43	46	48	50	52	54
66	43	47	48	51	53	55
68	44	48	49	52	54	56
70	45	49	50	52	55	57
72	46	50	51	53	56	58
73	46	50	52	54	57	59
74	46	51	52	54	57	59
75	47	51	53	55	58	60
76	47	52	53	55	58	60
77	48	52	53	56	59	61
78	48	53	54	56	59	61
79	48	53	54	57	60	62
80	49	53	55	57	60	62
81	49	54	55	58	61	63
82	50	54	56	58	61	63
83	50	55	56	59	62	64
84	50	55	57	59	62	64
85	51	56	57	60	63	65
86	51	56	57	60	63	65
87	52	57	58	61	64	66
88	52	57	58	61	64	66
89	52	57	59	61	64	67
90	53	58	59	62	65	67
91	53	58	60	62	65	68
92	54	59	60	63	66	68
93	54	59	61	63	66	69
94	54	60	61	64	67	70
95	55	60	61	64	67	70
96	55	60	62	65	68	71
97	56	61	62	65	68	71
98	56	61	63	66	69	72
99	56	62	63	66	69	72
100	57	62	64	67	70	73
101	57	63	64	67	70	73
102	58	63	65	68	71	74
103	58	63	65	68	71	74
104	58	64	65	68	72	75
105	59	64	66	69	72	75
106	59	65	66	69	73	76
108	60	66	67	70	74	77
110	61	66	68	71	75	78
112	61	67	69	72	76	79
114	62	68	70	73	77	79
116	63	69	71	74	78	80

28 Factors (*Continued*)

n	Probability (p) levels					
	.50	.10	.05	.01	.001	.0001
118	64	70	72	75	78	81
120	65	71	72	76	79	82
122	65	72	73	77	80	83
124	66	72	74	78	81	84
126	67	73	75	78	82	85
128	68	74	76	79	83	86
130	69	75	77	80	84	87
132	69	76	78	81	85	88
134	70	77	79	82	86	89
136	71	78	79	83	87	90
138	72	78	80	84	88	91
140	73	79	81	85	89	92

A.7. Tables for Significance of Congruence Coefficients in Factor Matching

The values are derived from many actual factor analyses by Monte Carlo methods, as described in Schneewind and Cattell (1970), where the full distributions are given. The values in the regular r_c column were calculated in the ordinary way, and those in the r_{cs} (s = symmetrized) by entering both the loadings and the reflected sign loadings, i.e., they are literally on twice the number of variables, whereupon they converge on one value. Probably most users will employ simply r_c. The values are very significantly different from correlation coefficients between factors.

X = number of variables common to the two factor analyses, and used for comparison; p is the usual significance value for resemblance of the two factors. The r_c and r_{cs} values are the required lower bound for that value. Korth and Tucker (1975, p. 361) present differently derived significances which run lower (p. 253).

X	p	r_c	r_{cs}
10	.001	−.87 through +.91	±.90
	.01	−.75 through +.78	±.77
	.025	−.67 through +.70	±.69
	.05	−.59 through +.63	±.62
	.10	−.50 through +.53	±.52
20	.001	−.76 through +.82	±.78
	.01	−.63 through +.68	±.64
	.025	−.53 through +.57	±.54

Table A.7. (*Continued*)

X	p	r_c	r_{cs}
	.05	− .47 through +.50	±.48
	.10	− .38 through +.41	±.39
30	.001	− .68 through +.73	±.70
	.01	− .55 through +.58	±.57
	.025	− .48 through +.51	±.49
	.05	− .38 through +.43	±.41
	.10	− .31 through +.36	±.33
40	.001	− .51 through +.65	±.60
	.01	− .38 through +.46	±.42
	.025	− .33 through +.39	±.35
	.05	− .27 through +.32	±.28
	.10	− .20 through +.24	±.22
50	.001	− .42 through +.54	±.52
	.01	− .34 through +.39	±.38
	.025	− .28 through +.32	±.29
	.05	− .23 through +.27	±.24
	.10	− .17 through +.19	±.18

A.8. Tables for Significance of Salient Variable Similarity Index *s*

The following significance values are appropriate for the salient variable similarity index as calculated by the formula on p. 255 above and the methods in Cattell, Balcar, Horn, and Nesselroade (1969) where the derivation of these distributions is described. Two parameters enter here: (1) The number of variables *n* that are the same in the two studies (all others are eliminated from any calculation) and (2) the percentage of variables lying in the hyperplane of the given factors being compared. The cut between hyperplane and salient variables should be the best statistical estimate for the boundary, often ±.10, but should be adjusted slightly if necessary from one study to the other to yield the same percentage in the hyperplane, and therefore of salients, in both. The distribution is clearer also if variables are reflected in sign (in both studies of course) to give about as many negative as positive salients. To yield the given *p* value the *s* value should be equal to or greater than that shown. As to percentages on the hyperplane the experimenter may need to interpolate between the values here.

60% in Hyperplane

n						
10	s:	.76	.51	.26	.01	.00
	p:	.001	.020	.138	.364	.500
20	s:	.63	.51	.26	.13	.01	.00
	p:	.000	.004	.086	.207	.393	.500

60% in Hyperplane (*Continued*)

n													
30	s:	.59	.51	.42	.34	.26	.17	.09	.01	.00			
	p:	.000	.001	.005	.019	.054	.131	.256	.411	.500			
40	s:	.51	.44	.38	.32	.26	.19	.13	.07	.01	.00		
	p:	.000	.001	.005	.016	.041	.090	.169	.282	.426	.500		
50	s:	.46	.41	.36	.31	.26	.21	.16	.11	.06	.01	.00	
	p:	.000	.001	.004	.011	.026	.061	.115	.200	.301	.428	.500	
60	s:	.42	.38	.34	.30	.26	.21	.17	.13	.09	.05	.01	.00
	p:	.000	.001	.003	.008	.018	.039	.078	.132	.215	.318	.440	.500
80	s:	.35	.32	.29	.26	.22	.19	.16	.13	.10	.07	.04	.01
	p:	.000	1.001	.003	.008	.018	.034	.063	.107	.170	.243	.338	.447
	s:	.00											
	p:	.500											
100	s:	.36	.31	.28	.26	.23	.21	.18	.16	.13	.11	.08	.06
	p:	.000	.001	.003	.006	.011	.019	.034	.058	.091	.136	.196	.271
	s:	.03	.01	.00									
	p:	.353	.449	.500									

70% in Hyperplane

n													
10	s:	.67	.34	.01	.00								
	p:	.002	.052	.316	.500								
20	s:	.67	.51	.34	.17	.01	.00						
	p:	.000	.002	.027	.135	.357	.500						
30	s:	.56	.45	.34	.23	.12	.01	.00					
	p:	.000	.002	.016	.064	.190	.383	.500					
40	s:	.51	.42	.34	.26	.17	.09	.01	.00				
	p:	.000	.002	.007	.034	.098	.222	.403	.500				
50	s:	.47	.41	.34	.27	.21	.14	.07	.01	.00			
	p:	.000	.001	.004	.018	.052	.123	.247	.407	.500			
60	s:	.39	.34	.28	.23	.17	.12	.06	.01	.00			
	p:	.000	.002	.007	.025	.066	.142	.262	.415	.500			
80	s:	.38	.34	.30	.26	.21	.17	.13	.09	.05	.01	.00	
	p:	.000	.001	.002	.007	.019	.046	.092	.174	.283	.425	.500	
100	s:	.34	.31	.27	.24	.21	.17	.14	.11	.07	.04	.01	.00
	p:	.000	.001	.002	.006	.016	.037	.069	.125	.207	.318	.438	.500

80% in Hyperplane

n						
10	s:	.51	.01	.00		
	p:	.012	.187	.500		
20	s:	.76	.51	.26	.01	.00
	p:	.000	.003	.041	.279	.500
30	s:	.67	.51	.34	.17	.01	.00
	p:	.000	.001	.011	.083	.316	.500

80% in Hyperplane (*Continued*)

n									
40	s:	.51	.38	.26	.13	.01	.00		
	p:	.000	.003	.024	.115	.347	.500		
50	s:	.51	.41	.31	.21	.11	.01	.00	
	p:	.000	.001	.006	.036	.142	.361	.500	
60	s:	.42	.34	.26	.17	.09	.01	.00	
	p:	.000	.002	.012	.052	.164	.369	.500	
80	s:	.38	.32	.26	.19	.13	.07	.01	.00
	p:	.000	.001	.004	.022	.076	.199	.391	.500
100	s:	.31	.26	.21	.16	.11	.06	.01	.00
	p:	.000	.002	.010	.036	.105	.220	.402	.500

90% in Hyperplane

n						
10	s:	.01	.00			
	p:	.052	.500			
20	s:	.51	.01	.00		
	p:	.003	.099	.500		
30	s:	.67	.34	.01	.00	
	p:	.000	.007	.133	.500	
40	s:	.51	.26	.01	.00	
	p:	.000	.012	.167	.500	
50	s:	.41	.21	.01	.00	
	p:	.000	.018	.198	.500	
60	s:	.51	.34	.17	.01	.00
	p:	.000	.002	.029	.217	.500
80	s:	.38	.26	.13	.01	.00
	p:	.000	.004	.045	.251	.500
100	s:	.31	.21	.11	.01	.00
	p:	.000	.007	.061	.286	.500

Professor J. Brennan of the Psychology Department, University of Hawaii, Honolulu, HI 96822, has kindly agreed to supply computer programs for the salient variable similarity index s and the congruence coefficient r_c at cost of mailing cards.

A.9. Recovery of V_0 and R_v from a Given V_{fp} and L Matrix

If space permitted, this Appendix might helpfully be extended to a number of practical needs arising in research through certain exigencies. Most of these would be accidental losses of particular matrices or data sheets! Others would be attempting to check published researches where insufficient data are presented. In this connection, the practice is to be deplored of publishing V_{fp}'s with only salient loadings, which prevents evaluation of goodness of simple structure or rerotation to a possibly better solution. If economies of presentation are genuinely necessary, the sending of basic R_v, V_0- and L_r matrices for microfilming at ASIS is an excellent idea, since everything—up through higher orders—except factor scores, can be obtained from this minimal set (or, even from R_v, if no claims are made for a particular L_r).

The following brief reminder of what can be done if only a V_{rs} and a transformation matrix L_r (or a factor correlation matrix R_f) exist and one wishes to return to V_0 and R_v (for new extractions or rotations) may be helpful. One proceeds:

$$V_0 = V_{rs} L_r^{-1} \qquad (A.15)$$

From this the R (reduced) is obtainable as

$$R = V_0 V_0^T \qquad (A.16)$$

Often published articles contain only the V_{fp} and the R_f. Then the experimenter who wants either to recover his own V_0 and R, or the un-published ones of some writer, e.g., to try alternative rotations, must either obtain R_v by

$$R_v = V_{fp} R_f V_{fp}^T \qquad (A.17)$$

or find R_{rv} by

$$R_{rv} = D_f R_f^{-1} D_f \qquad (A.18)$$

D_f being as usual the diagonal of roots of the inverses of the diagonals in R_f^{-1}. Taking the principal components of R_{rv}, one obtains what is L_T. From this one gets V_0 by

$$V_0 = V_{rs} L_r^{-1} \qquad (A.19)$$

This V_0 is not necessarily that from which the investigator began his rotations, since there are many possible equivalents of L_r from (A.18) but it will always be an orthogonally rotated adequate equivalent of it, and the R_v will, of course, be identical.

References

Ahmavaara, Y. The mathematical theory of factorial invariance under selection. *Psychometrika*, 1954, *19*, 27–38.

Ahmavaara, Y., & Markkanen, T. *The unified factor model.* Helsinki: Finnish Foundation for Alcohol Studies, 1958.

Aiken, L. S. Simultaneous processing of typal and dimensional variation among multidimensional events. *Multivariate Behavioral Research*, 1972, *3*, 305–316.

Aitkin, A. C. The evaluation of the latent roots and vectors of a matrix. *Proceedings of the Royal Society of Edinburgh*, 1937, *57*, 269–304.

Albert, A. A. The minimum rank of a correlation matrix. *Proceedings, National Academy of Sciences*, 1944, *30*, 144–146.

Allport, G. W., & Odbert, H. S. Trait-names: A psycholexical study. *Psychological Monographs*, 1936, *47*, 171.

Anderberg, A. *Cluster analysis with applications.* New York: Academic Press, 1973.

Anderson, H. E. Regression, discriminant analysis, and a standard notation for basic statistics. In R. B. Cattell (Ed.), *Handbook of multivariate experimental psychology.* Chicago: Rand McNally, 1966. Chap. 5.

Anderson, T. W. *An introduction to multivariate statistical analysis.* New York: Wiley, 1958.

Anderson, T. W., & Rubin, H. Statistical inference in factor analysis. *Proceedings of the 3rd Berkeley Symposium on Mathematical Statistics and Probability*, 1956, *5*, 111.

Archer, C. O., & Jennrich, R. I. Standard errors for rotated factor loadings. *Psychometrika*, 1973, *38*, 581–592.

Baggaley, A. R., & Cattell, R. B. A comparison of exact and approximate linear function estimates of oblique factor scores. *British Journal of Statistical Psychology*, 1956, *9*, 83–86.

Baltes, P. B. Longitudinal and cross sectional sequences in the study of age and generation effects. *Human Development*, 1968, *11*, 145–177.

Baltes, P. B., & Nesselroade, J. R. The developmental analysis of individual differences on multiple measures. In J. R. Nesselroade & H. W. Reese (Eds.), *Life span developmental psychology.* New York: Academic Press, 1973. Chapter 11.

Banks, C., & Broadhurst, P. B. (Eds.). *Studies in psychology, in honour of Sir Cyril Burt.* London: University of London Press, 1965.

Barcikowski, R. S., & Stevens, J. P. A Monte Carlo study of the stability of canonical correlations, canonical weights and canonical variate-variable correlations. *Multivariate Behavioral Research*, 1975, *10*, 353–364.

Bargmann, R. *Signifikantz-untersuchungen der einfachen Struktur in der Factoren analyse. Mitteilungsblatt fur Mathematische Statistik.* Wurzburg: Physica Verlag, 1954.

Bargmann, R., & Bock, R. D. Analysis of covariance structures. *Psychometrika*, 1966, *31*, 507–534.

Barker, R. G. The streams of behavior. New York: Appleton-Century-Crofts, 1963.

Barker, R. G. Explorations in ecological psychology. *American Psychologist*, 1965, *20*, 6–18.

Barlow, J. A., & Burt, C. L. The identification of factors from different experiments. *British Journal of Statistical Psychology, Statistical Section*, 1954, 7, 52–56.

Bartlett, H. W., & Cattell, R. B. An *R-dR*-technique operational distinction of the states of anxiety, stress, fear, etc. *Australian Journal of Psychology*, 1971, *2*, 279–287.

Bartlett, M. S. The statistical significance of canonical correlations. *Biometrika*, 1941, *32*, 29–38.

Bartlett, M. S. Tests of significance in factor analysis. *British Journal of Psychology*, 1950, *3*, 77.

Bartlett, M. S. A further note on tests of significance in factor analysis. *British Journal of Psychology*, 1951, *4*, 1.

Barton, K., & Cattell, R. B. An investigation of the common factor space of some well known questionnaire scales: The Eysenck EPI, the Comrey scales and the IPAT central trait-state kit. *Journal of Multivariate, Experimental and Clinical Psychology*, 1975, *1*, 268–277.

Bartsck, T. W., & Nesselroade, J. R. Test of the trait-state distinction using a manipulative factor analytic design. *Journal of Personality and Social Psychology*, 1973, *27*, 58–64.

Bentler, P. M. Alpha-maximized factor analysis (alphamax): Its relation to alpha and canonical factor analysis. *Psychometrika*, 1968, *3*, 335.

Bentler, P. M. A regression model for factor analysis. *Proceedings of the 78th Annual Convention of the APA*, Washington, D.C., Amer. Psychol. Assoc.,1970, 109–110.

Bentler, P. M. Monotonicity, an alternative to linear factor and test analysis. In D. R. Green *et al.* (Eds.), *Measurement and Piaget.* New York: McGraw-Hill, 1971.

Bentler, P. M. Assessment of developmental factor change at the individual and group level. In J. R. Nesselroade & H. W. Reese (Eds.), *Life span developmental psychology.* New York: Academic Press, 1973. Chapter 7.

Bentler, P. M. Multistructure statistical model applied to factor analysis. *Multivariate Behavioral Research*, 1976, *11*, 1-18.

Bentler, P. M., & Lee, S. Y. *Newton–Raphson approach to exploratory and confirmatory maximum likelihood factor analysis.* Manuscript submitted for publication, 1976.

Bereiter, C. Some persisting dilemmas in measurement of change. In C. W. Harris (Ed.), *Problems in measuring change.* Madison: University of Wisconsin Press, 1963.

Binder, A. Considerations of the place of assumptions in correlational analysis. *American Psychology*, 1959, *14*, 504–510.

Birkett, H., & Cattell, R. B. Diagnosis of the dynamic roots of a clinical symptom by P technique. *Multivariate Experimental Clinical Research*, 1978, *3*, 173–194.

Blalock, H. M. Path coefficients versus regression coefficients. *American Journal of Sociology*, 1967, *72*, 675–676.

Blalock, H. M. *Causal models in the social sciences.* Chicago: Aldine-Atherton, 1971.

Blalock, H. M., & Blalock, A. B. *Methodology in social research.* New York: McGraw-Hill, 1968.

Bock, R. D. Contributions of multivariate experimental designs to educational research. In R. B. Cattell (Ed.), *Handbook of multivariate experimental psychology.* Chicago: Rand McNally, 1966. Chapter 28.

Bock, R. D., & Bargmann, R. E. Analysis of covariance structures. *Psychometrika,* 1966, *31,* 507–534.

Bolz, C. R. Types of personality. In R. M. Dreger (Ed.), *Multivariate personality research.* Baton Rouge: Claitor, 1972. Chapter 5.

Bolz, C. R., Cattell, R. B., & Barton, K. A P-technique study of motivations in a married couple (in preparation).

Borgatta, E. F. Difficulty factors and the use of r_{phi}. *Journal of General Psychology,* 1965, *73,* 321.

Borgatta, E. F. *Sociological methodology.* San Francisco: Jossey-Bass, 1969.

Borgatta, E. F., & Bohrnestedt, G. W. (Eds.). *Sociological methods.* San Francisco: Jossey Bass, 1970.

Borgatta, E. F., & Cottrell, L. S., Jr. On the classification of groups. *Sociometry,* 1955, *18,* 665–678.

Boruch, R. F., & Dutton, J. E. A program for testing hypotheses about correlation arrays. *Educational and Psychological Measurement,* 1970, *30,* 719.

Boruskowski, A., & Stevens, S. B. On stability of canonical solutions. *Psychometrika,* 1976, *41,* 10–14.

Brennan, J. A test of confactor rotational resolutions on orthogonal and oblique cases. Ph.D. Thesis, University of Hawaii, Honolulu, Hawaii, 1977.

Brogden, H. E. Pattern, structure, and the interpretation of factors. *Psychological Bulletin,* 1969, *72,* 375.

Broverman, D. M. Effects of score transformations in Q and R factor analysis techniques. *Psychological Review,* 1961, *68,* 68–80.

Broverman, D. M. Normative and ipsative measurement in psychology. *Psychological Review,* 1962, *69,* 295–305.

Browne, M. W. On oblique procrustes rotation. *Psychometrika,* 1967, *32,* 125–132.

Browne, M. W. Fitting the factor analysis model. *Psychometrika,* 1969, *34,* 375–394.

Browne, M. W., & Kristof, W. On the oblique rotation of a factor matrix to a specified pattern. *Psychometrika,* 1969, *34,* 237.

Brunswik, E. *Perception and the representative design of psychological experiments.* Berkeley: University of California Press, 1956.

Burdsal, C., & Macdonald, C. A check on the factor structure of form E of the 16P.F. *Multivariate Experimental Clinical Research.* In press, 1977.

Burdsal, C., & Vaughan, D. M. A comparison of the personality structure of college students found in the questionnaire medium by items compared to parcels. *Journal of Genetic Psychology,* 1974, *125,* 219–224.

Buros, O. K. *The seventh mental measurements yearbook.* New Jersey: Gryphon Press, 1972.

Burt, C. L. *Distribution and relations of educational abilities.* London: P. S. King, 1917.

Burt, C. L. Correlations between persons. *British Journal of Psychology,* 1937, *28,* 56–96.

Burt, C. L. *The factors of the mind: An introduction to factor-analysis in psychology.* New York: Macmillan, 1941.

Burt, C. L. The factorial analysis of qualitative data. *British Journal of Psychology, Statistical Section,* 1950, *3,* 166–185.

Burt, C. L. Tests of significance in factor analysis. *British Journal of Psychology, Statistical Section*, 1952, *5*, 109–133.

Burt, C. L. Scale analysis and factor analysis. *British Journal of Statistical Psychology*, 1953, *6*, 5-23.

Burt, C. L. The early history of multivariate techniques in psychological research. *Multivariate Behavioral Research*, 1966, *1*, 24–42. (a)

Burt, C. L. The appropriate use of factor analysis and analysis of variance. In R. B. Cattell (Ed.), *Handbook of multivariate experimental psychology*. Chicago: Rand-McNally, 1966. Pp. 267–287. (b)

Burt, C. L., & Banks, C. A factor analysis of body measurements for British adult males. *Ann. Eugen.*, 1947, *13*, 238–256.

Burt, C. L., & Howard, M. The multifactorial theory of inheritance. *British Journal of Statistical Psychology*, 1956, *9*, 115–125.

Burt, C. L., & Stephenson, W. Alternative views on correlations between persons. *Psychometrika*, 1939, *4*, 269–281.

Buss, A. An inferential strategy for determining factor invariance across different individuals and different variables. *Multivariate Behavioral Research*, 1975, *10*, 365–372.

Buss, A. R., & Royce, J. R. Detecting cross cultural communalities and differences: Intergroup factor analysis. *Psychological Bulletin*, 1975, *82*, 128–136.

Campbell, D. F., & Fiske, D. W. Convergent and discriminant validation by the multitrait–multimethod matrix. *Psychological Bulletin*, 1959, *56*, 81–105.

Carroll, J. B. The effect of difficulty and chance success on correlations between items or between tests. *Psychometrika*, 1945, *10*, 1–19.

Carroll, J. B. Biquartimin criterion for rotation to oblique simple structure in factor analysis. *Science*, 1957, *126*, 1114.

Carroll, J. B. Oblimin rotation solution in factor analysis. Mimeographed Computing Program for the IBM-704, 1958.

Carroll, J. B. The nature of the data, or how to choose a correlation coefficient. *Psychometrika*, 1961, *26*, 347–372.

Carroll, J. D. Individual differences and multidimensional scaling. In R. N. Shephard et al. (Eds.), *Multidimensional scaling: Theory and applications in the behavioral sciences*. New York: Academic Press, 1972.

Carroll, J. D., & Wish, M. Models and methods in three-way dimensional scaling. In D. H. Krantz et al. (Eds.), *Contemporary developments in mathematical psychology* (Vol. II). New York: Freeman & Co., 1974.

Carroll, R. M., & Field, J. A comparison of the classificatory accuracy of profile similarity measures. *Multivariate Behavioral Research*, 1974, *9*, 373–380.

Cartwright, D. S. A misapplication of factor analysis. *American Sociological Review*, 1965, *30*, 249–251.

Cattell, R. B. The measurement of interest. *Character and Personality*, 1935. *4*, 147–169.

Cattell, R. B. The concept of social status. *Journal of Social Psychology*, 1942, *15*, 293–308.

Cattell, R. B. Parallel proportional profiles and other principles for determining the choice of factors by rotation. *Psychometrika*, 1944, *9*, 267–283. (a)

Cattell, R. B. Interpretation of the twelve primary personality factors. *Character and Personality*, 1944, *13*, 55–91. (b)

Cattell, R. B. A note on correlation clusters and cluster search methods. *Psychometrika*, 1944, *9*, 169–184. (c)

Cattell, R. B. Psychological measurement: Normative, ipsative, interactive. *Psychological Review*, 1944, *51*, 292–303. (d)

Cattell, R. B. "Parallel proportional profiles" and other principles for determining the choice of factors by rotation. *Psychometrika*, 1944, *9*, 267–283. (e)

Cattell, R. B. The description of personality: Principles and findings in a factor analysis. *American Journal of Psychology*, 1945, *58*, 69–90. (a)

Cattell, R. B. The diagnosis and classification of neurotic states: A re-interpretation of Eysenck's factors. *Journal of Nervous and Mental Disease*, 1945, *14*, 85–92. (b)

Cattell, R. B. *The description and measurement of personality.* New York: Harcourt, Brace & World, 1946. (a)

Cattell, R. B. Simple structure in relation to some alternative factorizations of the personality sphere. *Journal of General Psychology*, 1946, *35*, 225–238. (b)

Cattell, R. B. Oblique, second order, and cooperative factors in personality analysis. *Journal of General Psychology*, 1947, *36*, 3–22.

Cattell, R. B. The integration of factor analysis with psychology. *Journal of Educational Psychology*, 1948, *39*, 227–236.

Cattell, R. B. The dimensions of culture patterns by factorization of national characters. *Journal of Abnormal and Social Psychology*, 1949, *44*, 443–469. (a)

Cattell, R. B. r_p and other coefficients of pattern similarity. *Psychometrika*, 1949, *14*, 279–298. (b)

Cattell, R. B. A note on factor invariance and the identification of factors. *British Journal of Psychology*, 1949, *II*, 134–138. (c)

Cattell, R. B. *Personality: A theoretical and factual study.* New York: McGraw-Hill, 1950. (a)

Cattell, R. B. The principal culture patterns discoverable in the syntal dimensions of existing nations. *Journal of Social Psychology*, 1950, *32*, 215–253. (b)

Cattell, R. B. *P*-technique factorization and the determination of individual dynamic structure. *Journal of Clinical Psychology*, 1952, *8*, 5–10. (a)

Cattell, R. B. *Factor analysis.* New York: Harper, 1952. (b)

Cattell, R. B. The three basic factor-analytic research designs—their interrelations and derivatives. *Psychological Bulletin*, 1952, *49*, 499–520. (c)

Cattell, R. B. Personality, role, mood, and situation-perception: A unifying theory of modulators. *Psychological Review*, 1953, *70*, 1–18.

Cattell, R. B. Growing points in factor analysis. *Australian Journal of Psychology*, 1954, *6*, 105–140. (a)

Cattell, R. B. Personality structures as learning and motivation patterns: A theme for the integration of methodologies. In *Learning theory, personality theory, and aclinical research.* New York: Wiley, 1954. (b)

Cattell, R. B. Psychiatric screening of flying personnel: Personality structure in objective tests. *USAF School of Aviation Medicine*, Rept. No. 9, *Project* 21-0202-0007, 1955, pp. 1-50.

Cattell, R. B. Validation and intensification of the sixteen personality factor questionnaire. *Journal of Clinical Psychology*, 1956, *12*, 205–214.

Cattell, R. B. A universal index for psychological factors. *Psychologia*, 1957, *1*, 74–85. (a)

Cattell, R. B. Formulae and table for obtaining validities and reliabilities of extended factor scales. *Educational and Psychological Measurement*, 1957, *17*, 491–498. (b)

Cattell, R. B. *Personality and motivation structure and measurement.* New York: World Book, 1957. (c)

Cattell, R. B. Extracting the correct number of factors in factor analysis. *Educational and Psychological Measurement*, 1958, *18*, 791–838.

Cattell, R. B. Evaluating interaction and non-linear relations by factor analysis. *Psychological Reports*, 1960, *7*, 69–70.

Cattell, R. B. Group theory, personality and role: A model for experimental researches. In J. Geldard (Ed.), *Defence psychology*. Oxford: Pergamon Press, 1961. Pp. 209–258. (a)

Cattell, R. B. Theory of situational, instrument, second order and refraction factors in personality structure research. *Psychological Bulletin*, 1961, *58*, 160–174. (b)

Cattell, R. B. The relational simplex theory of equal interval and absolute scaling. *Acta Psychologica*, 1962, *20*, 139–158. (a)

Cattell, R. B. The basis of recognition and interpretation of factors. *Educational and Psychological Measurement*, 1962, *22*, 667–697. (b)

Cattell, R. B. Formulating the environmental situation and its perception, in behavior theory. In S. B. Sells (Ed.), *Stimulus determinants of behavior*. New York: Ronald Press, 1963. Pp. 46–75. (a)

Cattell, R. B. The structuring of change by *P*-technique and incremental *R*-technique. In C. W. Harris (Ed.), *Problems in measuring change*. Madison, Wisconsin: University of Wisconsin Press, 1963. (b)

Cattell, R. B. Personality, role, mood and situation perception: A unifying theory of modulators. *Psychological Review*, 1963, *70*, 1–18. (c)

Cattell, R. B. The nature and measurement of anxiety. *Scientific American*, 1963, *208*, 96–104. (d)

Cattell, R. B. Validity and reliability: A proposed more basic set of concepts. *Journal of Educational Psychology*, 1964, *55*, 1-22. (a)

Cattell, R. B. Beyond validity and reliability: Some further concepts and coefficients for evaluating tests. *Journal of Experimental Education*, 1964, *33*, 133–143. (b)

Cattell, R. B. Factor analysis: An introduction to essentials. I. The purpose and underlying models. *Biometrics*, 1965, *21*, 190–215. (a)

Cattell, R. B. Factor analysis: An introduction to essentials. II. The role of factor analysis in research. *Biometrics*, 1965, *21*, 405–435. (b)

Cattell, R. B. The configurative method for surer identification of personality dimensions, notably in child study. *Psychological Reports*, 1965, *16*, 269–270. (c)

Cattell, R. B. Higher order factor structures and reticular vs. hierarchical formula for their interpretation. In C. Banks & P. L. Broadhurst (Eds.), *Studies in psychology*. (Presented to Cyril Burt) London: University of London Press, 1965. (d)

Cattell, R. B. *Handbook of multivariate experimental psychology*. Chicago: Rand-McNally, 1966. (a)

Cattell, R. B. The scree test for the number of factors. *Multivariate Behavioral Research*, 1966, *1*, 140–161. (b)

Cattell, R. B. The principles of experimental design and analysis in relation to theory building. In R. B. Cattell (Ed.), *Handbook of multivariate experimental psychology*. Chicago: Rand-McNally, 1966. Chapter 2, pp. 19–66. (c)

Cattell, R. B. Psychological theory and scientific method. In R. B. Cattell (Ed.), *Handbook of multivariate experimental psychology*. Chicago: Rand-McNally, 1966. Chapter 1, pp. 1–18. (d)

Cattell, R. B. Patterns of change: Measurement in relation to state-dimension, trait change, lability and process concepts. In R. B. Cattell (Ed.), *Handbook of multivariate experimental psychology*. Chicago: Rand-McNally, 1966. Chapter 11. (e)

Cattell, R. B. The theory of fluid and crystallized intelligence. *British Journal of Educational Psychology*, 1967, *37*, 209–224.

Cattell, R. B. Taxonomic principles for locating and using types (and the derived taxonome computer program). In B. Kleinmuntz (Ed.), *Formal representation of human judgment*. Pittsburgh: Pittsburgh University Press, 1968.

Cattell, R. B. Comparing factor trait and state scores across ages and cultures. *Journal of Gerontology*, 1969, *24*, 348–360.

Cattell, R. B. The isopodic and equipotent principles for comparing factor scores across different populations. *British Journal of Mathematical and Statistical Psychology*, 1970, *23*, 23–41.

Cattell, R. B. *Abilities: Their structure, growth, and action*. Boston: Houghton Mifflin, 1971. (a)

Cattell, R. B. Estimating modulator indices and state liabilities. *Multivariate Behavioral Research*, 1971, *6*, 7–33. (b)

Cattell, R. B. Real base, true zero factor analysis. *Multivariate Behavioral Research Monographs*, 1972, No. 72-1. Fort Worth: Texas Christian University Press, 1–162. (a)

Cattell, R. B. The nature and genesis of mood states: A theoretical model with experimental measurements concerning anxiety, depression, arousal, and other mood states. In C. B. Spielberger (Ed.), *Current trends in theory and research* (Vol. 1). New York: Academic Press, 1972. (b)

Cattell, R. B. *Personality and mood by questionnaire*. San Francisco: Jossey–Bass, 1973.

Cattell, R. B. Radial parcel factoring versus item factoring in defining personality structure in questionnaires: Theory and experimental checks. *Australian Journal of Psychology*, 1974, *26*, 103–119.

Cattell, R. B. Matched determiners vs. factor invariance: a reply to Korth. *Multivariate Behavioral Research*, 1978, *13*, 431–448.

Cattell, R. B. *Personality and learning theory*. New York: Springer, 1979.

Cattell, R. B., & Adelson, M. The dimensions of social change in the USA as determined by *P*-technique. *Social Forces*, 1951, *30*, 190–201.

Cattell, R. B., & Baggaley, A. R. The salient variable similarity index for factor matching. *British Journal of Statistical Psychology*, 1960, *13*, 33–46.

Cattell, R. B., & Bartlett, H. W. An *R-dR*-technique operational distinction of the states of anxiety, stress, fear, etc. *Australian Journal of Psychology*, 1971, *23*, 105–123.

Cattell, R. B., & Bolton, L. S. What pathological dimensions lie beyond the normal dimensions of the 16 P.F. *Journal of Consulting and Clinical Psychology*, 1969, *33*, 18–29.

Cattell, R. B. & Brennan, J. The practicality of an orthogonal confactor rotation for the approximate resolution of oblique factors. *Multivariate Experimental Clinical Research*, 1977, *3*, 95–104.

Cattell, R. B., & Burdsal, C. A. The radial parcel double factoring design: A solution to the item-vs-parcel controversy. *Multivariate Behavioral Research*, 1975, *10*, 165–179.

Cattell, R. B., & Butcher, H. J. *The prediction of achievement and creativity*. Indianapolis: Bobbs Merrill, 1968.

Cattell, R. B., & Cattell, A. K. S. Factor rotation for proportional profiles: Analytical solution and an example. *British Journal of Statistical Psychology*, 1955, *8*, 83–92.

Cattell, R. B., & Child, D. *Motivation and dynamic structure*. New York: Halstead Press, 1975.

Cattell, R. B., & Cross, K. P. Comparison of the ergic and self sentiment structure found by dynamic traits by *R*- and *P*-techniques. *Journal of Personality*, 1952, *21*, 250–271.

Cattell, R. B., & De Khanna, A. Principles and procedures for unique rotation in factor analysis. In Enslein, Ralston and Wilf (Eds.), *Statistical methods for digital computers. Mathematical methods for digital computers* (Vol. III). New York: Wiley, 1976.

Cattell, R. B., & Dickman, K. A dynamic model of physical influences demonstrating the necessity of oblique simple structure. *Psychological Bulletin*, 1962, *59*, 389–400.

Cattell, R. B., & Digman, J. M. A theory of the structure of perturbations in observer ratings and questionnaire data in personality research. *Behavioral Science*, 1964, *9*, 341–358.

Cattell, R. B., & Finkbeiner, C. *An empirical determination of hyperplane widths (standard error of a zero loading) at simple structure positions.* Manuscript submitted for publication, 1977.

Cattell, R. B., & Foster, M. J. The rotoplot program for multiple, single-plane, visually-guided rotation. *Behavioral Science*, 1963, *8*, 156–165.

Cattell, R. B., & Gorsuch, R. L. The uniqueness and significance of simple structure demonstrated by contrasting organic "natural structure" and "random structure" data. *Psychometrika*, 1963, *28*, 55–67.

Cattell, R. B., & Jaspars, J. A general plasmode (No. 30-10-5-2) for factor analytic exercises and research. *Multivariate Behavioral Research Monographs*, 1967, *67-3*, 1–212.

Cattell, R. B., & Muerle, J. L. The "maxplane" program for factor rotation to oblique simple structure. *Educational and Psychological Measurement*, 1960, *20*, 569–590.

Cattell, R. B., & Nichols, K. E. An improved definition from ten researches of second order personality factors in Q-data (with cross cultural checks). *Journal of Social Psychology*, 1972, *86*, 187–203.

Cattell, R. B., & Radcliffe, J. A. Reliabilities and validities of simple and extended weighted and buffered unifactor scales. *British Journal of Statistical Psychology*, 1962, *15*, 113–128.

Cattell, R. B., & Saunders, D. R. Inter-relation and matching of personality factors from behavior rating, questionnaire and objective test data. *Journal of Social Psychology*, 1950, *31*, 243–260.

Cattell, R. B., & Scheier, I. H. *The meaning and measurement of neuroticism and anxiety.* New York: Ronald Press, 1961.

Cattell, R. B., & Schuerger. J. M. Objective personality testing: The handbook for the O-A Personality Measurement Kit. Champaign, Ill., *Institute for Personality and Ability Testing*, 1977.

Cattell, R. B., & Stice, G. F. *The dimensions of groups and their relations to the behavior of members.* Champaign, Illinois: IPAT, 1960.

Cattell, R. B., & Sullivan, W. The scientific nature of factors: A demonstration by cups of coffee. *Behavioral Science*, 1962, *7*, 184–193.

Cattell, R. B., & Tsujioka, B. The importance of factor-trueness and validity, versus homogeneity and orthogonality, in test scales. *Educational and Psychological Measurement*, 1964, *24*, 3–30.

Cattell, R. B., & Vogelman, S. An empirical test with varying variable and factor numbers of the relative accuracies of the scree and the K-G test for the number of factors. *Multivariate Behavioral Research*, 1977, *12*, in press.

Cattell, R. B., & Warburton, F. W. *Objective personality and motivation tests. A theoretical introduction and practical compendium.* Champaign: University of Illinois Press, 1967.

Cattell, R. B., & Watterson, D. *A check on the "seven missing personality factors" in normals beyond the 16 P.F.* Manuscript submitted for publication, 1978.

Cattell, R. B., & White, O. The method of confactor rotation illustrated in the oblique case on the cups of coffee plasmode (Advance Publication No.-17). Laboratory of Personality Assessment and Group Behavior, University of Illinois, 1962.

Cattell, R. B., & Williams, H. F. *P*-technique, a new statistical device for analyzing functional unities in the intact organism. *British Journal of Preventive and Social Medicine*, 1953, *7*, 141–153.

Cattell, R. B., & Wispe, L. G. The dimensions of syntality in small groups. *Journal of Social Psychology*, 1948, *28*, 57–78.

Cattell, R. B., Balcar, K. R., Horn, J. L., & Nesselroade, J. R. Factor matching procedures: An improvement of the *s* index; with tables. *Educational and Psychological Measurement*, 1969, *29*, 781–792.

Cattell, R. B., Bolz, C., & Korth, B. Behavioral types in pure bred dogs objectively determined by taxonome. *Behavior and Genetics*, 1973, *3*, 205–216. (a)

Cattell, R. B., Cattell, A. K. S., & Rhymer, R. M. *P*-technique demonstrated in determining psycho-physiological source traits in a normal individual. *Psychometrica*, 1947, *12*, 267–288.

Cattell, R. B., Coulter, M. A., & Tsujioka, B. The taxonometric recognition of types and functional emergents. In R. B. Cattell (Ed.), *Handbook of multivariate experimental psychology*. Chicago: Rand-McNally, 1966, Chapter 9.

Cattell, R. B., Eber, H. W., & Tatsuoka, M. *The 16 personality factor questionnaire and handbook*. Champaign, Illinois: IPAT, 1949, new edition, 1971.

Cattell, R. B., Horn, J. L., Radcliffe, J. A., & Sweney, A. B. The nature and measurement of components of motivation. *Genetic Psychology Monographs*, 1963, *68*, 49–211.

Cattell, R. B., Horn, J. L., Radcliffe, J. A., & Sweney, A. B. *The Motivation Analysis Test*. Champaign, Illinois: IPAT, 1964.

Cattell, R. B., Graham, R. K., & Wolliver, W. E. A re-assessment of the factorial cultural dimensions of modern nations. In preparation.

Cattell, R. B., Pichot, P., & Rennes, P. Constance interculturelle des facteurs de personalité, mesuré, par le test 16 P.F. *Revue de Psychologie Appliquée*, 1961, *11*, 165–196.

Cattell, R. B., Pierson, G., & Finkbeiner, C. Alignment of personality source trait factors from questionnaires and observer ratings: The theory of instrument-free patterns. *Multivariate Experimental Clinical Research*, 1976, *2*, 67–88. (a)

Cattell, R. B., Schmidt, L. R., & Bjersted, A. Clinical diagnosis by the objective analytic personality batteries. *Journal of Clinical Psychology Monographs*. Vermont: Brandon, 1972.

Cattell, R. B., Schmidt, L. R., & Pawlik, K. Cross cultural comparison (U.S.A., Japan, Austria) of the personality factor structures of 10–14 year olds in objective tests. *Social Behavior and Personality*, 1973, *1*, 182–211. (b)

Cattell, R. B., Schuerger, J. M., Klein, T. W., & Finkbeiner, C. A definitive large sample factoring of personality structure in objective measures, as a basis for the HSOA. *Journal of Research in Personality*, 1976, *10*, 22–41. (b)

Child, D. *The essentials of factor analysis*. London: Holt, Rinehart & Winston, 1970.

Christofferson, A. Factor analysis of dichotomized variables. *Psychometrika*, 1975, *40*, 5–32.

Clarke, M. R. B. A rapidly convergent method for maximum likelihood factor analysis. *British Journal of Mathematical and Statistical Psychology*, 1970, *23*, 43–52.

Clemans, W. V. An analytical and empirical examination of some properties of ipsative scores (unpublished doctoral dissertation, University of Washington, Seattle, Washington, 1956). *Psychometry Monographs*, 1966.

Cliff, N. Orthogonal rotation to congruence. *Psychometrika*, 1939, *4*, 149–162.

Cliff, N., & Hamburger, C. D. The study of sampling error in factor analysis by means of artificial experiments. *Psychological Bulletin*, 1967, *68*, 430–445.

Cliff, N., & Pennell, R. The influence of communality, factor strength, and loading size on the sampling characteristics of factor loadings. *Psychometrika*, 1967, *32*, 309.

Coan, R. W. A comparison of oblique and orthogonal factor solutions. *Journal of Experimental Education*, 1959, *27*, 151–166.

Coan, R. W. Basic forms of covariation and concomitance designs. *Psychological Bulletin*, 1961, *58*, 317–324.

Coan, R. W. Facts, factors and artifacts: The quest for psychological meaning. *Psychological Review*, 1964, *71*, 123–140.

Coan, R. W., & Cattell, R. B. Reproducible personality factors in middle childhood. *Journal of Clinical Psychology*, 1958, *14*, 339–345.

Cohen, J. The impact of multivariate research in clinical psychology. In R. B. Cattell (Ed.), *Handbook of multivariate experimental psychology*. Chicago: Rand-McNally, 1966. Chapter 30.

Cohen, J. Multiple regression as a general data-analytic system. *Psychological Bulletin*, 1968, *70*, 426.

Cohen, J. A profile similarity coefficient invariant over variable reflection. *Psychological Bulletin*, 1969, *71*, 281.

Cohen, J., & Cohen, P. *Applied multiple regression correlation analysis for the behavior sciences*. Hillsdale, N. J.: Erlbaum Press, 1975.

Coleman, J. S. *Introduction to mathematical sociology*. Glencoe: Free Press, 1964.

Comrey, A. L. The minimum residual method of factor analysis. *Psychological Reports*, 1962, *11*, 15.

Comrey, A. L. *A manual for the Comrey personality scales*. San Diego: Educational and Industrial Testing Service, 1968.

Comrey, A. L. *A first course in factor analysis*. New York: Academic Press, 1973.

Comrey, A. L., & Duffy, K. E. Cattell and Eysenck factor scores related to Comrey personality factors. *Multivariate Behavioral Research*, 1968, *4*, 379–392.

Comrey, A. L., & Levonian, E. A comparison of three point coefficients in factor analysis of MMPI items. *Educational and Psychological Measurement*, 1958, *18*, 739–755.

Conger, A. J. Estimating profile reliability and maximally reliable composites. *Multivariate Behavioral Research*, 1974, *9*, 85–104.

Cooley, W. W., & Lohnes, P. R. *Multivariate data analysis*. New York: Wiley, 1971.

Coombs, C. H. Mathematical models in psychological scaling. *Journal of the American Statistical Association*, 1951, *46*, 480–489.

Coombs, C. H. *A theory of data*. New York: Wiley, 1964.

Coombs, C. H., & Kao, R. C. *Nonmetric factor analysis*. Ann Arbor: Engineering Research Institute, University of Michigan, 1955.

Coombs, C. H., & Kao, R. C. On a connection between factor analysis and multidimensional unfolding. *Psychometrika*, 1960, *25*, 219–231.

Corballis, M. C., & Traub, R. E. Longitudinal factor analysis. *Psychometrika*, 1970, *35*, 79–93.

Costner, H. L., & Schönberg, R. Diagnosing indicator ills in multiple indicator models. In A. S. Goldberger (Ed.), *Structural equation models in the social sciences*. New York: Seminar Press, 1973, pp. 167–199.

Coulter, M. A., & Cattell, R. B. Principles of behavioral taxonomy and the mathematical basis of the Taxonome computer program. *British Journal of Mathematical and Statistical Psychology*, 1966, *19*, 237–269.

Cramer, E. M. On Browne's solution for oblique procrustes rotation. *Psychometika*, 1974, *39*, 159–164.

Crawford, L. B. Determining the number of interpretable factors. *Psychological Bulletin*, 1975, *82*, 226–237.

Cronbach, L. J. Coefficient alpha and the internal structure of tests. *Psychometrika*, 1951, *16*, 297–334.

Cronbach, L. J. Validity. In R. L. Thorndike (Ed.), *Educational measurement*. Washington: American Council on Education, 1971.

Cronbach, L. J., & Furby, L. How should we measure "change"—or should we? *Psychologial Bulletin*, 1970, *74*, 68–80.

Cronbach, L. J., & Gleser, G. C. Assessing similarity between profiles. *Psychological Bulletin*, 1953, *50*, 456–473.

Cureton, E. E. Note on ϕ max. *Psychometrika*, 1959, *24*, 89-91.

Cureton, E. E., & Mulaik, S. A. The weighted varimax rotation and the promax rotation. *Psychometrika*, 1975, *40*, 183–196.

Damarin, F., Tucker, L. R., & Messick, S. A base free measure of change. *Psychometrika*, 1966, *31*, 457–473.

Davies, M. The general factor in correlations between persons. *British Journal of Psychology*, 1939, *29*, 404–421.

Delhees, K. H., & Cattell, R. B. The dimensions of pathology: Proof of their projection beyond the normal traits. *Journal of Genetic Psychology*, 1971, *2*, 149–173.

DeYoung, G. *The combination of variance–covariance matrices across sets of the data box*. Manuscript submitted for publication, 1975.

Dickman, K. W. *Factorial validity of a rating instrument*. Unpublished doctoral dissertation, University of Illinois, Urbana, Illinois, 1960.

Dielman, T. E., Cattell, R. B., & Wagner, A. Evidence on the simple structure and factor invariance achieved by five rotational methods on four types of data. *Multivariate Behavioral Research*, 1972, *7*, 223.

Digman, J. M. Interaction and non-linearity in multivariate experiments. In R. B. Cattell (Ed.), *Handbook of multivariate experimental psychology*. Chicago: Rand-McNally, 1966. Chapter 15.

Digman, J. M. The procrustes class of factor-analytic transformation. *Multivariate Behavioral Research*, 1967. *2*, 89.

Digman, J. M. A criticism of Peterson's oversimplified factor analysis. American Psychological Association Annual Meeting, 1973.

Digman, J. M., & Woods, G. A comparison among five methods of factor analysis. Manuscript submitted for publication, 1976.

Dingman, H. F., Miller, C. R., & Eyman, R. K. A comparison between two analytic rotational solutions where the number of factors is indeterminate. *Behavioral Science*, 1964, *9*, 76.

Dreger, R. M. *Multivariate personality research*. Baton Rouge: Claitor, 1972.

DuBois, P. H. *Multivariate correlational analysis*. New York: Harper & Row, 1957.

Duncan, O. D. Path analysis: Sociological examples. *American Journal of Sociology*, 1966, *72*, 1–16.

Dunn-Rankin, P. The visual characteristics of words. *Scientific American*, 1977, *222*, 90–98.

Dwyer, P. S. The determination of the factor loadings of a given test from the known factor loadings of other tests. *Psychometrika*, 1937, *2*, 173.

Easley, J. A., & Tatsuoka, M. M. *Scientific thought: Cases from classical physics*. Boston: Allyn & Bacon, 1968.

Eber, H. W. Toward oblique simple structure: Maxplane. *Multivariate Behavioral Research*, 1966, *1*, 112–125.

Eckhart, C., & Young, G. The approximation of one matrix by another of lower rank. *Psychometrika*, 1936, *1*, 211–218.

Edwards, A. L. *The social desirability variable in personality assessment and research*. New York: Holt, 1957.

Elashoff, J. D. Analysis of covariance: A delicate instrument. *American Educational Research Journal*, 1969, *6*, 383.

Evans, G. T. Transformation of factor matrices to achieve congruence. *British Journal of Mathematical and Statistical Psychology*, 1971, *24*, 22.

Eysenck, H. J. On the choice of personality tests for research and prediction. *Journal of Behavioral Science*, 1971, *1*, 85–90.

Eysenck, H. J. Multivariate analysis and experimental psychology. In R. M. Dreger (Ed.), *Multivariate personality research*. Baton Rouge: Claitor, 1972. Chapter 3.

Ezekiel, M., & Fox, K. A. *Methods of correlation and regression analysis*. New York: Wiley, 1959.

Ferguson, G. A. The factorial interpretation of test difficulty. *Psychometrika*, 1941, *6*, 323–329.

Finkbeiner, C. T. The effects of three parameters on hyperplane widths found in real data. Unpublished Masters thesis, University of Illinois, Urbana, Illinois, 1973.

Finkbeiner, C. T. Estimation for the multiple common factor model when data are missing. Ph.D. thesis, University of Illinois, Urbana, Illinois, 1976.

Fisher, D. R. A comparison of various techniques of multiple factor analysis applied to biosystemic data. *University of Kansas Science Bulletin*, 1973, *50*, 127–162.

Fleishman, E. A. A comparative study of aptitude patterns in unskilled and skilled motor performances. *Journal of Applied Psychology*, 1957, *41*, 263–272.

Foa, U. G. New developments in facet design and analysis. *Psychological Review*, 1965, *72*, 262–274.

French, J. W. *The description of personality measurements in terms of rotated factors*. Princeton, New Jersey: ETS, 1953.

French, J. W. Application of T-technique factor analysis to the stock market. *Multivariate Behavioral Research*, 1972, *3*, 279.

Fruchter, B. *Introduction to factor analysis*. New York: Van Nostrand, 1954. (Revised, 1964.)

Fruchter, B. Manipulative and hypothesis-testing factor analytic experimental designs. In R. B. Cattell (Ed.), *Handbook of multivariate experimental psychology*. Chicago: Rand-McNally, 1966. Chapter 10.

Fuller, E. L., Jr., & Hemmerle, W. F. Robustness of the maximum-likelihood estimation procedure in factor analysis. *Psychometrika*, 1966, *31*, 255.

Gibb, C. A. Changes in the culture pattern of Australia, 1906–1946, as determined by P-technique. *Journal of Social Psychology*, 1956, *43*, 225–238.

Gibson, W. A. Remarks on Tucker's interbattery method of factor analysis. *Psychometrika*, 1960, *25*, 19.

Gibson, W. A. On the least squares orthogonalization of an oblique transformation. *Psychometrika*, 1962, *27*, 193–196.

Gibson, W. A. On the symmetric treatment of an asymmetric approach to factor analysis. *Psychometrika*, 1963, *28*, 423.

Glass, G. V. Alpha factor analysis of infallible variables. *Psychometrika*, 1966, *31*, 545.

Glass, G. V., & Hakstian, A. R. Measures of association in comparative experiments: Their development and interpretation. *American Educational Research Journal*, 1969, *6*, 403–414.

Glass, G. V., & McGuire, F. Abuses of factor scores. *American Educational Research Journal*, 1966, *3*, 297.

Gleason, F. C., & Staehlin, R. A proposal for handling missing data. *Psychometrika*, 1975, *40*, 229–252.

Goldberg, L. *Toward a universal taxonomy of personality traits*. Paper presented at the annual meeting of the Society of Multivariate Experimental Psychology, Oregon, November, 1975.

Goldberger, A. S., & Duncan, O. D. (Eds.). *Structural equation models in the social sciences*. New York: Seminar Press, 1973.

Goldfarb, N. *An introduction to longitudinal statistical analysis*. Glencoe: Free Press, 1960.

Gollob, H. F. Confounding of sources of variation in factor-analytic techniques. *Psychological Bulletin*, 1968, *70*, 330. (a)

Gollob, H. F. Rejoinder to Tucker's "Comments on 'Confounding of sources of variation in factor-analytic techniques.'" *Psychological Bulletin*, 1968, *70*, 355. (b)

Gollob, H. F. A statistical model which combines features of factor analytic and analysis of variance techniques. *Psychometrika*, 1968, *33*, 73. (c)

Gorsuch, R. L. (Ed.). *Computer programs for statistical analysis*. Nashville: Vanderbilt University Computer Center, 1968.

Gorsuch, R. L. A comparison of Biquartimin, Maxplane, Promax, and Varimax. *Educational and Psychological Measurement*, 1970, *30*, 861.

Gorsuch, R. L. Bartlett's significance test to determine the number of factors to extract. *Educational and Psychological Measurement*, 1973, *33*, 361.

Gorsuch, R. L. *Factor analysis*. Philadelphia: Saunders, 1974.

Gorsuch, R. L., & Cattell, R. B. Second stratum personality factors defined in the questionnaire realm by the 16 P.F. *Multivariate Behavioral Research*, 1967, *2*, 211-224.

Gourlay, N. Difficulty-factors arising from the use of tetrachoric correlations. *British Journal of Statistical Psychology*, 1951, *4*, 65–76.

Gower, J. C. Generalized procrustes analysis. *Psychometrika*, 1975, *40*, 33–52.

Gruvaeus, G. T. A general approach to procrustes pattern rotation. *Psychometrika*, 1970, *35*, 493–505.

Guertin, W. H., & Bailey, J. *Introduction to modern factor analysis*. Ann Arbor: Edwards Brothers, Inc., 1970.

Guilford, J. P. The minimal phi coefficient and the maximal phi. *Educational and Psychological Measurement*, 1965, *25*, 3–8. (a)

Guilford, J. P. *Fundamental statistics in psychology and education*. (4th ed.). New York: McGraw-Hill, 1965. (b)

Guilford, J. P., & Hoepfner, R. Comparisons of varimax rotations with rotations to theoretical targets. *Educational and Psychological Measurement*, 1969, *29*, 3.

Guilford, J. P., & Michael, W. B. Approaches to univocal factor scores. *Psychometrika*, 1948, *13*, 1–22.

Guttman, L. Image theory for the structure of quantitative variates. *Psychometrika*, 1953, *18*, 277–296.

Guttman, L. Some necessary conditions for common factor analysis. *Psychometrika*, 1954, *19*, 149.

Guttman, L. The determinancy of factor score matrices with implications for five other basic problems of common factor theory. *British Journal of Statistical Psychology*, 1955, *8*, 65. (a)

Guttman, L. A new approach to factor analysis. In P. F. Lazarsfeld (Ed.), *Mathematical thinking in the social sciences*. New York: Columbia University Press, 1955. (b)

Guttman, L. "Best possible" systematic estimates of communalities. *Psychometrika*, 1956, *21*, 273–289.

Guttman, L. Metricizing rank-ordered or unordered data for a linear factor analysis. *Sankhya*, 1959, *21*, 257–268.

Guttman, L. Order analysis of correlation matrices. In R. B. Cattell (Ed.), *Handbook of multivariate experimental psychology*. Chicago: Rand-McNally, 1966. Chapter 14.

Guttman, L. *A general nonmetric technique for finding the smallest Euclidean space for a configuration of points*. Unpublished manuscript, 1967.

Hakstian, A. R. A computer program for oblique factor transformation using the generalized Harris–Kaiser procedure. *Educational and Psychological Measurement*, 1970, *30*, 703.

Hakstian, A. R. A comparative evaluation of several prominent methods of oblique factor transformation. *Psychometrika*, 1971, *36*, 175.

Hakstian, A. R. Procedures for the factor analytic treatment of measures obtained on different occasions. *British Journal of Mathematical and Statistical Psychology*, 1973, *26*, 219–239.

Hakstian, A. R. The development of a class of oblique factor solutions. *British Journal of Mathematics and Statistical Psychology*. 1974, *27*, 100–114. (a)

Hakstian, A. R. Comparative assessment of multivariate association in psychological research. *Psychological Bulletin*, 1974, *81*, 1049–1052. (b)

Hakstian, A. R. Procedures for Φ-constrained confirmatory factor transformation. *Multivariate Behavioral Research*, 1975, *10*, 245–253.

Hakstian, A. R. *A method of estimating scores from items to produce factor estimate correlations equal to those of the true factors*. Manuscript submitted for publication, 1976.

Hakstian, A. R., & Abell, R. A. A further comparison of oblique factor transformation methods. *Psychometrika*, 1974, *39*, 429–444.

Hakstian, A. R., & Cattell, R. B. *The comprehensive ability battery and handbook*. Champaign, Illinois: Institute for Personality and Ability Testing, 1976.

Hakstian, A. R., & Cattell, R. B. *An examination of inter-domain relationships among some ability and personality traits*. Manuscript submitted for publication, 1977.

Hakstian, A. R., & Muller, V. J. Some notes on the number of factors problem. *Multivariate Behavioral Research*, 1973, *8*, 461–475.

Hakstian, A. R., & Skakun, E. N. Sampling and factor transformation. *Multivariate Behavioral Research*, 1976, *11*, 119–124.

Hakstian, A. R., Rogers, W., & Cattell, R. B. *A test by a sample of plasmodes stratified as to N, n and k of the scree and K-G tests for the number of factors*. Manuscript submitted for publication, 1978.

Hamburger, C. D. Factorial stability as a function of analytical rotation method, type of simple structure, and size of sample. Unpublished doctoral dissertation, University of Southern California, 1965.

Hammer, A. G. *Elementary matrix algebra for psychologists and social scientists*. New York: Pergamon Press, 1971.

Harman, H. H. *Modern factor analysis*. Chicago: University of Chicago Press, 1967. (Revised, 1976).

Harman, H. H., & Jones, W. H. Factor analysis by minimizing residuals (minres). *Psychometrika*, 1966, *31*, 351–368.

Harris, C. H., & Harris, M. *A structure of concept attainment abilities*. Madison, Wisconsin: Wiseman R & D Center, 1973.

Harris, C. W. Relationships between two systems of factor analysis. *Psychometrika*, 1956, *21*, 185–190.

Harris, C. W. (Ed.). *Problems in measuring change*. Madison: University of Wisconsin Press, 1963.

Harris, C. W. On factors and factor scores. *Psychometrika*, 1967, *32*, 363.

Harris, C. W., & Kaiser, H. F. Oblique factor analytic solutions by orthogonal transformations. *Psychometrika*, 1964, *29*, 347.

Hartley, H. O. Maximum likelihood estimation from incomplete data. *Biometrics*, 1958, *14*, 174–194.

Hartmann, W. Uber ein Verfahren der numerischen Taxonomie von Cattell und Coulter. *Biometrische Zeitschrift*. 1976, *18*, 273–290.

Haynes, J. R. The effect of double standardized scoring on the semantic differential. *Educational and Psychological Measurement*, 1975, *35*, 107–114.

Heeler, R. M., & Whipple, F. W. A Monte Carlo aid to the evaluation of maximum likelihood factor analysis solutions. *British Journal of Mathematical and Statistical Psychology*, 1976, *29*, 253–256.

Heerman, E. F. Univocal or orthogonal estimators of orthogonal factors. *Psychometrika*, 1963, *28*, 161.

Heerman, E. F. The geometry of factorial indeterminacy. *Psychometrika*, 1964, *29*, 371–381.

Hendrickson, A. E., & White, P. O. Promax: A quick method for rotation to oblique simple structure. *British Journal of Statistical Psychology*, 1964, *17*, 65.

Henrysson, S. *Applicability of factor analysis in the behavioral sciences: A methodological study*. Stockholm: Almquist & Wiksell, 1957.

Henrysson, S. *Applicability of factor analysis in the behavioral sciences*. Stockholm: Almquist & Wiksell, 1960.

Henrysson, S., & Thurberg, P. *Tetrachoric or phi coefficients in factor analysis*. (Report No. 27). Department of Psychology, University of Uppsala, Sweden, 1965.

Hocking, R. R. Selection of the best subset of regression variances. In K. Enslein *et al.* (Eds.) *Statistical methods for digital computers*, Vol. 3, New York: Wiley, 1977, pp. 39–57.

Hoel, P. G. A significance test for component analysis. *Annals of Mathematical Statistics*, 1937, *8*, 149–158.

Hohn, F. E. *Elementary matrix algebra* (2nd ed.). New York: Macmillan, 1958.

Holley, J. W. On the Burt reciprocity principle: A final generation. *Psychological Research Bulletin*, 1970, *9*, 9. (a)

Holley, J. W. On the generalization of the Burt reciprocity principle. *Multivariate Behavioral Research*, 1970, *5*, 241–250. (b)

Holley, J. W., & Guilford, J. P. A note on the *G* index of agreement. *Educational and Psychological Measurement*, 1964, *24*, 749.

Holley, J. W., & Sjöberg, L. Some characteristics of the *G* index of agreement. *Multivariate Behavioral Research*, 1968, *3*, 107.

Holman, E. W. The relations between hierarchical and Euclidean models for psychological distances. *Psychometrika*, 1972, *37*, 417–424.

Holzinger, K. J., & Harman, H. H. *Factor Analysis*. Chicago: University of Chicago Press, 1941.

Horn, J. L. An empirical comparison of methods for estimating factor scores. *Educational and Psychological Measurement*, 1965, *25*, 313–322. (a)

Horn, J. L. A rationale and test for the number of factors in factor analysis. *Psychometrika*, 1965, *30*, 179. (b)

Horn, J. L. Fluid and crystallized intelligence: A factor analytic study of the structure among primary mental abilities. Unpublished doctoral dissertation, University of Illinois, 1965. (c)

Horn, J. L. Integration of structural and developmental concepts in the theory of fluid and crystallized intelligence. In R. B. Cattell (Ed.), *Handbook of multivariate experimental psychology*, Chicago: Rand-McNally, 1966.

Horn, J. L. Factor analyses with variables of different metric. *Educational and Psychological Measurement*, 1969, *29*, 753. (a)

Horn, J. L. On the internal consistency reliability of factors. *Multivariate Behavioral Research*, 1969, *4*, 115. (b)

Horn, J. L. *Concepts and methods of correlational analyses*. New York: Holt, Rinehart & Winston, 1973.

Horn, J. L. On extension analysis and its relation to correlations between variables and factor scores. *Multivariate Behavioral Research*, 1976, *11*, 320–331.

Horn, J. L., & Cattell, R. B. Age differences in primary mental abilities. *Journal of Gerontology*, 1966, *21*, 210–220.

Horn, J. L., & Little, K. B. Isolating change and invariance in patterns of behavior. *Multivariate Behavioral Research*, 1966, *1*, 219–228.

Horn, J. L., & Miller, W. C. Evidence on problems in estimating common factor scores. *Educational and Psychological Measurement*, 1966, *26*, 617.

Horst, P. Relations among M sets of variables. *Psychometrika*, 1961, *26*, 129–149.

Horst, P. Generalized canonical correlations. *Journal of Clinical Psychology*. Monog. Suppl. No. 4, 1962.

Horst, P. *Matrix algebra for social scientists*. New York: Holt, Rinehart & Winston, 1963.

Horst, P. *Factor analysis of data matrices*. New York: Holt, Rinehart & Winston, 1965.

Horst, P. The missing data matrix. *Journal of Clinical Psychology*, 1968, *24*, 286.

Hotelling, H. Analysis of a complex of statistical variables into principal components. *Journal of Educational Psychology*, 1933, *24*, 417–441; 498–520.

Hotelling, H. The most predictable criterion. *Journal of Educational Psychology*, 1935, *26*, 134–142.

Hotelling, H. The relations of the newer multivariate statistical methods to factor analysis. *British Journal of Statistical Psychology*, 1957, *10*, 69–79.

Howard, K. I., & Cartwright, D. S. An empirical note on the communality problem in factor analysis. *Psychological Reports*, 1962, *10*, 797.

Howard, K. I., & Diesenhaus, H. Intra-individual variability, response set and response uniqueness in a personality questionnaire. *Journal of Clinical Psychology*, 1965, *21*, 392–396.

Howard, K. I., & Gordon, R. A. Empirical note on the "number of factors" problem in factor analysis. *Psychological Reports*, 1963, *12*, 247.

Howarth, H. E., & Browne, J. A. An item analysis of the 16 P.F. personality questionnaire. *Personality: An International Journal*, 1971, *2*, 117–139.

Hubert, L., & Schulz, J. Hierarchical clustering and the concept of space distortion. *Multivariate Behavioral Research*, 1975, *28*, 121–133.

Humphreys, L. G. Number of cases and number of factors: An example where N is very large. *Educational and Psychological Measurement*, 1964, *24*, 457–460.

Humphreys, L. G., Ilgen, D., McGrath, D., & Montanelli, R. Capitalization on chance in rotation of factors. *Educational and Psychological Measurement*, 1969, *29*, 259.

Humphreys, L. G., Tucker, L. R., & Dachler, P. Evaluating the importance of factors in any given order of factoring. *Multivariate Behavioral Research*, 1970, *5*, 209.

Hundleby, J., Pawlik, K., & Cattell, R. B. *Personality factors in objective test devices*. San Diego: Educational and Industrial Testing, 1965.

Hunka, S. Alpha factor analysis. *Behavioral Science*, 1966, *11*, 80.

Hurley, J. R., & Cattell, R. B. The Procrustes program: Producing direct rotation to test a hypothesized factor structure. *Behavioral Science*, 1962, *7*, 258–262.

Jackson, D. N., & Moof, M. E. Testing the null hypothesis for rotation to a target. *Multivariate Behavioral Research*, 1974, *9*, 303–310.

Jacobi, C. G. J. Über ein leichter Verfakren die in der Theorie der Säcularstörungen vookommenden Gleichuugen numerisch aufzulösen. *J. rein u augewandte Mathematik*, 1846, *30*, 51–94.

Jenkins, T. N. Limitations of iterative procedures for estimating communalities. *Journal of Psychological Studies*, 1962, *13*, 69.

Jennrich, R. I. *Simplified formulas for standard errors in maximum likelihood factor analysis*. (Research Bulletin 73–40) Princeton: E.T.S., 1973.

Jennrich, R. I. Standard errors for oblique factor loadings. *Psychometrika*, 1974, *39*, 593–604.

Jennrich, R. I. Stepwise regression. In K. Enstein *et al.* (Eds.) *Statistical methods for digital computers*. New York: Wiley, 1977.

Jennrich, R. I., & Robinson, S. M. A Newton–Raphson algorithm for maximum likelihood factor analysis. *Psychometrika*, 1969, *34*, 111–120.

Jennrich, R. I., & Thayer, D. T. A note on Lawley's formula for standard error in maximum likelihood. *Psychometrika*, 1974, *39*, 571–580.

Johnston, J. *Econometric methods*. New York: McGraw-Hill, 1963.

Jones, K. J. Relation of R-analysis factor scores and Q-analysis factor loadings. *Psychological Reports*, 1967, *20*, 247.

Jones, L. V. Analysis of variance in its multivariate developments. In R. B. Cattell (Ed.), *Handbook of multivariate experimental psychology*. Chicago: Rand-McNally, 1966.

Jöreskog, K. G. On the statistical treatment of residuals in factor analysis. *Psychometrika*, 1962, *27*, 335.

Jöreskog, K. G. Testing a simple structure hypothesis in factor analysis. *Psychometrika*, 1966, *31*, 165.

Jöreskog, K. G. A computer program for unrestricted maximum likelihood factor analysis. *Research Bulletin*. Princeton: Educational Testing Service, 1967. (a)

Jöreskog, K. G. Some contributions to maximum likelihood factor analysis. *Psychometrika*, 1967. *32*, 443–482. (b)

Jöreskog, K. G. A general approach to confirmatory maximum likelihood factor analysis. *Psychometrika*, 1969, *34*, 183.

Jöreskog, K. G. A general method of analysis of covariance structures. *Biometrika*, 1970, *57*, 239–251.

Jöreskog, K. G. Simultaneous factor analysis in several populations. *Psychometrika*, 1971, *36*, 409–426.

Jöreskog, K. G. General methods for estimating a linear structural equation system. In A. S. Goldberger and O. D. Duncan (Eds.), *Structural equation models in the social sciences*. New York: Seminar Press, 1973.

Jöreskog, K. G. Factor analysis by least squares and maximum likelihood methods. In K. Enstein *et al.* (Eds.) *Statistical methods for digital computers*. New York: Wiley, 1977.

Jöreskog, K. G., & Goldberger, A. S. Factor analysis by generalized least squares. *Psychometrika*, 1972, *37*, 243–260.

Jöreskog, K. G., & Lawley, D. N. New methods in maximum likelihood factor analysis. *British Journal of Mathematical and Statistical Psychology*, 1968, *21*, 85–97.

Jöreskog, K. G., & Van Thillo, M. New rapid algorithms for factor analysis by unweighted least squares, generalized least squares, and maximum likelihood. (Research Memo RM-71-5). Princeton: E.T.S., 1971.

Jöreskog, K. G., Gruvaeus, G., & Van Thillo, M. ACOUS: A general computer program for analysis of covariance structures. *Research Bulletin*. Princeton: Educational Testing Service, 1970.

Kaiser, H. F. The varimax criterion for analytic rotation in factor analysis. *Psychometrika*, 1958, *23*, 187–200.

Kaiser, H. F. Image analysis. In C. W. Harris (Ed.), *Problems in measuring change*. Madison: University of Wisconsin Press, 1963.

Kaiser, H. F. An index of factorial simplicity. *Psychometrika*, 1974, *39*, 31–36. (a)

Kaiser, H. F. A computational starting point for Rao's canonical factor analysis: Implications for computerized procedures. *Educational and Psychological Measurement*, 1974, *34*, 691–692. (b)

Kaiser, H. A note on the Equamax criterion. *Multivariate Behavioral Research*, 1974, *9*, 501–504. (c)

Kaiser, H. F. Factor analysis of the image correlation matrix. 1977, in press.

Kaiser, H. F., & Caffrey, J. Alpha factor analysis. *Psychometrika*, 1965, *30*, 1.

Kaiser, H. F., & Cerny, B. A. A study of a measure of sampling adequacy for factor analytic correlation matrices. *Multivariate Behavioral Research*, 1977, *12*, 401–417.

Kaiser, H. F., & Hunka, S. Some empirical results with Guttman's stronger lower bound for the number of common factors. *Educational and Psychological Measurement*, 1973, *33*, 99–102.

Kaiser, H. & Madow, 1976. Private communication.

Kaiser, H., Hunka, S., & Bianchini, J. Relating factors between studies based upon different individuals. *Multivariate Behavioral Research*, 1971, *6*, 409.

Kameoka, V. State and trait pattern congruences on the Clinical Analysis Questionnaire Scales, in preparation.

Kameoka, V., & Sine, L. An extension of Bargmann's tables and a computer program for testing the statistical significance of simple structure in factor analysis. Manuscript submitted for publication, 1977.

Katz, J. O., & Rohlf, F. J. Function-plane—a new approach to simple structure rotation. *Psychometrika*, 1974, *39*, 37–51.

Katz, J. O., & Rohlf, F. J. Primary product function plane: An oblique rotation to simple structure. *Multivariate Behavioral Research*, 1975, *10*, 219–232.

Kelley, T. L. *Crossroads in the mind of man.* Stanford: Stanford University Press, 1928.

Kempthorne, O. *et al.* (Eds.). *Statistics and mathematics in biology.* Ames, Iowa: State College Press, 1954.

Kendall M. G. *A course in multivariate analysis.* London: Griffin, 1961.

Kendall, M. G., & Stuart, A. *The advanced theory of statistics.* London: Griffin, 1966.

Kerlinger, F. N., & Pedhazur, E. J. Multivariate regression in behavioral research. New York: Holt, Rinehart & Winston, 1973.

Kiel, D., & Wrigley, C. *Effects upon the factorial structure of rotating varying numbers of factors.* Paper presented at the annual convention of the American Psychological Association, September 1–7, 1960.

Kline, P., & Grindley, J. A 28 day case study with MAT. *Journal of Multivariate, Experimental, Personality and Clinical Psychology*, 1974, *1*, 13–22.

Knapp, R. R. Objective personality test and sociometric correlates of frequency of sick bay visits. *Journal of Applied Psychology*, 1961, *45*, 104–110.

Knapp, T. R., & Swoyer, V. H. Some empirical results concerning the power of Bartlett's test of the significance of a correlation matrix. *American Educational Research Journal*, 1967, *4*, 13.

Koopman, R. F. Fitting a multidimensional model to binary data. Unpublished doctoral dissertation, University of Illinois, Urbana, Illinois, 1968.

Koopmans, T. C. *Linear regression analyses in economic time series.* Haarlem: Bohn, 1936.

Korth, B., & Tucker, L. The distribution of chance congruence coefficients from simulated data. *Psychometrika*, 1975, *40*, 361–372.

Kraemer, H. C. An estimation and hypothesis testing problems for correlation coefficients. *Psychometrika*, 1975, *40*, 473–485.

Krause, M. S. Ordinal scale construction for convergent validity, object discrimination and resolving power. *Multivariate Behavioral Research*, 1966, *1*, 379–385.

Krause, M. S., & Vaitkus, A. Codimensionality without high correlation. *Multivariate Behavioral Research*, 1970, *5*, 125–131.

Kristof, W. Orthogonal interbattery factor analysis. *Psychometrika*, 1967, *32*, 199–221.

Kristof, W. Testing a linear relation between true scores of two measures. *Psychometrika*, 1973, *38*, 101–112.

Kristof, W., & Wingersky, B. A generalization of the orthogonal procrustes rotation procedure to more than two matrices. *Proceedings of the 79th Annual Convention of the American Psychological Association*, 1971, *6*, 89.

Kruskal, J. B. *Non-metric analysis of factorial experiments*. (Research Report). Murray Hill, New Jersey: Bell Telephone Labs Inc., 1964. (a)

Kruskal, J. B. Multidimensional scaling by optimizing goodness of fit to a nonmetric hypothesis. *Psychometrika*, 1964, *29*, 1–28. (b)

Kruskal, J. B. Nonmetric multidimensional scaling: A numerical method. *Psychometrika*, 1964, *29*, 115–129. (c)

Kruskal, J. B., & Shepard, R. N. A nonmetric variety of linear factor analysis. *Psychometrika*, 1974, *39*, 123–158.

Kuhn, T. S. *The structure of scientific revolutions*. Chicago: University of Chicago Press, 1964.

Landahl, H. D. Centroid orthogonal transformations. *Psychometrika*, 1938, *3*, 219–223.

Lawley, D. N. The estimation of factor loadings by the method of maximum likelihood. *Proceedings of the Royal Society of Edinburgh*, 1940, *60*, 64–82.

Lawley, D. N. The application of the maximum likelihood method to factor analysis. *British Journal of Psychology*, 1943, *33*, 172–175.

Lawley, D. N., & Maxwell, A. E. *Factor analysis as a statistical method*. London: Butterworth, 2nd ed., 1971.

Lawley, D. N., & Swanson, Z. Tests of significance in a factor analysis of artificial data. *British Journal of Statistical Psychology*, 1954, *7*, 75–79.

Lawlis, G. F., & Chatfield, D. *Multivariate approaches for the behavioral sciences: A brief text*. Lubbock, Texas: Texas Tech. University Press, 1974.

Lawlis, G. F., & Rubin, S. E. 16 P.F. study of personality patterns in alcoholism. *Quarterly Journal of Studies in Alcoholism*, 1971, June, 318–327.

Lazarsfeld, P. F., & Rosenberg, M. *The language of social research*. Glencoe: Free Press, 1955.

Ledermann, W. On a shortened method of estimation of mental factors by regression. *Psychometrika*, 1939, *4*, 109–116.

Levin, J. Three-mode factor analysis. *Psychological Bulletin*, 1965, *64*, 442–452.

Levin, J. A rotational procedure for separation of trait method and interaction factors in multi-trait multi-method factoring. *Multivariate Behavioral Research*, 1974, *9*, 231–240.

Li, C. C. *Path analysis—a primer*. Pacific Grove, Boxwood Press, 1975.

Lingoes, J. C. An IBM-7090 program for Guttman–Lingoes smallest space analysis—IV. *Behavioral Science*, 1966, *11*, 407.

Lingoes, J. C. The multivariate analysis of qualitative data. *Multivariate Behavioral Research*, 1968, *3*, 61–94.

Lingoes, J. C., & Guttman, L. Nonmetric factor analysis: A rank reducing alternative to linear factor analysis. *Multivariate Behavioral Research*, 1967, *2*, 485.

Linn, R. L. A Monte Carlo approach to the number of factors problem. *Psychometrika*, 1968, *33*, 37.

Linn, R. L., Centra, J. A., & Tucker, L. Between, within, and total group factor analyses of student ratings of instruction. *Multivariate Behavioral Research*, 1975, *10*, 277–288.

Lord, F. M. Further problems in the measurement of growth. *Educational and Psychological Measurement*, 1958, *18*, 437–454.

Lord, F. M. Elementary models for measuring change. In C. W. Harris (Ed.), *Problems in measuring change*. Madison: University of Wisconsin Press, 1963.

Lord, F. M., & Novick, M. R. *Statistical theories of mental test scores*. Reading, Mass.: Addison Wesley, 1968.

Luborsky, L., & Mintz, J. The contribution of P-technique to personality, psychotherapy and psychosomatic research. In R. M. Dreger (Ed.), *Multivariate personality research*. Baton Rouge: Claitor, 1972. Chapter 10.

Mahalanobis, P. C. On the generalized distance in statistics. *Proceedings of the National Institute of Science, Calcutta*, 1936, *12*, 49–55.

Maxwell, A. E. Statistical methods in factor analysis. *Psychological Bulletin*, 1959, *56*, 228.

May, D. R. Psychiatric syndrome classifications checked by Taxonome. Unpublished doctoral dissertation, University of Illinois, Urbana, Illinois, 1971.

May, D. R. An application of the taxonome method to a plasmode. *Multivariate Behavioral Research*, 1973, *8*, 503–510.

McAndrew, C., & Forgy, E. A note on the effect of score transformations in Q and R factor analysis techniques. *Psychological Review*, 1963, *70*, 116–118.

McDonald, R. P. A general approach to nonlinear factor analysis. *Psychometrika*, 1962, *27*, 397–415.

McDonald, R. P. Difficulty factors and non-linear factor analysis. *British Journal of Mathematical and Statistical Psychology*, 1965, *18*, 11–23.

McDonald, R. P. The common factor analysis of multicategory data. *British Journal of Mathematical and Statistical Psychology*, 1969, *22*, 165.

McDonald, R. P. The theoretical foundations of principal factor analysis, canonical factor analysis, and alpha factor analysis. *British Journal of Mathematical and Statistical Psychology*, 1970, *23*, 1.

McDonald, R. P. Testing pattern hypotheses for covariance matrices. *Psychometrika*, 1974, *39*, 189–201.

McDonald, R. P. Descriptive axioms for common factor theory, image theory and component theory. *Psychometrika*, 1975, *40*, 137–152.

McDonald, R. P., & Burr, E. J. A comparison of four methods of constructing factor scores. *Psychometrika*, 1967, *32*, 381.

McKeon, J. J. Canonical analysis: Some relations between canonical correlation factor analysis, discriminant function analysis, and scaling theory. *Psychometric Monographs*, 1965, No. 13.

McNemar, Q. Lost: Our intelligence? Why? *American Psychologist*, 1964, *19*, 871–882.

McQuitty, L. L. Rank order typal analysis. *Educational and Psychological Measurement*, 1963, *23*, 55–61.

McQuitty, L. Multiple rank order typal analysis for the isolation of independent types. *Educational and Psychological Measurement*, 1966, *26*, 3–11.

Meredith, W. Notes on factorial invariance. *Psychometrika*, 1964, *29*, 177. (a)

Meredith, W. Rotation to achieve factorial invariance. *Psychometrika*, 1964, *29*, 187–206. (b)

Meredith W. Canonical correlations with fallible data. *Psychometrika*, 1964, *29*, 55–69. (c)

Merrifield, P. R. Factor analysis in educational research. *Annual Review of Research in Education*, 1976, Chap. 10, 394–436.

Meyer, E. P. On the relationship between ratio of number of variables to number of factors and factorial determinacy. *Psychometrika*. 1973, *38*, 375–380.

Meyer, E. P., Kaiser, H. F., Cerny, B. G., & Grun, B. G. MSA for a special Spearman matrix. *Psychometrika*. 1977, *42*, 153–156.

Morton, N. E. Analysis of family resemblance, I. Introduction. *American Journal of Human Genetics*, 1974, *26*, 318-330.

Moseley, E. C., & Klett, C. J. An empirical comparison of factor scoring methods. *Psychological Reports*, 1964, *14*, 179.

Mosier, C. I. Influence of chance error on simple structure. *Psychometrika*, 1939, *4*, 33-44. (a)

Mosier, C. I. Determining a simple structure when loadings for certain tests are known. *Psychometrika*, 1939, *4*, 149-162. (b)

Mosier, C. I. On the reliability of a weighted composite. *Psychometrika*, 1943, *8*, 161-168.

Mosteller, F. A., & Tukey, J. Data analysis statistics. In G. Lindzey & R. Aaronson (Eds.), *Handbook of social psychology* (Vol. 2, 2nd ed.). Reading, Mass.: Addison-Wesley, 1968.

Mote, T. A. An artifact of the rotation of too few factors: Study orientation vs. trait anxiety. *Revista Interamericana de Psicologia*, 1970, *4*, 171.

Mulaik, S. A. *The foundations of factor analysis*. New York: McGraw-Hill, 1972.

Nesselroade, J. R. The separation of state and trait factors by dR-technique. Unpublished doctoral dissertation, University of Illinois, Urbana, 1966.

Nesselroade, J. R. A Note on the "longitudinal factor analysis" model. *Psychometrika*, 1972, *37*, 187-191.

Nesselroade, J. R. Faktoren-analyse von Kreuzprodukten zur Bescreibung von Veränderangs phanomene. *Zeitschrift fuer Experimentelle und Angewandte Psychologie*, 1973, *20*, 92-106.

Nesselroade, J. R., & Baltes, P. B. On a dilemma of comparative factor analysis: A study of factor matching based on random data. *Educational and Psychological Measurement*, 1970, *30*, 935.

Nesselroade, J. R., & Bartsch, T. W. Multivariate experimental perspectives on the construct validity of the trait state distinction. In R. B. Cattell & R. M. Dreger (Eds.), *Handbook of modern personality study*. New York: Century, 1976.

Nesselroade, J. R., & Cable, D. G. Sometimes it's okay to factor difference scores—the separation of state and trait anxiety. *Multivariate Behavioral Research*, 1974, *9* 273-282.

Nesselroade, J. R., & Reese, J. (Eds.) *Lifespan developmental psychology*. New York: Academic Press, 1975.

Nesselroade, J. R., Baltes, P. B., & Labouvie, E. W. Evaluating factor invariance in oblique space: Baseline data generated from random numbers. *Multivariate Behavioral Research*, 1971, *6*, 233.

Nesselroade, J. R., Schaie, K. W., & Baltes, P. B. Ontogenetic and generational components of structural and quantitative change in adult behavior. *Journal of Gerontology*, 1972, *27*, 222-228.

Norman, W. F. *2800 personality trait descriptions: Normative operating characteristics for a university population.* (NIMH Grant No. MH 07195). Michigan: University of Michigan, 1967.

Nosal, M. A note on the minres method. *Psychometrika*, 1977, *42*, 149-151.

Nunnally, J. C. *Psychometric theory*. New York: McGraw-Hill, 1967.

Olkin, J. Correlations re-visited. In J. C. Stanley (Ed.) *Improving experimental design and statistical analysis*. Chicago: Rand-McNally, 1966.

Olkin, I., & Siotani, M. Testing for the equality of correlation coefficients for various multivariate models. (Tech Rep. No. 4) Stanford: Stanford University Laboratory for Quantitative Research in Education, 1964.

Osgood, C. E., & Miron, M. S. Language behavior: The multivariate structure of qualification. In R. B. Cattell (Ed.), *Handbook of Multivariate Experimental Psychology*, Chicago, Rand McNally, 1966.

Overall, J. E. Note on the scientific status of factors. *Psychological Bulletin*, 1964, *61*, 270–276.

Overall, J. E. Marker variable factor analysis: A regional principal axes solution. *Multivariate Behavioral Research*, 1974, *9*, 149–164.

Overall, J. E., & Klett, E. J. *Applied multivariate analysis*. New York: McGraw-Hill, 1972.

Overall, J. E., & Woodward, J. A. Unreliability of difference scores: A paradox for measurement of change. *Psychological Bulletin*, 1975, *82*, 85–86.

Pawlik, K., & Cattell, R. B. Third-order factors in objective personality tests. *British Journal of Psychology*, 1964, *55*, 1–18.

Peay, E. R. Non-metric grouping: Clusters and cliques. *Psychometrika*, 1975, *40*, 297–313.

Pennell, R. The influence of communality and N on the sampling distributions of factor loadings. *Psychometrika*, 1968, *33*, 423.

Pennell, R. Routinely computable confidence intervals for factor loadings using the "jacknife." *British Journal of Mathematical and Statistical Psychology*, 1972, *25*, 107–120.

Pieszko, H. J. Global perceptiveness as a process common to sensitization-repression and the factors independence, comentive supergo and asthenia. Unpublished Masters thesis, University of Illinois, Urbana, 1967.

Pinneau, S. R., & Newhouse, A. Measures of invariance and comparability in factor analysis for fixed variables. *Psychometrika*, 1964, *29*, 271.

Pinzka, C., & Saunders, D. R. *Analytic rotation to simple structure, II: Extension to an oblique solution* (Research Bulletin RB-54-31). Princeton, N.J.: Educational Testing Service, 1954.

Porter, A. C. How errors of measurement affect ANOVA, regression analyses, ANCOVA and factor analyses (Report No. 14). Office of Research Consultation, School for Advanced Studies, College of Education, Michigan State University, 1971.

Porter, A. C. The effect of correction for attenuation upon factor analyses (Report No. 20). Office of Research Consultation, School for Advanced Studies, College of Education, Michigan State University, 1976.

Quenouille, M. H. *The analysis of multiple time series*. New York: Hafner, 1957.

Rao, C. R. Advanced statistical methods in biometric research. New York: Wiley, 1952.

Rao, C. R. Estimation and tests of significance in factor analysis. *Psychometrika*, 1955, *20*, 92–111.

Rao, C. R. *Linear statistical inference and its applications*. New York: Wiley, 1965.

Rao, C. R., Morton, N. E., & Yee, S. Analysis of family resemblance, II. A linear model for familial correlation. *American Journal of Human Genetics*, 1974, *26*, 331–359.

Rao, D. C. Morton, N. E., & Yee, S. Resolution of cultural and biological inheritance by path analysis. *American Journal of Human Genetics*. 1976, *28*, 228–242.

Rao, D. C., Morton, N. E., Elston, R. C., & Yee, S. Causal analysis of academic performance. *Behavior Genetics*, 1977, *7*, 147–159.

Rickard, S. The assumptions of causal analysis for incomplete causal sets of two multilevel variables. *Multivariate Behavioral Research*, 1972, *3*, 317–359.

Rickels, K., Cattell, R. B., Weise, C. *et al.* The effects of psychotherapy upon measured anxiety and regression. *American Journal of Psychotherapy*, 1966, *20*, 261–269.

Rickels, K. *et al.* Controlled psychopharmacological research in private psychiatric practice. *Psychopharmacologia*, 1966, *9*, 288–306.

Roff, M. Some properties of the communality in multiple factor theory. *Psychometrika*, 1935, *1*, 1.

Ross, J. The relation between test and person factors. *Psychological Review*, 1963, *70*, 432–443.

Ross, J. Mean performance and the factor analysis of learning data. *Psychometrika*, 1964, *29*, 67.

Ross, J., & Cliff, N. A generalization of the interpoint distance model. *Psychometrika*, 1964, *29*, 167–176.

Royce, J. R. Factors as theoretical constructs. *American Psychologist*, 1963, *18*, 522–528.

Royce, J. R. (Ed.). *Multivariate analysis and psychological theory*. New York: Academic Press, 1973.

Rozeboom, W. W. Linear correlations between sets of variables. *Psychometrika*, 1965, *30*, 57–71.

Rummel, R. J. Dimensions of dyadic war. *Journal of Conflict Resolution*, 1967, *11*, 176–183.

Rummel, R. J. *Applied Factor Analysis*. Evanston: Northwestern University Press, 1970.

Rummel, R. J. Dimensions of error in cross-national data. In R. Naroll & R. Cohen (Eds.), *A handbook of method in cultural anthropology*. New York: Halsted, 1972. (a)

Rummel, R. J. *The dimensions of nations*. Beverley Hills: Sage Publications, 1972 (b)

Russell, J. The relationship between college freshman withdrawal and certain critical personality and study orientation factors. Ph.D. dissertation. Penn State University, 1969.

Sarason, I. G. Experimental approaches to test anxiety: Attention and the uses of information. In C. Spielberger (Ed.), *Anxiety: Current trends in theory and research* (Vol. 2). New York: Academic Press, 1972. Chapter 11.

Saunders, D. R. Factor analysis I: Some effects of chance error. *Psychometrika*, 1948, *13*, 251–257.

Saunders, D. R. Practical methods in the direct factoring of psychological score matrices. Unpublished doctoral dissertation, University of Illinois Library, Urbana, 1950.

Saunders, D. R. The rationale for an "oblimax" method of transformation in factor analysis. *Psychometrika*, 1961, *26*, 317.

Saunders, D. R. *Transvarimax: Some properties of the ratiomax and equamax criteria for blind orthogonal rotation*. Paper delivered at the American Psychological Association meeting, 1962.

Schaie, 'K. W. Tests of hypotheses about differences between two correlation matrices. *Journal of Experimental Education*, 1958, *26*, 241–245.

Schmid, J., & Leiman, J. The development of hierarchical factor solutions. *Psychometrika*, 1957, *22*, 53–61.

Schmidt, F. L. The relative efficiency of regression and simple unit predictor weights in applied psychology. *Educational and Psychological Measurement*, 1971, *31*, 699.

Schneewind, K., & Cattell, R. B. Zum Problem der Faktoridentifikation: Verteilungen und Vertranensintervalle von Kongruentzkoeffizienten. *Psychol. Beiträge*, 1970, *12*. 214–226.

Schönemann, P. H. Varisim: A new machine method for orthogonal rotation. *Psychometrika*, i966, *31*, 235–254. (a)

Schönemann, P. H. The generalized solution of the orthogonal procrustes problem. *Psychometrika*, 1966, *31*, 1. (b)

Schönemann, P. H., & Wong, M. M. Some new results on factor indeterminacy. *Psychometrika*, 1972, *37*, 61.

Schröder, G., Cattell, R. B., & Wagner, A. Verification of the structure of the 16 P.F. Questionnaire in German. *Psychologische Forschung*, 1969, *32*, 369–386.

Sells, S. B. *Stimulus determinants of behavior*. New York: Ronald, 1963.

Shepard, R. N. *Metric structures in nonmetric data*. Murray Hill, N.J.: Bell Telephone Laboratories, 1965.

Shepard, R. N., & Carroll, J. D. Parametric representation of non-linear data structures. In P. R. Krishmaiah (Ed.), *Multivariate analysis*. New York: Academic Press, 1966.

Shepard, R. N., & Kruskal, J. B. Non-metric methods for scaling and for factor analysis. *American Psychology*, 1964. (Abstract)

Siegel, S. *Non-parametric statistics for the behavioral sciences*. New York: McGraw-Hill, 1956.

Simon, H. A. *Models of man*. New York: Wiley, 1957.

Snyder, T. W. *A unique variance for three-mode factor analysis*. Department of Psychology, University of Illinois, Urban, 1968.

Sokal, R. R. A comparison of five tests for completeness of factor extraction. *Transactions of the Kansas Academy of Science*, 1959, *62*, 141–152.

Sokal, R. R., & Sneath, P. H. A. *Principles of numerical taxonomy*. San Francisco: Freeman, 1963.

Solomon, H. A survey of mathematical models in factor analysis. In H. Solomon (Ed.), *Mathematical thinking in the measurement of behavior*. Glencoe: Free Press, 1960, pp. 269–314.

Sörbom, D. Detection of correlated errors in longitudinal data. *Multivariate Behavioral Research*, 1975, *28*, 138–151.

Spearman, C. General intelligence objectively determined and measured. *American Journal of Psychology*, 1904, *15*, 201–293.

Spearman, C. *The abilities of man*. New York: Macmillan, 1927.

Spence, I., & Graef, J. The determination of the underlying dimensionality of an empirically obtained matrix of proximities. *Multivariate Behavioral Research*, 1974, *9*, 337–342.

Spreen, O. The position of time estimation in a factor analysis and its relation to some personality variables. *Psychological Record*, 1963, *13*, 455–464.

Srivastava, J. N. On testing hypotheses regarding a class of covariance structures. *Psychometrika*, 1966, *31*, 147–164.

Stewart, D., & Love, W. A general canonical correlation index. *Psychological Bulletin*, 1958, *70*, 160–163.

Suits, D. B. Use of dummy variables in regression equations. *Journal of American Statistical Association*, 1957, *52*, 548–551.

Swain, A. J. A class of factor analysis estimation procedures with common asymptotic sampling procedures. *Psychometrika*, 1975, *40*, 315–336.

Sweney, A. B. See chapters on motivation in R. B. Cattell & F. W. Warburton (Eds.), *Objective personality and motivation tests. A theoretical introduction and practical compendium*. Champaign: University of Illinois Press, 1967.

Sweney, A. B., & Cattell, R. B. Components measurable in manifestations of mental conflicts. *Journal of Abnormal and Social Psychology*, 1964, *68*, 479–490.

Tatsuoka, M. M. *Selected topics in advanced statistics: An elementary approach. Standardized scales: Linear and area transformations* (Report No. 2) Champaign, Illinois, Institute for Personality and Ability Testing, 1969. (a)

Tatsuoka, M. M. *Selected topics in advanced statistics. An elementary approach. Vali-*

dation studies: The use of multiple regression equations (Report No. 5). Champaign, Illinois, Institute for Personality and Ability Testing, 1969. (b)

Tatsuoka, M. M. *Selected topics in advanced statistics: An elementary approach. Discriminating analysis: The study of group differences* (Report No. 6). Champaign, Illinois, Institute for Personality and Ability Testing, 1970.

Tatsuoka, M. M. *Selected topics in advanced statistics: An elementary approach. Significance tests: Univariate and multivariate.* (Report No. 4). Champaign, Illinois, Institute for Personality and Ability Testing, 1971. (a)

Tatsuoka, M. M. *Multivariate analysis.* New York: Wiley, 1971. (b)

Tatsuoka, M. M. *Selected topics in advanced statistics: An elementary approach. Classification procedures: Profile similarity* (Report No. 3). Champaign, Illinois, Institute for Personality and Ability Testing, 1974.

Tatsuoka, M. M. *Selected topics in advanced statistics. The general linear model* (Report No. 7). Champaign, Illinois, Institute for Personality and Ability Testing, 1975.

Taylor, J. E., & Coyne, L. Varimax and Maxplane rotational methods under different conditions of sampling error and hierarchical structure. *Multivariate Behavioral Research*, 1973, *8*, 365-378.

Thomson, G. H. The influence of univariate selection on the factorial analysis of ability. *British Journal of Psychology*, 1938, *28*, 451-459.

Thomson, G. H. On estimating oblique factors. *British Journal of Psychology, Statistical Section*, 1949, *2*, 1-2.

Thomson, G. H. *The factorial analysis of human ability* (5th ed.). Boston: Houghton Mifflin, 1951.

Thomson, G. H., & Ledermann, W. The influence of multivariate selection on the factorial analysis of ability. *British Journal of Psychology*, 1939, *29*, 288-306.

Thorndike, R. M. Canonical analysis and predictor selection. *Multivariate Behavioral Research*, 1977, *12*, 75-87.

Thorndike, R. M., & Weiss, D. J. A study of the stability of canonical correlations and canonical components. *Educational and Psychological Measurement*, 1973, *33*, 123-134.

Thurstone, L. L. The absolute zero in intelligence measurement. *Psychological Review*, 1928, *35*, 175-197.

Thurstone, L. L. *Primary mental abilities.* Chicago: University of Chicago Press, 1938.

Thurstone, L. L. *Multiple factor analysis.* Chicago: University of Chicago Press, 1947.

Thurstone, L. L. A method of factoring without communalities. Invitational Conference on Testing Problems. Princeton, N.J.: ETS, 1955, 59-62, 64-66.

Thurstone, L. L., & Thurstone, T. G. *Primary mental abilities test.* Chicago: S. R. A., 1948.

Torgerson, W. S. *Theory and methods of scaling.* New York: Wiley, 1958.

Toynbee, A. J. *A study of history.* New York: Oxford University Press, 1947.

Trites, D. K., & Sells, S. B. A note on alternative methods for estimating factor scores. *Journal of Applied Psychology*, 1955, *39*, 455.

Tryon, R. C. *Cluster analysis.* Chicago: University of Chicago Press, 1939.

Tryon, R. C. General dimensions of individual differences: Cluster analysis vs. multiple factor analysis. *Educational and Psychological Measurement*, 1958, *18*, 477-495.

Tryon, R. C. Salient dimensionality vs. the fallacy of 'minimal rank' in factor analysis. *American Psychologist*, 1961, *16*, 467.

Tsujioka, B., & Cattell, R. B. A cross cultural comparison of second stratum questionnaire personality factor structures—anxiety and extraversion—in America and Japan. *Journal of Social Psychology*, 1965, *65*.205-219.

Tucker, L. R. *A method for synthesis of factor analysis studies* (Personnel Research Section Report No. 984; Contract DA-49-083, Department of the Army). Princeton, N.J.: E.T.S., 1951.

Tucker, L. R. *Factor analyses of double centered score matrices.* Paper presented to the American Psychological Association and Psychometric Society, September, 1956.

Tucker, L. R. An inter-battery method of factor analysis. *Psychometrika*, 1958, *23*, 111.

Tucker, L. R. *Formal models for a central prediction system* (Research Bulletin, 60–14). Princeton, N.J.: Educational Testing Service, 1960.

Tucker, L. R. Implications of factor analysis of three-way matrices for measurement of change. In C. W. Harris (Ed.), *Problems in measuring change.* Madison: University of Wisconsin Press, 1963, Pp. 122–137. (a)

Tucker, L. R. The extension of factor analysis to three-dimensional matrices. In N. Frederiksen & H. Gulliksen (Eds.), *Contributions to mathematical psychology.* New York: Holt, Rinehart & Winston, 1963. Pp. 109–127. (b)

Tucker, L. R. Some mathematical notes on three-mode factor analysis. *Psychometrika*, 1966, *31*, 279–311. (a)

Tucker, L. R. Learning theory and multivariate experiment. In R. B. Cattell (Ed.), *Handbook of multivariate experimental psychology.* Chicago: Rand McNally, 1966. Chapter 16. (b)

Tucker, L. R. Comments on "confounding of sources of variation in factor-analytic techniques. *Psychological Bulletin*, 1968, *70*, 345.

Tucker, L. R. Relations of factor score estimates to their use. *Psychometrika*, 1971, *36*, 427.

Tucker, L. R. Relations between multidimensional scaling and three mode factor analysis. *Psychometrika*, 1972, *37*, 3–27.

Tucker, L. R., & Lewis, C. A reliability coefficient for maximum likelihood factor analysis. *Psychometrika*, 1973, *38*, 1–10.

Tucker, L. R., Koopman, R. F., & Linn, R. L. Evaluation of factor analytic research procedures by means of simulated correlation matrices. *Psychometrika*, 1969, *34*, 421.

Tukey, J. W. Causation, regression and path analysis. In O. Kempthorne *et al.* (Eds.), *Statistics and mathematics in biology.* Ames, Iowa: State College Press, 1954. Pp. 34–45.

Turner, M. E., & Stevens, C. D. The regression analysis of causal paths. *Biometrics*, 1959, *15*, 236–258.

Uhr, L., & Miller, J. G. (Eds.), *Drugs and Behavior.* New York: Wiley, 1960.

Ullman, L. P., & Krasner, L. *A psychological approach to abnormal behavior.* Englewood Cliffs, New Jersey: Prentice-Hall, 1969.

Van de Geer, J. P. *Introduction to multivariate analysis for the social sciences.* San Francisco: Freeman, 1971.

Van Egeren, L. F. Experimental determination by P-technique of functional unities of depression and other psychological states. Unpublished Masters thesis, University of Illinois, Urbana, 1963.

Van Egeren, L. F. Multivariate statistical analysis. *Psychophysiology*, 1973, *10*, 517–532.

Vaughan, D. S. The relative methodological soundness of several major personality factor analyses. *Journal of Behavioral Science*, 1973, *1*, 305–313.

Veldman, D. J. Simple structure and the number of factors problem. *Multivariate Behavioral Research*, 1974, *9*, 191–200.

Veldman, D. J. COMMAP: A computer program for commonality analysis. Dept. of Project PRIME, Austin, Texas: University of Texas, 1975.

Velicer, W. F. An empirical comparison of the similarity of principal component, image and factor patterns. *Multivariate Behavioral Research*, 1977, *12*, 3–22.

Vernon, P. E. *Intelligence and cultural environment*. London: Methuen, 1969.

Wackwitz, J. H., & Horn, J. L. On obtaining the best estimates of factor scores within an ideal simple structure. *Multivariate Behavioral Research*, 1971, *6*, 389.

Wardell, D., & Royce, J. R. Relation between cognitive and temperament traits and the concept of "style." *Journal of Multivariate, Experimental, Peronality and Clinical Psychology*, 1975, *1*, 244–267.

Werts, C. E. & Linn, R. L. Path analysis: Psychological examples. *Psychological Bulletin*, 1970, *74*, 193–212.

Wherry, R. J., & Gaylord, R. H. Factor pattern of test items and tests as a function of the correlation coefficient: Content, difficulty, and constant error factors. *Psychometrika*, 1944, *9*, 237–244.

White, O. Some properties of three factor contribution matrices. *Multivariate Behavioral Research*, 1966, *1*, 373.

Wiggins, J. S. *Personality and predicition: Principles of personality assessment*. New York: Addison-Wesley, 1973.

Wiley, D. E., Smith, W. H., & Bramble, W. J. Studies of a class of covariance structure models. *Journal of American Statistical Association*, 1973, *68*, 317–323.

Woodrow, H. Intelligence and improvement in school subjects. *Journal of Educational Psychology*, 1945, *36*, 155–166.

Woods, G. *A computer program for the scree test for number of factors*. Manuscript submitted for publication, 1976.

Woods, G. F. *Computer sub-routines for Cattell's Scientific Use of Factor Analysis*. New York: Plenum, in press.

Woodward, J. A., & Overall, J. E. *Factor analysis of rank ordered data*. Book in preparation, 1976.

Wright, S. The method of path coefficients. *Annals of Mathematic Statistics*, 1934, *5*, 161–215.

Wright, S. The interpretation of multivariate systems. In O. Kempthorne, T. A. Bancroft, J. W. Gowen, & J. L. Lush (Eds.), *Statistics and mathematics in biology*. Ames: Iowa State College Press, 1954.

Wright, S. The treatment of reciprocal interaction, with or without lab, in path analysis. *Biometrics*, 1960, *16*, 423–425.

Wrigley, C. S., & Neuhaus, J. O. The matching of two sets of factors. *American Psychologist*, 1955, *10*, 418.

Wrigley, C. S., Saunders, D. R., & Neuhaus, J. O. Application of the quartimax method of rotation to Thurstone's primary mental abilities study. *Psychometrika*, 1958, *23*, 151–170.

Young, G., & Householder, A. S. Discussion of a set of points in terms of their mutual distances. *Psychometrika*, 1938, *3*, 19–22.

Yule, C. W. Why we sometimes get nonsense correlations between time series. *Journal of Royal Statistical Association*, 1926, *89*, 1–64.

Author Index

Subject Index